# Advancements in Wastewater Treatment

## Cutting Edge Techniques and Sustainable Solutions

nova
science publishers
www.novapublishers.com

## Environmental Science, Engineering and Technology

**More information about this series can be found at**
https://novapublishers.com/product-category/series/environmental-science-engineering-and-technology/

## Waste and Waste Management

**More information about this series can be found at**
https://novapublishers.com/product-category/series/waste-and-waste-management/

**Neha Agarwal**
**Maulin P. Shah**
**Vijendra Singh Solanki**
**Sankara Rao Miditana**
Editors

# Advancements in Wastewater Treatment

## Cutting Edge Techniques and Sustainable Solutions

nova
science publishers
www.novapublishers.com

DOI: https://doi.org/10.52305/JVPT2231



## NOTICE TO THE READER



Additional color graphics may be available in the e-book version of this book.

## Library of Congress Cataloging-in-Publication Data

Names: Agarwal, Neha (Professor of chemistry) editor
Title: Advancements in wastewater treatment : cutting edge techniques and sustainable solutions / Neha Agarwal ... [and 3 others], editors.
Other titles: Cutting edge techniques & sustainable solutions
Description: New York : Nova Science Publishers, Inc, [2026] | Series: Waste and waste management | Includes bibliographical references and index. |
Identifiers: LCCN 2026001358 (print) | LCCN 2026001359 (ebook) | ISBN 9798901340073 hardcover | ISBN 9798901340936 adobe pdf
Subjects: LCSH: Sewage--Purification | Sewage--Environmental aspects
Classification: LCC TD745 .A29 2026 (print) | LCC TD745 (ebook) | DDC 628.3--dc23/eng/20260202
LC record available at https://lccn.loc.gov/2026001358
LC ebook record available at https://lccn.loc.gov/2026001359

*Published by Nova Science Publishers, Inc. † New York*

# Contents

# Foreword

Water is one of our planet's most precious resources and yet it remains one of the most threatened due to unchecked industrialization, rapid urban development, and escalating environmental pollution. Wastewater, once seen solely as a byproduct of human activity, has now become a pivotal element in the global pursuit of sustainable development, environmental protection, and public health. In this context, *Advancements in Wastewater Treatment: Cutting Edge Techniques and Sustainable Solutions* emerges as a timely and significant contribution to the ongoing dialogue and innovation within the field of environmental science.

This book presents a wide-ranging and forward-thinking exploration of both traditional and cutting-edge technologies in wastewater treatment. What makes this volume particularly valuable is its balanced integration of interdisciplinary approaches including microbial, phytoremediation, nanomaterial, and resource recovery frameworks with clear scientific reasoning and practical relevance. Each chapter offers in-depth insights, supported by up-to-date research, that reflect not only the complexity of wastewater contaminants but also the creativity and commitment of the global scientific community in seeking sustainable solutions.

As someone deeply engaged in environmental biotechnology and green engineering, I am particularly pleased to see the strong emphasis this book places on ecological sensitivity, scalability, and circular economy models. It is an essential reference for researchers, professionals, and students who wish to better understand the dynamic and rapidly evolving landscape of wastewater treatment technologies.

I commend the editors and contributing authors for their dedication to creating a comprehensive and impactful volume. I am confident that this book will inspire continued innovation and action toward a cleaner, healthier, and more sustainable world.

Susana Rodríguez-Couto
Full Professor (tenured) in Biological Water Treatment
Lappeenranta-Lahti University of Technology LUT, Finland
14.04.2025

# Preface

In an era marked by escalating environmental degradation, freshwater scarcity, and rapid industrialization, the sustainable management of wastewater has become a global imperative. The book titled "*Advancements in Wastewater Treatment: Cutting Edge Techniques and Sustainable Solutions*" emerges from the pressing need to explore, innovate, and implement diverse approaches that can transform wastewater from a burden into a valuable resource.

This volume presents an integrated and interdisciplinary examination of contemporary wastewater treatment strategies. It brings together contributions from leading researchers and practitioners to address not only the scientific and technical dimensions of wastewater remediation but also the socio-environmental and industrial contexts in which these technologies must operate. From fundamental studies to futuristic insights, this book offers a comprehensive overview of existing challenges and progressive solutions.

The journey begins with an in-depth analysis of the sources, fate, and nature of wastewater contaminants (Chapter 1), followed by an exploration of the critical hurdles and promising avenues in treatment technologies (Chapter 2). Chapters 3 through 6 delve into the core mechanisms of contaminant removal using physical, chemical, adsorptive, phytoremediation, and microbial processes, laying the technical foundation for more specialized discussions.

Subsequent chapters (7–9) introduce readers to the removal of trace-level micropollutants, emerging technologies, and biotechnological breakthroughs that are reshaping the field. Addressing sector-specific concerns, Chapter 10 examines industrial wastewater management through tailored treatment systems, while Chapter 11 discusses the transformative role of MXene-integrated carbon materials in structural engineering for remediation.

Chapters 12 to 14 emphasize the integration of ecological sustainability with technology through the use of microalgae, green photocatalysis, and resource recovery strategies aligned with circular economy principles. The focus on sustainability continues in Chapter 15, which elaborates on bioremediation techniques for heavy metal removal.

Chapters 16 to 18 advance the discussion with cutting-edge approaches such as biochar-mediated advanced oxidation, pesticide-specific treatment methods, and nanotechnology-driven solutions for industrial effluents. Chapter 19 brings the reader to the frontier of research, summarizing current trends, unresolved challenges, and future directions in wastewater treatment.

The final chapter (Chapter 20) presents a unique case study that intersects environmental science with social entrepreneurship, showcasing an innovative model of temple waste management through green initiatives in Uttar Pradesh, India underscoring the importance of community participation in sustainable development.

This book is intended for a diverse audience, including environmental scientists, chemical engineers, biotechnologists, policy makers, industry professionals, and graduate students. Its multi-faceted approach bridges fundamental science, applied engineering, and socio-environmental strategies, providing readers with the tools and insights necessary for both academic advancement and practical implementation.

We believe this compilation not only reflects the current state of the field but also charts a path forward one that redefines wastewater as a resource to be recovered, reused, and respected. It is our sincere hope that this book fosters continued innovation and responsible stewardship of our planet's most precious resource: water.

# Acknowledgments

We extend our heartfelt gratitude to all those who contributed to the successful completion of this volume, "*Advancements in Wastewater Treatment: Cutting Edge Techniques and Sustainable Solutions*." This book is the result of the combined efforts of a dedicated group of researchers, scholars, and practitioners from diverse domains who have shared their knowledge, expertise, and insights to enrich this compilation.

First and foremost, we thank all the contributing authors whose in-depth research and thoughtful analyses form the backbone of this work. Their commitment to scientific rigor and sustainability has added immense value to this publication. Without their enthusiasm and timely contributions, this project would not have been possible. Special thanks are due to the institutions and research centres that supported the contributors through resources, infrastructure, and an environment conducive to research and innovation. Their role in nurturing scholarly work is invaluable.

We are also thankful to the technical support team and publication coordinators who handled the formatting, compilation, and communication processes with professionalism and efficiency. Their behind-the-scenes efforts were crucial in transforming a vision into a polished publication. We acknowledge the encouragement and support of our families, colleagues, and mentors who continuously inspired us throughout the editing process. Their patience, motivation, and understanding provided strength during the demanding phases of this work.

Lastly, we express our sincere gratitude to Nova Science Publishers for their overall support and co-operation. It is our sincerc hope that this book will contribute meaningfully to all the efforts and serve as a foundation for future innovations in wastewater treatment.

# Chapter 1

# Origin of Wastewater: Various Sources of Contamination, Fate, and Occurrence

**S. S. Kirthiga***
**and R. Dhinesh**
Department of Aquatic Environment Management, Faculty of Fisheries Science,
Kerala University of Fisheries and Ocean Studies, Kochi, Kerala, India

## Abstract

The present era is grappling with a growing number of environmental and health issues caused by an enormous volume of wastewater which is generated from a wide range of human activities. The wastewater contains hazardous pollutants and contaminants that pose significant ecological and health risks upon their release into the environment. Consequently, there is a pressing need to assess the effectiveness of both existing and emerging remediation and treatment technologies in mitigating the harmful effects of wastewater pollutants. In this context, the chapter highlights the urgent need to address wastewater contamination as a critical step toward safeguarding ecosystem integrity and human health in an increasingly industrialized and urbanized world.

This chapter provides a comprehensive discussion on the fate of wastewater contaminants, emphasizing their physical, chemical, and biological transformations within water bodies, sediments, and the atmosphere. It also explores the mechanisms that govern their behaviour, including the co-occurrence and interactions of multiple contaminants within complex environmental matrices. Key transformation and transport processes—such as dilution, volatilization, sorption, and biodegradation—are examined, along with the bioaccumulation of persistent pollutants within the food chain. It further explores the capabilities of both established and innovative treatment methods in addressing the harmful effects of these pollutants. Ultimately, it highlights the urgent need to address wastewater contamination to protect ecosystem integrity and human health in an increasingly industrialized and urbanized world.

**Keywords:** wastewater, contaminants, hazardous, occurrence, atmosphere, environment, ecosystem

---

* Corresponding Author's Email: sskkirthiga@gmail.com

In: Advancements in Wastewater Treatment
Editors: Neha Agarwal, Maulin P. Shahl, Vijendra Singh Solankil et al.
ISBN: 979-8-90134-007-3

## 1. Introduction

Wastewater, an unavoidable by-product of human activities, represents a significant environmental and public health challenge due to its complex composition and vast sources. It comprises a wide array of contaminants, including organic matter, inorganic compounds, nutrients, heavy metals, pathogens, and emerging pollutants [1]. The exponential growth of population, urbanization, industrialization, and agricultural intensification has amplified the generation of wastewater, creating immense pressure on natural water systems [2]. Globally, untreated or inadequately treated wastewater discharges contribute to the deterioration of water quality, loss of biodiversity, and heightened risks to human health [3]. Domestic wastewater constitutes a major portion of global wastewater production, originating primarily from households, urban settlements, and small-scale industries [4]. It contains a complex mixture of organic and inorganic matter, nutrients such as nitrogen and phosphorus, pathogens, and emerging contaminants like pharmaceuticals, personal care products (PCPs), and microplastics [5]. The growing use of synthetic compounds in households has significantly increased the chemical load in domestic wastewater as many of these compounds resist degradation and persist in aquatic environments. Untreated or partially treated domestic wastewater often results in the eutrophication of water bodies, algal blooms, and hypoxic conditions that disrupt aquatic ecosystems [3]. Furthermore, domestic wastewater is a major vector for the spread of waterborne diseases, underscoring the urgent need for improved wastewater collection, treatment, and reuse systems.

The industrial sector, being one of the largest contributors to wastewater contamination, introduces an array of hazardous pollutants into water systems [6]. Industrial wastewater varies significantly in composition depending on the type of industry, often containing high concentrations of heavy metals, dyes, solvents, hydrocarbons, and persistent organic pollutants (POPs). For instance, textile industries discharge wastewater rich in dyes and chemicals, while mining and metal processing activities introduce toxic heavy metals such as lead, mercury, and cadmium into the environment [7]. These contaminants not only pose direct toxicity to aquatic organisms but also accumulate in sediments and bioaccumulate through food chains, posing severe risks to ecological and human health. Additionally, industrial wastewater is frequently characterized by high thermal loads and fluctuating pH levels, which further exacerbate its environmental impacts [8]. Agriculture is another significant contributor to wastewater contamination, primarily through surface runoff and effluents from farming activities. Agricultural wastewater is typically laden with nutrients, pesticides, herbicides, and animal waste [9]. The excessive use of fertilizers and pesticides on farmlands results in nutrient-enriched runoff, which is a primary driver of eutrophication in freshwater and marine systems [10]. Livestock farming contributes to agricultural wastewater through the release of nutrient-rich effluents containing ammonia, organic matter, and pathogens [11]. These contaminants not only alter aquatic ecosystems but also contaminate groundwater, which is a crucial source of drinking water in many regions [12].

The fate of agricultural pollutants is further complicated by climate variability, which influences runoff patterns and increases pollutant loads in water bodies [13]. Sustainable farming practices, coupled with effective runoff management strategies, are essential to mitigate agricultural wastewater impacts. Urban stormwater runoff is a unique and growing source of wastewater contamination, particularly in rapidly urbanizing regions [14]. The proliferation of impervious surfaces such as roads, pavements, and rooftops accelerates the flow of stormwater, which carries a diverse range of pollutants into nearby water bodies [15]. Urban runoff often contains hydrocarbons from vehicles, heavy metals from industrial and construction activities, sediments, and plastic debris. Additionally, urban areas with combined sewer systems frequently experience combined sewer overflows (CSOs) during heavy rainfall events, which discharge untreated domestic and industrial wastewater directly into the environment. The episodic and unpredictable nature of urban stormwater pollution presents a significant challenge for water resource management. Integrated urban planning, low-impact development techniques, and green infrastructure solutions are vital for addressing urban runoff and its associated impacts.

This chapter delves into the intricate web of wastewater sources, highlighting the diverse contaminants introduced through domestic, industrial, agricultural, and urban processes. It also explores the pathways by which these pollutants enter natural systems, their interactions and transformations within the environment, and their eventual persistence or removal. A comprehensive understanding of these factors is critical for developing effective wastewater management strategies, mitigating environmental impacts, and safeguarding both human and ecological health.

## 2. Sources of Wastewater Contaminants

Wastewater contaminants originate from various sources, including domestic effluents, industrial discharges, agricultural runoff, and urban storm water (Figure 1). Domestic wastewater contains organic matter, nutrients, pathogens, and pharmaceuticals, contributing significantly to water pollution [16]. Industrial discharges introduce heavy metals, synthetic chemicals, and POPs [17]. Agricultural runoff carries nutrients like nitrogen and phosphorus, pesticides, and animal waste, exacerbating water quality issues [18]. Additionally, urban stormwater runoff transports oils, metals, and plastics from impervious surfaces, further contaminating water bodies [14]. These sources collectively contribute to complex environmental challenges.

### 2.1. Domestic Effluents

#### *2.1.1. Organic Matter and Nutrients*

Domestic effluents, primarily derived from residential and commercial activities, are a significant source of organic pollutants and nutrients. The organic matter in domestic wastewater comes mainly from human waste, food residues, detergents, and household chemicals [19]. When untreated or insufficiently treated, this organic load can lead to severe environmental issues, particularly eutrophication of water bodies. Nutrients such as nitrogen (from ammonia, urea) and phosphorus (from detergents and human waste) are present in high concentrations in domestic effluents that contribute to the enrichment of aquatic ecosystems [16]. This nutrient overload causes algal blooms, oxygen depletion, and subsequent degradation of aquatic life. Moreover, the biochemical oxygen demand (BOD) of organic matter consumes oxygen in the water, further exacerbating oxygen stress in aquatic ecosystems [20].

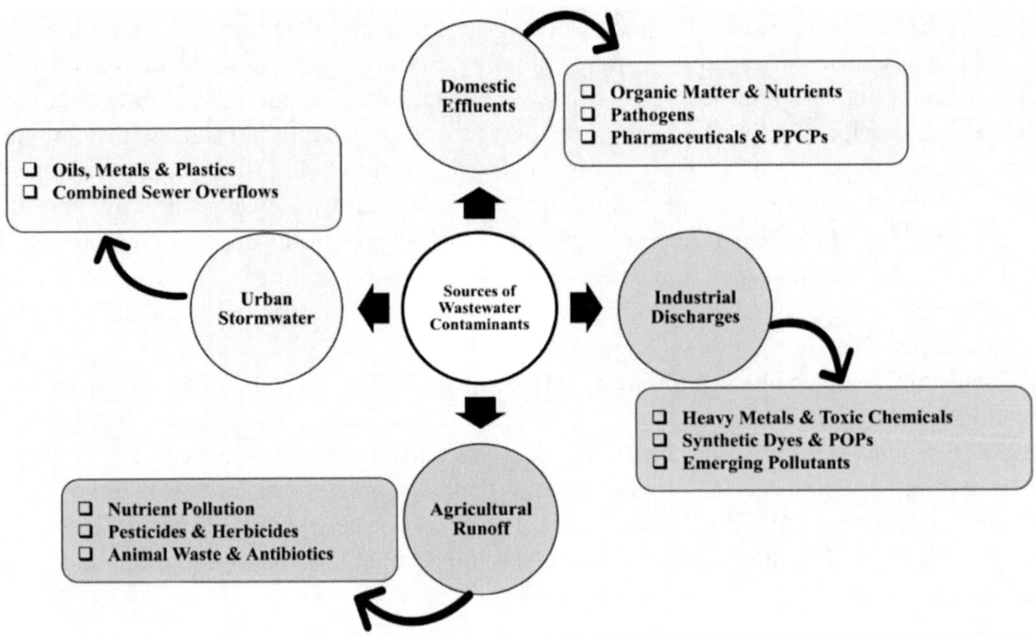


**Figure 1.** Major sources of wastewater and the major contaminants present in it.

### *2.1.2. Pathogens and Microbial Contaminants*

Domestic effluents are also a major source of pathogenic microorganisms, including bacteria, viruses, and protozoa, which can cause waterborne diseases. Pathogens are introduced into wastewater primarily through human waste and household discharges containing contaminated materials [21]. Improperly treated wastewater from domestic sources has been linked to the spread of diseases like cholera, dysentery, and typhoid fever [22]. These pathogens, when released into natural water bodies, can contaminate drinking water supplies and recreational water, posing serious public health risks. Therefore, the proper treatment and disinfection of domestic effluents is crucial to preventing microbial contamination in aquatic ecosystems.

### *2.1.3. Pharmaceuticals and Personal Care Products (PPCPs)*

PPCPs are emerging contaminants that have been detected in significant concentrations in domestic wastewater. These compounds include prescription medications, over-the-counter drugs, cosmetics, and fragrances. PPCPs are primarily excreted by humans through urine and feces, subsequently entering wastewater systems [23]. Many PPCPs, including antibiotics, hormones, and painkillers, are not effectively removed by conventional wastewater treatment processes and thus remain in the environment [24]. The presence of PPCPs in aquatic ecosystems can disrupt aquatic organisms' endocrine systems, leading to developmental and reproductive abnormalities [25]. Furthermore, the accumulation of pharmaceuticals in the food chain has raised concerns regarding human exposure to low levels of these chemicals through the consumption of contaminated water and seafood [26].

## 2.2. Industrial Discharges

### *2.2.1. Heavy Metals and Toxic Chemicals*

Industrial discharges are a major source of pollutants, particularly heavy metals and toxic chemicals, which are released from manufacturing processes, mining operations, and chemical production [27]. Heavy metals such as mercury, lead, cadmium, and arsenic are persistent pollutants that pose severe ecological and health risks due to their toxicity and bioaccumulation potential [28]. These metals can contaminate water bodies, sediments, and aquatic organisms, leading to chronic poisoning and long-term environmental damage [29]. Industrial effluents often contain high concentrations of these metals, which can enter water bodies through direct discharges or through runoff from industrial sites. The presence of these toxic metals in the environment is linked to various health issues, including neurological disorders, kidney damage, and cancer [29]. Efforts to mitigate heavy metal contamination have primarily focused on advanced treatment technologies such as adsorption, ion exchange, and membrane filtration.

### *2.2.2. Synthetic Dyes and POPs*

Industries such as textiles, leather, and paper manufacturing release a wide variety of synthetic dyes and POPs into wastewater [6]. These pollutants are characterized by their stability and resistance to degradation, leading to their accumulation in water bodies and sediments. Synthetic dyes, commonly used in the textile industry, can color water bodies and interfere with light penetration, inhibiting photosynthesis and disrupting aquatic food webs [30]. Similarly, POPs like polychlorinated biphenyls (PCBs) and pesticides are known to persist in the environment for decades, bioaccumulate in organisms, and pose significant risks to aquatic life and human health [31]. Industrial discharges containing these contaminants often exceed regulatory limits, contributing to environmental pollution.

### *2.2.3. Emerging Pollutants from Industrial Processes*

Emerging pollutants, including pharmaceuticals, endocrine-disrupting chemicals endocrine disrupting chemicals (EDCs), and nanomaterials, are increasingly being detected in industrial effluents [32]. These pollutants, often used in the production of goods or by-products from various industries, can pass through wastewater treatment systems and enter the environment. Nanomaterials, for instance, are widely used in electronics, textiles, and coatings, but their release into wastewater can cause environmental toxicity due to their small size and high reactivity [33]. EDCs, such as bisphenol A (BPA), interfere with hormone systems in wildlife and humans, leading to developmental and reproductive issues [34]. The challenge of addressing these emerging pollutants lies in the development of treatment technologies that can efficiently remove these substances from industrial effluents.

## 2.3. Agricultural Runoff

### *2.3.1. Nutrient Pollution*

Agricultural runoff is a significant contributor to nutrient pollution in water bodies, particularly nitrogen and phosphorus, which originate from fertilizers and animal manure. The excessive use of chemical fertilizers in agricultural practices leads to high concentrations of nitrogen (as nitrate) and phosphorus in runoff, which are then carried into rivers, lakes, and coastal zones [35]. These nutrients, when introduced in excess into aquatic ecosystems, cause eutrophication, promoting the growth of harmful algal blooms that deplete oxygen levels and degrade water quality [36]. This process can result in "dead zones," areas of water where oxygen concentration is so low that aquatic life cannot survive [37]. Moreover, the presence of nitrogen and phosphorus in drinking water poses health risks to humans, including methemoglobinemia, commonly known as "blue baby syndrome" [38].

#### *2.3.2. Pesticides and Herbicides*

Pesticides and herbicides, widely used in agriculture to protect crops from pests and weeds, are another major component of agricultural runoff. These chemicals are highly toxic to non-target organisms and can persist in the environment long after their application. When these chemicals wash into nearby water bodies, they can harm aquatic life, reduce biodiversity, and impair ecosystem functions [39]. Studies have shown that pesticides such as glyphosate, atrazine, and chlorpyrifos are detectable in water sources around the world and can cause toxicity in aquatic organisms, affecting reproduction and survival [39, 40]. Additionally, pesticide residues in water bodies can bioaccumulate in the food chain, posing risks to both wildlife and humans [41].

#### *2.3.3. Animal Waste and Antibiotics*

Animal waste, including manure from livestock operations, is another significant contributor to agricultural runoff. This waste contains a variety of contaminants, including pathogens, nutrients, and pharmaceuticals like antibiotics [42]. Additionally, the high nutrient content in animal waste, particularly nitrogen and phosphorus, exacerbates the eutrophication process and impacts water quality [43].

### 2.4. Urban Stormwater

#### *2.4.1. Pollutants from Impervious Surfaces*

Urban stormwater runoff is another critical source of pollution, particularly in cities with extensive impervious surfaces such as roads, parking lots, and rooftops [44]. These surfaces prevent the natural infiltration of rainwater, leading to the rapid runoff of pollutants into water bodies. Urban stormwater typically carries a mixture of oils, heavy metals, plastics, and other debris [45]. Oils and greases, often from vehicles, contribute to water contamination by forming thin films on the surface of water bodies, disrupting oxygen exchange and harming aquatic life [46]. Heavy metals such as lead, zinc, and copper are commonly found in stormwater due to their widespread use in construction, automotive, and industrial processes [47]. Plastics, including microplastics, are another growing concern, as they are carried into water bodies during rainstorms and persist in the environment for long periods [48].

#### *2.4.2. Combined Sewer Overflows (CSOs) and Their Impacts*

CSOs occur when stormwater and wastewater systems (typically designed to handle both types of discharge) become overwhelmed during heavy rainfall [49]. When this happens, untreated sewage, along with stormwater, is directly discharged into rivers, lakes, and coastal waters.

CSOs are a major source of water pollution, as they release pathogens, organic matter, nutrients, and hazardous chemicals into the environment [50]. These events can lead to the contamination of drinking water sources, increase the incidence of waterborne diseases, and degrade water quality, making it unsafe for recreational use. Addressing CSOs requires upgrading urban infrastructure, increasing treatment capacity, and implementing sustainable urban drainage systems to manage stormwater more effectively [51].

## 3. Types of Wastewater Contaminants

Wastewater contaminants can be broadly classified into organic, inorganic, biological, and emerging pollutants. Organic contaminants, such as BOD contributors, hydrocarbons, and synthetic organic compounds, often originate from domestic and industrial sources. Inorganic contaminants include heavy metals such as lead, cadmium, and mercury, which are toxic and persistent in the environment [52]. Biological contaminants, including pathogens, pose direct health risks to humans [53]. Emerging contaminants, such as pharmaceuticals, PCPs, and microplastics, are increasingly recognized for their environmental persistence and potential harm [54].

### 3.1. Organic Contaminants

Organic contaminants in wastewater are a key driver of environmental degradation, particularly in aquatic ecosystems. One of the most significant organic pollutants is Biochemical Oxygen Demand (BOD), which represents the amount of oxygen consumed by microorganisms in the process of decomposing organic matter [55]. High BOD levels in wastewater indicate the presence of significant organic matter, including human waste, food residues, and biodegradable industrial by-products [56]. These organic pollutants lead to oxygen depletion in receiving water bodies, adversely affecting aquatic life. The decomposition process consumes oxygen from the water, causing hypoxic conditions that can result in fish kills and a decrease in biodiversity [57]. Consequently, high BOD concentrations are used as an indicator of water quality, and reducing BOD levels is critical to the protection of aquatic ecosystems [58]. Hydrocarbons and synthetic organic compounds, which are common contaminants in wastewater from industrial and urban sources, are another major category of organic pollutants. Hydrocarbons, including petroleum-derived substances, are typically introduced into wastewater through industrial discharges, vehicle runoff, and accidental spills [59]. These compounds are highly hydrophobic and tend to accumulate in sediments, where they can be toxic to aquatic organisms over prolonged exposure. In addition to hydrocarbons, synthetic organic chemicals such as solvents, pesticides, and plasticizers are present in wastewater and contribute to its toxicity [60]. These synthetic chemicals are often persistent and resistant to biodegradation, leading to their accumulation in the environment and potential risks to both wildlife and human health. Many of these compounds are also classified as POPs, which are bio accumulative and can have long-term adverse effects on the food chain [61].

### 3.2. Inorganic Contaminants

Heavy metals, including lead, cadmium, mercury, and arsenic, are some of the most toxic inorganic contaminants found in wastewater [28]. These metals are introduced into the environment through industrial discharges, mining operations, and agricultural runoff [65]. Once in the aquatic environment, heavy metals can be absorbed by aquatic organisms, leading to bioaccumulation and biomagnification in the food chain. Lead, for example, is a potent neurotoxin, particularly harmful to young children, and can cause developmental delays and cognitive impairments [62]. Similarly, mercury, when methylated by microorganisms, becomes highly toxic and can result in neurological disorders and reproductive toxicity in humans and wildlife [63]. Treatment of heavy metals in wastewater often requires specialized methods such as adsorption, membrane filtration, and chemical precipitation to remove these pollutants effectively and prevent their harmful impacts on ecosystems. The salinity of wastewater is another important inorganic contaminant, especially in regions with high industrial or agricultural activities that discharge brine or saline water. Excessive salinity in wastewater can have detrimental effects on the quality of freshwater resources and on the health of aquatic organisms, particularly freshwater species that are sensitive to changes in salinity levels [64]. In addition to salinity, trace elements, including elements like zinc, copper, and nickel, are commonly found in wastewater. These trace elements, although present in small concentrations, can be toxic to aquatic life and have the potential to accumulate in the food chain, leading to long-term ecological and health risks [65]. Effective management of salinity and trace elements in wastewater requires advanced treatment processes such as reverse osmosis or ion-exchange techniques, which can remove dissolved salts and metals from water [66].

### 3.3. Biological Contaminants

Biological contaminants, including pathogens and disease-causing microorganisms, are a major concern in wastewater due to their potential to cause widespread health problems. These pathogens, which include bacteria, viruses, protozoa, and helminths, enter wastewater primarily through human feces and urine, particularly in regions without adequate sanitation systems [21]. When wastewater is not treated or is inadequately treated, these pathogens can contaminate drinking water supplies and recreational waters, leading to the transmission of diseases such as cholera, dysentery, typhoid, and hepatitis [29]. The presence of these microorganisms in wastewater poses significant public health risks, especially in developing countries, where access to proper sanitation and wastewater treatment is limited. Effective treatment of wastewater for pathogen removal involves processes such as filtration, ultraviolet (UV) radiation, and chlorine disinfection, which are essential to ensure safe water quality for public health [67]. Antimicrobial resistance (AMR) is an emerging threat associated with the presence of PPCPs in wastewater, particularly antibiotics [68]. AMR occurs when microorganisms such as bacteria evolve resistance to the drugs designed to kill them, rendering treatments ineffective [69]. Wastewater systems act as reservoirs for AMR bacteria, as treated effluents often contain trace amounts of antibiotics and other antimicrobial agents that promote the selection of resistant strains [70]. These resistant bacteria can spread through environmental media such as water, soil, and sediments, leading to a wider distribution of AMR across ecosystems. The role of wastewater in the dissemination of AMR has prompted calls for improved wastewater treatment processes that can effectively remove antimicrobial agents and prevent the development of resistance [71]. Strategies to tackle AMR in wastewater include enhancing disinfection protocols and utilizing advanced technologies such as ozonation and membrane filtration [70].

### 3.4. Emerging Contaminants

PPCPs are a class of emerging contaminants that have been increasingly recognized for their persistence in wastewater and their potential impacts on the environment [72]. These compounds include a wide range of substances such as antibiotics, hormones, pain relievers, and cosmetics, which enter wastewater systems primarily through human excretion or the disposal of unused products. Conventional wastewater treatment methods are often ineffective at removing PPCPs, resulting in their continued presence in treated effluent, surface waters, and sediments [73]. The presence of PPCPs in aquatic ecosystems can disrupt endocrine systems in wildlife, leading to altered reproductive behaviors and developmental abnormalities [74]. Furthermore, the long-term environmental and human health impacts of PPCPs remain poorly understood, highlighting the need for further research and more effective treatment technologies to address this issue.

EDCs are another group of emerging contaminants found in wastewater, known for their ability to interfere with hormone systems in both humans and wildlife. EDCs include substances such as (BPA, phthalates, and certain pesticides, which are commonly found in wastewater due to industrial discharges, agricultural runoff, and household waste [75]. EDCs can mimic or block hormones, leading to a range of adverse effects such as reproductive dysfunction, developmental disorders, and increased risk of certain cancers [76]. The widespread presence of EDCs in water bodies has raised concerns about their potential to affect aquatic ecosystems, including the reproductive health of fish and amphibians. As EDCs are resistant to biodegradation, they persist in the environment and can accumulate in the food chain, posing risks to both wildlife and human populations [77].

Microplastics and (NPs) are two types of emerging contaminants gaining increasing attention due to their widespread occurrence and potential harmful effects. Microplastics, defined as plastic particles smaller than 5 mm, are commonly found in wastewater due to the breakdown of larger plastic items, personal care products, and synthetic textiles [78]. These particles can accumulate in aquatic environments and are ingested by marine organisms, potentially causing physical harm or chemical contamination. NPs, which are engineered particles with sizes ranging from 1 to 100 nm, are increasingly used in various industrial applications, including in electronics, cosmetics, and medicine [79]. Due to their small size and unique properties, NPs can easily enter biological systems, potentially leading to toxicity and long-term health effects. Both microplastics and NPs are resistant to traditional wastewater treatment methods, and their presence in the environment necessitates the development of novel treatment technologies to mitigate their impact.

## 4. Fate and Behavior of Contaminants in the Environment

The fate and behavior of wastewater contaminants in the environment are governed by various physical, chemical, and biological processes that influence their persistence, transformation, and potential for harm. The major transformation and transport mechanisms are represented in Figure 2. Physical processes such as sedimentation, sorption, and volatilization play a significant role in contaminant distribution. Chemical processes like hydrolysis, oxidation, and reduction further alter contaminant properties. Biological processes, especially biodegradation and biotransformation, involve microorganisms that break down organic pollutants into simpler compounds [80]. The fate of the contaminants is also influenced by their interactions with other substances. Co-occurrence and synergistic effects can alter their persistence, toxicity, and transport in the environment. For instance, the presence of other pollutants like heavy metals can affect the degradation rates of organic contaminants. Understanding these processes is crucial for predicting the environmental impact and designing effective treatment strategies for wastewater contaminants.

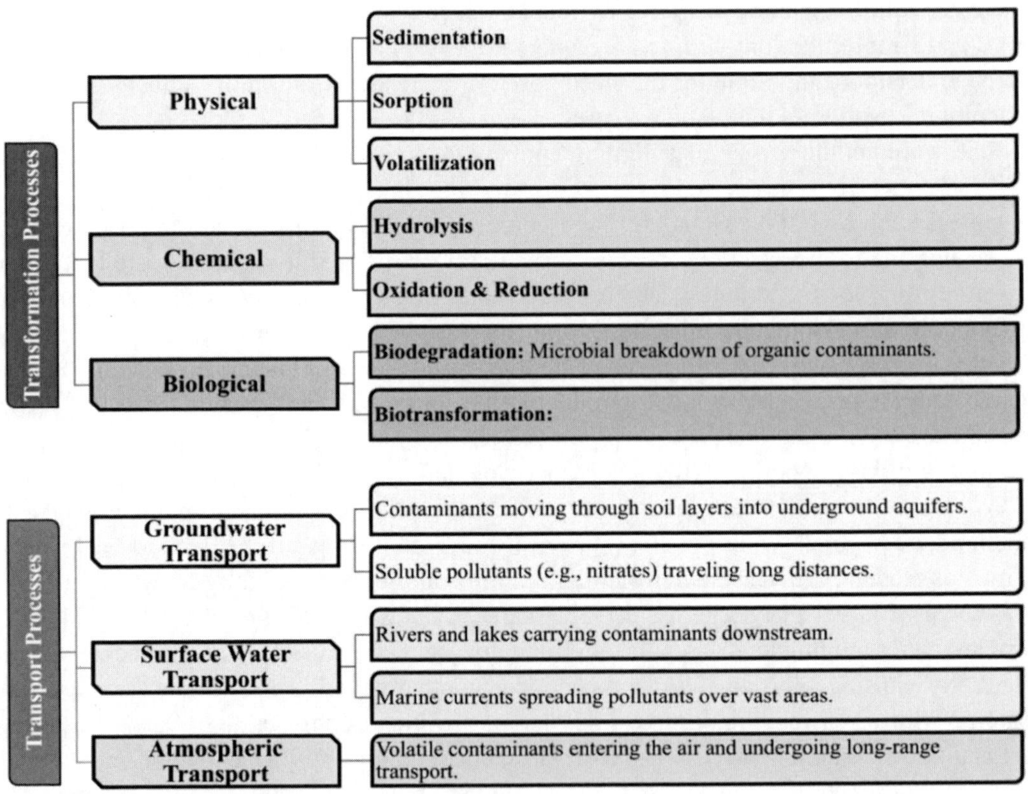


**Figure 2.** Transformation mechanisms and transport processes of wastewater contaminants.

## 4.1 Transformations in Aquatic and Terrestrial Ecosystems

### *4.1.1. Physical Processes*

The fate and behavior of contaminants in the environment are significantly influenced by physical processes such as sedimentation, sorption, and volatilization. Sedimentation occurs when suspended particles, including contaminants such as metals and organic matter, settle out of the water column and accumulate in sediments [81]. This process is particularly important for particulate-bound contaminants such as heavy metals and microplastics, which can become trapped in sediment layers for extended periods, potentially causing long-term environmental impacts [82]. Sorption refers to the binding of contaminants to soil, sediment, or organic matter, and is a critical process influencing the transport and bioavailability of pollutants [83]. Hydrophobic contaminants, such as hydrocarbons and some pesticides, often undergo sorption to particles, which reduces their mobility in water bodies and affect their availability to aquatic organisms. Volatilization, on the other hand, is the process by which volatile contaminants, such as certain organic solvents or pesticides, transition from water to the atmosphere. Volatilization is especially important in shallow water bodies and during warmer conditions, where temperature-driven evaporation can lead to the release of contaminants into the air, affecting atmospheric quality and contributing to long-range transport [84].

### *4.1.2. Chemical Processes*

Chemical transformations also play a major role in determining the fate of contaminants in the environment. Hydrolysis, oxidation, and reduction reactions can alter the structure and toxicity of pollutants. Hydrolysis is the process by which a contaminant reacts with water, breaking down into more water-soluble products. For instance, many pesticides and pharmaceuticals undergo hydrolysis in aquatic environments, altering their persistence and toxicity [85]. Oxidation and reduction reactions, which are commonly mediated by environmental factors such as light, oxygen, or microbial activity, can also influence the chemical stability and reactivity of pollutants. For example, the oxidation of organic pollutants in water bodies can result in the formation of reactive intermediates that may be more toxic than the parent compounds [86]. Conversely, reduction reactions, such as the reduction of chromium (VI) to chromium (III), can alter the mobility and toxicity of metals in aquatic ecosystems [87]. These chemical processes are critical in determining the persistence, bioavailability, and overall environmental impact of contaminants.

### *4.1.3. Biological Processes: Biodegradation and Biotransformation*

Biological processes, such as biodegradation and biotransformation, are key mechanisms by which contaminants are broken down or modified in the environment. Biodegradation refers to the microbial breakdown of organic contaminants, such as pesticides, pharmaceuticals, and hydrocarbons, through enzymatic activity [88]. This process is often the most important pathway for the removal of organic pollutants from water and soil. Biodegradation rates are influenced by factors such as temperature, pH, and the availability of oxygen or nutrients. In some cases, microorganisms can completely mineralize contaminants, converting them into harmless by-products like carbon dioxide, water, and inorganic nutrients. Biotransformation, on the other hand, refers to the enzymatic alteration of contaminants within organisms, which can lead to the formation of metabolites that are either more or less toxic than the parent compound [89]. In aquatic ecosystems, fish and other organisms may metabolize contaminants such as pharmaceuticals and heavy metals, leading to changes in their toxicity and bioaccumulation potential. These biological processes play a vital role in reducing the persistence of contaminants and mitigating their impact on the environment.

## 4.2. Transport Mechanisms

### *4.2.1. Groundwater Transport*

Groundwater transport is a crucial process in the movement of contaminants through aquifers. The speed and distance of contaminant movement depend on factors such as hydraulic conductivity, which reflects the ability of soil or rock to transmit water, and the solubility and density of the contaminants [90]. Contaminants that are highly soluble in water, such as nitrates, can travel long distances through aquifers, while less soluble substances tend to remain closer to their source [91]. The density of the contaminant also influences its downward movement, with denser substances, such as certain metals, sinking deeper into the groundwater system [90].

### *4.2.2. Surface Water Transport*

Surface water transport refers to the movement of contaminants through rivers, lakes, and oceans. This process is primarily influenced by water currents, tides, and dispersion, which help carry contaminants over long distances [92]. In rivers and lakes, the flow velocity and water volume determine how quickly contaminants are transported downstream [93]. In coastal and marine environments, tides and ocean currents can spread pollutants across vast areas, influencing the distribution of contaminants. Dispersion, the spreading of contaminants as they move through water, plays a key role in diluting pollutants, but can also result in the widespread contamination of large aquatic areas.

### *4.2.3. Mechanisms of Co-occurrence and Interaction Among Contaminants*

The co-occurrence and interaction of multiple contaminants in wastewater and the environment are complex processes that can influence the fate and toxicity of pollutants. Contaminants often interact with one another in ways that can enhance or mitigate their individual effects, posing challenges for both treatment and ecological risk assessment. For instance, heavy metals such as cadmium, lead, and mercury may interact with organic contaminants, such as polycyclic aromatic hydrocarbons (PAHs) or pesticides, leading to synergistic toxicity [94]. These interactions can result in increased bioaccumulation in organisms, as certain contaminants may alter the uptake or detoxification pathways of others. Additionally, chemical interactions between contaminants can lead to the formation of more toxic or persistent compounds. For example, the presence of high concentrations of organic matter in wastewater can facilitate the formation of toxic disinfection by-products, such as trihalomethanes, during chlorination processes [95]. The presence of one contaminant may also influence the environmental behavior of others. For example, the sorption of metals to sediment particles may reduce their bioavailability, but the addition of organic pollutants can displace metals from particles, increasing their bioavailability and toxicity [96]. Understanding the mechanisms of co-occurrence and interaction among contaminants is essential for accurately assessing the ecological risks associated with wastewater contamination and for developing effective remediation strategies.

## 5. Occurrence of Contaminants in Various Environments

The contaminants in the water travel through different ecosystem and ultimately end up in aquatic ecosystems. The larger and bigger contaminants originating from the terrestrial ecosystems undergo various breakdown and transformation process which result in accumulation in the aquatic ecosystems as smaller particles. The contaminants in the wastewater from the land are washed up by the freshwater ecosystems like rivers and lakes to the marine ecosystem (Figure 3).

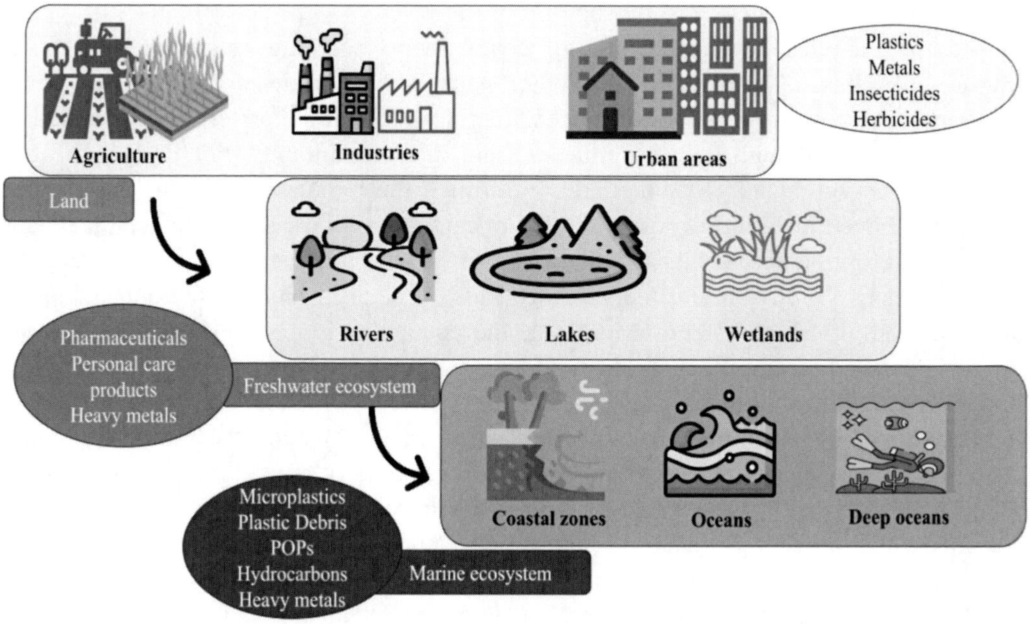


**Figure 3.** Occurrence of different wastewater contaminants in different ecosystems.

### 5.1. Freshwater Bodies

Freshwater bodies, including rivers, lakes, reservoirs, and wetlands, are critical components of the hydrological cycle, supporting biodiversity, providing drinking water, and serving as recreational spaces. However, these ecosystems are increasingly subjected to contamination due to anthropogenic activities. The primary sources of contaminants in freshwater bodies are agricultural runoff, urban wastewater discharge, and industrial effluents [97]. These pollutants often include nutrients, pesticides, PPCPs, which are transported via runoff from agricultural fields, industrial sites, and residential areas. One of the most significant contaminants in freshwater systems is nutrient pollution, especially nitrogen and phosphorus. These nutrients typically originate from fertilizers used in agriculture and urban areas, sewage discharge, and animal manure. When excessive nutrients enter freshwater bodies, they can cause eutrophication, a process that leads to an overgrowth of algae. Algal blooms can deplete dissolved oxygen levels in the water, creating hypoxic zones or "dead zones" where aquatic life, particularly fish, cannot survive [36, 37]. In addition to oxygen depletion, some algal blooms produce toxins that can be harmful to both aquatic organisms and human health. For example, the cyanobacteria blooms in freshwater systems have been linked to livestock deaths and the contamination of drinking water sources [98]. Another major concern is the introduction of pesticides into freshwater ecosystems, which often come from agricultural runoff. These chemicals, designed to protect crops from pests, can persist in the environment and accumulate in water bodies, leading to long-term ecological impacts. Pesticides like atrazine, chlorpyrifos, and glyphosate have been detected in freshwater systems worldwide and are known to adversely affect aquatic life, particularly fish and amphibians. The presence of these substances in freshwater bodies also poses significant risks to human health, especially when they contaminate drinking water supplies [99].

PPCPs are another emerging class of contaminants that are frequently detected in freshwater ecosystems. These compounds enter water bodies through the disposal of unused medications, effluents from wastewater treatment plants, and runoff from residential and hospital facilities. Although they are often present in trace amounts, PPCPs can still have disruptive effects on aquatic life, particularly through their endocrine-disrupting properties [100]. Studies have shown that exposure to low concentrations of certain pharmaceuticals, such as estrogens and antibiotics, can lead to alterations in the reproductive systems of fish and other aquatic organisms [101]. These effects can lead to a decline in biodiversity and disrupt the stability of freshwater ecosystems. Moreover, the contamination of freshwater bodies with heavy metals such as mercury, cadmium, and lead has been documented in many regions due to industrial activities, mining, and urban runoff. Heavy metals are persistent in the environment and can accumulate in sediments, where they may enter the food chain, posing risks to aquatic organisms and humans. For instance, mercury contamination in freshwater systems can lead to the bioaccumulation of methylmercury in fish, which can then biomagnify up the food chain, affecting predators such as birds and mammals, including humans [102]. The impacts of these pollutants are not limited to the biota of freshwater ecosystems; they also extend to the ecosystem services these systems provide. For instance, eutrophication and the degradation of water quality can impair the ability of freshwater bodies to serve as sources of potable water, thus affecting human health and socio-economic development. Additionally, contaminated freshwater systems can affect recreational activities, tourism, and fisheries, leading to significant economic losses.

### 5.2. Marine Systems

Marine ecosystems are critical for the functioning of the Earth's biosphere, providing a wide range of ecosystem services, including climate regulation, carbon sequestration, food resources, and biodiversity maintenance. However, these ecosystems are increasingly vulnerable to contamination from various anthropogenic sources, which have significant implications for the health of marine species and the functioning of oceanic processes. Contaminants in marine systems can originate from multiple sources, including industrial discharges, agricultural runoff, sewage, and atmospheric deposition [103]. One of the primary pollutants in marine ecosystems is microplastics, which have emerged as a major environmental concern in recent decades. Microplastics are small plastic particles, typically less than 5mm in size, that originate from the breakdown of larger plastic debris or from products such as synthetic textiles, PCPs, and cleaning agents [78]. These particles are carried by ocean currents and accumulate in various marine environments, from surface waters to deep-sea ecosystems. Microplastics can be ingested by marine organisms, leading to bioaccumulation and potential harm to their health, including physical blockages, impaired feeding, and exposure to toxic chemicals attached to the particles. In addition to microplastics, heavy metals such as mercury, lead, and cadmium are prevalent contaminants in marine systems. These metals enter the marine environment through industrial activities, mining operations, and atmospheric deposition, and they can persist in the environment for long periods. Heavy metals are often toxic to marine life, even at low concentrations, and can accumulate in the tissues of marine organisms, leading to bioaccumulation and biomagnification through the food web. For example, mercury contamination in marine systems is particularly concerning due to its ability to transform into methylmercury, a highly toxic form that biomagnifies in marine organisms, especially fish and marine mammals [102]. The accumulation of such metals can result in reproductive failures, impaired growth, and even mortality in marine species. Oil spills are another significant source of contamination in marine environments. These spills, often caused by tanker accidents, offshore drilling, and industrial discharges, release large quantities of hydrocarbons into the ocean, where they can have devastating effects on marine life. Oil slicks can coat the feathers of seabirds and the fur of marine mammals, leading to loss of insulation, dehydration, and, ultimately, death. In addition to physical damage, the toxic components of oil can disrupt the health of marine organisms by affecting their respiratory, digestive, and reproductive systems [104]. The persistence of oil in marine environments can cause long-term ecological damage, including the degradation of coastal habitats, such as mangroves and coral reefs, which are vital for biodiversity and coastal protection.

POPs, such as PCBs and organochlorine pesticides, are another group of contaminants that have significant effects on marine ecosystems. These chemicals are highly stable and remain in the marine environment for decades, accumulating in sediments and marine organisms. POPs can interfere with the endocrine systems of marine organisms, leading to reproductive problems, developmental abnormalities, and decreased immune function [105]. Due to their lipophilic nature, POPs accumulate in the fatty tissues of marine organisms and biomagnify up the food chain, ultimately impacting apex predators, including humans, who rely on seafood as a primary protein source. Nutrient pollution, particularly from nitrogen and phosphorus, also poses significant risks to marine ecosystems. These nutrients often derived from agricultural runoff, wastewater discharge, and industrial effluents, can lead to eutrophication in coastal areas. The excessive availability of nutrients can trigger harmful algal blooms, which deplete oxygen levels in the water, resulting in hypoxic or anoxic conditions that are detrimental to marine life. Furthermore, some species of algae produce toxins that can harm marine organisms and, in some cases, affect human health, especially through the consumption of contaminated seafood [16]. The impacts of these contaminants on marine ecosystems are profound and far-reaching, affecting biodiversity, ecosystem services, and the socio-economic value of marine resources. To mitigate these effects, integrated management strategies are required, including stricter regulations on pollutant discharges, the reduction of plastic waste, the implementation of oil spill response plans, and the restoration of damaged habitats. Additionally, advancing technologies for pollution detection, such as remote sensing and molecular monitoring techniques, could provide valuable tools for assessing the extent and impacts of contamination in marine ecosystems.

## 6. Processes Affecting Contaminant Dynamics

### 6.1. Dilution and Volatilization

Dilution and volatilization are key processes that influence the distribution and fate of contaminants in aquatic systems. Dilution refers to the reduction in the concentration of pollutants due to the mixing of contaminated water with cleaner water, usually occurring in large water bodies, rivers, or lakes. This process is driven by hydrodynamic factors such as current velocities, turbulence, and water flow, which can disperse contaminants over a larger volume. However, the effectiveness of dilution depends on the size of the receiving water body, the pollutant's initial concentration, and the physical properties of the contaminant. For example, pollutants with high solubility in water, like salts or some organic compounds, tend to dilute more effectively than hydrophobic substances [106]. Volatilization, the process by which contaminants transition from water to the atmosphere, is influenced by the volatility of the pollutant, temperature, wind speed, and the surface area of the water body. Volatile organic compounds, such as solvents and certain pesticides, are particularly prone to volatilization [107]. Both dilution and volatilization are important mechanisms that affect the persistence and spread of contaminants, with dilution reducing concentrations and volatilization potentially contributing to atmospheric pollution.

### 6.2. Sorption and Desorption

Sorption and desorption processes play a crucial role in determining the fate and mobility of contaminants in aquatic environments. Sorption refers to the binding of contaminants to solid surfaces, such as sediments, organic matter, or particles suspended in water, which reduces their bioavailability and mobility [96]. The partitioning of contaminants between sediment and water is influenced by several factors, including the chemical properties of the contaminant, the composition and organic content of the sediment, and the pH and temperature of the water. Hydrophobic contaminants, such as many pesticides and metals, tend to bind strongly to sediments, limiting their transport through the water column [108]. In contrast, more soluble contaminants may remain in the dissolved phase, increasing their mobility and bioavailability. Desorption, the reverse process, occurs when contaminants bound to sediments are released back into the water phase. This process is influenced by factors such as changes in water chemistry, sediment disturbance, and microbial activity [109]. Sorption and desorption processes are fundamental for understanding the environmental behavior of contaminants, as they affect both the short-term and long-term dynamics of pollutant distribution in aquatic ecosystems.

### 6.3. Biodegradation

Biodegradation is a natural process through which microorganisms break down organic contaminants into simpler, less harmful compounds. Microbial communities, including bacteria, fungi, and algae, play a central role in the degradation of different categories of pollutants such as hydrocarbons, pesticides, and pharmaceuticals [110]. These microorganisms possess a wide range of enzymatic pathways that enable them to degrade complex organic molecules into smaller molecules that can be assimilated or mineralized. Factors such as temperature, oxygen availability, pH, and nutrient concentrations can significantly influence the rate of biodegradation [111]. For example, in aerobic environments, bacteria may utilize oxygen to oxidize organic compounds, converting them into carbon dioxide and water. In anaerobic conditions, different microbial communities may engage in reductive dechlorination or other specialized metabolic pathways. The ability of microbial communities to degrade contaminants also depends on the composition of the community, which can vary with environmental conditions. Additionally, co-metabolism, where microorganisms degrade contaminants in the presence of other substances, can accelerate biodegradation processes [112, 113]. Understanding the role of microbial communities in biodegradation is essential for developing effective bioremediation strategies to mitigate the impact of contaminants on the environment.

### 6.4. Bioaccumulation in Food Chains

Bioaccumulation and biomagnification are critical processes that influence the fate and toxicity of contaminants in ecosystems, especially in the context of aquatic food webs. Bioaccumulation refers to the accumulation of contaminants in an organism's tissues over time, usually through ingestion, respiration, or absorption from the surrounding environment [114]. Persistent pollutants, such as heavy metals, POPs, and certain pesticides, tend to bioaccumulate in organisms because they are not easily metabolized or excreted. These pollutants can build up in the tissues of organisms at the base of the food chain, such as algae or small fish. As these organisms are consumed by predators, the contaminants are transferred and accumulated in higher trophic levels, a process known as biomagnification. This results in higher concentrations of pollutants in organisms at the top of the food chain, such as large fish, birds, and mammals [115]. Biomagnification can lead to toxic effects in higher trophic levels, including reproductive failure, immune system impairment, and developmental issues. The extent of bioaccumulation and biomagnification depends on several factors, including the chemical properties of the contaminant, the organism's feeding behavior, and the contaminant's persistence in the environment. Understanding the processes of bioaccumulation and biomagnification is crucial for assessing the long-term ecological risks of contaminants and for developing strategies to protect both human and ecological health.

# 7. Impacts of Wastewater Contaminants

## 7.1. Health Risks to Humans

Human exposure to wastewater contaminants occurs through various pathways, including direct contact with contaminated water, ingestion of contaminated food, and inhalation of pollutants from the air. Waterborne transmission is one of the most common routes, especially in regions with inadequate sanitation or wastewater treatment facilities [116]. Contaminants in wastewater, such as pathogens, heavy metals, and pharmaceuticals, can enter drinking water supplies or recreational waters, leading to a variety of health risks. Additionally, pollutants in wastewater can enter the food chain through irrigation of crops or consumption of contaminated aquatic organisms [117]. For example, heavy metals such as lead and mercury, as well as POPs, can bioaccumulate in fish and shellfish, posing risks to human health when consumed. Airborne exposure is another pathway, particularly for volatile organic compounds (VOCs) and gases like ammonia, which can be released from untreated wastewater or sludge [118]. These pollutants can be inhaled, leading to respiratory problems and other health concerns.

Overall, the health risks associated with wastewater contaminants are multifaceted and can affect individuals through multiple exposure routes, underscoring the importance of effective wastewater management and treatment practices. Wastewater contaminants pose both acute and chronic health risks to humans, depending on the type and concentration of the pollutant, as well as the duration of exposure. Acute health effects generally occur following short-term exposure to high levels of contaminants and can lead to symptoms such as gastrointestinal illnesses, respiratory problems, skin rashes, and eye irritation [1]. For instance, pathogens present in wastewater, such as bacteria, viruses, and protozoa, can cause waterborne diseases like cholera, dysentery, and typhoid fever upon ingestion or contact. In contrast, chronic health effects result from long-term exposure to low concentrations of contaminants, which may accumulate in the body over time, causing more serious health problems. Heavy metals, such as cadmium, lead, and arsenic, are notorious for their chronic toxicity, leading to organ damage, developmental disorders, neurotoxicity, and cancer [29]. Additionally, endocrine-disrupting chemicals (EDCs) present in wastewater, such as phthalates and BPA, can interfere with hormonal systems, causing reproductive and developmental problems, particularly in vulnerable populations like children and pregnant women [119]. The long-term impact of these contaminants requires ongoing monitoring and mitigation strategies to protect public health.

## 7.2. Ecological Risks in Aquatic and Terrestrial Ecosystems

Wastewater contamination poses significant ecological risks to both aquatic and terrestrial ecosystems, particularly in terms of biodiversity loss and habitat degradation. In aquatic ecosystems, the influx of organic matter, nutrients, and toxic chemicals from wastewater can lead to eutrophication, a process that depletes oxygen levels in water bodies and creates dead zones where most aquatic organisms cannot survive [120]. The nutrient overload, particularly nitrogen and phosphorus, fuels the excessive growth of algae, which can block sunlight and disrupt the balance of aquatic ecosystems, leading to a decline in biodiversity. Moreover, the presence of toxic substances such as heavy metals, POPs, and pharmaceuticals in wastewater can directly affect aquatic life by impairing the reproductive health, growth, and survival of species [121]. Terrestrial ecosystems are also at risk, particularly through the deposition of contaminants on soil or vegetation. The bioaccumulation of heavy metals and other pollutants in plants and animals can affect the health of terrestrial species and disrupt food webs, reducing biodiversity and ecosystem functionality. Ecosystem services benefits provided by ecosystems to humans, such as clean water, air purification, soil fertility, and pollination can be severely disrupted by wastewater contamination. In aquatic ecosystems, the degradation of water quality due to pollutants such as nutrients, pathogens, and chemicals can impair the ability of water bodies to provide clean drinking water and support aquatic biodiversity [120]. For example, the presence of heavy metals and synthetic chemicals in water bodies can contaminate drinking water sources, affecting the availability of safe water for human consumption. Additionally, the accumulation of pollutants in soil and sediments can hinder soil fertility and agricultural productivity, which disrupts the provisioning services of ecosystems. In coastal and marine ecosystems, wastewater contamination can alter habitats essential for fish, shellfish, and other marine species, undermining fisheries and aquaculture industries [122]. Thus, wastewater contamination not only affects biodiversity but also threatens critical ecosystem services that support human livelihoods and well-being.

## 8. Possible Strategies for Mitigating Wastewater Contamination

Mitigating wastewater contamination requires a multi-faceted approach that integrates advanced treatment technologies, effective policies, and public awareness (Figure 4). One key strategy is the implementation of advanced wastewater treatment methods, such as membrane bioreactors, constructed wetlands, and electrochemical processes, which offer higher removal efficiencies for contaminants, including micropollutants like heavy metals, microplastics, and pharmaceutical residues. Bio-based technologies, such as the use of biochar, algae, or microbial consortia, have also gained traction due to their sustainability and ability to simultaneously treat wastewater and recover valuable resources. For industrial and agricultural sectors, adopting source-segregation measures and pre-treatment techniques can significantly reduce the pollutant load entering centralized treatment facilities. Coupled with these approaches, real-time monitoring systems using sensors and machine learning algorithms can help detect contamination hotspots, optimize treatment processes, and prevent environmental release of untreated wastewater.

Policy frameworks and community engagement are equally critical in mitigating wastewater contamination. Governments and regulatory bodies need to enforce stringent wastewater discharge standards, incentivize green technologies, and support the transition to circular economy models where wastewater is treated as a resource for energy, water, and nutrient recovery. Public-private partnerships can fund infrastructure development and encourage innovations in decentralized wastewater treatment systems tailored to local contexts. Additionally, public awareness campaigns emphasizing the environmental and health risks of untreated wastewater can drive behavioral changes, such as reducing the use of harmful chemicals and promoting proper disposal practices. Together, these strategies create a comprehensive approach to mitigate wastewater contamination, protecting aquatic ecosystems and ensuring sustainable water resources for future generations.

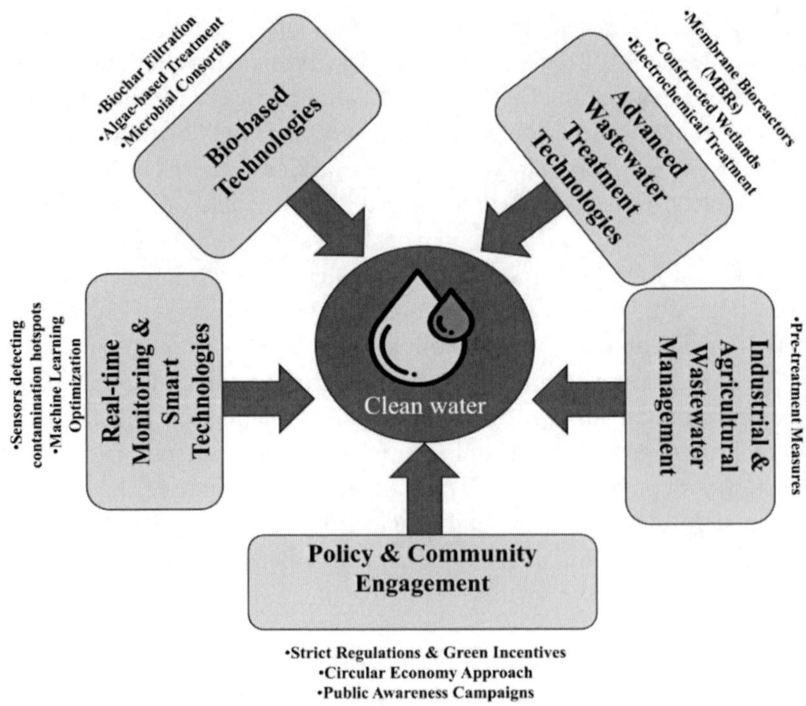


**Figure 4.** Possible Strategies for Mitigating Wastewater Contamination.

## Conclusion

The origin of wastewater and its diverse contamination sources present a significant challenge to global water sustainability. Domestic, industrial, agricultural, and urban activities introduce a wide range of pollutants that vary in composition, toxicity, and environmental persistence. These pollutants, including nutrients, pathogens, heavy metals, POPs, and emerging contaminants like microplastics and pharmaceuticals, pose severe risks to aquatic ecosystems and human health. Their dynamic transformations in the environment further complicate mitigation efforts, emphasizing the urgent need for innovative and adaptive management approaches. A deeper understanding of wastewater's fate and occurrence, particularly the pathways and mechanisms by which contaminants persist or are transformed in natural systems, is essential to tackle this multifaceted issue. To address these challenges, a holistic and integrative approach to wastewater management must be prioritized. This includes the adoption of advanced treatment technologies, stricter enforcement of regulatory frameworks, and the implementation of sustainable practices across all sectors. Furthermore, enhancing public awareness and fostering collaboration among stakeholders and governments, industries, researchers, and local communities can play a pivotal role in mitigating wastewater impacts. By acknowledging wastewater not only as a waste product but also as a resource for water, energy, and nutrient recovery, we can transition toward a circular water economy that supports environmental protection and resource sustainability. Only through these concerted efforts can we achieve the dual goals of safeguarding water quality and ensuring a sustainable future for generations to come.

## References


[1] Akpor, O. B., Otohinoyi, D. A., Olaolu, D. T., & Aderiye, B. I. (2014). Pollutants in wastewater effluents: Impacts and remediation processes. *International Journal of Environmental Research and Earth Science*, 3(3), 50–59.

[2] du Plessis, A. (2023). Water resources from a global perspective. In *South Africa's water predicament: Freshwater's unceasing decline* (pp. 1–25). Cham: Springer International Publishing.

[3] Bhat, S. U., & Qayoom, U. (2021). *Implications of sewage discharge on freshwater ecosystems. (Details such as journal/book title, volume, and pages are missing*. Please provide them for a complete APA citation).

[4] Karki, B. K., Baniya, S., Kharel, H. L., Angove, M. J., & Paudel, S. R. (2024). Urban wastewater management in Nepal: Generation, treatment, engineering and policy perspectives. *$H_2$Open Journal,* 7(2), 222–242.

[5] Chaturvedi, M., Mishra, A., Sharma, K., Sharma, G., Saxena, G., & Singh, A. K. (2021). Emerging contaminants in wastewater: Sources of contamination, toxicity, and removal approaches. In *Emerging treatment technologies for waste management* (pp. 103–132).

[6] Kishor, R., Purchase, D., Saratale, G. D., Saratale, R. G., Ferreira, L. F. R., Bilal, M., Chandra, R., & Bharagava, R. N. (2021). Ecotoxicological and health concerns of persistent coloring pollutants of textile industry wastewater and treatment approaches for environmental safety. *Journal of Environmental Chemical Engineering*, 9(2), 105012.

[7] Velusamy, S., Roy, A., Sundaram, S., & Kumar Mallick, T. (2021). A review on heavy metal ions and containing dyes removal through graphene oxide-based adsorption strategies for textile wastewater treatment. *The Chemical Record*, 21(7), 1570–1610.

[8] Sar, T., Marchlewicz, A., Harirchi, S., Mantzouridou, F. T., Hosoglu, M. I., Akbas, M. Y., Hellwig, C., & Taherzadeh, M. J. (2024). Resource recovery and treatment of wastewaters using filamentous fungi. *Science of the Total Environment*, 951, 175752. https://doi.org/10.1016/j.scitotenv.2022.175752.

[9] Abonyi, M. N., Obi, C. C., Nwabanne, J. T., & Aniagor, C. O. (2024). Emerging and ecofriendly biological methods for agricultural wastewater treatment. *Environmental Systems Research*, 13(1), 45.

[10] Altamira-Algarra, B., Puigagut, J., Day, J. W., Mitsch, W. J., Vymazal, J., Hunter, R. G., & García, J. (2022). A review of technologies for closing the P loop in agriculture runoff: Contributing to the transition towards a circular economy. *Ecological Engineering*, 177, 106571.

[11] Vaishnav, S., Saini, T., Chauhan, A., Gaur, G. K., Tiwari, R., Dutt, T., & Tarafdar, A. (2023). Livestock and poultry farm wastewater treatment and its valorization for generating value-added products: Recent updates and way forward. *Bioresource Technology*, 382, 129170.

[12] Mukhopadhyay, A., Duttagupta, S., & Mukherjee, A. (2022). Emerging organic contaminants in global community drinking water sources and supply: A review of occurrence, processes and remediation. *Journal of Environmental Chemical Engineering*, 10(3), 107560.

[13] Pericherla, S., Karnena, M. K., & Vara, S. (2020). A review on impacts of agricultural runoff on freshwater resources. *International Journal of Emerging Technologies*, 11, 829–833.

[14] Müller, A., Österlund, H., Marsalek, J., & Viklander, M. (2020). The pollution conveyed by urban runoff: A review of sources. *Science of the Total Environment*, 709, 136125.

[15] Yang, Y. Y., & Lusk, M. G. (2018). Nutrients in urban stormwater runoff: Current state of the science and potential mitigation options. *Current Pollution Reports*, 4, 112–127.

[16] Manasa, R. L., Mehta, A. (2020). Wastewater: Sources of Pollutants and Its Remediation. In: Gothandam, K., Ranjan, S., Dasgupta, N., Lichtfouse, E. (eds) Environmental Biotechnology Vol. 2 (pp.197–219). *Environmental Chemistry for a Sustainable World*, vol 45. Springer.

[17] Ajiboye, T. O., Oyewo, O. A., & Onwudiwe, D. C. (2021). Simultaneous removal of organics and heavy metals from industrial wastewater: A review. *Chemosphere,* 262, 128379. https://doi.org/10.1016/j.chemosphere.2020.128379.

[18] Zahoor, I., & Mushtaq, A. (2023). Water pollution from agricultural activities: A critical global review. *International Journal of Chemical and Biochemical Sciences*, 23(1), 164–176.

[19] Agarwal, S., Darbar, S., & Saha, S. (2022). Challenges in management of domestic wastewater for sustainable development. In *Current Directions in Water Scarcity Research* (Vol. 6, pp. 531–552). Elsevier.

[20] Yang, H., Tan, T., Du, X., Feng, Q., Liu, Y., Tang, Y., Bai, G., Liu, Z., Xia, S., Song, S., & Quintana, M. (2024). Advancements in freshwater aquaculture wastewater management: A comprehensive review. *Aquaculture,* 741346.

[21] Cui, Q., Huang, Y., Wang, H., & Fang, T. (2019). Diversity and abundance of bacterial pathogens in urban rivers impacted by domestic sewage. *Environmental Pollution,* 249, 24–35. https://doi.org/10.1016/j.envpol.2019.03.023.

[22] Tariq, A., & Mushtaq, A. (2023). Untreated wastewater reasons and causes: A review of most affected areas and cities. *International Journal of Chemical and Biochemical Sciences*, 23(1), 121–143.

[23] Nannou, C., & Lambropoulou, D. (2020). Chiral pharmaceuticals: Source, human risk, and future studies. In *Chiral Organic Pollutants* (pp. 249–278). CRC Press.

[24] Samal, K., Mahapatra, S., & Ali, M. H. (2022). Pharmaceutical wastewater as emerging contaminants (EC): Treatment technologies, impact on environment and human health. *Energy Nexus*, 6, 100076.

[25] Wydro, U., Wołejko, E., Luarasi, L., Puto, K., Tarasevičienė, Ž., & Jabłońska-Trypuć, A. (2023). A review on pharmaceuticals and personal care products residues in the aquatic environment and possibilities for their remediation. *Sustainability*, 16(1), 169.

[26] Nilsen, E., Smalling, K. L., Ahrens, L., Gros, M., Miglioranza, K. S., Picó, Y., & Schoenfuss, H. L. (2019). Critical review: Grand challenges in assessing the adverse effects of contaminants of emerging concern on aquatic food webs. *Environmental Toxicology and Chemistry*, 38(1), 46–60.

[27] Rafique, M., Hajra, S., Tahir, M. B., Gillani, S. S. A., & Irshad, M. (2022). A review on sources of heavy metals, their toxicity and removal technique using physico-chemical processes from wastewater. *Environmental Science and Pollution Research,* 29(11), 16772–16781.

[28] Raj, D., & Maiti, S. K. (2020). Sources, bioaccumulation, health risks and remediation of potentially toxic metal(loid)s (As, Cd, Cr, Pb and Hg): An epitomised review. *Environmental Monitoring and Assessment,* 192, 108.

[29] Ali, M. M., Hossain, D., Al-Imran, A., Khan, M. S., Begum, M., & Osman, M. H. (2021). Environmental pollution with heavy metals: A public health concern. In *Heavy Metals – Their Environmental Impacts and Mitigation* (pp. 771–783).

[30] Islam, M. T., Islam, T., Islam, T., & Repon, M. R. (2022). Synthetic dyes for textile colouration: Process, factors and environmental impact. *Textile and Leather Review,* 5, 327–373.

[31] Krithiga, T., Sathish, S., Renita, A. A., Prabu, D., Lokesh, S., Geetha, R., Namasivayam, S. K. R., & Sillanpaa, M. (2022). Persistent organic pollutants in water resources: Fate, occurrence, characterization and risk analysis. *Science of The Total Environment*, 831, 154808.

[32] Vieira, W. T., de Farias, M. B., Spaolonzi, M. P., da Silva, M. G. C., & Vieira, M. G. Ạ. (2021). Endocrine-disrupting compounds: Occurrence, detection methods, effects and promising treatment pathways—A critical review. *Journal of Environmental Chemical Engineering*, 9(1), 104558.

[33] Zahmatkesh, S., Hajiaghaei-Keshteli, M., Bokhari, A., Sundaramurthy, S., Panneerselvam, B., & Rezakhani, Y. (2023). Wastewater treatment with nanomaterials for the future: A state-of-the-art review. *Environmental Research*, 216, 114652.

[34] Encarnação, T., Pais, A. A., Campos, M. G., & Burrows, H. D. (2019). Endocrine disrupting chemicals: Impact on human health, wildlife and the environment. *Science Progress*, 102(1), 3–42.

[35] Rashmi, I., Roy, T., Kartika, K. S., Pal, R., Coumar, V., Kala, S., & Shinoji, K. C. (2020). Organic and inorganic fertilizer contaminants in agriculture: Impact on soil and water resources. In *Contaminants in Agriculture: Sources, Impacts and Management* (pp. 3–41).

[36] Wurtsbaugh, W. A., Paerl, H. W., & Dodds, W. K. (2019). Nutrients, eutrophication and harmful algal blooms along the freshwater to marine continuum. *Wiley Interdisciplinary Reviews: Water,* 6(5), e1373.

[37] Kirchman, D. L. (2021). *Dead zones: The loss of oxygen from rivers, lakes, seas, and the ocean.* Oxford University Press.

[38] Nieder, R., Benbi, D. K., & Reichl, F. X. (2018). Reactive water-soluble forms of nitrogen and phosphorus and their impacts on environment and human health. In *Soil Components and Human Health* (pp. 223–255). Springer.

[39] AbuQamar, S. F., El-Saadony, M. T., Alkafaas, S. S., Elsalahaty, M. I., Elkafas, S. S., Mathew, B. T., Aljasmi, A. N., Alhammadi, H. S., Salem, H. M., Abd El-Mageed, T. A., & Zaghloul, R. A. (2024). Ecological impacts and management strategies of pesticide pollution on aquatic life and human beings. *Marine Pollution Bulletin*, 206, 116613.

[40] Huang, X., Cui, H., & Duan, W. (2020). Ecotoxicity of chlorpyrifos to aquatic organisms: A review. *Ecotoxicology and Environmental Safety*, 200, 110731.

[41] Ray, S., & Shaju, S. T. (2023). Bioaccumulation of pesticides in fish resulting toxicities in humans through food chain and forensic aspects. *Environmental Analysis, Health and Toxicology*, 38. https://doi.org/10.5620/eaht.2023012 (Add DOI if available)

[42] Zalewska, M., Błażejewska, A., Czapko, A., & Popowska, M. (2021). Antibiotics and antibiotic resistance genes in animal manure–consequences of its application in agriculture. *Frontiers in Microbiology*, 12, 610656.

[43] Khan, M. N., & Mohammad, F. (2014). Eutrophication: Challenges and solutions. In *Eutrophication: Causes, Consequences and Control* (Vol. 2, pp. 1–15). Springer.

[44] Yang, Y. Y., & Lusk, M. G. (2018). Nutrients in urban stormwater runoff: Current state of the science and potential mitigation options. *Current Pollution Reports*, 4, 112–127.

[45] Haque, S. E. (2022). Urban water pollution by heavy metals, microplastics, and organic contaminants. In *Current Directions in Water Scarcity Research* (Vol. 6, pp. 21–43). Elsevier.

[46] Sudan, A., Ghosh, A., Verma, M., & Tomar, P. (2024). Impact of water pollution & perspective techniques to mitigate it: An overview. In *Handbook of Water Pollution* (pp. 29–64).

[47] Liu, A., Liu, L., Li, D., & Guan, Y. (2015). Characterizing heavy metal build-up on urban road surfaces: Implication for stormwater reuse. *Science of the Total Environment*, 515, 20–29.

[48] Hitchcock, J. N. (2020). Storm events as key moments of microplastic contamination in aquatic ecosystems. *Science of the Total Environment*, 734, 139436.

[49] Ghodsi, S. H. (2021). Urban stormwater management and green infrastructure planning: Predicting and abating combined sewer overflows (Doctoral dissertation, State University of New York at Buffalo). ProQuest Dissertations Publishing.

[50] Crocetti, P., Eusebi, A. L., Bruni, C., Marinelli, E., Darvini, G., Carini, C. B., Bollettini, C., Recanati, V., Akyol, Ç., & Fatone, F. (2021). Catchment-wide validated assessment of combined sewer overflows (CSOs) in a Mediterranean coastal area and possible disinfection methods to mitigate microbial contamination. *Environmental Research*, 196, 110367.

[51] Botturi, A., Ozbayram, E. G., Tondera, K., Gilbert, N. I., Rouault, P., Caradot, N., Gutierrez, O., Daneshgar, S., Frison, N., Akyol, Ç., & Foglia, A. (2021). Combined sewer overflows: A critical review on best practice and innovative solutions to mitigate impacts on environment and human health. *Critical Reviews in Environmental Science and Technology*, 51(15), 1585–1618.

[52] Singh, G., Singh, A., Singh, P., Shukla, R., Tripathi, S., & Mishra, V. K. (2021). The fate of organic pollutants and their microbial degradation in water bodies. In *Pollutants and water management: Resources, strategies and scarcity* (pp. 210–240). Springer.

[53] Coelho, A. C., & García Díez, J. (2015). Biological risks and laboratory-acquired infections: A reality that cannot be ignored in health biotechnology. *Frontiers in Bioengineering and Biotechnology*, 3, 56.

[54] Khan, S., Naushad, M., Govarthanan, M., Iqbal, J., & Alfadul, S. M. (2022). Emerging contaminants of high concern for the environment: Current trends and future research. *Environmental Research*, 207, 112609.

[55] Lv, Z., Ran, X., Liu, J., Feng, Y., Zhong, X., & Jiao, N. (2024). Effectiveness of chemical oxygen demand as an indicator of organic pollution in aquatic environments. *Ocean-Land-Atmosphere Research*, 3, 0050.

[56] Gil, A., Toledo, M., Siles, J. A., & Martín, M. A. (2018). Multivariate analysis and biodegradability test to evaluate different organic wastes for biological treatments: Anaerobic co-digestion and co-composting. *Waste Management*, 78, 819–828.

[57] Altieri, A. H., & Diaz, R. J. (2019). Dead zones: Oxygen depletion in coastal ecosystems. In *World seas: An environmental evaluation* (Vol. 2, pp. 453–473). Academic Press.

[58] Nayar, R. (2020). Assessment of water quality index and monitoring of pollutants by physico-chemical analysis in water bodies: A review. *International Journal of Engineering Research and Technology*, 9(1), 178-185.

[59] Jain, M., Noor, A., Sahoo, A., Pant, K. K., Ziora, Z. M., & Blaskovich, M. A. (2024). Biodegradation of petrochemical refinery wastewater. In *Biodegradation of toxic and hazardous chemicals* (pp. 1–13). CRC Press.

[60] Postigo, C., & Barceló, D. (2015). Synthetic organic compounds and their transformation products in groundwater: Occurrence, fate and mitigation. *Science of the Total Environment*, 503, 32–47.

[61] Akhtar, A. B. T., Naseem, S., Yasar, A., Naseem, Z. (2021). Persistent Organic Pollutants (POPs): Sources, Types, Impacts, and Their Remediation. In: Prasad, R. (eds) *Environmental Pollution and Remediation. Environmental and Microbial Biotechnology*. (pp. 213–246), Springer.

[62] Santa Maria, M. P., Hill, B. D., & Kline, J. (2019). Lead (Pb) neurotoxicology and cognition. *Applied Neuropsychology: Child*, 8(3), 272–293.

[63] Wu, Y. S., Osman, A. I., Hosny, M., Elgarahy, A. M., Eltaweil, A. S., Rooney, D. W., Chen, Z., Rahim, N. S., Sekar, M., Gopinath, S. C., & Mat Rani, N. N. I. (2024). The toxicity of mercury and its chemical compounds: Molecular mechanisms and environmental and human health implications: A comprehensive review. *ACS Omega*, 9(5), 5100–5126.

[64] Herbert, E. R., Boon, P., Burgin, A. J., Neubauer, S. C., Franklin, R. B., Ardón, M., Hopfensperger, K. N., Lamers, L. P., & Gell, P. (2015). A global perspective on wetland salinization: Ecological consequences of a growing threat to freshwater wetlands. *Ecosphere,* 6(10), 1–43.

[65] Ali, H., & Khan, E. (2019). Trophic transfer, bioaccumulation, and biomagnification of non-essential hazardous heavy metals and metalloids in food chains/webs—Concepts and implications for wildlife and human health. *Human and Ecological Risk Assessment: An International Journal*, 25(6), 1353–1376.

[66] Al-Obaidi, M., Kara-Zaitri, C., & Mujtaba, I. M. (2020). *Wastewater treatment by reverse osmosis* process. CRC Press.

[67] Chahal, C., Van Den Akker, B., Young, F., Franco, C., Blackbeard, J., & Monis, P. (2016). Pathogen and particle associations in wastewater: Significance and implications for treatment and disinfection processes. *Advances in Applied Microbiology*, 97, 63–119.

[68] Sengar, A., & Vijayanandan, A. (2022). Human health and ecological risk assessment of 98 pharmaceuticals and personal care products (PPCPs) detected in Indian surface and wastewaters. *Science of the Total Environment*, 807, 150677.

[69] Moo, C. L., Yang, S. K., Yusoff, K., Ajat, M., Thomas, W., Abushelaibi, A., Lim, S. H. E., & Lai, K. S. (2020). Mechanisms of antimicrobial resistance (AMR) and alternative approaches to overcome AMR. *Current Drug Discovery Technologies*, 17(4), 430–447.

[70] Riaz, L., Anjum, M., Yang, Q., Safeer, R., Sikandar, A., Ullah, H., Shahab, A., Yuan, W., & Wang, Q. (2020). Treatment technologies and management options of antibiotics and AMR/ARGs. In *Antibiotics and Antimicrobial Resistance Genes in the Environment* (pp. 369–393). Elsevier.

[71] Hiller, C. X., Hübner, U., Fajnorova, S., Schwartz, T., & Drewes, J. E. (2019). Antibiotic microbial resistance (AMR) removal efficiencies by conventional and advanced wastewater treatment processes: A review. *Science of the Total Environment*, 685, 596–608.

[72] Chaturvedi, P., Shukla, P., Giri, B. S., Chowdhary, P., Chandra, R., Gupta, P., & Pandey, A. (2021). Prevalence and hazardous impact of pharmaceutical and personal care products and antibiotics in environment: A review on emerging contaminants. *Environmental Research*, 194, 110664.

[73] Shahid, M. K., Kashif, A., Fuwad, A., & Choi, Y. (2021). Current advances in treatment technologies for removal of emerging contaminants from water–A critical review. *Coordination Chemistry Reviews*, 442, 213993.

[74] Kanda R. (2019). Reproductive Impact of Environmental Chemicals on Animals. *Advances in experimental medicine and biology,* 1200, 41–70.

[75] Ismanto, A., Hadibarata, T., Kristanti, R. A., Maslukah, L., Safinatunnajah, N., & Kusumastuti, W. (2022). Endocrine disrupting chemicals (EDCs) in environmental matrices: Occurrence, fate, health impact, physio-chemical and bioremediation technology. *Environmental Pollution*, 302, 119061.

[76] Kabir, E. R., Rahman, M. S., & Rahman, I. (2015). A review on endocrine disruptors and their possible impacts on human health. *Environmental Toxicology and Pharmacology*, 40(1), 241–258.

[77] Jasrotia, R., Langer, S., & Dhar, M. (2021, December). Endocrine disrupting chemicals in aquatic ecosystem: An emerging threat to wildlife and human health. In *Proceedings of the Zoological Society*, 74(4), 634–647. New Delhi: Springer India.

[78] Habib, R. Z., Thiemann, T., & Al Kendi, R. (2020). Microplastics and wastewater treatment plants—a review. *Journal of Water Resource and Protection*, 12(1), 1–18.

[79] Marouzi, S., Sabouri, Z., & Darroudi, M. (2021). Greener synthesis and medical applications of metal oxide nanoparticles. *Ceramics International*, 47(14), 19632–19650.

[80] Ahmed, S. F., Mofijur, M., Nuzhat, S., Chowdhury, A. T., Rafa, N., Uddin, M. A., Inayat, A., Mahlia, T. M. I., Ong, H. C., Chia, W. Y., & Show, P. L. (2021). Recent developments in physical, biological, chemical, and hybrid treatment techniques for removing emerging contaminants from wastewater. *Journal of Hazardous Materials*, 416, 125912.

[81] Akcil, A., Erust, C., Ozdemiroglu, S., Fonti, V., & Beolchini, F. (2015). A review of approaches and techniques used in aquatic contaminated sediments: Metal removal and stabilization by chemical and biotechnological processes. *Journal of Cleaner Production*, 86, 24–36.

[82] Ottander, N. (2022). Distribution and transport of microplastics and metals in sewage sludge-amended agricultural soil. [Master's thesis, Linköping University].

[83] Durães, N., Novo, L. A., Candeias, C., & Da Silva, E. F. (2018). Distribution, transport and fate of pollutants. In *Soil Pollution* (pp. 29–57). Academic Press.

[84] Vasiljevic, T., Su, K., & Harner, T. (2021). A first look at atmospheric concentrations and temporal trends of phthalates in distinct urban sectors of the Greater Toronto Area. *Atmospheric Pollution Research,* 12(2), 173–182.

[85] Narayanan, M., El-Sheekh, M., Ma, Y., Pugazhendhi, A., Natarajan, D., Kandasamy, G., Raja, R., Kumar, R. S., Kumarasamy, S., Sathiyan, G., & Geetha, R. (2022). Current status of microbes involved in the degradation of pharmaceutical and personal care products (PPCPs) pollutants in the aquatic ecosystem. *Environmental Pollution*, 300, 118922.

[86] Shi, J., Jiang, J., Chen, Q., Wang, L., Nian, K., & Long, T. (2023). Production of higher toxic intermediates of organic pollutants during chemical oxidation processes: A review. *Arabian Journal of Chemistry,* 16(7), 104856.

[87] Jiang, B., Gong, Y., Gao, J., Sun, T., Liu, Y., Oturan, N., & Oturan, M. A. (2019). The reduction of Cr (VI) to Cr (III) mediated by environmentally relevant carboxylic acids: State-of-the-art and perspectives. *Journal of Hazardous Materials*, 365, 205–226.

[88] Bose, S., Kumar, P. S., Vo, D. V. N., Rajamohan, N., & Saravanan, R. (2021). Microbial degradation of recalcitrant pesticides: A review. *Environmental Chemistry Letters*, 19, 3209–3228.

[89] Roohi, Bano, K., Kuddus, M., Zaheer, M. R., Zia, Q., Khan, M. F., Ashraf, G. Md., Gupta, A., & Aliev, G. (2017). Microbial enzymatic degradation of biodegradable plastics. *Current Pharmaceutical Biotechnology,* 18(5), 429–440.

[90] Al-Hashimi, O., Hashim, K., Loffill, E., Marolt Čebašek, T., Nakouti, I., Faisal, A. A., & Al-Ansari, N. (2021). A comprehensive review for groundwater contamination and remediation: Occurrence, migration and adsorption modelling. *Molecules,* 26(19), 5913.

[91] Sharma, P. K., Mayank, M., Ojha, C. S. P., & Shukla, S. K. (2020). A review on groundwater contaminant transport and remediation. *ISH Journal of Hydraulic Engineering*, 26(1), 112–121.

[92] González, C. J., Reyes, E., Álvarez, Ó., Izquierdo, A., Bruno, M., & Mañanes, R. (2019). Surface currents and transport processes in the Strait of Gibraltar: Implications for modeling and management of pollutant spills. *Ocean & Coastal Management*, 179, 104869. https://doi.org/10.1016/j.ocecoaman.2019.104869.

[93] Ji, Z. G. (2017). *Hydrodynamics and water quality: Modeling rivers, lakes, and estuaries* (2nd ed.). John Wiley & Sons.

[94] Mukherjee, A. G., Wanjari, R., Eladl, M. A., El-Sherbiny, M., Elsherbini, D. M. A., Sukumar, A., Kannampuzha, S., Ravichandran, M., Renu, K., Vellingiri, B., & Kandasamy, S. (2022). Mixed contaminants: Occurrence, interactions, toxicity, detection, and remediation. *Molecules,* 27(8), 2577.

[95] Liu, C., Ersan, M. S., Plewa, M. J., Amy, G., & Karanfil, T. (2018). Formation of regulated and unregulated disinfection byproducts during chlorination of algal organic matter extracted from freshwater and marine algae. *Water Research*, 142, 313–324.

[96] Zhang, C., Yu, Z. G., Zeng, G. M., Jiang, M., Yang, Z. Z., Cui, F., Zhu, M. Y., Shen, L. Q., & Hu, L. (2014). Effects of sediment geochemical properties on heavy metal bioavailability. *Environment International,* 73, 270–281.

[97] Mushtaq, N., Singh, D. V., Bhat, R. A., Dervash, M. A., & Hameed, O. B. (2020). Freshwater contamination: Sources and hazards to aquatic biota. In *Fresh Water Pollution Dynamics and Remediation* (pp. 27–50). Springer.

[98] Hilborn, E. D., & Beasley, V. R. (2015). One Health and cyanobacteria in freshwater systems: Animal illnesses and deaths are sentinel events for human health risks. *Toxins,* 7(4), 1374–1395.

[99] Bhat, S. U., Akhter, Z., Neelofar, M. R., & Qayoom, U. (2022). Towards understanding the impact of pesticides on freshwater ecosystem. In *Pesticides in the Natural Environment* (pp. 121–138). Elsevier.

[100] Kasonga, T. K., Coetzee, M. A., Kamika, I., Ngole-Jeme, V. M., & Momba, M. N. B. (2021). Endocrine-disruptive chemicals as contaminants of emerging concern in wastewater and surface water: A review. *Journal of Environmental Management*, 277, 111485.

[101] Badamasi, I., Odong, R., & Masembe, C. (2020). Threats posed by xenoestrogenic chemicals to the aquatic ecosystem, fish reproduction and humans: A review. *African Journal of Aquatic Science*, 45(3), 243–258.

[102] de Almeida Rodrigues, P., Ferrari, R. G., Dos Santos, L. N., & Junior, C. A. C. (2019). Mercury in aquatic fauna contamination: A systematic review on its dynamics and potential health risks. *Journal of Environmental Sciences*, 84, 205–218.

[103] Weldeslassie, T., Naz, H., Singh, B., & Oves, M. (2018). Chemical contaminants for soil, air and aquatic ecosystem. In *Modern Age Environmental Problems and Their Remediation* (pp. 1–22). Springer.

[104] Dupuis, A., & Ucan-Marin, F. (2015). *A literature review on the aquatic toxicology of petroleum oil: An overview of oil properties and effects to aquatic biota* (p. 52). Canadian Science Advisory Secretariat.

[105] Islam, R., Kumar, S., Karmoker, J., Kamruzzaman, M., Rahman, M. A., Biswas, N., Tran, T. K. A., & Rahman, M. M. (2018). Bioaccumulation and adverse effects of persistent organic pollutants (POPs) on ecosystems and human exposure: A review study on Bangladesh perspectives. *Environmental Technology & Innovation,* 12, 115–131.

[106] Walker, D. B., Baumgartner, D. J., Gerba, C. P., & Fitzsimmons, K. (2019). Surface water pollution. In M. L. Brusseau, I. Pepper, & C. Gerba (Eds.), *Environmental and Pollution Science* (3rd ed., pp. 261–292). Academic Press.

[107] Li, Z., Chen, H., Zhan, L., Zheng, S., Wu, H., & Yang, L. (2023). Review of migration, transformation, and control of volatile components during desulfurization wastewater evaporation: Advances and perspectives. *Energy & Fuels*, 37(20), 15248–15266.

[108] Christensen, E. R., Wang, Y., Huo, J., & Li, A. (2022). Properties and fate and transport of persistent and mobile polar organic water pollutants: A review. *Journal of Environmental Chemical Engineering*, 10(2), 107201.

[109] Yang, X., Dai, X., Zhang, Y., Lin, H., Wang, J., He, Z., & Li, Y. (2020). Sorption, desorption, and transformation of synthetic progestins in soil and sediment systems. *Geoderma*, 362, 114141.

[110] Parte, S. G., Mohekar, A. D., & Kharat, A. S. (2017). Microbial degradation of pesticide: A review. *African Journal of Microbiology Research*, 11(24), 992–1012.

[111] Fouad, F. A., Youssef, D. G., Shahat, F. M., & Abd El-Ghany, M. N. (2022). Role of microorganisms in biodegradation of pollutants. In *Handbook of biodegradable materials* (pp. 1–40). Springer International Publishing.

[112] Maier, R. M., & Gentry, T. J. (2015). Microorganisms and organic pollutants. In *Environmental microbiology* (pp. 377–413). Academic Press.

[113] Lü, H., Wei, J. L., Tang, G. X., Chen, Y. S., Huang, Y. H., Hu, R., Mo, C. H., Zhao, H. M., Xiang, L., Li, Y. W., & Cai, Q. Y. (2024). Microbial consortium degrading of organic pollutants: Source, degradation efficiency, pathway, mechanism and application. *Journal of Cleaner Production*, 141913.

[114] Agbugui, M. O., & Abe, G. O. (2022). Heavy metals in fish: Bioaccumulation and health. *British Journal of Earth Sciences Research*, 10(1), 47–66.

[115] Rodríguez-Jorquera, I. A., Vitale, N., Garner, L., Perez-Venegas, D. J., Galbán-Malagón, C. J., Duque-Wilckens, N., & Toor, G. S. (2017). Contamination of the upper class: Occurrence and effects of chemical pollutants in terrestrial top predators. *Current Pollution Reports*, 3, 206–219.

[116] Ravindra, K., Mor, S., & Pinnaka, V. L. (2019). Water uses, treatment, and sanitation practices in rural areas of Chandigarh and its relation with waterborne diseases. *Environmental Science and Pollution Research,* 26, 19512–19522.

[117] Minhas, P. S., Saha, J. K., Dotaniya, M. L., Sarkar, A., & Saha, M. (2022). Wastewater irrigation in India: Current status, impacts and response options. *Science of the Total Environment*, 808, 152001.

[118] Arora, K., Kumar, V., Nayak, S. J., Surya, A., & Sugumar, S. (2023). Recent advances in biological remediation of volatile organic compounds (VOCs) and heavy metals. In *Bio-Nano Filtration in Industrial Effluent Treatment* (pp. 49–68). Elsevier.

[119] Hassan, S., Thacharodi, A., Priya, A., Meenatchi, R., Hegde, T. A., Thangamani, R., Nguyen, H. T., & Pugazhendhi, A. (2024). Endocrine disruptors: Unravelling the link between chemical exposure and women's reproductive health. *Environmental Research*, 241, 117385.

[120] Ogidi, O. I., & Akpan, U. M. (2022). Aquatic biodiversity loss: Impacts of pollution and anthropogenic activities and strategies for conservation. In *Biodiversity in Africa: Potentials, threats and conservation* (pp. 421–448). Springer Nature Singapore.

[121] Singh, P. K., Kumar, U., Kumar, I., Dwivedi, A., Singh, P., Mishra, S., Seth, C. S., & Sharma, R. K. (2024). Critical review on toxic contaminants in surface water ecosystem: Sources, monitoring, and its impact on human health. *Environmental Science and Pollution Research*, 31(45), 56428–56462.

[122] Wear, S., Cunningham, S., Feller, I. C., Fiorenza, E. A., Frielaender, A., Halpern, B. S., Hirashiki, C., Lamb, J., Lovelock, C. E., McLean, J., & Nichols, R. C. (2024). 6.13 Wastewater pollution impacts on estuarine and marine environments. In *Treatise on Estuarine and Coastal Science* (2nd ed., pp. 434–466). Academic Press.

## Chapter 2

# Wastewater Treatment: Arduous Challenges and Avenues

**Sudakshina Trivedi**
**and Amrita Srivastava***
Department of Chemistry, University of Lucknow, Lucknow, India

## Abstract

The economy of a developing country is significantly driven by agriculture, but as the nation progresses towards becoming a developed country, industrial growth plays a crucial role in accelerating economic expansion. Industries such as pharmaceuticals, textiles, petroleum, slaughterhouses, oil mills, and refineries contribute substantially to the economy but also generate large volumes of waste. Much of this waste is discharged into nearby water bodies, leading to severe water pollution. Additionally, domestic wastewater further exacerbates the issue, posing a serious threat to both the environment and human health. Several key challenges need urgent attention, including inadequate sanitation facilities, lack of awareness regarding domestic wastewater reuse, and inefficient wastewater treatment technologies that require high energy consumption. Effective wastewater management extends beyond purificationit also involves proper sludge treatment. While spreading awareness through social initiatives is crucial, research into cost-effective and sustainable treatment methods is equally essential.

Various treatment techniques, including membrane, chemical, and biological processes, play vital roles in removing different pollutants. Recent advancements, such as the integration of nanotechnology and artificial intelligence, have shown great promise in enhancing treatment efficiency. This chapter explores the major challenges in wastewater treatment and presents innovative approaches to tackle this critical issue. It highlights potential remedial measures to mitigate water contamination, ultimately benefiting public health and preserving aquatic ecosystems.

**Keywords:** wastewater, environment, nanotechnology, contaminants, aquatic ecosystems, membrane technologies

---

* Corresponding Author's Email: amrita.sri18@gmail.com

In: Advancements in Wastewater Treatment
Editors: Neha Agarwal, Maulin P. Shahl, Vijendra Singh Solankil et al.
ISBN: 979-8-90134-007-3

## 1. Introduction

The population increase has caused a rise in anthropogenic activities, which disturb the environment by releasing harmful products, leading to environmental pollution. Human activities are exploiting two major natural resources energy and water—contributing to issues such as global warming and water contamination [1]. Rapid population growth and urbanization are generating large amounts of wastewater, polluting the environment. It is expected that this rapid increase in population will make water one of the scarcest resources in future [2]. India comprises merely 2.4% of the world's total land area, yet sustains over 15% of the global population. The total utilizable water resources in India account for only 4% of the global availability. Collectively, these factors are exerting increasing pressure on water resources, both in India and globally, leading to a decline in per capita water availability. . When water quality is adulterated by agricultural practices, household activities, mining, energy production, deforestation, industrial activities and urban settlements, it is termed as 'wastewater'. Industrial wastewater contains chemicals, dyes, metals, metalloids, volatile compounds and other harmful substances whereas domestic wastewater contains pathogens. Infections caused by water-based pathogens have become a significant threat to public health. Pathogens present in domestic wastewater can include viruses, parasites, helminthes, or bacteria. A common source of domestic wastewater is inadequate sanitation facilities. For example, the presence of *E. coli* (an intestinal bacteria excreted by all warm-blooded animals) in fecal matter is used as an indicator to measure the extent of fecal contamination [3].

Industrial wastewater varies in composition depending on its source of origin, posing treatment challenges due to the need for varied methods based on composition. Such polluted water causes numerous ailments in living beings including humans, animals and plants. Some reported health problems caused by wastewater exposure include haemorrhage, nervous problems, ulcers, kidney damage and reproductive damage. Contaminated water has decreased oxygen levels, affecting aquatic fauna. Wastewater exhibits high levels of Biological Oxygen Demand (BOD) and Chemical Oxygen Demand (COD). Elevated BOD and COD levels reduce dissolved oxygen, thereby impairing the sustainability of aquatic life [4]. The most harmful substances present in wastewater include pesticides, fertilizers, heavy metals, dyes, micro-plastics, pharmaceutically active compounds from hospital wastes, cosmetic products, and waste from the paper industry. Inorganic pollutants are also a major source of water contamination. Natural water that comes into contact with rocks and minerals components, may contain inorganic pollutants in the form of dissolved inorganic matter. However, a major proportion of inorganic matter, such as nitrite ions, is introduced through human activities [5]. Some inorganic pollutants include trace elements, inorganic salts, mineral acids, nitrates, phosphates, sulphates, fluorides, chlorides, cyanides, oxalates etc. An increased amount of phosphates may lead to eutrophication of water bodies which leads to toxin production. Inorganic pollutants such as sulphates are generally reduced to sulphides, biologically. These sulphides react with hydrogen to form hydrogen sulphide, which is highly toxic to aquatic flora and fauna.

The exponential global population increase has led to a corresponding rise in food-related demands. To enhance both the quality and quantity of crops, pesticides are applied in substantial quantities. Trace amounts of pesticides are absorbed by plants and accumulate in the soil, which subsequently leaches them into nearby water streams. . Due to their high toxicity and hazardous nature, pesticides are regarded as one of the most dangerous and persistent categories of chemical contaminants. Fields treated with chemicals, fertilizers and pesticides also serve as a source of wastewater. When dissolved in rainwater, these substances run off into nearby freshwater streams and contaminate the aquatic environment. The most commonly used chemicals are insecticides, herbicides, fungicides, acaricides, molluscicides, nematicides and rodenticides. Based on their chemical composition, pesticides are mainly classified into four types: organochlorines, organophosphates, carbamates and substituted urea, all of which are major contributors to water pollution. As per the available world report, most of the used pesticides are employed without giving much attention to their possible consequences [5].

Cooling operations in power generation plants also release large quantities of wastewater. Wastewater is also generated in the form of bilge water by ships and cargo vessels. Fire extinguishers consume large quantities of water, which is often not treated after use. As a result, this water becomes contaminated. Depending on the materials burnt, water may become contaminated with various elements and can run off into nearby streams as wastewater [6]. Leaching of landfills also release waste, which when dissolved in rainwater, runs off as wastewater.

Synthetic dyes being non-biodegradable, are more harmful and are mostly carcinogenic. These dyes decrease the solubility of necessary gases in water and also hinders light penetration into water bodies, rendering them unfit for aquatic flora and fauna. Accumulation of heavy metals present in dyes can impart carcinogenic and mutagenic properties. Dyes removal can be carried out using different methods such as biological methods, physicochemical methods, chemical methods, and advanced oxidation processes (AOPs).

Pollutants that have recently emerged and require modern detection and treatment methods are known as "emerging contaminants". These emerging contaminants include endocrine-disrupting chemicals (EDCs) which damage not only aquatic life but also the entire ecosystem [7]. These detrimental pollutants stay for a longer period in water and are absorbed by plants and animals causing various ailments. Reports of United Nations Development Program (UNDP) have shown a scarcity of clean water and inadequate sanitation practices [5]. Water treatment systems also vary significantly between rural and urban areas due to financial constraints. In India, only approximately 24% of wastewater is treated [8]. A Safe water supply is not only a basic societal need but also a fundamental right for everyone. However, the quality of water supplied in most countries and continents remain poor due to inadequate hygiene and unsatisfactory treatment practices.

Some water borne diseases such as cholera, typhoid, and hepatitis are caused by detrimental pathogens present in contaminated water. Their entry into main-water streams occurs primarily through sewage discharge or waste from slaughterhouses [9]. Diseases caused by such pathogens vary in their fatality rates. For Example, some viruses can cause hepatitis, while protozoa, fungi, algae, plankton, and worms can also pose serious threats to human health.

According to a 2010 WHO report, the annual average mortality from waterborne diseases exceeds 5 million people, and approximately more than 50% of the affected population suffers from intestinal microbial inflammation [10]. Globally, only 24% of wastewater is treated before release, with the remainder discharged untreated. Very few countries have efficient primary treatment systems, and even fewer have both primary and secondary treatment schemes to treat wastewater effectively. Consequently, this chapter discusses the existing challenges in treating wastewater and also explores emerging opportunities for improvement and innovation.

## 2. Avenues and Challenges

While working for science and development, proper treatment of wastewater is one of the most pressing issues of current time. Wastewater treatment has undergone appreciable transformations from rudimentary methods previously employed to newly developed sophisticated systems, aimed at protecting public health and environment.

### 2.1. Early Methods and Evolution of Technology

Wastewater treatment initially began with simple methods such as filtration. However, these methods were insufficient to treat contaminated water, leading to serious public health issues from the consumption of polluted water. Wastewater treatment methods can be broadly categorized into three main types as energy - intensive methods, chemical methods and physical methods. Among these, physical methods are the most commonly employed for wastewater treatment. They involve recirculation technologies, which can be classified as conventional or non-conventional. The recirculation method is particularly effective for purifying urban and industrial wastewater through various unit processes [11]. Conventional methods of wastewater treatment are primarily based on physical and chemical methods. Commonly used physical methods are sedimentation, filtration, boiling, distillation, desalination, reverse osmosis, adsorption, and ion exchange. Physical adsorption is an exothermic process, as it is accompanied by a decrease in both free energy and entropy of the adsorption system [12]. However, physical methods of wastewater treatment have two major limitations. The first is clogging of filter media due to accumulation of solid pollutants. The second is poor interaction between heavy metals and filter media, which results in limited pollutant removal efficiency. For drinking water applications, energy intensive methods are employed, where an electric current is used to aid the treatment.

An ancient wastewater treatment method developed in 1970s was the Biological Activated Carbon (BAC) method, which combined activated carbon adsorption with the bio-degradation of organic pollutants. This process significantly supported the development of aerobic biomass [13]. Some merits of BAC method include:

1. Fast kinetics and flexibility of adsorption process
2. Low initial cost of investment
3. Removal of various target contaminants
4. Regeneration of adsorbents makes it economically beneficial

On the other hand, some demerits associated with this method are:

1. This is a non-selective method
2. Different kind of adsorbents are needed based on the nature of wastewater
3. Efficiency of adsorption is decreased on repeated use of adsorbent
4. Regeneration process is expensive and may also cause adsorbent loss.

Municipal wastewater treatment involves primary, secondary and tertiary treatment methods. Primary treatment includes the removal of grit through screening, grinding and sedimentation. Sedimentation is a settling process in which the aqueous solution is left undisturbed, allowing suspended, dense metal pollutants to settle at the bottom due to gravity, based on their density and solubility. Major limitations of this process include low efficiency of removal of fine smaller sized particles. Thus, complete removal of metal pollutants cannot be achieved employing this method. Secondary treatment operates by oxidation of organic matter dissolved in water-streams by means of biologically active sludge. Tertiary treatment includes nitrogen removal and other steps such as granular filtration and activated carbon absorption. Significant progress was made with the introduction of biological methods in this field. These advancements improved the ability to address complex contaminants, enhancing treatment outcomes.

Initially, the screening process is used to extract suspended solids from contaminated water. Usually industrial and sewage wastewater contains more suspended solid contaminants than domestic wastewater. Therefore, efficiency of treatment plant is also based on its effectiveness in eliminating suspended solid contaminants [14]. Adsorption and coagulation are among the most commonly employed methods; however, they only transfer pollutants from aqueous phase to solid phase, generating sludge and creating disposal challenges, which makes both methods environmentally non-friendly.

Desalination helps in removing dissolved minerals from water. However, traditional desalination methods are ineffective in removing organic compounds and toxic metal ions. Various studies have shown that nearly 30% of organic matter in wastewater is non-biodegradable and, therefore, must be disposed of as sludge [15]. Deep percolation- mediated groundwater recharge also has the capacity to remove microorganisms. Although chlorination is used as the final step in water treatment process to prevent the regrowth of pathogens, chlorine can produce by-products that act as chemical contaminants [16]. This side effect of chlorination needs to be thoroughly investigated and addressed. In addition to pathogens, chemical pollutants are also a major concern, especially for nations prioritizing industrial development. In such cases, industrial effluents often flow into and mix with domestic water supplies and nearby water bodies. Wastewater generated from chemical industries contains various carcinogenic compounds such as Polychlorinated Biphenyls (PCBs), in high concentrations [17]. PCBs, being fat-soluble and easily transportable in the environment, have been widely used in industrial applications for a long time. They also exhibit high stability within tissues and cells [18]. A recent survey conducted along the Musi River in India revealed that metal ions originating from wastewater were detected in cow's milk which was likely transferred through grass fodder irrigated with contaminated water [8].

Limitations associated with traditional methods gave rise to advanced methods of wastewater treatment. These advanced wastewater treatment methods are primarily focused on the removal of additional organic and suspended solids, nitrogenous oxygen demand, nutrients and toxic materials [19]. Heavy metals are also removed through hydrochar adsorption, which exhibits excellent heavy metal removing capacity due to the presence of oxygen based functional groups on its surface [20].

Another extensively used eco technology to treat water contaminants is Constructed Wetlands (CWs) [21]. They function as ecosystem stabilizers, particularly in rural areas. Their notable features include low maintenance, long lifespan and robust operation. Emergent pollutants are treated using a recently introduced technology of involving 'hydrogels [22].

These are porous materials with a 3-D flexible network containing hydrophilic groups. Due to these groups hydrogels swell in water and other fluids. Hydrogels deserve much attention in wastewater treatment due to their peculiar intrinsic features such as re-usability, adsorption capacity and fast kinetics. Specific disadvantages associated with chitosan hydrogels are weak mechanical strength, poor stability which limits their use. Chitosan hydrogels are recently developed technique but they lack mechanical strength in thermal and acidic mediums. Although extensive work has been done, the current design of wastewater treatment plants is still not efficient in eliminating contaminants and their metabolites from wastewater [23]. Nanomaterials (NMs) due to their large surface area are effective at adsorbing toxic pollutants and aiding in pollution detection. After evaluating all the available techniques, it may be said that CWs can act as tertiary treatment systems or standalone units for wastewater treatment. They also present a cost-effective alternative to engineered NMs.

## 2.2. Prevailing Challenges in Wastewater Treatment

There are multiple factors that prevent many developing countries from implementing comprehensive wastewater treatment methods. It is evident that the efficiency of wastewater treatment depends primarily on two major factors: the composition of dissolved contaminants and the mechanism adopted for pollutant removal. Activated carbon has great adsorption efficiency which makes it superior to coagulation-flocculation method. Activated carbon has better catalytic efficiency than commercially available catalysts. However, the performance of activated carbon is subjected to certain limitations, primarily due to pore blocking and competition for adsorption sites and treating the filtrate. Activated biochar needs to be discarded after a year of usage, which can lead to secondary pollution. Although biochar can be regenerated by various methods, the regeneration cost is often higher than producing new biochar. Undoubtedly, modified activated biochar is very efficient in wastewater treatment; however, its production cost is a significant concern. On the other hand, biomass derived engineered activated biochar can efficiently catalyse the wastewater treatment process.

The increasing volume of wastewater generates additional challenges in minimizing impacts on human health and the environment. Therefore, the foremost challenge is to assess the scale at which wastewater treatment is both feasible and viable. Emphasis should be on the reuse of industrial and domestic wastewater. With wastewater discharge increasing at a tremendous rate globally, establishing and maintaining large wastewater treatment plants remain difficult for many developing nations. Commonly promoted wastewater treatment plants are decentralized, offering flexibility for long term use, economic stability and compatibility with local effluent treatment needs [24]. Despite the availability of numerous wastewater treatment methods, it is important to focus on techniques that are eco-friendly and support long-term treatment with minimal expenditure.

### *2.2.1. Aging Infrastructure*

Rapid urbanization and industrialization are significantly increasing the complexity of contaminants. Urban water supply systems are often interconnected with various other networks, leading to indirect impacts that contribute to water contamination. Existing technologies have not advanced sufficiently to keep pace with the rising pollution levels. To address these challenges effectively, development of modern treatment facilities with advanced infrastructure is essential. Conventional wastewater treatment plants have proven inadequate in removing pathogens and other micro-pollutants [25]. This creates a demand for advanced methods that can effectively treat micro-pollutants while remaining economically viable. The main drawback of conventional water treatment systems is their limited ability to remove dyes, which often pass untreated. Therefore, conventional methods are not suitable for textile-based water treatment plants. While the ion exchange method can be used for dye removal, its effectiveness is limited due to its narrow range of action. It cannot treat all types of dyes, restricting its applicability as a comprehensive water treatment solution.

### *2.2.2. Complex Contaminants*

Increasing industrialization is leading to the generation of complex pollutants that are difficult to degrade. Among these, some of the most challenging contaminants are pharmaceuticals and personal care products. Various pharmaceutical compounds have been detected in water bodies worldwide, highlighting the growing concern over water contamination [26]. Their extensive use in the 21$^{st}$ century has led to the presence of their traces in surface and ground water in the form of emerging contaminants [27]. These chemical compounds are found to be physiologically active in nature, slightly water soluble and consisting of stubborn organic and inorganic components [28]. Such pharmaceutical substances are termed evolving organic pollutants due to their existence in aquatic habitat. These pharmaceutically active components have shown toxicity even in very minute quantity (such as $0.04gL^{-1}$) due to their high rate of persistence, obstinacy and incomplete metabolism. Their bioactive nature can also cause endocrine disruption in living beings. Commonly used drugs like Paracetamol, Diclofenac, and Atenolol have been detected in municipal wastewater, posing a significant toxic threat. Therefore, their eradication is essential before using treated water.

### *2.2.3. Economic Constraints*

Major limitations in implementing efficient technologies lie in financial aspects. All available technologies for wastewater treatment are costly in terms of energy except solar powered or activated carbon treatment. High cost of infrastructure, operation procedure and maintenance of technologies is difficult to afford and manage. Lack of enough funding resources also makes it tough to meet financial needs. Thus, financial barriers; comprising high costs and inadequate funding, pose significant challenges to establishing proper wastewater treatment plants. Due to financial restrictions, local municipalities are unable to adopt recently evolved methods to treat wastewater. Integral cost of wastewater treatment plants include sum of operating expenditures (OPEX) and capital expenditures (CAPEX) [29]. For a wastewater treatment plan to be economical, OPEX must be lower than CAPEX. Factors leading to lower OPEX are:

a. Saved disposing cost due to less sludge production.
b. Extraction of value-added substances from photosynthetic bacteria used in wastewater treatment.
c. Saved electricity cost since photosynthetic bacteria do not need aeration while performing treatment activity.

Whereas, factors leading to higher value of CAPEX are:

a. Photosynthetic bacteria need additional facilities like dryer and extraction column.

b. Photo-bioreactors are much complex but less developed than conventionally used aeration tanks.

### *2.2.4. Environmental Impact*

Impact of treatment methods on surrounding environment is also a significant challenge. Among the various environmental impacts, one of the major concerns is the proper management of generated sludge. Bio-solids generated after decontamination also need to be treated and disposed of properly. Managing this sludge requires substantial energy consumption, which again is exhaustive and harmful to the environment. Pollutants removed from wastewater often get adsorbed onto the surfaces, which in turn poses an environmental concern [30].

Another advanced wastewater treatment method is Granular Activated Carbon (GAC), which is highly effective for removing odorous organic pollutants from industrial wastewater. This method involves adsorbing pollutants from water streams onto GAC in fixed beds [31]. These materials can be easily re-activated and re-used, thereby, reducing their carbon footprint. However, the regeneration and reuse processes need exhaustive amount of energy which limits its applications in wastewater treatment [32]. These problems can be resolved by working in two directions: use of optimal energy to treat and dispose wastewater as well as sludge and finding alternative methods for sludge disposal and recycling.

### *2.2.5. Regulatory Issues*

Policy gaps and inadequate compliance are the primary regulatory challenges for wastewater treatment plants. There is a lack of consistent and comprehensive policy frameworks for wastewater treatment. For many facilities, it is a challenge to fully comply with the existing regulations properly. This gap is further widened due to lack of cohesive policies for protection of environment particularly in terms of the most vital resource 'water'. Wastewater disposal and recycling should be regulated through national and international legislation. Although, in many cases, sufficient data and information are available, government policies make the plans inaccessible. While framing policies, the biggest concern is regular monitoring and control of generated contaminated water [33]. This issue can be addressed by combining incentives with monitoring plans to ensure compliance. Even when legislation and policies are there to treat wastewater effectively, their implementation is often hindered by unrealistic criteria.

### *2.2.6. Social Factors*

Society plays a vital role in addressing major issues, still, there is widespread negligence regarding the treatment and reuse of wastewater. The limited societal involvement reflects poor political will, inadequate investments, and a lack of coordination among agencies. A significant challenge is that public authorities often lack sufficient knowledge of available technologies to mitigate environmental and health risks. However, a shift in their approach can greatly enhance the effectiveness of treatment policies. Strong coordination among agencies, institutions, and stakeholders is essential for achieving success in wastewater management.

## 3. Emerging Innovations and Technologies

Prevailing challenges in wastewater treatment have created a need for advanced methodologies. As an initial step towards improving water quality, primary treatment can be implemented. Additionally, secondary treatments such as CWs and infiltration-percolation methods can bring good results. The effectiveness of secondary treatment is evaluated based on the removal of BOD and suspended solids (SS). A well-designed and properly operated secondary treatment system is characterized by high efficiency in BOD and SS removal [19]. Some chemical methods of water purification include chlorination, coagulation and flocculation. Another widely used physicochemical method for treating wastewater is photo-catalytic degradation of dissolved contaminants under irradiation. Physicochemical process involves a combination of biological and physical-chemical methods to achieve desired results [19]. Such methods-namely ozone treatment, incineration, and irradiation with ultraviolet light are often effective, but considerably more costly.

Ozonation is another effective method for the removal of dye contaminants [34]. In this method chromophores are oxidized, and ozone attacks these oxidized sites leading to the degradation of dye molecules [35]. Ozonation generates by-products like Nitroso-dimethylamine (NDMA) which requires post treatment. When comparing Ozonation and activated carbon (AC) treatment, the latter one demonstrates better performance in eliminating neutral contrast media like iopromide. However, both ozonation and AC treatments have edge over one another and thus can't be compared conclusively. Some recently evolved techniques to treat wastewater are looked upon with hope to get an efficient solution. Some of these rising innovations are:

1. AOPs
2. Membrane Technologies
3. Biological Treatment Innovations
4. Green and Sustainable Technologies
5. Smart Technologies and Automation
6. Artificial Intelligence

### 3.1. AOPs

AOP offers promising solution for removing complex pollutants that are challenging to treat with conventional methods. It utilizes strong oxidants, making it an effective approach for treating complex water streams. Additionally, this method efficiently removes dyes from contaminated water. However, both UV radiation and AOPs can generate unwanted disinfection byproducts, posing potential risks to both human and environmental health [36]. Among different AOPs, photo-catalysis and sonocatalysis with metal organic framework (MOF) of NMs have gained much attention for degrading toxic contaminants [37]. Semiconductor assisted AOP is a trending approach towards detoxification of wastewater by eliminating pollutants. Photo-induced process occurring on semiconductors under UV or visible light have the potential to produce numerous free radical species, such as superoxide radicals, hydroperoxyl radicals, and hydroxide radicals. Among these, hydroxide radicals are considered to play a crucial role in AOPs [38]. On the other hand, photo-sensitization facilitates the degradation of pollutants under visible light exposure. This process occurs when an electron from the excited state of pollutant molecule is transferred to the conduction band (CB) of the employed catalyst. However, catalyst-mediated degradation occurs only when the radiation intensity matches or exceeds the band-gap energy of the semiconductor [39]. However, these methods are costly and require aggressive applications. Another challenge is sustainability of UV-based processes; as about only 4% of solar light is in the UV range and artificial UV sources are quite expensive. Compared to conventional methods, AOPs have proven to be highly effective due to their non-selective nature and ability to generate hydroxyl radicals [40]. Some notable Benefits of AOP over conventional methods are:

a. They can mineralize dye aggregates completely
b. Sludge formation is reduced to a greater extent
c. Non selective nature of hydroxide radical can mineralize broad range of dyes
d. Being resistant to textile-based toxicity, they do not produce fouling.

### 3.2. Membrane Technologies

Different types of methods are based on membrane-dependent mechanisms, and all such methods can be collectively classified under membrane technologies. Various membrane processes such as micro-filtration, ultra-filtration, nano-filtration (NF), reverse osmosis (RO) and forward osmosis, when performed with some pre-treatment operations, help in removing colors, ions, microbes and some organic compounds from wastewater. The most advantageous feature of these methodologies is their high efficiency of pollutants removal. However, despite their usefulness, these methods have certain limitations, such as need of high pressure and sensitivity to fouling. Also, when used for treating pesticides, pathogens and volatile organic compounds; deterioration in membrane is observed. The need for regular replacement in membranes makes their adoption more costly [41].

Filtration methods using tight membranes are economically feasible for RO and NF techniques. Cross-flow filtration is a widely adopted design for membrane filtration, both from an engineering perspective and according to industrial standards. Due to less sludge production, membrane technology is considered one of the most efficient methods; however, it has certain limitations, such as high energy consumption and high maintenance requirements [42]. Enhanced reduction of concentration of FRNA coliphage using membrane bioreactors (MBRs) or bio-sand systems has also been reported. However; recently introduced technologies like MBRs have a drawback of generating sludge [36]. Thus, current wastewater treatment schemes should also focus on decreased sludge production and its proper disposal.

### 3.3. Biological Treatment Innovations

The most commonly used advanced biological treatment methods are anaerobic digestion and Enhanced Biological Phosphorus Removal abbreviated as EBPR. Both these methods are effective in nutrients removal and biogas production. Since anaerobic treatment systems are not yet efficient in removing pollutants, aerobic methods cannot be entirely replaced with anaerobic technologies. Anaerobic treatment works as a step in the pre-treatment process. It has emerged as a cost effective alternative to the costly activated sludge process and is widely employed in developing countries such as Colombia, Brazil and India.

Wastewater, rich in nutrients like carbon, phosphorus, and nitrogen, serves as a suitable medium for the growth of photosynthetic bacteria. Photosynthetic bacteria associated with MBRs were used in earlier applications and represented a unique approach in water purification by retaining biomass [43]. This method also contributes to the goal of reducing carbon emissions by generating cellular molecules which absorb carbon, present in wastewater. The dye removal can also be successfully achieved using biological treatment methods. Although, biological methods are economical and eco-friendly, there effectiveness is limited due to their dependence on fluctuating parameters like temperature and pH.

### 3.4. Green and Sustainable Technologies

Efficient and sustainable green methods for wastewater treatment include wetlands and algal-based systems. These eco-friendly approaches serve as a viable alternative for addressing wastewater challenges. Similar to biological treatment methods, these technologies not only purify wastewater but also contribute to biofuel production. Artificial wetlands are considered an effective option for wastewater treatment. The mobility of engineered nanoparticles (ENPs) can be significantly reduced due to their interaction with sedimented organic contaminants in wetlands. Additionally, plants absorb ENPs from the wetlands, leading to their accumulation in roots, which may ultimately pose toxicity risks to the plants. Phytoremediation is a recently evolved technique to remove heavy metals from wastewater. The basic mechanism involves the exchange of ions such as $Ca^{2+}$, $Na^{+}$ and $K^{+}$ present on algal cell surface with heavy metal ions in polluted water [44]. Phytoremediation is preferred over bioremediation due to advantages such as:

a. Algal biomass can be easily used throughout the year
b. This is a cost-effective method
c. Nontoxic chemicals are generated
d. No sludge is produced
e. Algal biomass synthesis is not needed
f. Easy application for heavy metals removal
g. Feasible generation and reuse of algal biomass

### 3.5. Smart Technologies and Automation

Modern smart technologies play a crucial role in analyzing real-time data for improved decision-making. Integrating sensors and real-time data collection can enhance the efficiency of wastewater treatment systems. Nanotechnology is emerging as a cost-effective tool for water pollution sensing and treatment. NMs have significant potential not only for removing water contaminants but also for monitoring pollution levels. However, a major drawback of this approach is the simultaneous risk of environmental toxicity associated with NMs. During wastewater purification, NMs reduce bacterial activity, ultimately leading to a decreased rate of pollutant removal. Nanotechnology can be commercialized for wastewater treatment by reducing energy consumption. Other recent techniques for treating wastewater include two-dimensional transitional metal carbides known as MXenes [45]. Exclusive features of this material are high chemical stability, good electrical conductivity and hydrophilicity which make them ideal for treating wastewater [46].

### 3.6. Artificial Intelligence

Artificial Intelligence (AI) simulates human cognitive processes through machine-based systems [47]. At present, AI is creating a noteworthy impact across key areas such as education, agriculture, healthcare, transportation, space exploration, home automation and many other fields. AI operates within different model frameworks. These different types of AI models, such as neural networks and hybrid models are developed to better understand the non -linear patterns of pollutant removal during wastewater treatment. These AI models have numerous redundant processes, methodologies, conceptual ideas, activities, and tasks. AI technologies have already displayed their efficiency in managing complex, interactive and dynamic wastewater treatment processes. These models can be employed to regulate and predict the effluent quality of the treatment process. Continuous monitoring of employed methods with the help of sensors can aid in improved operation efficiency, regulating energy consumption, and managing cost [48]. AI has already been employed in predicting the performances of water treatment plant, testing the quality of effluents discharged, optimizing operational parameters, constructing the treatment units, designing of sensors, regulating maintenance plans, optimization of energy consumption and system automation [47]. Major steps in wastewater treatment, such as BOD measurement, are being executed using multi-layer network and link network analysis methods governed by AI [49]. Various parameters used to assess wastewater quality such as BOD, COD and TSS have also been evaluated using different AI based models. COD removal efficiency has also been predicted using artificial neural network (ANN) model [50]. AI has also been effectively applied in modelling the degradation of pharmaceuticals belonging to the Penicillin class. It has also been used to manage the activated sludge process through Kohonen mapping combined with ANN and a fuzzy interference system, demonstrating superior viability and flexibility compared to traditional methods [47]. The mapping system has also been proven to be useful in BOD prediction. Performance of wastewater treatment plant can also be supervised with the help of integrated computational intelligence system and multilevel architecture. Such AI based model is helpful in developing a better understanding in system-based deviations for both single as well as multiple operational units. AI has also helped in reducing energy consumption while performing complex operations thereby increasing efficiency of the treatment plants [51].

AI systems and associated automation techniques are more cost effective than conventional engineering-based methods [50]. Despite its advantages, certain drawbacks associated with this technique limit its ability to achieve the expected performance. Some of the major drawbacks include high cost and the potential for unemployment. Additionally, the lack of structured and unified framework hinders the effective management, sharing, and re-use of AI applications in wastewater treatment. In the coming years, AI is expected to be a game-changer in the field of wastewater treatment.

## 4. Policy and Regulatory Framework

Effective implementation of wastewater treatment schemes depends on a comprehensive understanding of existing regulations and policies, coupled with their proper enforcement. Addressing point source pollution—wastewater discharged from identifiable sources—can be achieved by either promoting incentives or enforcing regulatory principles. For instance, upgrading wastewater treatment plants with advanced filtration systems can significantly reduce the discharge of harmful chemicals into nearby water bodies. Additionally, policies designed to safeguard the livelihoods of farmers and traders, who utilize wastewater for irrigation, should be supported by adequate investments. Adhering to current regulations enables informed decision making regarding future needs and plays a significant role in addressing emerging issues.

### 4.1. Existing Policies

Comparative analysis of existing policies is mandatory to compare issues that vary from one place to another. Policies need to be improved, financial cycles must be examined, and institutional dialogues should be carefully considered. Together, these efforts can significantly contribute to the effective management of wastewater treatment plants and processes. World Health Organization (WHO) has also released guidelines for treatment of wastewater and its minimal exposure to mankind.

### 4.2. Future Regulatory Changes

Policy changes are expected in response to advancement of technology. These may include wastewater pricing, implementation of stricter regulatory measures, introduction of tradable property rights, and incorporation of religious considerations within WHO guidelines. When designing wastewater management plans and programs, it is essential to prioritize low-income communities. These communities often rely on wastewater for irrigation, forming the foundation of their livelihoods. Engaging stakeholders—including government agencies, community groups, and the private sector—in the planning process can significantly enhance the success rate of wastewater management projects. This collaborative approach ensures that diverse perspectives are considered, leading to more effective and sustainable outcomes.

Regulating the use of wastewater in agriculture is essential to mitigate health risks associated with such practices. Public awareness campaigns should inform communities about potential hazards, including exposure to pathogens and contaminants. Prioritizing the management of domestic wastewater is crucial. Implementing proper sanitation facilities such as pit latrines, composting toilets, and septic tanks can significantly reduce health risks. These facilities should remain separate from household water pipelines to prevent cross-contamination. These measures can significantly improve public health and ensure safer agricultural practices.

Thus, substantial investment in sanitation is needed. Low-cost solution to sanitation facilities should be engineered that are based on the safety of health and environment. Effluent standards, taxes and tradable permits can be implemented to encourage advancements in water management. Currently, there is a noticeable backlog in addressing the Millennium Development Goal (MDG) related to improved sanitation. Incorporating meaningful criteria that reflect local, technical, economic, social and cultural contexts is much needed. Dividing the city into small, manageable units can also be a good strategy allowing for tailored schemes and approaches based on specific needs. This can lead to step-by-step success of the planned projects.

## 5. Social and Economic Implications

Effective wastewater treatment is essential not only for safeguarding public health but also for supporting sustainable economic development. By mitigating health risks, generating employment through treatment-related initiatives, and promoting public awareness, wastewater treatment plants offer a broad spectrum of social benefits. Conversely, inadequate treatment can lead to serious health consequences, with contaminated water contributing to the spread of diseases such as hemorrhage, and kidney and liver disorders.

Water sanitation remains a global challenge in reducing the burden of diseases on human populations. The implementation of advanced wastewater treatment methods offers significant economic advantages. These technologies can be cost-efficient, leading to substantial savings and bolstering a nation's economy. Proper wastewater management safeguards community health, stimulates local economies through employment, and enhances environmental quality. Engaging communities through awareness programs ensures the sustainability and acceptance of these initiatives. Effective collaboration between stakeholders and different groups is also important to make any scheme successful.

## 6. Future Outlook

Developing sustainable and effective wastewater treatment projects requires a multifaceted approach that integrates technological advancements, policy reforms, and strategic investments. Embracing innovative technologies, such as membrane filtration and artificial intelligence, enhance the efficiency of treatment processes. On the other hand, strategic investments comprise of priorities and practices. Prioritizing investments in wastewater treatment should be guided by the specific needs and funding capacities of each area. Engaging stakeholders in this process ensures that resources are allocated effectively, addressing the most pressing issues first. Treated wastewater can be utilized for various purposes; however; its quality check is essential before use. If reclaimed water is deemed non-potable, it can still be used in multiple ways such as irrigation, industrial applications, power-plants, and as fire extinguisher.

## Conclusion

As discussed, water recycling offers environmental benefits and helps meet human needs; however, future research should focus not only on treatment technologies but also on effectively engaging and convincing the public. A circular economy is required in order to reduce human footprint on earth. The concept of circular economy, combined with growing social awareness and environmental concerns, has contributed to several paradigm shifts in wastewater treatment. Consequently, a centralized management system is urgently needed to ensure efficient water conveyance and distribution, utilizing both conventional as well as unconventional resources. Moreover, such rapid evolution in this direction has made it clear that wastewater treatment is no longer a local issue, but a global concern. Innovative techniques that integrate risk management and decision support systems with advanced technologies are emerging in wastewater treatment. Society now stands at a pivotal juncture, confronting significant challenges alongside promising opportunities in this field. By addressing current issues and embracing emerging innovations, we can adopt sustainable and resilient approaches to managing Earth's most vital resource.

## Reeferences


[1] Nishat, A., Yusuf, M., Qadir, A., Ezaier, Y., Vambol, V., Khan, M. I., Moussa, S. B., Kamyab, H., Sehgal, S. S., Prakash, C., Yang, H. H., Ibrahim, H., & Eldin, S. M. (2023). Wastewater treatment: a short assessment on available techniques. *Alexandria Engineering Journal*, 76:505-516. https://doi.org/10.1016/j.aej.2023.06.054.

[2] Day, D. G. (1996). How Australian social policy neglects water environments. *Australian Journal of Soil and Water Conservation*, 9, 4 : 3-9. https://doi.org/10.3316/ielapa.970505017.

[3] Devane, M. L., Moriarty, E., Weaver, L., Cookson, A., & Gilpin, B. (2020). Fecal indicator bacteria from environmental sources; strategies for identification to improve water quality monitoring. *Water Research*, 185: 116204. https://doi.org/10.1016/j.watres.2020.116204.

[4] Krishnan, S., Rawindran, H., Sinnathambi, C. M., & Lim, J. W. (2017, June). Comparison of various AOPs used in remediation of industrial wastewater laden with recalcitrant pollutants. In IOP Conference Series: *Materials Science and Engineering*, 206, 1: 012089. https://doi.org/10.1088/1757-899X/206/1/012089.

[5] Altowayti, W. A. H., Allozy, H. G. A., Shahir, S., Goh, P. S., & Yunus, M. A. M. (2019). A novel nanocomposite of aminated silica nanotube (MWCNT/Si/NH 2) and its potential on adsorption of nitrite. *Environmental Science and Pollution Research*, 26 : 28737-28748. https://doi.org/10.1007/s11356-019-06059-0.

[6] Plucinski, M. P., Sullivan, A. L., & Hurley, R. J. (2017). A methodology for comparing the relative effectiveness of suppressant enhancers designed for the direct attack of wildfires. *Fire Safety Journal*, 87: 71-79. https://doi.org/10.1016/j.firesaf.2016.12.005.

[7] Mohapatra, D. P., & Kirpalani, D. M. (2019). Advancement in treatment of wastewater: fate of emerging contaminants. *The Canadian Journal of Chemical Engineering*, 97, 10: 2621-2631. https://doi.org/10.1002/cjce.23533.

[8] Qadir, M., Wichelns, D., Raschid-Sally, L., McCornick, P. G., Drechsel, P., Bahri, A., & Minhas, P. S. (2010). The challenges of wastewater irrigation in developing countries. *Agricultural water management*, 97, 4: 561-568. https://doi.org/10.1016/j.agwat.2008.11.004.

[9] Izah, S. C., Ngun, C. T., & Richard, G. (2022). Microbial quality of groundwater in the Niger Delta region of Nigeria: Health implications and effective treatment technologies. *In Current Directions in Water Scarcity Research Elsevier*, 6: 149-172. https://doi.org/10.1016/B978-0-323-91838-1.00010-5.

[10] Cabral, J. P. (2010). Water microbiology. Bacterial pathogens and water. *International journal of environmental research and public health*, 7, 10: 3657-3703. https://doi.org/10.3390/ijerph7103657.

[11] Bakar, N. A., Othman, N., Yunus, Z. M., Altowayti, W. A. H., Al-Gheethi, A., Asharuddin, S. M., & Mohd-Salleh, S. N. A. (2021). Nipah (Musa Acuminata Balbisiana) banana peel as a lignocellulosic precursor for activated carbon: Characterization study after carbonization process with phosphoric acid impregnated activated carbon. *Biomass Conversion and Biorefinery*, 1-14. https://doi.org/10.1007/s13399-021-01937-5.

[12] Chai, W. S., Cheun, J. Y., Kumar, P. S., Mubashir, M., Majeed, Z., Banat, F., Ho, S. H. & Show, P. L. (2021). A review on conventional and novel materials towards heavy metal adsorption in wastewater treatment application. *Journal of Cleaner Production*, 296: 126589. https://doi.org/10.1016/j.jclepro.2021.126589.

[13] Xiaojian, Z., Zhansheng, W., & Xiasheng, G. (1991). Simple combination of biodegradation and carbon adsorption—the mechanism of the biological activated carbon process. *Water research*, 25, 2: 165-172. https://doi.org/10.1016/0043-1354(91)90025-L.

[14] Pitás, V., Somogyi, V., Kárpáti, Á., Thury, P., & Fráter, T. (2020). Reduction of chemical oxygen demand in a conventional activated sludge system treating coke oven wastewater. *Journal of Cleaner Production*, 273: 122482. https://doi.org/10.1016/j.jclepro.2020.122482.

[15] Whitton, R., Fane, S., Jarvis, P., Tupper, M., Raffin, M., Coulon, F., & Nocker, A. (2018). Flow cytometry-based evaluation of the bacterial removal efficiency of a blackwater reuse treatment plant and the microbiological changes in the associated non-potable distribution network. *Science of the Total Environment*, 645: 1620-1629. https://doi.org/10.1016/j.scitotenv.2018.07.121.

[16] Altowayti, W. A. H., Shahir, S., Othman, N., Eisa, T. A. E., Yafooz, W. M., Al-Dhaqm, A., & Ali, A. (2022). The role of conventional methods and artificial intelligence in the wastewater treatment: a comprehensive review. *Processes*, 10, 9: 1832. https://doi.org/10.3390/pr10091832.

[17] Zhang, H., Zhang, H., Zhao, L., Zhou, B., Li, P., Liu, B., & Zhang, C. (2021). Ecosystem impact and dietary exposure of polychlorinated biphenyls (PCBs) and heavy metals in Chinese mitten crabs (Eriocheir sinensis) and their farming areas in Jiangsu, China. *Ecotoxicology and Environmental Safety*, 227: 112936. https://doi.org/10.1016/j.ecoenv.2021.112936.

[18] Sonune, A., & Ghate, R. (2004). Developments in wastewater treatment methods. *Desalination*, 167: 55-63. https://doi.org/10.1016/j.desal.2004.06.113.

[19] Saber, M., Takahashi, F., & Yoshikawa, K. (2018). Characterization and application of microalgae hydrochar as a low-cost adsorbent for Cu (II) ion removal from aqueous solutions. *Environmental Science and Pollution Research*, 25, 32: 32721-32734. https://doi.org/10.1007/s11356-018-3106-8.

[20] Matamoros, V., Arias, C., Brix, H., & Bayona, J. M. (2007). Removal of pharmaceuticals and personal care products (PPCPs) from urban wastewater in a pilot vertical flow constructed wetland and a sand filter. *Environmental science & technology*, 41, 23: 8171-8177. https://doi.org/10.1021/es071594.

[21] Hoffman, A. S. (2012). Hydrogels for biomedical applications. *Advanced drug delivery reviews*, 64: 18-23. https://doi.org/10.1016/j.addr.2012.09.010.

[22] Van der Bruggen, B., Schaep, J., Wilms, D., & Vandecasteele, C. (1999). Influence of molecular size, polarity and charge on the retention of organic molecules by nanofiltration. *Journal of Membrane Science*, 156, 1: 29-41. https://doi.org/10.1016/S0376-7388(98)00326-3.

[23] Luo, Y., Guo, W., Ngo, H. H., Nghiem, L. D., Hai, F. I., Zhang, J.,Liang, S. & Wang, X. C. (2014). A review on the occurrence of micropollutants in the aquatic environment and their fate and removal during wastewater treatment. *Science of the total environment*, 473: 619-641. https://doi.org/10.1016/j.scitotenv.2013.12.065.

[24] Luo, Y., Guo, W., Ngo, H. H., Nghiem, L. D., Hai, F. I., Zhang, J., Liang, S. & Wang, X. C. (2014). A review on the occurrence of micropollutants in the aquatic environment and their fate and removal during wastewater treatment. *Science of the total environment,* 473: 619-641. https://doi.org/10.1016/j.scitotenv.2013.12.065.

[25] Killedar, L., Ilager, D., Shetti, N. P., Aminabhavi, T. M., & Reddy, K. R. (2021). Synthesis of ruthenium doped titanium dioxide nanoparticles for the electrochemical detection of diclofenac sodium. *Journal of Molecular Liquids*, 340: 116891. https://doi.org/10.1016/j.molliq.2021.116891.

[26] Talwar, S., Verma, A. K., & Sangal, V. K. (2021). Synergistic degradation employing photocatalysis and photo-Fenton process of real industrial pharmaceutical effluent utilizing the Iron-Titanium dioxide composite. *Process Safety and Environmental Protection, 146:* 564-576. https://doi.org/10.1016/j.psep.2020.11.029.

[27] Domingo-Echaburu, S., Dávalos, L. M., Orive, G., & Lertxundi, U. (2021). Drug pollution & sustainable development goals. *Science of The Total Environment, 800:* 149412. https://doi.org/10.1016/j.scitotenv.2021.149412.

[28] Judd, S. J. (2017). Membrane technology costs and me. *Water research, 122:* 1-9. https://doi.org/10.1016/j.watres.2017.05.027.

[29] Rajasulochana, P. and Preethy, V., 2016. Comparison on efficiency of various techniques in treatment of waste and sewage water–A comprehensive review. *Resource-Efficient Technologies, 2,* 4 :175-184. https://doi.org/10.1016/j.reffit.2016.09.004.

[30] Ademiluyi, F. T., Amadi, S. A., & Amakama, N. J. (2009). Adsorption and Treatment of Organic Contaminants using Activated Carbon from Waste Nigerian Bamboo. *Journal of Applied Sciences and Environmental Management, 13,* 3. https://doi.org/10.4314/jasem.v13i3.55351.

[31] Bui, X. T., Vo, T. P. T., Ngo, H. H., Guo, W. S., & Nguyen, T. T. (2016). Multicriteria assessment of advanced treatment technologies for micropollutants removal at large-scale applications. *Science of the Total Environment, 563 :* 1050-1067. https://doi.org/10.1016/j.scitotenv.2016.04.191.

[32] Gentile, A., Piccolo, P., Iannece, P., Cicatelli, A., Castiglione, S., & Guarino, F. (2024). Reduction of antimicrobial resistance: Advancements in nature-based wastewater treatment. *Journal of Hazardous Materials, 471:* 134330. https://doi.org/10.1016/j.coesh.2017.12.004.

[33] Theron, J., Walker, J. A., & Cloete, T. E. (2008). Nanotechnology and water treatment: applications and emerging opportunities. *Critical reviews in microbiology, 34,* 1: 43-69. https://doi.org/10.1080/10408410701710442.

[34] Moore, G., Griffith, C., & Peters, A. (2000). Bactericidal properties of ozone and its potential application as a terminal disinfectant. *Journal of food protection, 63,* 8: 1100-1106. https://doi.org/10.4315/0362-028X-63.8.1100.

[35] Bradley, I., Straub, A., Maraccini, P., Markazi, S., & Nguyen, T. H. (2011). Iron oxide amended biosand filters for virus removal. *Water research, 45,* 15: 4501-4510. https://doi.org/10.1016/j.watres.2011.05.045.

[36] Zhang, X., Peng, T., & Song, S. (2016). Recent advances in dye-sensitized semiconductor systems for photocatalytic hydrogen production. *Journal of Materials Chemistry A, 4,* 7: 2365-2402. https://doi.org/10.1039/C5TA08939E.

[37] Wang, J. L., & Xu, L. J. (2012). Advanced oxidation processes for wastewater treatment: formation of hydroxyl radical and application. *Critical reviews in environmental science and technology, 42,* 3: 251-325. https://doi.org/10.1080/10643389.2010.507698.

[38] Borthakur, S., Basyach, P., Kalita, L., Sonowal, K., Tiwari, A., Chetia, P., & Saikia, L. (2020). Sunlight assisted degradation of a pollutant dye in water by a WO 3@ gC 3 N 4 nanocomposite catalyst. *New Journal of Chemistry, 44,* 7 : 2947-2960. https://doi.org/10.1039/C9NJ05142B.

[39] Vilar, V. J., Pinho, L. X., Pintor, A. M., & Boaventura, R. A. (2011). Treatment of textile wastewaters by solar-driven advanced oxidation processes. *Solar Energy, 85,* 9: 1927-1934. https://doi.org/10.1016/j.solener.2011.04.033.

[40] Fersi, C., & Dhahbi, M. (2008). Treatment of textile plant effluent by ultrafiltration and/or nanofiltration for water reuse. *Desalination, 222 ,* 1-3 : 263-271. https://doi.org/10.1016/j.desal.2007.01.171.

[41] Ben Aim, R. M., & Semmens, M. J. (2003). Membrane bioreactors for wastewater treatment and reuse: a success story. *Water science and technology, 47,* 1 : 1-5. https://doi.org/10.2166/wst.2003.0001.

[42] Okuda, A., & Kobayashi, M. (1961). Production of slime substance in mixed cultures of Rhodopseudomonas capsulatus and Azotobacter vinelandii. *Nature, 192,* 4808: 1207-1208. https://doi.org/10.1038/1921207a0.

[43] Salama, E. S., Roh, H. S., Dev, S., Khan, M. A., Abou-Shanab, R. A., Chang, S. W., & Jeon, B. H. (2019). Algae as a green technology for heavy metals removal from various wastewater. *World Journal of Microbiology and Biotechnology, 35 :* 1-19. https://doi.org/10.1007/s11274-019-2648-3.

[44] Crini, G., & Lichtfouse, E. (2019). Advantages and disadvantages of techniques used for wastewater treatment. *Environmental chemistry letters, 17 :* 145-155. https://doi.org/10.1007/s10311-018-0785-9.

[45] Cooper, P. F. (2001). Historical aspects of wastewater treatment. *Decentralised Sanitation and Reuse.*

[46] Malviya, A., & Jaspal, D. (2021). Artificial intelligence as an upcoming technology in wastewater treatment: a comprehensive review. *Environmental Technology Reviews, 10,* 1: 177-187. https://doi.org/10.1080/21622515.2021.1913242.

[47] Shi, S., & Xu, G. (2018). Novel performance prediction model of a biofilm system treating domestic wastewater based on stacked denoising auto-encoders deep learning network. *Chemical Engineering Journal, 347:* 280-290. https://doi.org/10.1016/j.cej.2018.04.087.

[48] Oliveira-Esquerre, K. P., Seborg, D. E., Mori, M., & Bruns, R. E. (2004). Application of steady-state and dynamic modeling for the prediction of the BOD of an aerated lagoon at a pulp and paper mill: Part II. Nonlinear approaches. *Chemical Engineering Journal, 105,* 1-2: 61-69. https://doi.org/10.1016/j.cej.2004.06.012.

[49] Tarke, P. D., Sarda, P. R., & Sadgir, P. A. (2016). Performance of ANNs for prediction of TDS of Godavari river, India. *International Journal of Engineering Research, 5,* 2: 115-118. https://doi.org/10.17950/ijer/v5s2/209.

[50] Sànchez, M., Cortés, U., Lafuente, J., Roda, I. R., & Poch, M. (1996). DAI-DEPUR: an integrated and distributed architecture for wastewater treatment plants supervision. *Artificial Intelligence in Engineering, 10,* 3: 275-285. https://doi.org/10.1016/0954-1810(96)00004-0.

[51] Bongards, M., Gaida, D., Trauer, O., & Wolf, C. (2014). Intelligent automation and IT for the optimization of renewable energy and wastewater treatment processes. *Energy, Sustainability and Society, 4:* 1-12. https://doi.org/10.1186/s13705-014-0019-3.

## Chapter 3

# Physical and Chemical Processes for the Removal of Contaminants from Wastewater

**Sankara Rao Miditana[1,*]**
**Neha Agarwal[2]**
**Joy Sarkar[3]**
**A. Ramesh Babu[4]**
**and Jaya Pandey[5]**

[1]Department of Chemistry, Government Degree College, Puttur, Tirupati, Andhra Pradesh, India
[2]Department of Chemistry, Navyug Kanya Mahavidyalaya, University of Lucknow, India
[3]Department of Chemistry, University of Lucknow, India
[4]Department of Chemistry, SVA Government Degree College (M), Srikalahasti, A.P, India
[5]Department of Chemistry, Amity School of Applied Sciences Amity University, Lucknow Campus, Uttar Pradesh, India

## Abstract

Global water consumption has more than doubled in recent decades, driven by its widespread use in agriculture, industry, households, and transportation. Climate change exacerbates water scarcity, highlighting the urgent need for advanced water treatment technologies to enhance reuse and improve effluent quality. The presence of chemical and particulate contaminants, including nutrients, heavy metals, microorganisms, and other toxic pollutants significantly influences water quality. In this context, wastewater treatment is essential for protecting public health and the environment by effectively removing physical, chemical, and biological contaminants. Physical processes such as screening, sedimentation, filtration, flotation, and adsorption eliminate suspended solids and organic pollutants. Dissolved air flotation (DAF) removes oils and grease, while activated carbon adsorption captures dissolved contaminants.

Chemical treatment enhances pollutant removal through coagulation and flocculation, which promotes particle aggregation. Oxidation techniques, including chlorine, ozone, and advanced oxidation processes (AOPs), degrade organic matter, while chemical precipitation removes heavy metals and excess nutrients. Ion exchange is used to extract dissolved ions, and disinfection methods such as chlorination, ultraviolet (UV) irradiation,

* Corresponding Author's Email: sraom90@gmail.com

In: Advancements in Wastewater Treatment
Editors: Neha Agarwal, Maulin P. Shahl, Vijendra Singh Solankil et al.
ISBN: 979-8-90134-007-3

and ozonation effectively eliminate pathogens. Hybrid technologies, including electrocoagulation and membrane bioreactors, integrate physical and chemical processes to enhance treatment efficiency. These advanced approaches contribute to sustainable wastewater management, ensuring cleaner water and improved environmental safety.
This chapter presents an overview of key physical and chemical treatment methods and hybrid technologies that enhance the effectiveness of wastewater treatment.

**Keywords:** adsorption, oxidation process, contaminants, coarse screens, dissolved air flotation, wastewater

## 1. Introdution

Wastewater treatment is a crucial process for maintaining environmental sustainability and public health. With rapid industrialization and urbanization, wastewater generation has significantly increased, leading to severe environmental and health concerns. Proper treatment helps prevent waterborne diseases, reduce pollution, and conserve water resources. It also ensures compliance with environmental regulations and promotes the reuse of treated water for various applications such as irrigation, industrial processes, and even potable water supply in certain cases. According to the United Nations, approximately 80% of the world's wastewater is discharged untreated into the environment, leading to the degradation of aquatic ecosystems and freshwater resources [1]. As water scarcity intensifies due to climate change and population growth, advanced treatment technologies are becoming a priority for policymakers and industries worldwide.

Wastewater contains a variety of contaminants that can be broadly classified into three categories: organic, inorganic, and biological pollutants. Organic contaminants include oils, fats, detergents, and industrial chemicals, while inorganic pollutants comprise heavy metals, nitrates, phosphates, and other dissolved minerals. Biological contaminants consist of bacteria, viruses, protozoa, and other pathogens that can pose serious health risks. The presence of these pollutants in untreated wastewater can lead to severe ecological and health hazards, making it imperative to implement effective treatment methods. Effective wastewater treatment not only prevents contamination of natural water bodies but also promotes water reuse, thereby reducing the pressure on freshwater resources.

Wastewater treatment involves the application of physical and chemical processes that play a vital role in contaminant removal, complementing biological methods to achieve comprehensive purification [2, 3]. Physical processes such as sedimentation, filtration, flotation, and adsorption effectively remove suspended solids, sediments, and certain dissolved substances by physically separating them from the water. These processes are typically the initial treatment stages and play a fundamental role in reducing pollutant loads before chemical treatments are applied. Physical methods also enhance the efficiency of subsequent treatment stages by minimizing the concentration of contaminants. Similarly, chemical treatment methods such as coagulation and flocculation, chemical precipitation, advanced oxidation, and ion exchange are used to target dissolved and complex contaminants that cannot be effectively removed by physical means [4]. The physical and chemical processes used in the wastewater treatment are depicted in Figure 1.

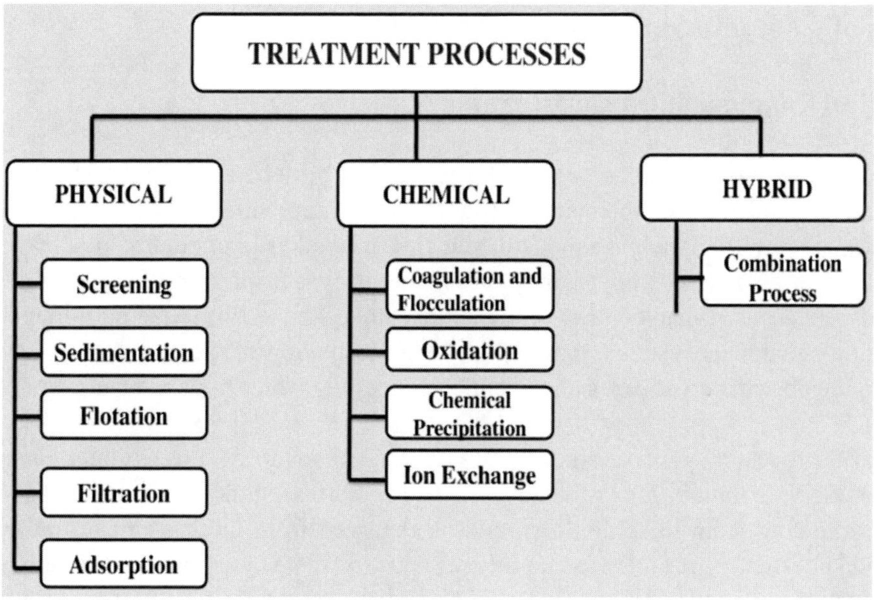


**Figure 1.** Different Processes in Wastewater Treatment.

Organic contaminants in wastewater include various synthetic chemicals, oils, and fats, which can have detrimental effects on aquatic ecosystems and pose serious health risks, if left untreated. Chemical processes, such as AOPs, have proven effective in breaking down and removing organic pollutants [4]. Inorganic contaminants, including heavy metals and salts, can be efficiently removed through membrane filtration and ion exchange [2]. Meanwhile, biological contaminants such as pathogens require disinfection techniques such as ultraviolet irradiation, ozonation, and chlorination to ensure the safety of treated water [3].

The integration of physical and chemical processes with biological treatment methods, such as activated sludge and constructed wetlands (CWs), results in a comprehensive wastewater treatment approach capable of addressing a broad spectrum of contaminants [5]. This holistic approach ensures that the treated water meets the necessary standards for safe discharge or reuse, contributing to the sustainable management of water resources. The selection of appropriate physical and chemical treatment processes depends on the characteristics of the wastewater and the specific contaminants it contains. Understanding these characteristics allow for optimized treatment efficiency, ensuring compliance with environmental regulations while supporting sustainability goals.

This chapter provides a comprehensive overview of the key physical and chemical processes involved in wastewater treatment, highlighting their advantages and limitations. The chapter emphasizes the importance of integrating multiple treatment methods to achieve high-quality treated water while ensuring environmental protection. As wastewater management continues to evolve, advancements in hybrid technologies that combine physical, chemical, and biological processes will play a crucial role in enhancing treatment efficiency and sustainability.

## 2. Types of Contaminants in Wastewater

### 2.1. Physical Contaminants in Wastewater

Physical contaminants in wastewater are those that can be separated from the water through physical means, such as sedimentation, filtration, and screening. These contaminants can include suspended solids, such as sand, silt, and clay, as well as larger debris like leaves, twigs, and plastic. The presence of physical contaminants in wastewater can have several adverse effects. Suspended solids can contribute to the turbidity of the water, reducing light penetration and affecting aquatic ecosystems. Additionally, the deposition of these solids in water bodies can lead to the obstruction of pipes and channels, impeding water flow and causing operational issues [6].

Physical processes play a crucial role in the removal of these contaminants from wastewater. Sedimentation, for instance, utilizes gravity to separate the suspended solids from the water, allowing them to settle down and be removed from the system. Filtration, on the other hand, physically traps the contaminants as the wastewater passes through a porous medium, such as sand or membrane filters. The effectiveness of physical processes in removing physical contaminants can be enhanced through the use of coagulants and flocculants [7]. Coagulants, such as alum or iron salts, work by neutralizing the electric charges on the suspended particles, allowing them to agglomerate and form larger, more easily settable flocs. Flocculants, such as polyacrylamides, further enhance the formation of these flocs, improving the efficiency of the sedimentation process [2].

In addition to sedimentation and filtration, screening is another important physical process used in wastewater treatment. Screens, which come in various sizes and configurations, are designed to physically remove larger debris and objects from the wastewater stream, protecting downstream equipments and processes from damage or clogging.

### 2.2. Chemical Contaminants in Wastewater

Chemical contaminants in wastewater can include a wide range of organic and inorganic compounds, each with their own unique properties and environmental impacts. Organic contaminants can include oils, fats, and various synthetic chemicals, while inorganic contaminants may include heavy metals and salts. The presence of these chemical contaminants in wastewater can pose significant threats to aquatic ecosystems and human health [8]. Organic compounds, for example, can exert a high biological oxygen demand (BOD) on the receiving water bodies, depleting the dissolved oxygen levels and suffocating aquatic life. Heavy metals, on the other hand, can bioaccumulate in the food chain, posing long-term health risks to both wildlife and human populations.

To address these chemical contaminants, a range of chemical processes can be employed in wastewater treatment. Chemical precipitation, for instance, can be used to remove heavy metals from wastewater. This process works by adjusting the $p^H$ of the wastewater to a level where the metal ions form insoluble precipitates, which can then be removed through sedimentation or filtration. Another important chemical process in wastewater treatment is adsorption, which involves the binding of contaminants to the surface of a solid material, such as activated carbon or zeolites [9]. Adsorption is particularly effective in the removal of organic compounds and heavy metals, and can be used as a standalone treatment or in conjunction with other processes.

AOPs such as ozonation and UV/hydrogen peroxide treatment represent a more advanced approach to chemical contaminant removal [10]. These processes utilize highly reactive oxygen species to break down and transform organic compounds into less harmful byproducts. The selection and application of these chemical processes in wastewater treatment is often a complex and site-specific endeavour, requiring a deep understanding of the contaminant profile, the treatment goals, and the overall system constraints.

### 2.3. Biological Contaminants in Wastewater

Biological contaminants in wastewater can include a wide range of microorganisms, such as bacteria, viruses, and protozoa, as well as other living organisms like algae and macrophytes. These contaminants can pose significant risks to human health and the environment, as they can spread pathogenic strains that can cause various infectious diseases [11]. The presence of these biological contaminants in wastewater is a major concern, as they can be transmitted through direct contact with the water or through the consumption of contaminated food or drinking water. To address this issue, wastewater treatment plants utilize various physical and chemical processes to disinfect and eliminate these contaminants.

UV radiation is one of the most commonly used and effective physical disinfection methods used in wastewater treatment. This process involves exposing the wastewater to high-intensity UV light, which damages the genetic material of the microorganisms, rendering them inactive and unable to replicate [12]. Chemical disinfection, on the other hand, involves the use of strong oxidizing agents, such as chlorine or ozone, to kill or inactivate the microorganisms. The choice of disinfection method depends on factors such as the type and concentration of the biological contaminants, the desired level of disinfection, and the potential for the formation of disinfection byproducts. The removal of biological contaminants is a critical step in the overall wastewater treatment process, as it helps to protect public health and the environment from the potential risks associated with these pollutants.

## 3. Physical Processes for Contaminant Removal

Physical processes play a fundamental role in wastewater treatment by removing contaminants through mechanical means. These processes are typically the first stage of treatment and are essential in reducing pollutant loads before applying chemical and biological treatments. They are particularly effective in eliminating suspended solids, sediments, oils, grease, and certain dissolved substances, ensuring the efficiency of subsequent treatment stages. The key physical processes used for contaminant removal in wastewater treatment include screening, sedimentation, flotation, filtration, and adsorption [13].

### 3.1. Screening and Grit Removal

Screening and grit removal are essential preliminary processes in wastewater treatment. Screening removes large solids such as plastics, wood, and rags to prevent clogging and equipment damage. It includes coarse screens, fine screens, and micro screens, which are listed in Table 1. Grit removal eliminates heavy inorganic particles like sand, gravel, and metal fragments that can cause abrasion in pumps and pipes. Common methods include gravity settling tanks, aerated grit chambers, vortex grit chambers, and hydrocyclones, which enhance grit separation efficiency [14]. These processes improve plant efficiency, reduce maintenance costs, and enhance downstream treatment performance. Modern wastewater treatment plants integrate mechanized screening with vortex grit chambers to optimize efficiency. Advanced technologies, such as ultrasonic grit detection and cyclonic separators, further improve removal processes, making wastewater treatment more sustainable [15].

**Table 1.** Classification of screens based on the size of the materials they remove

| Type of Screen | Opening Size | Function |
|---|---|---|
| Coarse Screens | >6 mm | Remove large objects like wood, plastics, and rags |
| Fine Screens | 1-6 mm | Capture smaller debris such as paper and food particles |
| Micro Screens | <1 mm | Used for advanced wastewater treatment to remove fine particles |

### 3.2. Sedimentation

Sedimentation is a critical physical process in wastewater treatment that relies on the gravitational settling of suspended solids and particles. In a sedimentation tank or clarifier, the wastewater flow is slowed down, allowing heavier particles to settle out, while the clarified water is then directed to the next stage of treatment [6]. The efficiency of the sedimentation process can be further enhanced through the use of coagulants and flocculants, which work by neutralizing the surface charge of suspended particles, promoting their aggregation into larger, more easily settleable floc [16]. Commonly used coagulants include aluminum sulfate (alum), ferric chloride, and organic polymers [17].

Sedimentation is particularly effective at removing suspended solids, organic matter, and certain heavy metals that are bound to the particulate matter. The removal efficiency of sedimentation can be influenced by factors such as the particle size and density, residence time in the sedimentation tank, and the presence of mechanical mixing or flocculation.

### 3.3. Filtration

Filtration is another important physical process in wastewater treatment, which involves the removal of suspended solids and particles by passing the wastewater through a porous medium, such as sand, gravel, or a membrane. The selection of the appropriate filtration technology depends on the specific contaminants present in the wastewater and the desired level of treatment.

One common filtration method is granular media filtration, where the wastewater is passed through a bed of sand, anthracite, or other granular materials. This process is effective in removing suspended solids, organic matter, and some microbial contaminants [18]. Another filtration approach is membrane filtration, which utilizes semipermeable membranes to separate contaminants from the water. Membrane filtration techniques, such as microfiltration, ultrafiltration, nanofiltration, and reverse osmosis, can remove a wide range of contaminants, including suspended solids, organic compounds, dissolved salts, and even some microorganisms [19, 20].

### 3.4. Flotation and Applications in Oil and Grease Removal

Flotation is a physical separation process that relies on the difference in density between the contaminants and the water to facilitate their separation. This process is particularly effective in the removal of oil and grease from wastewater [21]. In a flotation system, air bubbles are introduced into the wastewater, causing the oil and grease droplets to attach to the air bubbles and rise to the surface, where they can be skimmed off. This process, known as dissolved air flotation, is commonly used in the treatment of oily wastewater from industrial processes, such as petroleum refineries, metal finishing operations, and food processing industries [22].

The efficiency of the flotation process can be further enhanced through the use of coagulants and flocculants, which help to destabilize the oil emulsions and promote the agglomeration of the oil droplets, making them easier to separate [23].

### 3.5. Adsorption and Role in Organic Contaminant Removal

Adsorption is a physicochemical process in which contaminants in the wastewater are removed by accumulating on the surface of a solid adsorbent material. The choice of adsorbent material is crucial, as it determines the efficiency and selectivity of the adsorption process. Commonly used adsorbents in wastewater treatment include activated carbon, zeolites, clay minerals, and various types of functionalized nanomaterials [24, 25].

The adsorption process is particularly effective in the removal of organic contaminants, such as pesticides, pharmaceuticals, personal care products, and other persistent organic pollutants (POPs). The high surface area and porous structure of adsorbents like activated carbon enable effective contaminants binding, resulting in efficient removal [26].

### 3.6. Dissolved Air Flotation (DAF) Process

Dissolved Air Flotation is an advanced physico-chemical separation process widely utilized in wastewater treatment for the removal of suspended solids, oils, and organic contaminants. The process relies on the introduction of microbubbles, generated by dissolving air under pressure and then releasing it into the flotation chamber at atmospheric pressure. These microbubbles adhere to particulate matter, increasing their buoyancy and facilitating their separation from the liquid phase [27].

A standard DAF system consists of key components, including a flotation tank, an air saturation system, and a skimming mechanism that collects and removes the floated impurities. The method is particularly effective in treating high-solids wastewater streams and is commonly employed in industrial and municipal applications [28].

The primary advantages of DAF include high removal efficiency of fine suspended particles, grease, and organic matter, as well as its capability to handle fluctuating loads. However, operational challenges exist, including the need for precise control of air-to-solids ratio, hydraulic retention time, and coagulant dosing to maintain process stability [29]. Additionally, the system generates concentrated sludge, necessitating appropriate treatment and disposal strategies to comply with environmental regulations. Skilled operation and routine maintenance are essential to prevent fouling and ensure long-term efficiency.

## 4. Chemical Processes for Contaminant Removal

### 4.1. Coagulation and Flocculation

Coagulation and flocculation are chemical processes that play a crucial role in the removal of contaminants from wastewater. Coagulation involves the addition of chemical coagulants, such as aluminum sulfate, ferric chloride, or organic polymers, to the wastewater. These coagulants neutralize the negative charge of suspended particles, allowing them to come together and form larger, more easily settleable flocs [30-32].

The process of flocculation then promotes the formation of these larger, more stable flocs through the use of mechanical mixing or the addition of flocculants. Flocculants, which may beinorganic or organic, act as bridging agents, linking the smaller coagulated particles into larger, more easily removable flocs. The coagulation-flocculation process is effective in removing a wide range of contaminants, including suspended solids, organic matter, heavy metals, and some pathogens. The removal efficiency can be influenced by factors such as the type and dosage of coagulants and flocculants, $p^H$, temperature, and the characteristics of the wastewater. The coagulation-flocculation process is widely used in various wastewater treatment applications, including municipal sewage treatment, industrial wastewater treatment, and the treatment of surface water for drinking water production.

### 4.2. Oxidation and AOPs

Oxidation is a chemical process that involves the transfer of electrons, resulting in the transformation or removal of contaminants from wastewater. Conventional oxidation processes, such as chlorination, ozonation, and the use of potassium permanganate, have been used extensively in wastewater treatment to remove organic matter, oxidize inorganic compounds, and disinfect the water [33]. However, these traditional oxidation methods have limitations, as they may not effectively degrade recalcitrant and POPs, and they can also produce harmful byproducts.

To address these limitations, AOPs have emerged as a more effective alternative. AOPs are characterized by the generation of highly reactive hydroxyl radicals (OH•), which are capable of degrading a wide range of organic contaminants, including pharmaceuticals, personal care products, pesticides, and industrial chemicals [34]. Some common AOPs include:

- Photocatalysis, which utilizes semiconductor materials, such as titanium dioxide, to generate hydroxyl radicals under UV or visible light irradiation.
- Fenton and photo-Fenton processes, which involve the use of iron salts and hydrogen peroxide to produce hydroxyl radicals.
- Ozone-based processes, which combine ozone with other oxidants or catalysts to enhance the production of hydroxyl radicals.

These AOPs have demonstrated high efficiency in the removal of a wide range of organic contaminants, and they can be tailored to specific wastewater characteristics and treatment objectives.

### 4.3. Chemical Precipitation

Chemical precipitation is a process in which dissolved contaminants, such as heavy metals, are converted into insoluble solid precipitates through the addition of chemical reagents. The most common precipitation reactions involve the use of hydroxides, carbonates, or sulphides to form insoluble metal salts that can be subsequently removed from the wastewater through sedimentation or filtration. The choice of precipitating agent depends on the specific contaminants present in the wastewater, the desired removal efficiency, and the $p^H$ conditions [35]. For example, the addition of lime (calcium hydroxide) is often used to precipitate heavy metals as insoluble metal hydroxides. Chemical precipitation is particularly effective in the removal of heavy metals, such as lead, cadmium, copper, and zinc, from industrial wastewater. The process can also be used to remove other contaminants, such as phosphorus, fluoride, and some organic compounds, depending on the specific chemical reagents and conditions employed.

### 4.4. Ion Exchange

Ion exchange is a reversible process in which ions from a solution are exchanged for other ions that are bound to an ion exchange resin or material. In the context of wastewater treatment, ion exchange is commonly used to remove dissolved ionic contaminants, such as heavy metals, nutrients (e.g., nitrates and phosphates), and some organic compounds [36]. The ion exchange process involves the replacement of undesirable ions in the wastewater with more favourable ions from the ion exchange material. The ion exchange material, which can be either synthetic or natural, has a fixed charge and a high surface area, allowing for efficient ion exchange.

Ion exchange is particularly effective in the removal of heavy metals, such as cadmium, chromium, and lead from industrial wastewater [37]. It can also be used to remove other ionic contaminants, such as nitrates, sulphates, and phosphates, from municipal and agricultural wastewater. The main advantages of ion exchange include its high selectivity, efficiency, and ability to produce a high-quality effluent. However, the process can be sensitive to fouling and requires regular regeneration of the ion exchange material, which can be a costly.

## 5. Hybrid Physical and Chemical Processes

Hybrid physical and chemical processes combine mechanical separation with chemical reactions to enhance wastewater treatment efficiency. These methods integrate coagulation-flocculation with filtration, adsorption with oxidation, or membrane filtration with chemical precipitation [38]. The detailed account of the processes is discussed as follows.

### 5.1. Combination Processes

In practice, the most effective wastewater treatment strategies often involve the integration of multiple physical and chemical processes to achieve a higher level of contaminant removal [39]. These combination or hybrid processes leverage the strengths of different treatment methods to address the complex and diverse nature of wastewater contaminants. One example of a combination process is the use of coagulation-flocculation followed by adsorption or membrane filtration. The coagulation-flocculation step helps to remove suspended solids and larger organic molecules, while the subsequent adsorption or membrane filtration step can effectively target smaller, dissolved contaminants [41]. Another hybrid approach is the combination of chemical oxidation and ion exchange, where the oxidation process is used to transform and degrade organic contaminants, followed by ion exchange to remove the resulting inorganic ions.

### 5.2. Benefits of Hybrid Systems

The use of hybrid physical and chemical processes offers several benefits for wastewater treatment:

a. *Improved contaminant removal efficiency:* By combining complementary treatment methods, hybrid systems can achieve a higher overall removal rate for a broader range of contaminants.
b. *Enhanced process flexibility:* Hybrid systems allow for greater adaptability to changes in wastewater composition and can be tailored to meet specific treatment objectives.
c. *Increased robustness and reliability:* The redundancy built into hybrid systems can improve the overall resilience and stability of the wastewater treatment process.
d. *Improved contaminant removal efficiency:* By combining complementary treatment methods, hybrid systems can achieve a higher overall removal rate for a broader range of contaminants.

### 5.3. Emerging Technologies and Future Trends

The field of wastewater treatment is continuously evolving, with the development of new and more advanced physical and chemical processes. Some emerging technologies and future trends in this area include:

a. *Membrane bioreactors:* MBRs combine biological treatment with membrane filtration, offering high-quality effluent and a smaller footprint compared to traditional activated sludge systems.
b. *Electrochemical oxidation:* This process utilizes electrochemical reactions to generate strong oxidizing agents, such as hydroxyl radicals, for the degradation of organic contaminants.
c. *Adsorption using novel adsorbents:* The development of new adsorbent materials, such as nanomaterials, metal-organic frameworks, and biochar, can improve the efficiency and selectivity of adsorption-based treatment processes.
d. *Integrated hybrid systems:* The future of wastewater treatment may involve the integration of multiple hybrid processes, leveraging the strengths of various physical, chemical, and biological methods to achieve comprehensive contaminant removal and resource recovery.

## 6. Challenges and Limitations

### 6.1. Limitations of Physical Processes

Physical processes, such as coagulation-flocculation, adsorption, and membrane filtration, are widely employed in wastewater treatment due to their effectiveness in removing a variety of contaminants. However, these methods also face certain limitations that need to be considered. One of the primary limitations of physical processes is their inability to effectively target dissolved, non-biodegradable, or recalcitrant organic compounds [41, 42]. These pollutants can be difficult to remove using conventional physical techniques, as they often do not readily form flocs or precipitates that are easily separable [43, 44]. Moreover, the removal efficiency of physical processes may be limited for small, charged ions or low-molecular-weight organic compounds, which can pass through membrane filtration systems or remain adsorbed onto the surface of adsorbents.

Another limitation of physical processes is the generation of significant amount of waste sludge or concentrate. The sludge produced during coagulation-flocculation or the concentrate generated from membrane filtration requires proper disposal or further treatment, adding to both operational costs and environmental considerations. Furthermore, the performance of physical processes can be sensitive to factors such as temperature, $p^H$, and the presence of interfering substances in the wastewater. These parameters can affect the efficiency of the treatment, requiring careful monitoring and optimization to maintain the desired level of contaminant removal [45, 46].

### 6.2. Limitations of Chemical Processes

Chemical treatment methods, such as oxidation and chemical precipitation, can be highly effective in removing specific contaminants from wastewater. However, these processes also face their own set of limitations. One of the primary concerns with chemical processes is the potential for the formation of toxic by-products [47, 48]. The use of strong oxidizing agents or the addition of chemical reagents can lead to the generation of undesirable compounds, which may pose environmental and health risks if not properly managed.

Additionally, chemical processes often require careful $p^H$ control to ensure the optimal performance of the treatment. Maintaining the appropriate $p^H$ range can be challenging, especially when dealing with complex wastewater matrices containing various contaminants.

Another limitation of chemical processes is the high operating costs associated with the consumption of chemical reagents. The continuous need for chemical additives, such as coagulants, flocculants, or precipitating agents, can significantly increase the overall treatment costs, making these methods less economically viable for large-scale or long-term operations.

Furthermore, chemical processes may be less effective in dealing with complex mixtures of contaminants, as the interactions between different compounds can hinder the desired treatment outcomes. The removal efficiency of chemical methods can be influenced by the presence of competing ions, the solubility of the target contaminants, and the kinetics of the chemical reactions.

### 6.3. Environmental Concerns and Disposal of By-Products

The use of physical and chemical processes in wastewater treatment can lead to the generation of various by-products, such as sludge, concentrate, and chemical residues, which require proper handling and disposal. Improper management of these by-products can lead to environmental issues, such as the contamination of soil, groundwater, or surface water bodies [49, 50]. Measures must be taken to ensure the safe disposal or further processing of these by-products, considering their chemical composition and potential environmental impact.

To address the limitations of physical and chemical processes, the integration of these methods with biological treatment techniques, as well as the development of advanced hybrid systems, can provide a more comprehensive solution for the removal of a wide range of contaminants from wastewater. These hybrid approaches can leverage the strengths of different treatment technologies, leading to improved overall performance, reduced environmental impacts, and more cost-effective wastewater treatment solutions.

## Conclusion

Wastewater treatment is an indispensable process for safeguarding public health, preserving ecosystems, and ensuring the sustainable use of water resources. Various contaminants, including physical, chemical, and biological pollutants, degrade water quality, making their removal an urgent necessity. This can be achieved through physical and chemical methods, which play a crucial role in improving water quality and enabling its safe discharge or reuse.

Physical processes, including screening, sedimentation, filtration, flotation, and adsorption, effectively eliminate suspended solids and organic matter. Meanwhile, chemical treatments such as coagulation, oxidation, precipitation, ion exchange, and disinfection target dissolved pollutants, heavy metals, and pathogens. The combination of these processes ensures comprehensive wastewater purification, reducing environmental pollution and mitigating health risks.

However, conventional wastewater treatment methods face challenges such as high operational costs, energy consumption, and secondary waste production. Hybrid methods, such as DAF, have proven to be more effective in contaminant removal. To address existing limitations, hybrid technologies that integrate physical and chemical processes have gained prominence, enhancing efficiency and adaptability. As global water demand continues to rise and climate change exacerbates water scarcity, the development of innovative, cost-effective, and environmentally friendly treatment solutions becomes imperative.

The future of wastewater treatment lies in technological advancements, policy interventions, and sustainable management practices that balance efficiency with environmental responsibility. Investments in research and infrastructure, along with the adoption of emerging treatment techniques, will be crucial in ensuring long-term water security. By refining these approaches, wastewater treatment can evolve into a more resilient, energy-efficient, and sustainable system, ultimately; contributing to a cleaner environment and improved public health for future generations.

## References


[1] The 2017 edition of the United Nations World Water Development Report, entitled "Wastewater: The Untapped Resource."

[2] Jeon, C. (2012). Review for Physical and Chemical Process for Heavy Metal Treatment. In *Clean Technology*, 18(4): 341. https://doi.org/10.7464/ksct.2012.18.4.341.

[3] Ahmed, S. F., Mofijur, M., Nuzhat, S., Chowdhury, A. T., Rafa, N., Uddin, M. A., Inayat, A., Mahlia, T. M. I., Ong, H. C., Chia, W. Y., & Show, P. L. (2021). Recent developments in physical, biological, chemical, and hybrid treatment techniques for removing emerging contaminants from wastewater. *Journal of Hazardous Materials*, 416: 125912. https://doi.org/10.1016/j.jhazmat.2021.125912.

[4] Ray, S. S., Gusain, R., & Kumar, N. (2020). Water purification using various technologies and their advantages and disadvantages. *Elsevier eBooks* (p. 37). https://doi.org/10.1016/b978-0-12-821959-1.00003-9.

[5] Biec Nhu Ha, Thong Nguyen Tan (2025). Chapter 8 - Integration of bioremediation and physico-chemical methods for wastewater treatment and resource recovery, Biotechnologies for Wastewater Treatment and Resource Recovery, *Current Trends and Future Scope*, Pages 109-122, https://doi.org/10.1016/B978-0-443-27376-6.00002-5.

[6] Sathya, K., Nagarajan, K., Malar, G. C. G., Rajalakshmi, S., & Lakshmi, P. R. (2022). A comprehensive review on comparison among effluent treatment methods and modern methods of treatment of industrial wastewater effluent from different sources. *Applied Water Science*, 12(4). Springer Nature. https://doi.org/10.1007/s13201-022-01594-7.

[7] Saravanan, A., P. Senthil Kumar, S. Jeevanantham, S. Karishma, B. Tajsabreen, P.R. Yaashikaa, B. Reshma. (2021). Effective water/wastewater treatment methodologies for toxic pollutants removal: Processes and applications towards sustainable development. *Chemosphere,* 280: 130595. https://doi.org/10.1016/j.chemosphere.2021.130595

[8] Bashir, I., Lone, F. A., Bhat, R. A., Mir, S. A., Dar, Z. A., & Dar, S. A. (2020). Concerns and Threats of Contamination on Aquatic Ecosystems. *Bioremediation and Biotechnology: Sustainable Approaches to Pollution Degradation*,1–26. https://doi.org/10.1007/978-3-030-35691-0_1.

[9] Tapas Ranjan Sahoo, Benedicte Prelot, Chapter 7 - Adsorption processes for the removal of contaminants from wastewater: the perspective role of nanomaterials and nanotechnology. *Nanomaterials for the Detection and Removal of Wastewater Pollutants, Micro and Nano Technologies* 2020, Pages 161-222. https://doi.org/10.1016/B978-0-12-818489-9.00007-4.

[10] Monali Priyadarshini, Indrasis Das, Makarand M. Ghangrekar, Lee Blaney. (2022). Advanced oxidation processes: Performance, advantages, and scale-up of emerging technologies. *Journal of Environmental Management,* 316: 115295. https://doi.org/10.1016/j.jenvman.2022.115295.

[11] Chahal, C., van den Akker, B., Young, F., Franco, C., Blackbeard, J., & Monis, P. (2016). Pathogen and Particle Associations in Wastewater: Significance and Implications for Treatment and Disinfection Processes. *Advances in applied microbiology*, 97: 63–119. https://doi.org/10.1016/bs.aambs.2016.08.001.

[12] Collivignarelli, M. C., Abbà, A., Benigna, I., Sorlini, S., & Torretta, V. (2018). Overview of the Main Disinfection Processes for Wastewater and Drinking Water Treatment Plants. *Sustainability,* 10(1) : 86. https://doi.org/10.3390/su10010086.

[13] Metcalf, L., Eddy, H. P., & Tchobanoglous, G. (1991). *Wastewater engineering: treatment, disposal, and reuse* (Vol. 4). New York: McGraw-Hill.

[14] Tchobanoglous, G., Stensel, H. D., Tsuchihashi, R., Burton, F., Abu-Orf, M., Bowden, G., & Pfrang, W. (2014). *Wastewater engineering: treatment and resources recovery.* Metcalf and Eddy Inc.

[15] Spellman, F. R. (2008). Handbook of water and wastewater treatment plant operations. CRC press.

[16] Iwuozor, K. O. (2019). Prospects and challenges of using coagulation-flocculation method in the treatment of effluents. *Advanced Journal of Chemistry-Section A,* 2(2): 105-127. http://dx.doi.org/10.29088/SAMI/AJCA.2019.2.105127.

[17] Bakar, A. F. A., & Halim, A. A. (2013). Treatment of automotive wastewater by coagulation-flocculation using poly-aluminum chloride (PAC), ferric chloride ($FeCl_3$) and aluminum sulfate (alum). In *AIP conference proceedings, American Institute of Physics*, 1571(1): 524-529. https://doi.org/10.1063/1.4858708.

[18] Shon, H. K., Vigneswaran, S., & Snyder, S. A. (2006). Effluent Organic Matter (EfOM) in Wastewater: Constituents, Effects, and Treatment. In *Critical Reviews in Environmental Science and Technology*. Taylor & Francis. 36 (4): 327. https://doi.org/10.1080/10643380600580011.

[19] Bart Van der Bruggen. (2018). Chapter 2 - Microfiltration, ultrafiltration, nanofiltration, reverse osmosis, and forward osmosis, undamental Modelling of Membrane Systems, Membrane and Process Performance, Pages 25-70, https://doi.org/10.1016/B978-0-12-813483-2.00002-2.

[20] Aziz, S., Mazhar, A. R., Ubaid, A., Shah S. M. H., Riaz Y., Talha T., Jung D. (2024). A comprehensive review of membrane-based water filtration techniques. *Appl Water Sci* 14:169. https://doi.org/10.1007/s13201-024-02226-y.

[21] Zulkefli, N. F., Alias, N. H., Jamaluddin, N. S., Abdullah, N., Manaf, S. F. A., Othman, N. H., Marpani, F., Shayuti, M. S. M., & Kusworo, T. D. (2021). *Recent Mitigation Strategies on Membrane Fouling for Oily Wastewater Treatment. Membranes, Multidisciplinary Digital Publishing Institute*. 12(1): 26. https://doi.org/10.3390/membranes12010026.

[22] Muñoz-Alegría, J. A., Muñoz-España, E. ., & Flórez-Marulanda, J. F. (2021). Dissolved air flotation: a review from the perspective of system parameters and uses in wastewater treatment. *TecnoLógicas,* 24(52): e2111. https://doi.org/10.22430/22565337.2111.

[23] Makhtar, N. S. M., Idris, J., Musa, M., Andou, Y., Hamid, K. H. K., & Puasa, S. W. (2020). Plant-Based Tacca leontopetaloides Biopolymer Flocculant (TBPF) *Produced High Removal of Heavy Metal Ions at Low Dosage. Multidisciplinary Digital Publishing Institute*, 9(1): 37. https://doi.org/10.3390/pr9010037.

[24] Liu, Y., Wu, D., Wang, X., Yu, J., & Li, F. (2018). Fabrication of eco-friendly nanofibrous membranes functionalized with carboxymethyl-β-cyclodextrin for efficient removal of methylene blue with good recyclability. In *RSC Advances,* 8(66): 37715. https://doi.org/10.1039/c8ra07523a.

[25] Younas, F., Mustafa, A., Farooqi, Z. U. R., Wang, X., Younas, S., Mohy-Ud-Din, W., Ashir Hameed, M., Mohsin Abrar, M., Maitlo, A. A., Noreen, S., & Hussain, M. M. (2021). Current and Emerging Adsorbent Technologies for Wastewater Treatment: Trends, Limitations, and Environmental Implications. *Water,* 13(2) : 215. https://doi.org/10.3390/w13020215.

[26] Liu, W., DuShu, H., Yi, P., & Li, Y. (2012). Status and Prospects of Concentrated Organic Wastewater Degradation. In *InTech eBooks*. https://doi.org/10.5772/31990.

[27] Edzwald, J. K. (2010). Dissolved Air Flotation and Me. *Water Research*, 44(7) : 2077–2106. https://doi.org/10.1016/j.watres.2009.12.040.

[28] Crittenden, J. C., Trussell, R. R., Hand, D. W., Howe, K. J., & Tchobanoglous, G. (2012). *MWH's Water Treatment: Principles and Design* (3$^{rd}$ ed.). John Wiley & Sons.

[29] Antonella L. Piaggio, Lais A. Soares, Malini Balakrishnan, Tavishi Guleria, Merle K. de Kreuk, Ralph E.F. Lindeboom. (2022). High suspended solids removal of Indian drain water with a down-scaled Dissolved Air Flotation (DAF) for water recovery. Assessing water-type dependence on process control variables. *Environmental Challenges*, 8 : 100567. https://doi.org/10.1016/j.envc.2022.100567.

[30] Saien, J., & Nejati, H. (2007). Enhanced photocatalytic degradation of pollutants in petroleum refinery wastewater under mild conditions. In *Journal of Hazardous Materials* 148 (1) : 491. https://doi.org/10.1016/j.jhazmat.2007.03.001.

[31] Wang, L. K., Wang, MH. S., Shammas, N. K., Hahn, H. H. (2021). Physicochemical Treatment Consisting of Chemical Coagulation, Precipitation, Sedimentation, and Flotation. In: Wang, L. K., Wang, MH. S., Hung, YT. (eds) Integrated Natural Resources Research. Handbook of Environmental Engineering, vol 22. Springer, Cham. https://doi.org/10.1007/978-3-030-61002-9_6.

[32] Sahu, O. P., & Chaudhari, P. K. (2014). Physicochemical Treatment of Sugar Industry Wastewater: Coagulation Processes. Environmental Quality Management, 23(4) : 49-69. https://doi.org/10.1002/tqem.21373.

[33] Thombre, N. V. (2023). Oxidation in Water and Used Water Purification. In: Lahnsteiner, J. (eds) *Handbook of Water and Used Water Purification*. Springer, Cham. https://doi.org/10.1007/978-3-319-66382-1_174-1.

[34] Pandis, P. K., Kalogirou, C., Kanellou, E., Vaitsis, C., Savvidou, M. G., Sourkouni, G., Zorpas, A. A., & Argirusis, C. (2022). Key Points of Advanced Oxidation Processes (AOPs) for Wastewater, Organic Pollutants and Pharmaceutical Waste Treatment: A Mini Review. *Chem Engineering*, 6(1) : 8. https://doi.org/10.3390/chemengineering6010008.

[35] Benalia, M. C., Youcef, L., Bouaziz, M. G., Samia Achour S., Menasra H. (2022) Removal of Heavy Metals from Industrial Wastewater by Chemical Precipitation: Mechanisms and Sludge Characterization. *Arab J Sci Eng* 47 : 5587–5599. https://doi.org/10.1007/s13369-021-05525-7.

[36] Zhang, T., Payne, K., Zhang, J. Purswani P., Karpyn P.,, Wang M. (2024). Hybrid ion exchange and biological processes for water and wastewater treatment: a comprehensive review of process applications and mathematical modeling. *Rev Environ Sci Biotechnol,* 23: 163–188. https://doi.org/10.1007/s11157-023-09677-w.

[37] Dąbrowski, A., Z. Hubicki, P. Podkościelny, E. Robens. (2004). Selective removal of the heavy metal ions from waters and industrial wastewaters by ion-exchange method. *Chemosphere,* 56 (2) : 91-106. https://doi.org/10.1016/j.chemosphere.2004.03.006.

[38] Davoodbeygi, Y., Askari, M., Salehi, E., & Kheirieh, S. (2023). A review on hybrid membrane-adsorption systems for intensified water and wastewater treatment: Process configurations, separation targets, and materials applied. *Journal of Environmental Management,* 335 :117577. https://doi.org/10.1016/j.jenvman.2023.117577.

[39] Ahmed, S. F., Mofijur, M., Nuzhat, S., Chowdhury, A. T., Rafa, N., Uddin, M. A., ... & Show, P. L. (2021). Recent developments in physical, biological, chemical, and hybrid treatment techniques for removing emerging contaminants from wastewater. *Journal of hazardous materials*, 416 : 125912. https://doi.org/10.1016/j.jhazmat.2021.125912.

[40] Sillanpää, M., Ncibi, M. C., Matilainen, A., & Vepsäläinen, M. (2018). Removal of natural organic matter in drinking water treatment by coagulation: A comprehensive review. *Chemosphere,* 190 :54-71. https://doi.org/10.1016/j.chemosphere.2017.09.113.

[41] Hubbe, M. A., Metts, J. R., Hermosilla, D., Blanco, M., Yerushalmi, L., Haghighat, F., Lindholm-Lehto, P., Khodaparast, Z., Kamali, M., & Elliott, A. (2016). Wastewater Treatment and Reclamation: A Review of Pulp and Paper Industry Practices and Opportunities. *BioResources,* 11(3) : 7953. North Carolina State University. https://doi.org/10.15376/biores.11.3.hubbe.

[42] Rathi, B. S., Kumar, P. S., & Show, P. L. (2020). A review on effective removal of emerging contaminants from aquatic systems: Current trends and scope for further research. *Journal of Hazardous Materials*, 409: 124413. Elsevier BV. https://doi.org/10.1016/j.jhazmat.2020.124413.

[43] Nageeb, M. (2013). Adsorption Technique for the Removal of Organic Pollutants from Water and Wastewater. In *InTech eBooks*. https://doi.org/10.5772/54048.

[44] Anjaneyulu, Y., Chary, N. S., & Raj, D. (2005). Decolourization of Industrial Effluents – Available Methods and Emerging Technologies – A Review. *Reviews in Environmental Science and Bio/Technology, Springer Science+Business Media.* 4(4) : 245. https://doi.org/10.1007/s11157-005-1246-z.

[45] Lau, G. E., Abdullah, C. A. C., Ahmad, W. A. N. W., Assaw, S., & Zheng, A. L. T. (2020). Eco-Friendly Photocatalysts for Degradation of Dyes. In *Catalysts.* Multidisciplinary Digital Publishing Institute. 10(10): 1129. https://doi.org/10.3390/catal10101129.

[46] Mostafazadeh, A. K., Zolfaghari, M., & Drogui, P. (2016). Electrofiltration technique for water and wastewater treatment and bio-products management: A review. *Journal of Water Process Engineering*, 14: 28. https://doi.org/10.1016/j.jwpe.2016.10.003.

[47] Booton, A., Mayer, B. K. & Zitomer, D. H. (2024). Chemical oxidation as an alternative for municipal wastewater secondary treatment: a review. *Rev Environ Sci Biotechnol* 23: 43–65. https://doi.org/10.1007/s11157-024-09684-5.

[48] Bui, N., Kang, H., Teat, S. J., Su, G. M., Pao, C., Liu, Y., Zaia, E. W., Guo, J., Chen, J., Meihaus, K. R., Dun, C., Mattox, T. M., Long, J. R., Fiske, P. S., Kostecki, R., & Urban, J. J. (2020). A nature-inspired hydrogen-bonded supramolecular complex for selective copper ion removal from water. *Natu Commun,* 11(1): 3947. https://doi.org/10.1038/s41467-020-17757-6.

[49] Muthukumar, V., Manoharan, H. (2025). Introduction to Wastewater Treatment Using Various Electrochemical Methods. In: Kumar, S., Sarkar, I. J. R., Shah, M. P. (eds) Electrochemical Perspective Towards Wastewater Treatment. *Advances in Wastewater Research*, pp: 1-26. https://doi.org/10.1007/978-981-96-0652-8_1.

[50] Saha, B., & Orvig, C. (2010). Biosorbents for hexavalent chromium elimination from industrial and municipal effluents. *Coordination Chemistry Reviews*, 254 (23) : 2959). https://doi.org/10.1016/j.ccr.2010.06.005.

# Chapter 4

# Unraveling the Adsorptive Removal of Micropollutants from Wastewater

**Sheetal Kumari**[1]
**and Prabhat Kumar Patel**[2,*]
[1]Amity institute of Environmental Sciences, Amity University, Noida, India
[2]School of Sciences, Harbin Institute of Technology Shenzhen, Shenzhen, China

## Abstract

Every living thing on our planet is in grave danger of extinction due to an exceptional spike in pollution caused by the swift proliferation of technology, population expansion, and globalization. Sustainable development has thus emerged as a top priority for decision-makers everywhere. Water and environmental pollution by persistent hazardous substances poses serious risks to safety for people and animals, with varying effects on development and well-being. Global concerns have been generated by the release of several toxic substances into the environment as a result of industrialization and societal expansion. Micropollutants (MPs) are a growing problem because of their carcinogenic, irreversible, and accumulative nature, all of which have led to widespread alarm. One of the biggest ecological problems nowadays is getting rid of such pollutants from manufacturing operations. Numerous artificial and natural sorbents, each with special advantages, have been established to solve this problem. Polymers and related composite substances have drawn interest recently due to their ability for contaminant adsorption. Economic feasibility, accessibility, and compliance with regulatory requirements are among the many benefits of the sorption procedure. The effectiveness, incentives, and drawbacks of chitosan as an adsorbent are examined in this chapter, which concludes with an evaluation of its capability for dealing with the urgent problem of pollution.

**Keywords:** wastewater, pharmaceuticals, pfas, microplastics, adsorption, micropollutants

---

* Corresponding Author's Email: pprabhat1991@gmail.com

In: Advancements in Wastewater Treatment
Editors: Neha Agarwal, Maulin P. Shahl, Vijendra Singh Solankil et al.
ISBN: 979-8-90134-007-3

## 1. Introduction

Water is essential for human existence and societal advancement. Water is vital for food production, electricity generation, and medicinal treatments. Human activities have polluted water systems by the release of organic chemicals, dyes, medicines, and pesticides [1]. Harmfulcontaminants in aquatic environments can bioaccumulate via food chain, leading to negative impacts on species [2]. Toxic textile dyes, MPs, drugs, beauty products, and pesticides are the most often documented wastes. Dyes provide a significant ecological risk due to their elevated toxicity. Specific dyes and their reaction by-products may disrupt the basic functioning of aquatic species [3]. The eradication of metallic ions is more difficult compared to dyes, leading to their accumulation in water and soil. These substances can induce toxicity in people through water contamination and agricultural produce [1]. Moreover, plasticizers, and pesticides can interfere with endocrine systems, even in minimal quantities. Owing to their resistance to natural degradation, trace quantities of disruptors may persist in the environment. Given its critical role in the surroundings, it is imperative to purify sewage sources utilizing economical methods. Currently, researchers are engaged in several domains of materials science, engineering, and chemistry to create economical and eco-friendly water treatment materials. Additionally, an assessment of energy efficiency, along with the associated ecological risks and benefits is being considered. Numerous studies have been performed to develop innovative solutions for addressing water contamination resulting from organic and inorganic contaminants. Advanced materials, including polymers, have been effectively applied for clean-up at several scales. Recent studies have concentrated on using biomaterials to identify, trap, and degrade both new and persistent pollutants [4]. Among the various traditional methods for wastewater treatment, such as oxidation, coagulation, photodegradation, ion exchange, and biological treatment, adsorption is the most prevalent and economically efficient method for dye removal. Adsorption is an excellent method for treating MPs impurities in water and offers a viable approach for the renewal and reuse of adsorbents. Although, not the most recent technique for pollution removal, it is considered a highly efficient and reliable method for sewage purification due to its operational simplicity, versatility, and resistance to detrimental substances. Recent research has focussed on chitosan, biochar, covalent organic frameworks, and cyclodextrins (CDs) for the economical separation of pollutants from natural water and wastewater, targeting micro-polluted environments.

## 2. Types of MPs

Expelled water contains MPs that are described as inorganic and organic substances manufactured by non-natural sources. Although these contaminants do not have an immediate influence on the surroundings, they can have long-term consequences, particularly for marine habitats. A clear link can be established between the existence of MPs in surroundings and the increased toxicity of novel compounds formed through their interactions with other substances. Personal grooming items, medicinal residues, and endocrine-disrupting substances are among the three primary categories of MPs [5]. MPs include heavy metals (HMs), polychlorinated biphenyls (PCBs), and polycyclic aromatic hydrocarbons (PAHs). Medications and personal care products were found to be the most prevalent sources of microplastics in industrial waste streams, according to statistical measurements. A significant amount of the MPs in the generated wastewater are medicinal chemical substances. Investigators found substantial quantities of codeine, atenolol, gabapentin, paracetamol, and tramadol in raw effluents (>10 mg/L) [6].

## 2.1. Inorganic MPs

One important category of inorganic MPs is HMs. Metals with a density of 6 g/cm$^3$ or greater are classified as HMs [7]. The most commonly encountered HMs with multiple commercial implementations are Mn, Cd, Cr, Pb, and Hg. Furthermore, some of these metals are often considered potential essential nutrients for plant growth. However, despite their beneficial aspects, there are evidences of their detrimental impacts on both human and environmental health. Compared to other metals, Cr is the most lethal metal and has the most severe adverse impacts on human health. It has extensive applications in fabrics, paper and pulp, tanning plants, alloy and metal processing, paint and dyeing, and other industries. These companies discharge tons of untreated chromium (Cr) in their effluent. Excessive Cr intake through meals and beverages has been linked to serious illnesses including dermatitis, skin allergies, nausea and vomiting, and carcinogenic effects. When Comparing the two different forms of Cr, Cr(VI) is found to be hundreds of times more toxic and carcinogenic than its counterpart, Cr(III). Cr(VI) easily enter living cells and convert into receptive byproducts in the presence of enzymes and other human antigens. These byproducts can inhibit cell proliferation and impair the cellular performance. They can also damage DNA structure and essential biomolecules such as lipids and proteins. The human body may exhibit negative health conditions if it absorbs too much Cr(III), however, it may accelerate the metabolism during the phase of digestion (glycogenolysis).

Dyes and pigment production, electroplating, coatings in stabilizers, battery manufacturing, and alloy and metal polishing all contribute to the emission of Cd. Ingesting or inhaling Cd can lead to harmful conditions including carcinogenicity and endocrine dysfunction, as well as harmful impacts on nerves, bones, lungs, gastrointestinal tract, and the urinary tract. According to the US Government Agency for Toxic Substances and Disease Registry, Hg is the third-leading detrimental component on Earth. When inhaled, it can damage the nervous system, liver, and kidney cortex. It is found in electronic components, thermometers, sphygmomanometers, amalgams employed in dental care, barometers, colors, and coatings. Another harmful element Pb occurs in freshwater and it enters water bodies through industrial operations such as calligraphy, petrol manufacture, ceramics and vessel building, and the mordant pigment manufacturing. According to toxicological research, babies are more likely to be affected by lead poisoning due to their developing musculoskeletal and brain functions. At extremely low levels, Pb ions can cause neurological conditions, IQ deficiencies, and cognitive difficulties for children. They can also contribute to anemia, hypertension, heart disease, brain hemorrhage, renal failure, birth defects during pregnancy, and even death if their concentration exceeds the lethal dosage. On the other hand, manganese also plays a significant role in water pollution due to its ability to alter water color and smell, as well as its harmful effects on the central nervous and endocrine systems in humans following prolonged exposure.

In addition to HMs, other chemical micro-pollutants include nitrates ($NO_3^-$) and halides. Although these anions are also detectable in pharmaceutical wastewater, they are primarily found in urban water that is derived from household detergents and solvents [8]. The great dissolution of these inorganic anions makes them challenging to eliminate because they dissociate completely into solutions and produce chemically resilient water-soluble compounds. Nitrates are particularly detrimental to infants, as excessive accumulation can damage arteries and vascular system, impairing proper blood flow. This can lead to methemoglobinemia (also known as "blue baby syndrome") and other serious illnesses such as Type 2 diabetes, thyroid growth, and homeostasis of the veins. In addition, halides cause cognitive impairment, fatigue, deterioration of the cartilage, and impeded development.

## 2.2. Organic MPs

Remnants of PAHs, PCBs, EDCs, and medicinal products are the broad categories of organic MPs (OMPs). Attached benzene rings (numbers 2–8) make up PAHs, which can have either a homocyclic or heterocyclic molecular structure. PAHs tend to accumulate in marine food chains from sewage and deposition. They are often persistent and possess carcinogenic potential even at extremely low concentrations. Vehicle tar, oil spills, pesticides, fossil fuels, and pyrogenic hydrocarbon combustion are the sources of these PAHs. Prominent PAHs identified in various industrial runoffs include anthracene (used in the colorant, synthetic materials, and medicinal industries), fluorine (used in the drugs, thermoplastic, and insecticide industries), fluoranthene (used in the drug business), pyrene (used in the colorant industry), and 1-naphthalene. Depending on the amount and extent of exposure, PAHs have an adverse impact on both marine life and mammals. Researchers believe that PAHs can cause serious health issues including eye and skin irritation, memory impairment, and diarrhea, as well as long term conditions such as swelling, respiratory issues, and destruction of red blood cells (RBCs). However, the short term impacts of PAHs on humans have not been adequately defined in previous literature. Among the many types of OMPs, PCBs stand out due to their 209 chemical congeners, which are highly poisonous, non-biodegradable, bio accumulative, and transportable. According to research, the level of various PCBs in treated wastewater ranges from 1.01 to 8.12 ng/L (dissolved media) and 2.03 to 31.2 ng/L (suspended media). If PCBs are inhaled or consumed, they may trigger diseases like carcinogenicity, bones deformity, mental retardation, and reproductive dysfunction [9]. Endocrine disruptors (EDCs) are external simulated substances that mimic hormones and prevent animals' endocrine receptors from performing their regular activities. Many commercialized items, including plasticizers, antimicrobials, and medications, have been produced employing these compounds and their byproducts as chemical solvents. According to reports, the primary contributor of different EDCs is chemical waste from paper and pulp production, leather tanning, fabric manufacture, and distilleries. Popular EDCs that enter the body through food, water, or the air include dioxins, alkylphenols, bisphenol A (BPA), triclosan, phthalates, and nonylphenol (NP). These substances build up in fat tissue and have synergistic consequences on human physiopathology, including cancer of the respiratory tract, liver, and prostate, metabolic disorders, weight gain, thyroid issues, heart disease, digestive tract, neural, and muscular illnesses.

## 3. Sources of MPs

### 3.1. Soil

The widespread use of insecticides in farming, which are employed to control weeds and bugs that endanger grain yields and efficiency, is a major contributor to soil pollution from OMPs [10]. Not only does the use of these chemical compounds degrade soil quality and harm natural heritage, but it also facilitates the transfer toxicants to other parts of the ecosystem [11]. The detrimental ecological effects of pesticide-derived pollutants are exacerbated by farm drainage, which serves as a major pathway for their spread into nearby and remote environments. A common method of assessing this process is to look at the drainage capacity of farmland-derived streams that ultimately combine to form larger aquatic ecosystems [12]. The processes of OMP production, consumption, and discharge into the surroundings are intricate processes. Research conducted in Nigeria has brought attention to the ingestion of OMPs, including organochlorines and polychlorinated biphenyls, showing how drainage and leakage mechanisms spread soil and water contamination across a larger area [13]. Moreover, pesticides may enter the food chain through soil absorption and subsequently accumulate in plants and animals, further disrupting ecological diversity and endangering the environmental health. Interestingly, it has also been found that organophosphates may enter the food chain and harm animals. Furthermore, a considerable amount of building supplies, biosolids, and household items (such as cookware, packaged food, color, and fabrics) that contain perfluoroalkyl substances (PFAS) are immediately disposed of in landfills. The majority of PFAS are easily dissolved in landfill leachates due to their high-water dissolution, after which they are discharged into the surrounding environment [14].

### 3.2. Air

Air can also act as a carrier for toxicants; however, its role in immediate contact is significantly less important than that of water. The air predominantly serves as dispersal agent, enabling OMPs to travel from their point of origin to water sources, which serve as the primary pathways for environmental interaction. When released into the environment, OMPs can volatilize, persist in the surroundings, and be inhaled, leading to contamination [15]. As it eventually flows into rivers, streams, and aquifers; it helps maintain the waterways that OMPs use for their dissemination. Investigations into this process have revealed the presence of OMPs, such as chemicals, phthalate esters, polycyclic aromatic hydrocarbons (PAHs), and PCBs in precipitation [16]. The role of air condensation in the ambient dispersion of OMPs has been confirmed by research conducted on snowfall in Norway, which detected the presence of PAHs, fatty acids, phthalic acid, and ethyl esters [17].

### 3.3. Food Chain

Toxicants reach a variety of species directly through food webs within their own environments. Greater levels of trophic differentiation between marine and land-based animals are especially susceptible to OMP contamination for two main reasons: (i) eating lower trophic level creatures that have ingested OMPs from their surroundings, and (ii) OMP persistence. As a result, various species display diverse patterns of OMP absorption, largely influenced by the unique characteristics of the OMPs [18]. The uses of computational approaches allow researchers to juxtapose the rates of MP absorption by aquatic species with those occurring across the food chain. The studies have demonstrated that PCBs' high ecological perseverance, trophic transmission accounts for nearly all of the biological accumulation of PCBs in apex predators. These results highlight the multiple routes that OMPs take within ecological systems, which are determined by their chemical characteristics and relationships with living organisms. When evaluating the most effective ecological dispersion paths, a chemical's solubility plays a crucial role. In general, substances that are less soluble in water are more noxious, persistent, and capable of bioaccumulating than their more soluble counterparts [19]. This property enables the detection of OMPs far from their source of origin. In contrast, pollutants that are extremely soluble and mobile can move faster through the surroundings. These substances can undergo metabolic transformations that generate derivatives that are easier for species to ingest; however, they are not completely broken down, leading to the excretion of original compounds through feces and urine. Instead of entering through the food chain, these contaminants are likely to reach living organisms directly through ambient exposure [20].

### 3.4. Water

Since, nearly all the producers of OMPs eventually release their contaminants into aquatic environments, water serves as a primary medium for the exposure and spread of OMPs in the ecosystem. Minor tributaries that eventually merge into larger rivers carry residues of pesticides from farmlands, allowing the contaminants to enter larger bodies of water. The efficient movement of MPs from their source of origin to the affected species depends on the water cycle [21]. With respect to sewage disposal techniques, such as the discharge of urine, PCPs, and feces into waterways, medicinal OMPs are mostly detected in waterways. These OMPs subsequently enter the environment, where they continue to expose organisms. According to a research conducted in the U.S., pharmaceutical OMPs were detected in over 80% of the nation's potable water supplies [22]. These findings are alarming since it implies that a significant percentage of Americans are constantly exposed to these harmful OMPs. Similar research has also detected medicinal MPs in water-based samples from Asia, Europe, and Africa. Prolonged exposure to these microcontaminants, whether through breathing or skin contact, offers major health concerns to the impacted communities. Animals are also exposed to OMPs, as water bodies serve as a source of food for them. Through this interaction, OMPs have the possibility to enter the food chain, potentially affecting higher trophic level species as well [15].

## 4. Toxic Trails: Unraveling the Fate of MPs

Water pollution has been a global issue for a long time. Today, the worldwide efforts to protect the aquatic environments focus on identifying suitable and efficient approaches within wastewater purification plants to successfully eliminate MPs, preserve ecological balance, and reduce the detrimental impacts on human health [23]. Industrial effluents must be treated to lessen its adverse consequences and to comply with the permitted levels of the discharges. MPs are often introduced into the ecosystems via either dispersed or isolated routes. Reducing MPs at the source, during utilization, and after the end of their life cycle are three affordable ways to limit their environmental dissemination [24]. MPs are eliminated from wastewater by abiotic alteration, microbial breakdown, and binding onto materials. Assimilation and breakdown are the main processes in wastewater purification facilities. The main objective of sewage purification facilities is to eliminate compounds that endanger the environment; however, MPs cannot be completely removed with all the advancements in wastewater treatment plants [23]. When compared to other conventional physical and chemical approaches, many biological treatment techniques demonstrated better capacity to eliminate MPs [23]. Traditional therapeutic techniques have evolved into more sophisticated methods that may be applied independently or in combination with traditional techniques. Wastewater characteristics determine the most suitable treatment approaches to use and also influence the treatment system's limitations and overall effectiveness [25, 26]. According to Nasr et al., (2007), the utilization of rotating biological contactors (RBCs) or activated sludge in biological processes results in a discharge that is acceptable with legally restricted licenses [25]. Since, MPs undergo frequent transformations during treatment, a bioassay study of the entire effluent rather than an individual chemical assessment offers clearer insights into their potential impacts [27]. According to Nasr et al., construction-related chemical-based discharges did not meet the allowable limits following activated sludge treatment [25]. This was due to the lack of nitrogen and phosphorous concentrations, which decreased the effectiveness of sludge. In order to address this issue and achieve the best outcome, nitrogen and phosphorous compounds had to be added. Combining different types of effluents during wastewater treatment increases the contaminants load, however, occasionally it helps the treatment procedures and may be a more economical approach.

In plastic footwear production facilities, the salts required by organisms to carry out their activities and, by extension, to remove contaminants; are supplied by diluting commercial sewage with domestic sewage rather than by using chemical salts as an external source [25]. Notably, the anaerobic and aerobic settings implemented in the nitrogen elimination methods have an impact on the elimination rate of MPs. The findings demonstrated that some MPs were eliminated solely under toxic circumstances; while others were eliminated in both, anaerobic and aerobic conditions [28]. In contrast to hydrophilic substances, which showed a higher elimination percentage at 20°C compared to 10 °C, hydrophobic substances maintained a consistent level of elimination at 10–35°C. Additionally, it was also discovered that the elimination of the majority of MPs in the MBR cleanup system was declined at 45 °C. Since acidity or basicity influences MP dispersion and microbe proliferation, the pH of wastewaters regulates the elimination procedure [26]. Many MPs, including antibiotics, become ionized at this pH level. The ionization state influences their adsorption and biodegradation by affecting the degree of component separation [29]. Hydraulic retention time (HRT) [27] and precipitation [6] are other significant variables that affect MP amounts in effluents. It is challenging and requires expensive and time-consuming investigations to foresee the eventual outcome and elimination of MPs after wastewater treatment. Nonetheless, an approximation of the situation can be made using the physical properties of the substance; such as pKa and dissolution [27]. There is no statistical correlation between functioning and framework in MP biosynthesis. MPs with an adsorption coefficient (Kd) less than 300 L/kg were often unable to sorb onto the subsequent sludge; instead, their evolution could be easily evaluated via comparison of the amounts of both effluent and influent [27]. The substance's qualities determine its effects, which may also be prolonged or manifest in distant locations. Due to their short half-life or limited distribution, certain chemicals are notoriously hard to detect. However, their toxicity poses significant hazards to marine life and leads to noticeable consequences [24]. Additionally, understanding the fate of MPs in the sewage purification facilities require predictive modeling, it provides a quantitative description of the actual elimination of MPs, irrespective of their detection. By comparing the MPs data at the plant's inlet and outlet with macro pollution parameters at the outlet, the mathematical examination of the data enables the evaluation of wastewater treatment plant's performance [23].

## 5. Adsorptive Removal of MPs

A variety of sorbent materials, such as clay [30], mineral [31,32], carbonaceous [33], polymeric [34–36], and nanocomposite materials [37], have been investigated for the removal of OMPs. The adaptability and economy of sorption-based approaches have led to their widespread implementation in OMPs remediation. This section provides a concise overview of the fundamental aspects regarding the utilization of chemically crosslinked cyclodextrin-assisted polymeric substances for the eradication of various contaminants in wastewater such as, textile dyes, MPs, polycyclic aromatic hydrocarbons (PAHs), pharmaceutical compounds, and pesticides. The presence of these pollutants in wastewater and freshwater, along with their adverse environmental impacts, has resulted in a rise in studies focussing on the sorption of impurities by chitosan, biochar, covalent organic frameworks (COFs), and cyclodextrins (CDs).

### 5.1. Chitosan

A relatively common biological polymer, chitin is present in the shell of crustaceans, the outermost coverings of insects, algal cells, and the cell walls of fungi. It is less common in nature for chitosan to be found in some fungi belonging to the *Mucoraceae* family. In the past, chitin derived from crustacean origins was primarily chemically deacetylated for production of commercialized chitosan compounds [1, 2, 4]. Different chitosan blends for attracting OMPs from sewage are shown in Table 1.

**Table 1.** Summary of chitosan composites utilized in OMP adsorptive eradication in last 5 years

| Chitosan Composites | MPs | Adsorption capabilities ($mg\ g^{-1}$) | References |
|---|---|---|---|
| GTH-CTNPEI aerogel | Perfluorooctanoic acid (PFOA) | ~100% Removal | [38] |
| chitosan/polyurethane (PU/CH) | Perfluorohexanesulfonic and perfluorobutanesulfonic acids | 24.74<br>15.86 | [39] |
| carboxymethyl chitosan and sodium poly(acrylate) cryogels | Fluoxetine | 80.6 ± 3.4 | [40] |
| Chitosan- Graphene oxide (CSN@GO) | Ciprofloxacin | 35.36 | [41] |

| Chitosan Composites | MPs | Adsorption capabilities ($mg\ g^{-1}$) | References |
|---|---|---|---|
| Chitosan/polyethyleneimine/ $Ti_3C_2$ MXene composite | Chlortetracycline<br>Ibuprofen | 128.42<br>511.54 | [42] |
| Polyethyleneimine-grafted CBs (GCBs) | Potassium nonafluoro-1- Butane sulfonate (PNFBS)<br>Perfluoro octane sulfonate (PFOS)<br>Perfluoro butanoic acid (PFBA)<br>PFOA | 260.14<br>250<br>305.78<br>230.02 | [43] |
| graphene oxide/chitosan/ genipin sponges | Diclofenac<br>Triclosan | 48.58<br>7.08 | [44] |
| PEI-modified montmorillonite chitosan beads | PFBA<br>PNFBS<br>Pentadecafluorooctanic acid<br>Perfluorooctanesulfonate | 12.2<br>16.7<br>18.5<br>20.8 | [45] |
| hyper-cross-linked chitosan | Endocrine Disrupting Chemicals<br>Bisphenol F (BPF)<br>Bisphenol A (BPA)<br>bisphenol B (BPB)<br>p-tert-butylphenol (4-t-BP) | 178<br>262<br>375<br>141 | [46] |
| Cyrogel derived from chitosan | PFOA<br>Perfluorodecanoic Acid | 451.4<br>312.3 | [47] |
| Graphene oxide/chitosan/genipin sponges | Polystyrene | 7.42 | [44] |
| chitosan/polyethyleneimine xerogel | perfluorobutanesulfonic acid (PFBS) | 305±45 | [48] |
| Chitin-PANIs sponge (ChPANIs) | Polystyrene (PS) | 91.7% | [49] |
| Chitin-CB sponge | Polystyrene | 99.74 | [50] |
| Hardwood/chitosan sponge | Polystyrene | 0.26 | [51] |
| chitosan/imprinted hydrogel | Ibuprofen | 103.93 | [52] |
| mud-crab shells origin chitosan | Rifampicin<br>Streptomycin<br>Ibuprofen | 66.91<br>11<br>24.21 | [53] |
| PSSMA-functionalized-chitosan magnetic | Norfloxacin<br>Sulfadiazine | 412.63<br>214.07 | [54] |
| Bentonite-chitosan composite | Amoxicillin<br>Ampicillin<br>Doripenem | 86.1<br>83.3<br>96 | [55] |
| Polydopamine enhanced magnetic chitosan aerogel | Polyethylene terephthalate (PET)<br>Polyethylene (PE)<br>PS | 97.3%<br>94.6%<br>92.3% | [56] |
| Chitin/GO/chitosan sponge | PS<br>PS-COOH,<br>PS-$NH_2$ | 9.74<br>7.83<br>12.4 | [57] |
| Chitin/GO sponges | Polystyrene | 8.461 | [58] |
| Tricaprylmethylammonium chloride conjugated chitosan hydrogel | Tetracycline | 20.85 | [59] |
| $NiFe_2O_4$-COF-chitosan-terephthalaldehyde | Tetracycline<br>Cefotaxime | 388.52<br>309.26 | [60] |
| Cross-linked chitosan beads grafted by polyethylenimine | Diclofenac Sodium | 253.32 | [61] |
| Chitosan film | Furosemide | 3.5 | [62] |
| PMCC | Ciprofloxacin | 527.93 | [63] |

### 5.2. Biochar

Biochar is recognized as an effective sorbent for the elimination of MPs from aquatic environments and soil due to its porous and carbonaceous structure, facilitated by a biosorption process [64]. The term "pyrolysis" refers to a process in which biogenic waste is heated at temperatures ranging from 250$^0$C to 850$^0$C, either in the absence of oxygen or in a limited oxygen environment. Pyrolysis produces hydrocarbon-rich bio-oil and syngas as byproducts. Pyrolysis may be classified into three primary categories: slow pyrolysis, quick pyrolysis, and flash pyrolysis. The factors that have the most significant impact on the classification of pyrolysis and the distribution of its products are the particle size of the biomass, retention time, reaction temperature, heating rate, and the duration. In recent years, a variety of agricultural commodities or by-products have been employed as biomass feedstocks for the synthesis of biochar. Lignin, cellulose, and hemicellulose are the three major polymers that are found in these sources. Although cellulose and hemicellulose are more vulnerable to volatilization and undergo pyrolysis at temperatures around 350°C, lignin is more resistant to combustion, with its pyrolysis temperature scaling to 500°C [65]. The yields of pyrolysis products are influenced by the ratios of volatile matter, fixed carbon, ash content, and moisture in biomass feedstocks. The surface characteristics of biochar, its aromaticity, and the interactions among active functional groups significantly affect the removal of MPs. Pyrolysis temperatures influence surface area, pore size distribution, hydrophobicity, and stability. The pore network of biochar consists of macropores with sizes ≥50 nm, mesopores with widths ranging from 2 nm to 50 nm, and micropores with diameters <2 nm. Mesopores are mainly linked to the surface properties of biochar, augmenting the adsorption of small organic molecules [66]. The surface charge and polarity of biochar are essential factors affecting the adsorption of ionic molecules and organic compounds, respectively. Biochar generated at low temperatures is more effective for the elimination of inorganic and polar organic pollutants because of its elevated concentration of oxygen- and hydrogen-containing functional groups [67]. The increased hydrophobicity of biochar facilitates greater penetration of non-polar pollutants due to the lack of hydrogen bonding between water and oxygen functional groups. Moreover, the enduring chemical stability of biochar is ensured by reduced O:C and H:C ratios (<0.2), which signifies its aromaticity and maturity, respectively. Consequently, an increased resilience of biochar to chemical and biological degradation is anticipated.

#### 5.2.1. Adsorption Mechanisms

Biochar can bind to organic matter in a several ways, including particle splitting, pore blockage, electrostatic interactions, donating and accepting electrons, and interacting via water molecules. Splitting is a common phenomenon that takes place in biochar with a high volatile substance and at high concentrations of solutes. An adsorption mechanism for biochar called "pore filling" depends on the meso- and micropores that are found on its outermost layer. For the pore-filling stage of biochar sorption, biochar with a lower volatile concentration is preferred [68]. Electrostatic interaction is the main mechanism for the assimilation of ionic organic compounds. According to this process, the negatively charged surface of the biochar attracts ionic solutes and vice versa. When aromatic substances are sorbed onto the graphene-like structure of biochar, an association between electron donors and acceptors begins. The sorption of hydrophobic and neutral organic molecules is governed by hydrophobic interactions and the partitioning process. Compared to partitioning, hydrophobic interactions require less hydration energy. The sorption of various organic pollutants onto the graphene interface of biochar depends on this process [69].

Dong et al investigated the interaction processes between Cr(VI) and biochar derived from sugar beetroot, which resulted in the formation of Cr(III). The positive charges of the biochar were electrostatically attracted to the negatively charged Cr(VI) species [70]. Furthermore, it was demonstrated that hydrogen ions generated from the biochar participate in the reduction of Cr(VI) to Cr(III). The diminished Cr(III) interacts with the functional groups of the biochar in an aqueous solution to create a Cr(III) complex. Another study revealed the same findings regarding the adsorption of $Pb^{2+}$ from wastewater utilizing biochar derived from sewage sludge, demonstrating that the biochar's functional groups are essential for complex formation with Pb [71]. Similarly a study by Sun et al. demonstrated a comparable process in the adsorption of Hg(II) using biochar derived from soybean stalks, and the adsorption of Cd(II) using biochar manufactured from corn straw [72].

## 5.3. Covalent Organic Frameworks (COFs)

A family of crystallized porous polymeric substances with extremely organized architectures and persistent porosity are known as COFs. COFs are distinguished from other polymers by their physical predesignability, technological controllability, and operational accessibility. Theoretically, polymerization processes provide synthetic strategies for constructing predesigned fundamental and high-order frameworks, while the topographical layout plan offers geometrical guidance for the architectural tiling of prolonged porous quads [73]. Several COF blends for attracting OMPs from sewage are shown in Table 2.

## 5.4. Cyclodextrin

Cyclodextrins (CDs) are cyclic oligosaccharides with a hollow, truncated cone structure. CD is a cost-effective, biocompatible, readily accessible, and biodegradable substance. Enzymes break down starch, maize, and potatoes by connecting glucose residues via $\alpha$−1,4-glycosidic bonds, forming a macro-cycle array of linear, branching, and cyclic dextrin [95]. CDs are soluble in water; yet, their cavities are hydrophobic, enabling them to preferentially bind hydrophobic guest molecules that conform to their cavity size and structure, therefore generating inclusion complexes. This distinctive quality has rendered CDs important in both industry and academics. CDs are known for their complexation capabilities and potential uses in several fields, including pharmaceutical, food, biomedicine, and supramolecular chemistry [96]. The fundamental design principle for cyclodextrin-based materials is to reduce their solubility in water. Although the abundance of hydroxyl groups on the external structure of cyclodextrin imparts high water solubility, these same groups also allow for chemical modification with various functional groups, thereby reducing water solubility while enabling the development of novel functionalized derivatives [97].

**Table 2.** Summary of COF composites utilized in OMP adsorptive eradication in last 5 years

| COF Composites | MPs | Adsorption capabilities ($mg\ g^{-1}$) | References |
|---|---|---|---|
| chitosan/CCP-$NH_2$ aerogels | Sodium diclofenac | 501 | [74] |
| $NH_2$MIL-125(Ti)@TpPa-$SO_3$H | Ciprofloxacin<br>Enoxacin<br>Norfloxacin<br>Sparfloxacin | 457<br>445<br>456<br>429 | [75] |
| CS@COF | PFOA | 2.8 mmol/g | [76] |
| COF-300/PVDF | Naphthalene<br>1-Naphthylamine | 303<br>278 | [77] |
| $Fe_3O_4$@COF-300 | Ibuprofen<br>Bisphenol A | 303.03<br>173.31 | [78] |
| $Fe_3O_4$@COF@$NH_2$-MIL-125Ti | Tebuconazole<br>Tetracycline | 87.73<br>138.1 | [79] |
| Hydrazone/COFs (TpBch) | Doxorubicin hydrochloride | 121.9 | [80] |
| Chitosan/COF | sulfamerazine | 102.5 | [81] |
| UiO-67/CS-MCA | 2,4-dichlorphenoxyacetic acid<br>Glyphosate | 615<br>675 | [82] |
| COF–S7 | Naproxen | 35 | [83] |
| COF-300/GO/PPy | Diclofenac<br>Indomethacin | 138<br>115 | [84] |
| $NH_2$/COF | Bisphenol A | 190.22 | [85] |
| F-COF/Chitosan | GenX<br>Perfluoro-octanoic acid<br>Perfluoro-octanesulfonic acid | 4603.3<br>6177.1<br>8307.1 | [86] |
| BPDA-TAPB-Dva | Bisphenol A<br>Bisphenol S | 149.25<br>200 | [87] |
| M-TpDAB | Chlortoluron<br>Isoproturon<br>Monolinuron<br>Buturon<br>Monuron | 10.8<br>12.1<br>10.7<br>11.9<br>11.7 | [88] |
| PPOP | Bisphenol A | 653 | [89] |
| H–COF-$SO_3$H | Paraquat and Diquat | 129.9 | [90] |
| Tb@TpPa-$SO_3$H | Ciprofloxacin<br>Enrofloxacin<br>Norfloxacin | 956<br>998<br>989 | [91] |
| TpBD-(CF3)2 | Ibuprofen | 27 (River)<br>42 (Lake) | [92] |
| TPB-DMTP-COF | Sulfamerazine | 209 | [93] |
| $NiFe_2O_4$ COF- chitosan terephthalaldehyde | Cefotaxime<br>Tetracycline | 309.26<br>388.52 | [94] |

Glucanotransferase (CGTase) and amylase act as catalysts in the breakdown of starch, which results in the production of CDs. Due to their non-toxic nature, adequate cavity diameters, stable crystalline structure, and affordable price, they have attracted significant interest from researchers since 1980s. Each CD molecule is composed of several stable hydrophobic cavities and many hydroxyls groups that are capable of interacting with aromatic compounds. Rizzi et al., developed a CD sponge that was capable of isolating ciprofloxacin ($C_{17}H_{18}FN_3O_3$) [98]. The modification process involved components such as epichlorohydrin, glutaraldehyde, citric acid, ethylenediaminetetraacetic acid, and ethylenediamine. Consequently, the cross-linked CD exhibits an increased number of adsorption sites and significant adsorption capacity [95]. Conversely, CD immobilization offers an alternative approach to enhance water stability. Jia et al. [99] immobilized β-CD onto the surface of poly(N-isopropyl acrylamide) fibres to improve the regeneration performance of CDs. The immobilized β-CDs demonstrated powerful removal efficiency for crystal violet after four adsorption-desorption cycles.

### 5.4.1. Adsorption Mechanism

The ability of CD to form integrated complexes with a wide variety of molecules through non-covalent interactions is the most prominent feature of this molecule. It includes non-covalent interactions, including hydrophobic contacts, electrostatic forces, dipole-dipole interactions, Van der Waals forces, hydrogen bonding, and charge transfer interactions. Hydrophobic interactions and Van der Waals forces are considered as the primary driving forces behind the formation of inclusion complexes. On the other hand, hydrogen bonding and electrostatic interactions help stabilize the structure of the inclusion complex [100]. In CD polymers, the addition of CD molecules is facilitated by the chemically active hydroxyl groups on CD, which may be modified with different other functional groups. These altered reactive groups could interact with the adsorbents hydrophilically, via hydrogen bonding, electrostatic interactions, or complexation. Liu et al. used p-toluenesulfonyl to change β-CD, which facilitated the coordination of $Th^{4+}$ to combine with the oxygen atom in p-toluenesulfonyl group, thereby enabling adsorption. Crini et al. [101] created carboxymethyl cellulose/CD complexes with epichlorohydrin as a crosslinking agent. In alkaline conditions, the carboxymethyl group in the cross-linked compounds absorbed cationic dyes by electrostatic interactions.

## 6. Implications for Sustainability and Environmental Health

When guided by environmentally friendly chemistry principles, the implementation of sophisticated sorption technologies for water cleanup has great potential for enhancing durability and ecological health [102]. The creation of innovative "green" sorbents from waste materials or economically derived substances can reduce the carbon emissions and minimize environmental impacts associated with their production and lifetime administration [34]. For Instance, biochar made from wood or crop residue has demonstrated efficacy in the adsorptive eradication of water contaminants and in promoting circular economy. The widespread applications of optimized sorption technology driven by alternative energy resources offers hope for healthier water supply and preservation of marine environments worldwide. Standardized lifespan evaluation procedures consider toxicity effects are necessary to guise the development and implementation of adsorbents. Studies suggest that nano-adsorbents with reduced cytotoxicity and ecological risk may be produced by green manufacturing processes that incorporate plant based compounds. The prospective for green-designed adsorbents to efficiently and sustainably clean polluted water still needs to be explored and developed [103]. Adsorption methods aligned with environmentally friendly standards offer a promising approach to ensuring cleaner water supplies for the future generations, provided they are implemented and managed responsibly.

## 7. Limitations of Adsorptive Removal

Despite significant advancements in the development of sorption methods for sewage purification, certain limitations require further attention. The identified challenges can be organized into five groups: the capability to adsorb, the potential to selectively adhere, non-specific attachment, substance breakdown, and reliability. The effectiveness and reliability of sorption methods for wastewater treatment are largely determined by the sorption capability. Sorption capacity refers to the amount of sorbate that an identifiable quantity of sorbent can retain under specific conditions. The features of the sorbate (such as its size, orientation, and quantity), the sorbent (such as porousness, and effective groups), and the conditions of the sorption process (such as contact duration temperature, and acidity) can all influence sorption capability. The sorption capability of the majority of sorbents is still very poor in comparison to the substantial quantity and diversity of contaminants in sewage, which is a current limitation of sorption techniques. To obtain successful elimination rates, a significant sorbent dosage is needed, which raises expenses for treatment and complications. Moreover, the complete saturation of available binding pores on the sorbent surface affects the sorption capability of most sorbents to decrease as the concentration of sorbate increases [104]. Therefore, the development of novel sorbents with improved sorption capabilities and selectivity for various contaminants in wastewater is crucial.

Selectivity has a significant role in determining the effectiveness and suitability of adsorption technologies in wastewater treatment. Selectivity refers to how effectively a sorbent can identify and retain a specific sorbate from a mixture of other substances in water. Factors influencing a sorbent's selectivity include interfacial interactions, porosity, reactive functional groups, molecular dimension, orientation, and charges of the sorbate. Furthermore, operating attributes like pH, temperature, and concentration of ions also have an impact on the sorbent's selectivity. It is highly challenging to obtain outstanding selectivity, particularly when dealing with the complexity and heterogeneity of the pollutants in wastewater. Activated carbon can efficiently eliminate an array of emergent contaminants from water by utilizing a wide range of connections, including $\pi$–$\pi$ connection, electrostatic interactions, H-bonds, and hydrophobicity. However, these chemical connections cause activated carbon to draw polar molecules like water, compromising its uniqueness and selectivity for organic pollutants in effluent [105]. Another problem that arises in the process of developing new sorption technologies is the phenomenon known as the competitive sorption effect. This phenomenon takes place when numerous sorbate species fight for the restricted surface sites of the sorbent.

Reduced selectivity results from increased competition amongst sorbates as their initial amount in water increases. To achieve a balance between specificity and effectiveness, this consequence presents an operational dilemma that requires careful consideration of functional variables and thoughtful selection of sorbent materials [106]. Material degradation and stability refer to an adsorbent's capacity to preserve its architecture, operation, and capability over diverse ecological and executable situations.

Several factors influence how quickly and effectively a material degrades and remains stable. These include the sorbent and sorbate possessions, and the working conditions [107]. A major drawback of sorption techniques is that the majority of sorbents have a very low rate of disintegration and persistence in contrast to the severe and intricate environments that are present in sewage. Additionally, the surface or architecture of most sorbents deteriorates physically or chemically with prolonged contact or continuous utilization, leading to reduced durability and effectiveness over time.

## 8. Future Perspectives and Outlook

This chapter's analysis reveals that, until now, batch studies on the removal of MPs using inexpensive and environmentally friendly sorbents have been based on batch experiments.. Regeneration and the possibility of dynamic implementation are dependent on restoration techniques and fixed-bed readings, which have received less attention. Additionally, a significant portion of the research is limited to laboratory-scale analysis of sorption capabilities. Unfortunately, there is a lack of sufficient data on the practical implementation of sorption cycles and MPs removal, particularly at a pilot scale. The majority of MPs are released into the environment through effluent discharge and use of fertilizers as they're not completely eliminated by the conventional wastewater/compost treatment/sludge digestion cycle. Although increasing research has focused on unconventional, inexpensive sorbents for drug eradication, their functional application remains poorly understood. Despite ongoing research on fixed-bed sorption and sorbent redemption studies that incorporate realistic conditions and reflect actual regulatory guidelines are particularly important as many recent investigations use synthetic solutions and contaminant concentrations much higher than those typically found in the environment. As a result, several areas still require research for contemporary usage. Future research should include sustained sorption, regeneration, multi-component sorption, and the treatment of actual sewage. On the other hand, a number of possible medications could be created in a sequence, including multi-level sorption using various sorbents, sorption assisted with post-treatment, an amalgamation of various methods, sorption assisted with advanced treatment methods, etc. When many treatment methods are combined, weakly degraded pollutants, MPs, and other residual organic pollutants can be effectively removed, making the whole process highly efficient. Therefore, there are numerous opportunities to develop more effective treatment configurations in future.

## Conclusion

Recent years indicate substantial progress in the creation and use of novel adsorbents for the clean-up of impurities from surroundings. These advancements reflect the progress made in this field, driven by the need to develop more efficient, cost effective and environmentally friendly solutions to address the growing issue of water pollution. Numerous studies have demonstrated that these novel sorbents have outstanding sorption properties for a wide variety of impurities. These pollutants include metallic ions, rare metals, colorants, drugs, and other developing toxins. Robust interactions with target pollutants provided by their specific surfaces lead to higher removal efficiencies.

The implementation of these adsorbents poses many obstacles. Critical technological challenges encompass the amplification of synthesis processes, the assurance of stability under real-world situations, and the advancement of efficient regeneration strategies. Resolving these difficulties is essential for optimizing the possibilities offered by these substances. Cost related issues significantly affect the real-world implications of these sorbents for industrial purpose. Furthermore, legal and economic structure must facilitate the utilization of these new substances by assuring compliance with protection and ecological requirements and by offering enticements for their research and implementation. Continued advancements are needed to provide scalable and economical synthesis techniques for the industrial production of high-performance sorbents. Moreover, research must focus on improving adsorbent selectivity and stability to efficiently target particular contaminants in intricate environmental settings. This involves the development of multifunctional adsorbents capable of simultaneously targeting multiple pollutants, thereby enhancing the overall efficacy of water treatment systems.

The advancement of sustainable regeneration approaches is a vital domain for future inquiry. Employing eco-friendly and effective techniques for renewing exhausted adsorbents would lower operational expenses and decrease waste. Furthermore, to foster innovations and ensure that new developments align with regulations and policies, it will be necessary for scientists and policymakers to collaborate across multiple disciplines.

Emphasizing field experiments and real world applications is necessary in order to validate the effectiveness of novel sorbents under realistic settings. These studies offer significant insights into operating characteristics, robustness, and efficacy in various ecological conditions. These data are crucial for bridging the gap between laboratory research and large scale industry applications, facilitating the transition from innovative concepts to applied resolutions.

## References

[1] Patel P. K., Pandey L. M., Uppaluri R. V. S. (2024). "Highly effective removal of multi-s from simulated industrial effluent through an adsorption process employing carboxymethyl-chitosan composites." *Environmental Research* 240. https://doi.org/10.1016/j.envres.2023.117502.

[2] Patel P. K., Pandey L. M., Uppaluri R. V. S. (2023). "Synthesized carboxymethyl-chitosan variant composites for cyclic adsorption-desorption based removal of Fe, Pb, and Cu." *Chemosphere* 340. https://doi.org/10.1016/j.chemosphere.2023.139780.

[3] Saeed T., Naeem A., Din I. U., Farooq M., Khan I. W., Hamayun M., Malik T. (2022). "Synthesis of chitosan composite of metal-organic framework for the adsorption of dyes; kinetic and thermodynamic approach." *Journal of Hazardous Materials* 427. https://doi.org/10.1016/j.jhazmat.2021.127902.

[4] Patel P. K., Pandey L. M., Uppaluri R. V. S. (2023). "Adsorptive removal of Zn, Fe, and Pb from Zn dominant simulated industrial wastewater solution using polyvinyl alcohol grafted chitosan variant resins." *Chemical Engineering Journal* 459. https://doi.org/10.1016/j.cej.2023.141563.

[5] Patel P. K., Uppaluri R. V. S. (2025). "Adsorption of emerging pollutants utilizing chitosan derivatives: Recent advances and future perspective." *International Journal of Biological Macromolecules* 299. https://doi.org/10.1016/j.ijbiomac.2025.140203.

[6] Kasprzyk-Hordern B., Dinsdale R. M., Guwy A. J. (2009). "The removal of pharmaceuticals, personal care products, endocrine disruptors and illicit drugs during wastewater treatment and its impact on the quality of receiving waters." *Water Research* 43, 363–380. https://doi.org/10.1016/j.watres.2008.10.047.

[7] Patel P. K., Pandey L. M., Uppaluri R. V. S. (2025). "Conceptual techno-economic analysis of chitosan derivative resin and commercial sorbents for multi- removal." *Separation and Purification Technology*, 131732. https://doi.org/10.1016/j.seppur.2025.131732.

[8] Ahmed M. B., Zhou J. L., Ngo H. H., Guo W., Thomaidis N. S., Xu J. (2017). "Progress in the biological and chemical treatment technologies for emerging contaminant removal from wastewater: A critical review." *Journal of Hazardous Materials* 323, 274–298. https://doi.org/10.1016/j.jhazmat.2016.04.045.

[9] Yao M., Li Z., Zhang X., Lei L. (2014). "Polychlorinated Biphenyls in the Centralized Wastewater Treatment Plant in a Chemical Industry Zone: Source, Distribution, and Removal." *Journal of Chemistry*. https://doi.org/10.1155/2014/352675.

[10] Wołejko E., Jabłońska-Trypuć A., Wydro U., Butarewicz A., Łozowicka B. (2020). "Soil biological activity as an indicator of soil pollution with pesticides – A review." *Applied Soil Ecology* 147. https://doi.org/10.1016/j.apsoil.2019.09.006.

[11] Mahmood I., Imadi S. R., Shazadi K., Gul A., Hakeem K. R. (2016). "Effects of pesticides on environment." *Implications in Crop Science*, Springer, 253–269. https://doi.org/10.1007/978-3-319-27455-3_13.

[12] Kushwaha A., Goswami S., Hans N., Goswami L., Devi G., Deshavath N. N., Yadav M. K., Lall A. M. (2021). *"An Insight into Biological and Chemical Technologies for MP Removal from Wastewater."* Springer, 199–226. https://doi.org/10.1007/978-981-15-6564-9_11.

[13] Rose A., Kehinde O., Babajide A. (2013). "The level of persistent, bioaccumulative, and toxic (PBT) organic MPs contamination of Lagos soils." *Journal of Environmental Chemistry and Ecotoxicology* 5, 26–38. https://doi.org/10.5897/JECE12.045.

[14] Cui D., Li X., Quinete N. (2020). "Occurrence, fate, sources and toxicity of PFAS: What we know so far in Florida and major gaps." *Trends in Analytical Chemistry* 13. https://doi.org/10.1016/j.trac.2020.115976.

[15] Goswami L., Kushwaha A., Shabbir M., Dikshit P.K., Oh S., Bhan U. (2025). "Covalent organic frameworks (COFs) and COFs-based hybrid architectonics: A promising material platform for organic MPs detection and removal from wastewater." *Coordination Chemistry Reviews* 527. https://doi.org/10.1016/j.ccr.2024.216398.

[16] Guidotti M., Giovinazzo R., Cedrone O., Vitali M. (2000). "Determination of organic MPs in rain water for laboratory screening of air quality in urban environment." www.elsevier.com/locate/envint.

[17] Shubhankar B., Ambade B. (2016). "A review on deposition, distribution of polycyclic aromatic hydrocarbons in different environmental matrix and study its toxicity and carcinogenic effect." *Asian Journal of Chemistry* 28, 2341–2345. https://doi.org/10.14233/ajchem.2016.20096.

[18] Okeke E. S., Olagbaju O. A., Okoye C. O., Addey C. I., Chukwudozie K. I., Okoro J. O., Deme G. G., Ewusi-Mensah D., Igun E., Ejeromedoghene O., Odii E.C., Oderinde O., Iloh V. C., Abesa S. (2022). "Microplastic burden in Africa: A review of occurrence, impacts, and sustainability potential of bioplastics." *Chemical Engineering Journal Advances* 12. https://doi.org/10.1016/j.ceja.2022.100402.

[19] Goswami L., Vinoth Kumar R., Borah S. N., Manikandan N. A., Pakshirajan K., Pugazhenthi G. (2018). "Membrane bioreactor and integrated membrane bioreactor systems for MP removal from wastewater: A review." *Journal of Water Process Engineering* 26, 314–328. https://doi.org/10.1016/j.jwpe.2018.10.024.

[20] Humphries M. S., Myburgh J. G., Campbell R., Buah-Kwofie A., Combrink X. (2021). "Organochlorine pesticide bioaccumulation in wild Nile crocodile (Crocodylus niloticus) fat tissues: Environmental influences on changing residue levels and contaminant profiles." *Science of the Total Environment* 753. https://doi.org/10.1016/j.scitotenv.2020.142068.

[21] Yang J., Luo Y., Chen M., Lu H., Zhang H., Liu Y., Guo C., Xu J. (2023). "Occurrence, spatial distribution, and potential risks of organic MPs in urban surface waters from qinghai, northwest China." *Chemosphere* 318. https://doi.org/10.1016/j.chemosphere.2023.137819.

[22] Kolpin D. W., Furlong E. T., Meyer M. T., Thurman E. M., Zaugg S. D., Barber L. B., Buxton H. T. (2002). "Pharmaceuticals, hormones, and other organic wastewater contaminants in U.S. streams, 1999-2000: A national reconnaissance." *Environmental Science & Technology* 36, 1202–1211. https://doi.org/10.1021/es011055j.

[23] Hammoudani Y. E., Dimane F. (2021). "Assessing behavior and fate of MPs during wastewater treatment: Statistical analysis." *Environmental Engineering Research* 26. https://doi.org/10.4491/eer.2020.359.

[24] Ali E. M., Farhoud B. S. (2024). "Removal of Micro Pollutants from Industrial Wastewater by Conventional and Advanced Biological Treatment Process." *Environmental Science and Engineering*, Springer, 21–42. https://doi.org/10.1007/978-3-031-62898-6_2.

[25] Nasr F. A., Doma H. S., Abdel-Halim H. S., El-Shafai S. A. (2007). "Chemical industry wastewater treatment." *Environmentalist* 27, 275–286. https://doi.org/10.1007/s10669-007-9004-0.

[26] Cirja M., Ivashechkin P., Schäffer A., Corvini P. F. X. (2008). "Factors affecting the removal of organic MPs from wastewater in conventional treatment plants (CTP) and membrane bioreactors (MBR)." *Reviews in Environmental Science and Bio/Technology* 7, 61–78. https://doi.org/10.1007/s11157-007-9121-8.

[27] Das S., Ray N. M., Wan J., Khan A., Chakraborty T., Ray M. B. (2017). "MPs in Wastewater: Fate and Removal Processes." I*nTech*, 2017. https://doi.org/10.5772/65644.

[28] Falås P., Longrée P., La Cour Jansen J., Siegrist H., Hollender J., Joss A. (2013). "MP removal by attached and suspended growth in a hybrid biofilm-activated sludge process." *Water Research* 47, 4498–4506. https://doi.org/10.1016/j.watres.2013.05.010.

[29] Gulde R., Helbling D. E., Scheidegger A., Fenner K. (2014). "PH-dependent biotransformation of ionizable organic MPs in activated sludge." *Environmental Science & Technology* 48, 13760–13768. https://doi.org/10.1021/es5037139.

[30] Bellaj M., Aziz K., Achaby M. E., Haddad M. E., Gebrati L., Kurniawan T. A., Chen Z., Yap P. S., Aziz F. (2024). "Cationic and anionic dyes adsorption from wastewater by clay-chitosan composite: An integrated experimental and modeling study." *Chemical Engineering Science* 285. https://doi.org/10.1016/j.ces.2023.119615.

[31] Patel P. K., Pandey L. M., Uppaluri R. V. S. (2023). "Multi-metal Adsorption and Cyclic Desorption Characteristics of $Zn^{+2}$ and $Cu^{+2}$ Constituting Multi-component Synthetic Wastewater System Using *Commercial Resins*." Springer Nature Singapore, 3–27. https://doi.org/10.1007/978-981-19-8464-8_1.

[32] Patel P. K., Nagireddi S., Uppaluri R. V. S., Pandey L. M. (2022). "Batch adsorption characteristics of Dowex Marathon MSA commercial resin for Au(III) removal from synthetic electroless plating solutions." *Materials Today Proceedings*, Elsevier Ltd, 824–829. https://doi.org/10.1016/j.matpr.2022.06.258.

[33] Giannakoudakis D. A., Kyzas G. Z., Avranas A., Lazaridis N. K. (2016). "Multi-parametric adsorption effects of the reactive dye removal with commercial activated carbons." *Journal of Molecular Liquids* 213, 381–389. https://doi.org/10.1016/j.molliq.2015.07.010.

[34] Patel P. K., Pandey L. M., Uppaluri R. V. S. (2024). "Cyclic Adsorption and Desorption Characteristics of Citric Acid-chitosan Variant Resins for Pb, Fe, and Zn Removal from Simulated Mining and Agricultural Wastewater System." *Journal of Polymers and the Environment*. https://doi.org/10.1007/s10924-024-03343-0.

[35] Nagireddi S., Patel P. K., Golder A. K., Uppaluri R. V. S. (2025). "Evaluation of a triethylenetetramine-chitosan derivative for the removal of palladium ion from an intricate adsorbate system." *Journal of Molecular Structure* 1323. https://doi.org/10.1016/j.molstruc.2024.140782.

[36] Patel P. K., Pandey L. M., Uppaluri R. V. S. (2023). "Cyclic desorption based efficacy of polyvinyl alcohol-chitosan variant resins for multi heavy-metal removal." *International Journal of Biological Macromolecules* 242. https://doi.org/10.1016/j.ijbiomac.2023.124812.

[37] Chaudhary S., Sood A., Mehta S. K. (2016). "Comparative study of dye adsorption on silica nanoparticles: Effects of surface functionalization and other operational parameters." *Journal of Nanoscience and Nanotechnology* 16, 7433–7443. https://doi.org/10.1166/jnn.2016.11128.

[38] Ilango A. K., Arathala P., Musah R. A., Liang Y. (2024). "Experimental and density functional theory investigation of surface-modified biopolymer for improved adsorption of mixtures of per- and polyfluoroalkyl substances in water." *Water Research* 255. https://doi.org/10.1016/j.watres.2024.121458.

[39] Sellaoui L., Dhaouadi F., Deghrigue M., Bouzidi M., Khmissi H., Dotto G. L., Oliveira M. L. S., Silva L. F. O., Erto A., Ernst B., Badawi M. (2024). "A multilayer adsorption of perfluorohexanesulfonic and perfluorobutanesulfonic acids on bio-based polyurethane/chitosan foam: Advanced interpretation of the adsorption mechanism." *Chemical Engineering Journal* 489. https://doi.org/10.1016/j.cej.2024.151173.

[40] Nkana G. R. N., Chabot B., Nguyen-Tri P. (2024). "Adsorption of fluoxetine in aqueous environments: Development of macroporous cryogels based on sodium poly(acrylate) and carboxymethyl chitosan." *Reactive and Functional Polymers* 205. https://doi.org/10.1016/j.reactfunctpolym.2024.106069.

[41] Vamsi E. B., Reshma M., Devatha C. P. (2024). "Adsorption of ciprofloxacin antibiotic using chitosan graphene oxide hybrid beads." *Case Studies in Chemical and Environmental Engineering* 10. https://doi.org/10.1016/j.cscee.2024.100982.

[42] Li Y., Zhang H., Qu G., Xie L., Tang S., Lei H., Zhong Y., Zhang Y. F. (2024). "Efficient removal of antibiotics from wastewater by chitosan/polyethyleneimine/Ti3C2 MXene composite hydrogels: Synthesis, adsorption, kinetics and mechanisms." *Colloids and Surfaces A: Physicochemical and Engineering Aspects* 702. https://doi.org/10.1016/j.colsurfa.2024.135111.

[43] Shahrokhi R., Park J. (2024). "Enhanced removal of short- and long-chain per- and poly-fluoroalkyl substances from aqueous phase using crushed grafted chitosan beads: Performance and mechanisms." *Environmental Pollution* 340. https://doi.org/10.1016/j.envpol.2023.122836.

[44] Ko M., Jang T., Yoon S., Lee J., Choi J. H., Choi J. W., Park J. A. (2024). "Synthesis of recyclable and light-weight graphene oxide/chitosan/genipin sponges for the adsorption of diclofenac, triclosan, and microplastics." *Chemosphere* 356. https://doi.org/10.1016/j.chemosphere.2024.141956.

[45] Shahrokhi R., Rahman A., Hubbe M. A., Park J. (2024). "Aminated clay-polymer composite as soil amendment for stabilizing the short- and long-chain per- and poly-fluoroalkyl substances in contaminated soil." *Journal of Hazardous Materials* 472. https://doi.org/10.1016/j.jhazmat.2024.134470.

[46] Wang C., Wang Q., Wang J., Si K., Zhu H., Wu Q. (2024). "Well-designed chitosan-based cationic porous polymer: A robust material for effective adsorption of endocrine disrupting chemicals." *International Journal of Biological Macromolecules* 280. https://doi.org/10.1016/j.ijbiomac.2024.135801.

[47] García-Castrillo M., Barandika G., Lizundia E. (2024). "Low Environmental Impact Magnetic Chitosan and Chitin Cryogels for PFAS Remediation." *Advanced Functional Materials*. https://doi.org/10.1002/adfm.202405298.

[48] Kebria M. R. S., Bono L., Salestan S. K., Armirotti A., Carzino R., Athanassiou A., Fragouli D. (2023). "Efficient removal of perfluorobutanesulfonic acid from water through a chitosan/polyethyleneimine xerogel." *Chemical Engineering Journal* 466. https://doi.org/10.1016/j.cej.2023.143236.

[49] Zhu Z., Wu X., Wang Z. (2023). "Effect of polyaniline dispersibility in chitin sponge matrix controlled by hydrophilicity on microplastics adsorption." *International Journal of Biological Macromolecules* 253. https://doi.org/10.1016/j.ijbiomac.2023.127292.

[50] Zhu Z., Wu X., Wang C., Meng Z., Sun C., Wang Z. (2023). "Effect of carbon black particle size in chitin sponges on microplastics adsorption." *Separation and Purification Technology* 322. https://doi.org/10.1016/j.seppur.2023.124321.

[51] Xu J., Guo Y., Tang C., Qian Y., Guo C., Wang Z., Li L. (2023). "Hardwood vessel-inspired chitosan-based sponge with superior compressibility, superfast adsorption and remarkable recyclability for microplastics removal in water." *Chemical Engineering Journal* 475. https://doi.org/10.1016/j.cej.2023.146130.

[52] Stachowiak M., Cegłowski M., Kurczewska J. (2023). "Hybrid chitosan/molecularly imprinted polymer hydrogel beads doped with iron for selective ibuprofen adsorption." *International Journal of Biological Macromolecules* 251 (2023). https://doi.org/10.1016/j.ijbiomac.2023.126356.

[53] Shahrin E. W. E. S., Narudin N. A. H., Shahri N. N. M., Nur M., Lim J. W., Bilad M. R., Mahadi A. H., Hobley J., Usman A. (2023). "A comparative study of adsorption behavior of rifampicin, streptomycin, and ibuprofen contaminants from aqueous solutions onto chitosan: Dynamic interactions, kinetics, diffusions, and mechanisms." *Emerging Contaminants* 9 (2023). https://doi.org/10.1016/j.emcon.2022.100199.

[54] Wang M., You X. Y. (2023). "Efficient adsorption of antibiotics and s from aqueous solution by structural designed PSSMA-functionalized-chitosan magnetic composite." *Chemical Engineering Journal* 454 (2023). https://doi.org/10.1016/j.cej.2022.140417.

[55] Yeo J. Y. J., Khaerudini D. S., Soetaredjo F. E., Waworuntu G. L., Ismadji S., Putranto A., Sunarso J. (2023). "Experimental and modelling study of adsorption isotherms of amoxicillin, ampicillin and doripenem on bentonite-chitosan composite." *South African Journal of Chemistry* 43 (2023) 38–45. https://doi.org/10.1016/j.sajce.2022.09.013.

[56] Zheng B., Li B., Wan H., Lin X., Cai Y. (2023). "Coral-inspired environmental durability aerogels for micron-size plastic particles removal in the aquatic environment." *Journal of Hazardous Materials* 431 (2022). https://doi.org/10.1016/j.jhazmat.2022.128611.

[57] Sun C., Wang Z., Zheng H., Chen L., Li F. (2023). "Biodegradable and re-usable sponge materials made from chitin for efficient removal of microplastics." *Journal of Hazardous Materials* 420 (2021). https://doi.org/10.1016/j.jhazmat.2021.126599.

[58] Sun C., Wang Z., Chen L., Li F. (2023). "Fabrication of robust and compressive chitin and graphene oxide sponges for removal of microplastics with different functional groups." *Chemical Engineering Journal* 393 (2020). https://doi.org/10.1016/j.cej.2020.124796.

[59] Ranjbari S., Tanhaei B., Ayati A., Khadempir S., Sillanpää M. (2023). "Efficient tetracycline adsorptive removal using tricaprylmethylammonium chloride conjugated chitosan hydrogel beads: Mechanism, kinetic, isotherms and thermodynamic study." *International Journal of Biological Macromolecules* 155 (2020) 421–429. https://doi.org/10.1016/j.ijbiomac.2020.03.188.

[60] Li Z., Liu Y., Zou S., Lu C., Bai H., Mu H., Duan J. (2023). "Removal and adsorption mechanism of tetracycline and cefotaxime contaminants in water by $NiFe_2O_4$-COF-chitosan-terephthalaldehyde nanocomposites film." *Chemical Engineering Journal* 382 (2020). https://doi.org/10.1016/j.cej.2019.123008.

[61] Lu Y., Wang Z., Ouyang X. K., Ji C., Liu Y., Huang F., Yang L. Y. (2020). "Fabrication of cross-linked chitosan beads grafted by polyethylenimine for efficient adsorption of diclofenac sodium from water." *International Journal of Biological Macromolecules* 145, 1180–1188. https://doi.org/10.1016/j.ijbiomac.2019.10.044.

[62] Rizzi V., Gubitosa J., Fini P., Romita R., Nuzzo S., Gabaldón J. A., Gorbe M. I. F., Gómez-Morte T., Cosma P. (2020). "Chitosan film as recyclable adsorbent membrane to remove/recover hazardous pharmaceutical pollutants from water: the case of the emerging pollutant Furosemide." *Journal of Environmental Science and Health, Part A, Toxic/Hazardous Substances and Environmental Engineering* 56, 145–156. https://doi.org/10.1080/10934529.2020.1853985.

[63] Zheng C., Zheng H., Hu C., Wang Y., Wang Y., Zhao C., Ding W., Sun Q. (2020). "Structural design of magnetic biosorbents for the removal of ciprofloxacin from water." *Bioresource Technology* 296. https://doi.org/10.1016/j.biortech.2019.122288.

[64] Ahmad M., Rajapaksha A. U., Lim J. E., Zhang M., Bolan N., Mohan D., Vithanage M., Lee S. S., Ok Y. S. (2014). "Biochar as a sorbent for contaminant management in soil and water: A review." Chemosphere 99, 19–33. https://doi.org/10.1016/j.chemosphere.2013.10.071.

[65] Welker C. M., Balasubramanian V. K., Petti C., Rai K. M., Bolt S. D., Mendu V. (2015). "Engineering plant biomass lignin content and composition for biofuels and bioproducts." *Energies* (Basel) 8, 7654–7676. https://doi.org/10.3390/en8087654.

[66] Shimabuku K. K., Kearns J. P., Martinez J. E., Mahoney R. B., Moreno-Vasquez L., Summers R. S. (2016). "Biochar sorbents for sulfamethoxazole removal from surface water, stormwater, and wastewater effluent." *Water Research* 96, 236–245. https://doi.org/10.1016/j.watres.2016.03.049.

[67] Tong Y., McNamara P. J., Mayer B. K. (2019). "Adsorption of organic MPs onto biochar: A review of relevant kinetics, mechanisms and equilibrium." *Environ Science* (Camb) 5, 821–838. https://doi.org/10.1039/c8ew00938d.

[68] Kasozi G. N., Zimmerman A. R., Nkedi-Kizza P., Gao B. (2010). "Catechol and humic acid sorption onto a range of laboratory-produced black carbons (biochars)." *Environmental Science & Technology* 44, 6189–6195. https://doi.org/10.1021/es1014423.

[69] Ersan G., Apul O. G., Karanfil T. (2016). "Linear solvation energy relationships (LSER) for adsorption of organic compounds by carbon nanotubes." *Water Research* 98, 28–38. https://doi.org/10.1016/j.watres.2016.03.067.

[70] Dong X., Ma L. Q., Li Y. (2011). "Characteristics and mechanisms of hexavalent chromium removal by biochar from sugar beet tailing." *Journal of Hazardous Materials* 190, 909–915. https://doi.org/10.1016/j.jhazmat.2011.04.008.

[71] Lu H., Zhang W., Yang Y., Huang X., Wang S., Qiu R. (2012). "Relative distribution of Pb2+ sorption mechanisms by sludge-derived biochar." *Water Research* 46, 854–862. https://doi.org/10.1016/j.watres.2011.11.058.

[72] Sun Y., Gao B., Yao Y., Fang J., Zhang M., Zhou Y., Chen H., Yang L. (2014). "Effects of feedstock type, production method, and pyrolysis temperature on biochar and hydrochar properties." *Chemical Engineering Journal* 240, 574–578. https://doi.org/10.1016/j.cej.2013.10.081.

[73] Geng K., He T., Liu R., Dalapati S., Tan K. T., Li Z., Tao S., Gong Y., Jiang Q., Jiang D. (2020). "Covalent Organic Frameworks: Design, Synthesis, and Functions." *Chemical Reviews* 120, 8814–8933. https://doi.org/10.1021/acs.chemrev.9b00550.

[74] Ravi S., Han H. H., Choi Y., Choi H., Park H. H., Bae Y. S. (2024). "Covalent organic polymer/chitosan multifunctional granular aerogels for diclofenac sodium removal from water." *Chemical Engineering Journal* 492. https://doi.org/10.1016/j.cej.2024.152334.

[75] Gutiérrez-Sánchez P., Hrichi A., Garrido-Zoido J.M., Álvarez-Torrellas S., Larriba M., Gil M. V., Amor H. B., García J. (2023). "Natural clays as adsorbents for the efficient removal of antibiotic ciprofloxacin from wastewaters: Experimental and theoretical studies using DFT method." *Journal of Industrial and Engineering Chemistry* 134, 137–151. https://doi.org/10.1016/j.jiec.2023.12.044.

[76] Zhang X., Wang S., Zhu X., Zhu D., Wang W., Wang B., Deng S., Yu G. (2024). "Efficient removal of per/polyfluoroalkyl substances from water using recyclable chitosan-coated covalent organic frameworks: Experimental and theoretical methods." *Chemosphere* 356. https://doi.org/10.1016/j.chemosphere.2024.141942.

[77] Peng H., Minic A., Banjerdteerakul K., Li K. (2023). "COF-300/PVDF adsorbents with aligned microchannels for fast removal of polycyclic aromatic hydrocarbons (PAHs)." *Chemical Engineering Journal* 465. https://doi.org/10.1016/j.cej.2023.142901.

[78] Zhang B., Tian Y., Gao X., Zheng H., Niu Y., Liu J. (2023). "Adsorption Performance of Magnetic Covalent Organic Framework Composites for Bisphenol A and Ibuprofen." *Molecules* 28. https://doi.org/10.3390/molecules28135214.

[79] Zeng Z. Y., Zhou D. D., Yang S. T., Su M. M., Yang Z. H. (2023). "A novel Ti-based magnetic MOFs combined with COF for tetracycline and tebuconazole removal from water." *Journal of Environmental Chemical Engineering* 11. https://doi.org/10.1016/j.jece.2023.111336.

[80] Ji S. L., Xiao S. S., Wang L. L., Wang Q. P. (2023). "Synthesis of hydrazone-linked covalent organic framework for pH-mediated adsorption and removal of doxorubicin hydrochloride." *European Polymer Journal* 201. https://doi.org/10.1016/j.eurpolymj.2023.112572.

[81] Liu K., Yang J., Liu J., Shuai Q., Yamauchi Y., Han M., Huang L. (2023). "Robust self-floating covalent organic framework/chitosan aerogels for the efficient removal of sulfamerazine." *Chemical Engineering Journal* 472. https://doi.org/10.1016/j.cej.2023.144966.

[82] Luo X., Wang C., Huang G., Tan Y., Tang W., Kong J., Li Z. (2022). "Bio-inspired chitosan aerogel decorated with MOF-on-COF heterostructure hybrid as recyclable scavenger of herbicides in water." *Separation and Purification Technology* 298. https://doi.org/10.1016/j.seppur.2022.121616.

[83] Bala S., Abdullah C. A. C., Tahir M. I. M., Rahman M. B. A. (2022). "Adsorptive Removal of Naproxen from Water Using Polyhedral Oligomeric Silesquioxane (POSS) Covalent Organic Frameworks (COFs)." *Nanomaterials* 12. https://doi.org/10.3390/nano12142491.

[84] Feng J. B., Li Y. Y., Zhang Y., Xu Y. Y., Cheng X. W. (2022). "Adsorptive removal of indomethacin and diclofenac from water by polypyrrole doped-GO/COF-300 nanocomposites." *Chemical Engineering Journal* 429. https://doi.org/10.1016/j.cej.2021.132499.

[85] Xiong M., Wang B., Wang H., Xu F., Zeng Y., Ren H., Zeng H. (2022). "Probing the adsorption behavior and mechanism of NO2 and NH2 functionalized covalent organic frameworks (COFs) for removal of bisphenol A." *Microporous and Mesoporous Materials* 346. https://doi.org/10.1016/j.micromeso.2022.112299.

[86] He C., Yang Y., Hou Y. J., Luan T., Deng J. (2022). "Chitosan-coated fluoro-functionalized covalent organic framework as adsorbent for efficient removal of per- and polyfluoroalkyl substances from water." *Separation and Purification Technology* 294. https://doi.org/10.1016/j.seppur.2022.121195.

[87] Fu D., Zhang Q., Chen P., Zheng X., Hao J., Mo P., Liu H., Liu G., Lv W. (2021). "Efficient removal of bisphenol pollutants on imine-based covalent organic frameworks: Adsorption behavior and mechanism." *RSC Advances* 11, 18308–18320. https://doi.org/10.1039/d1ra02342j.

[88] Guo L., Liu J., Li J., Hao L., Liu W., Wang C., Wu Q., Wang Z. (2021). "A core-shell structured magnetic covalent organic framework as a magnetic solid-phase extraction adsorbent for phenylurea herbicides." *Journal of Chromatography* A 1651. https://doi.org/10.1016/j.chroma.2021.462301.

[89] Lee M. Y., Ahmed I., Yu K., Lee C. S., Kang K. K., Jang M. S., Ahn W. S. (2021). "Aqueous adsorption of bisphenol A over a porphyrinic porous organic polymer." *Chemosphere* 265. https://doi.org/10.1016/j.chemosphere.2020.129161.

[90] Tan W., Xu X., Lv Y., Lei W., Hu K., Ye F., Zhao S. (2021). "Sulfonic acid functionalized hierarchical porous covalent organic frameworks as a SALDI-TOF MS matrix for effective extraction and detection of paraquat and diquat." *Journal of Colloid and Interface Science* 603, 172–181. https://doi.org/10.1016/j.jcis.2021.06.077.

[91] Jiang W., Cui W.R., Liang R.P., Qiu J.D. (2021). "Difunctional covalent organic framework hybrid material for synergistic adsorption and selective removal of fluoroquinolone antibiotics." *Journal of Hazardous Materials* 413. https://doi.org/10.1016/j.jhazmat.2021.125302.

[92] Fernandes S. P. S., Mellah A., Kovár P., Sárri M. P., Pšenička M., Djamila H., Salonen L. M., Espiña B. (2020). "Extraction of ibuprofen from natural waters using a covalent organic framework." *Molecules* 25. https://doi.org/10.3390/molecules25143132.

[93] Zhuang S., Liu Y., Wang J. (2020). "Covalent organic frameworks as efficient adsorbent for sulfamerazine removal from aqueous solution." *Journal of Hazardous Materials* 383. https://doi.org/10.1016/j.jhazmat.2019.121126.

[94] Li Z., Liu Y., Zou S., Lu C., Bai H., Mu H., Duan J. (2020). "Removal and adsorption mechanism of tetracycline and cefotaxime contaminants in water by NiFe2O4-COF-chitosan-terephthalaldehyde nanocomposites film." *Chemical Engineering Journal* 382. https://doi.org/10.1016/j.cej.2019.123008.

[95] Sikder M. T., Rahman M. M., Jakariya M., Hosokawa T., Kurasaki M., Saito T. (2019). "Remediation of water pollution with native cyclodextrins and modified cyclodextrins: A comparative overview and perspectives." *Chemical Engineering Journal* 355, 920–941. https://doi.org/10.1016/j.cej.2018.08.218.

[96] Tian B., Hua S., Liu J. (2020). "Cyclodextrin-based delivery systems for chemotherapeutic anticancer drugs: A review." *Carbohydrate Polymers* 232. https://doi.org/10.1016/j.carbpol.2019.115805.

[97] Periasamy R. (2020). "A systematic review on the significant roles of cyclodextrins in the construction of supramolecular systems and their potential usage in various fields." *Journal of Carbohydrate Chemistry* 39, 189–216. https://doi.org/10.1080/07328303.2020.1792919.

[98] Rizzi V., Gubitosa J., Signorile R., Fini P., Cecone C., Matencio A., Trotta F., Cosma P. (2021). "Cyclodextrin nanosponges as adsorbent material to remove hazardous pollutants from water: The case of ciprofloxacin." *Chemical Engineering Journal* 411. https://doi.org/10.1016/j.cej.2021.128514.

[99] Jia S., Tang D., Peng J., Sun Z., Yang X. (2019). "β-Cyclodextrin modified electrospinning fibers with good regeneration for efficient temperature-enhanced adsorption of crystal violet." *Carbohydrate Polymers* 208, 486–494. https://doi.org/10.1016/j.carbpol.2018.12.075.

[100] Liu L., Guo Q. X. (2002). "*The Driving Forces in the Inclusion Complexation of Cyclodextrins*".

[101] Crini G., Ndongo Peindy H. (2006). "Adsorption of C.I. Basic Blue 9 on cyclodextrin-based material containing carboxylic groups." *Dyes and Pigments* 70, 204–211. https://doi.org/10.1016/j.dyepig.2005.05.004.

[102] Satyam S., Patra S. (2024). "Innovations and challenges in adsorption-based wastewater remediation: A comprehensive review." *Heliyon* 10. https://doi.org/10.1016/j.heliyon.2024.e29573.

[103] Kazemi A., Bahramifar N., Heydari A., Olsen S. I. (2018). "Life cycle assessment of nanoadsorbents at early stage technological development." *Journal of Cleaner Production* 174, 527–537. https://doi.org/10.1016/j.jclepro.2017.10.245.

[104] Momina, Ahmad K. (2023). "Feasibility of the adsorption as a process for its large scale adoption across industries for the treatment of wastewater: Research gaps and economic assessment." *Journal of Cleaner Production* 388. https://doi.org/10.1016/j.jclepro.2023.136014.

[105] Ali M. E., Hoque M. E., Hossain S. K. S., Biswas M. C. (2020). "Nanoadsorbents for wastewater treatment: next generation biotechnological solution." *International Journal of Environmental Science and Technology* 17, 4095–4132. https://doi.org/10.1007/s13762-020-02755-4.

[106] Cai J., Niu B., Xie Q., Lu N., Huang S., Zhao G., Zhao J. (2022). "Accurate Removal of Toxic Organic Pollutants from Complex Water Matrices." *Environmental Science & Technology* 56, 2917–2935. https://doi.org/10.1021/acs.est.1c07824.

[107] Crini G., Lichtfouse E. (2019). "Advantages and disadvantages of techniques used for wastewater treatment." *Environmental Chemistry Letters 17, 145–155.* https://doi.org/10.1007/s10311-018-0785-9.

# Chapter 5

# The Role of Phytoremediation in Wastewater Treatment

**Richa Sharma**[1]
**Rishi Rana**[1,*]
**and Raj Kumar**[2]
[1]Department of Civil Engineering, Jaypee University of Information Technology, Waknaghat, Himachal Pradesh, India
[2]Department of Biotechnology and Bioinformatics, Jaypee University of Information Technology, Waknaghat, Himachal Pradesh, India

## Abstract

Wastewater treatment has become increasingly important as global industrialization, population growth, and urbanization exacerbate water pollution. Current trends in wastewater treatments focus on sustainable and eco-friendly approaches, with special emphasis on phytoremediation and bioremediation. Phytoremediation leverages the ability of plants to absorb, accumulate, and detoxify pollutants, making it a low-cost, green solution for treating heavy metals, organic pollutants, and nutrients in contaminated water. Bioremediation, on the other hand, uses microorganisms to degrade or transform harmful pollutants into less toxic forms. This technique is effective in removing organic contaminants, oils, and other hazardous compounds, while minimizing secondary pollution. Both techniques offer promising alternatives to conventional chemical and physical treatment methods, which are often energy-intensive and expensive. However; some challenges including the slow rate of pollutant degradation in some environments, the specificity of certain plants and microbes, and the need for better understanding of the interactions between pollutants and biological agents are still found. Future perspectives in wastewater treatment involve enhancing these processes through genetic engineering, improving plant and microbial tolerance to pollutants, and integrating bioremediation and phytoremediation with other treatment technologies for greater efficiency. The study discusses the current trends, potential benefits, and challenges of phytoremediation and bioremediation in wastewater treatment, highlighting their role in addressing emerging pollutants, promoting environmental sustainability, and ensuring safe water resources for future generations.

---

* Corresponding Author's Email: rishi.rana@juitsolan.in

In: Advancements in Wastewater Treatment
Editors: Neha Agarwal, Maulin P. Shahl, Vijendra Singh Solankil et al.
ISBN: 979-8-90134-007-3

**Keywords:** phytoremediation, bioremediation, wastewater treatment, contaminants, environment, constructed wetlands

## 1. Introduction

In recent years, the global rise in industrial activities, coupled with significant population growth and rapid urbanization, has created an urgent need for effective wastewater treatment solutions. As freshwater resources dwindle and environmental regulations tighten, the challenge to mitigate water pollution intensifies, underscoring the critical role of innovative and sustainable treatment methods. Traditional approaches to wastewater management, often characterized by chemical and physical interventions, are increasingly viewed as inadequate due to their high energy requirements, cost, and potential for secondary pollution [1, 2]. Amid these concerns, eco-friendly alternatives such as phytoremediation and bioremediation have emerged as vital components of sustainable wastewater treatment strategies. These biological processes offer a green solution to the pervasive problem of water contamination, capitalizing on natural mechanisms to purify water without introducing further environmental hazards [3]. Phytoremediation utilizes the natural capabilities of plants to absorb, accumulate, and detoxify a wide range of pollutants, including heavy metals, organic compounds, and excess nutrients [4-6]. On the other hand, bioremediation employs various microorganisms to degrade or convert harmful pollutants into less toxic forms, effectively addressing contaminants such as organic waste, oils, and other hazardous substances. Figure 1 depicts the basic steps involved in the process of bioremediation.

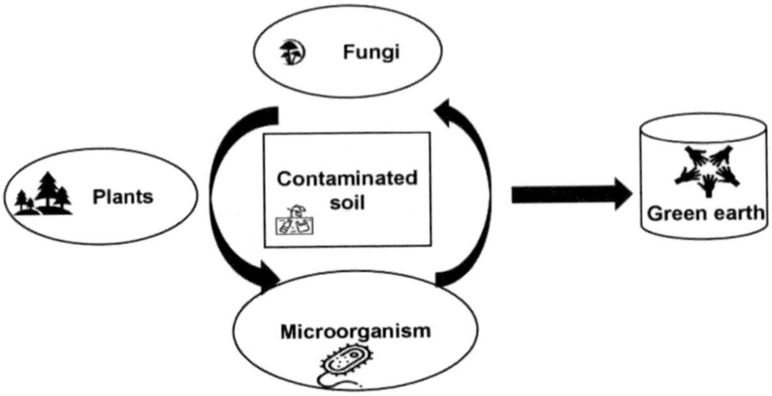


**Figure 1.** Process of bioremediation.

As the world grapples with the dual challenges of protecting its water bodies and promoting environmental sustainability, the adoption of phytoremediation and bioremediation not only aligns with global efforts to reduce ecological footprints but also offers a practical path forward in the management of wastewater. By exploring and enhancing these biological techniques, it is possible to meet the growing demands for clean water and contribute to the health and well-being of ecosystems and human populations alike. Phytoremediation utilizes the natural abilities of specific plants to absorb, sequester, and detoxify pollutants from water, soils, and air [5]. Plants involved in phytoremediation can degrade organic pollutants, stabilize metal contaminants, or extract and accumulate toxic elements from the environment [6]. It is particularly effective in treating water bodies contaminated with heavy metals, nitrates, phosphates, solvents, pesticides, and other organic compounds [7]. The versatility of phytoremediation makes it suitable for application in a variety of settings, from small-scale rural areas to larger, industrial sites. These sustainable methods offer significant advantages over conventional techniques. They are generally less disruptive to the environment and can be integrated into existing ecosystems with minimal ecological disturbance [8-10]. Additionally, both phytoremediation and bioremediation are cost-effective, especially in the long term, as they require less energy and fewer chemical inputs.

## 2. Role of Plants in Phytoremediation and Wastewater Treatment

Experimental studies have demonstrated that *Salvinia molesta* (*S. molesta*) exhibits a high capacity for removing toxic dyes, making it a promising tool for industrial wastewater treatment [11]. For heavy metal remediation, aquatic plants have shown remarkable efficiency, with *Duckweed* (*Lemna minor*) and other species proving to be highly effective metal accumulators [12]. Additionally, bio adsorbents derived from natural lignin residues isolated from black liquor are considered optimal for heavy metal removal [13]. Bio sorption, a widely studied technique, has been found effective in eliminating heavy metals and other toxic contaminants from the environment [14]. Various natural materials such as coconut waste and black oak bark have demonstrated the ability to remove lead (Pb), cadmium (Cd), and mercury (Hg) from contaminated environments [15].

Among plant-based bio sorption techniques, *Opuntia ficus-indica* has exhibited maximum sorption capacities for a wide range of pollutants [16]. Constructed Wetlands (CWs), known for their cost-effectiveness and eco-friendly approach, are widely used for large-scale wastewater treatment to eliminate hazardous chemicals [17]. *Water hyacinth* (*Eichhornia crassipes*) and other aquatic plants have proven efficient in removing both organic and inorganic contaminants [18]. Additionally, *Phragmites australis*, when combined with certain bacterial strains, has been investigated for its ability to remove inorganic pollutants from oilfields and industrial wastewater [19]. Several aquatic plants, including *Pistia stratiotes* and *S. molesta*, are being extensively introduced in large-scale industries due to their high biomass efficiency for pollutant absorption [20]. Beyond plants, fungi and bacteria also play a critical role in phytoremediation, facilitating the transformation and breakdown of organic and inorganic contaminants [21]. Certain plants, including *Lactuca sativa L.*, *Centaurea cyanus L.*, *Helianthus annuus* (sunflower), and *Silybum marianum Gareth*, have demonstrated extraordinary potential for removing hazardous radionuclide, which pose severe threats to living organisms [22]. Various aquatic plants, such as *E. crassipes*, *L. minor*, and *Elodea canadensis*, are commonly used for treating water pollution due to their high absorption capacity, rapid growth rates, and ease of harvesting [23]. Notably, *Elodea canadensis*, a fast-growing macrophyte native to North America, has proven effective in removing pesticide residues from wastewater. Researchers [24] investigated its potential in eliminating three major pesticides—dimethomorph (fungicide), copper sulfate (fungicide), and flazasulfuron (herbicide)—with the toxicity ranking in descending order: flazasulfuron> copper sulfate> dimethomorph. These findings highlight the versatility of aquatic plants, bacterial strains, and bio adsorbents in wastewater treatment and pollutant removal across diverse environmental conditions.

## 3. Phytoremediation Techniques in Wastewater Treatment

Phytoremediation, a green technology, harnesses the natural capabilities of plants to remediate environments polluted by hazardous substances such as heavy metals, hydrocarbons, and pesticides. This technique is appealing due to its cost-effectiveness, environmental friendliness, and the ability to be implemented in situ, thereby minimizing secondary pollution commonly associated with conventional remediation methods. The selection of a specific phytoremediation technique is contingent upon the nature of the contaminant and the prevailing environmental conditions. This strategic approach ensures that the most effective remediation method is employed, optimizing the cleanup process.

Phyto-extraction utilizes specific plants to remove contaminants from soil, groundwater, or surface water. These plants are selected based on their high capacity to accumulate toxic substances, and the process can be continuous or induced. This method is effective for extracting heavy metals and organic pollutants, and it requires plants that can tolerate high concentrations of these substances while still thriving and producing significant biomass [25]. As shown in Table 1, the range of plant species utilized in phytoremediation varies widely, with species like *Helianthus annuus* demonstrating significant capabilities in heavy metal uptake. The table outlines the phytoremediation potential of numerous plant species, emphasizing the targeted pollutants, removal efficiency, and essential properties. Sunflowers (*Helianthus annuus*) can remove heavy metals like lead and uranium with 85% effectiveness, making them appropriate for soil and water cleanup. Indian Mustard (*Brassica juncea*) is a fast-growing plant with high biomass output that can remove cadmium, chromium, and nickel with a 78% efficiency, making it an effective metal accumulator. Water hyacinths (*Eichhornia crassipes*) remove nutrients, heavy metals, and organic pollutants with 90% effectiveness, making them excellent for use in water bodies while reducing algal blooms. Poplar trees (*Populus spp.*), which degrade trichloroethylene and benzene at a 65% efficiency, are ideal for large-scale remediation due to their broad root systems. Alfalfa (*Medicago sativa*) removes organic pollutants with 75% effectiveness and improves soil quality. Chinese Brake Fern (*Pteris vittata*) stands out for its 95% arsenic removal effectiveness, making it one of the most effective plants for arsenic-contaminated soil. Finally, cattails (*Typha spp.*), which are typically found in manmade wetlands, may remove fertilizers and certain metals with an 80% effectiveness. These findings show the broad range of plant-based remediation techniques for heavy metal and organic pollutants removal.

Phyto-degradation involves plants in the breakdown of pollutants through metabolic processes. This can occur within the plant itself or externally if the plant releases enzymes into the environment [26]. These enzymes catalyze the degradation of contaminants in soils, river sediments, and sludge, as well as in ground and surface waters [27]. Phyto-degradation is particularly useful for the treatment of complex organic chemicals that are otherwise difficult to degrade, offering an environmentally friendly solution by leveraging natural plant processes. Rhizofiltration, or phytofiltration, is a method used primarily for the treatment of wastewater from industrial and agricultural activities [28]. In this technique, plants with extensive, fast-growing root systems are either grown directly in water or have their roots suspended in contaminated water. The plants used must exhibit high tolerance to toxic compounds and low oxygen environments. Rhizofiltration is particularly effective for the removal of heavy metals and radioactive elements from water, leveraging the plant's ability to absorb and sequester these pollutants within their root systems [29]. Table 2 illustrates the substantial changes in soil pH and contaminant concentrations before and after applying phytoremediation, underlining the method's effectiveness in altering soil chemical properties. The pH rose from 5.8 to 7.2 (+24%), indicating a shift to neutral. Heavy metal (Pb) concentrations were lowered by 80% (from 200 mg/kg to 40 mg/kg), while organic pollutants (benzene) dropped by 80% (from 150 mg/L to 30 mg/L). Soil conductivity decreased by 40% (from 25 dS/m to 15 dS/m), indicating a reduction in salinity or dissolved ions. Nitrate ($NO_3$) and phosphate ($PO_4$) levels dropped by 67% (from 300 mg/L to 100 mg/L) and 60% (from 50 mg/L to 20 mg/L), reducing eutrophication hazards. Furthermore, total organic carbon (TOC) levels fell by 30% (from 10,000 mg/kg to 7,000 mg/kg), showing a considerable decrease in organic pollutants. These findings show how the treatment method improves environmental quality by lowering contaminants in soil and water.

**Table 1.** Summary of plant species and their contaminant removal

| Plant Species | Pollutants Targeted | Removal Efficiency | Notes |
|---|---|---|---|
| Sunflowers (Helianthus annuus) | Heavy metals (e.g., lead, uranium) | 85% | Effective in soil and water environments. |
| Indian Mustard (Brassica juncea) | Metals like cadmium, chromium, nickel | 78% | Fast-growing, high biomass production. |
| Water Hyacinths (Eichhornia crassipes) | Nutrients, heavy metals, organic pollutants | 90% | Used in water bodies; controls algae growth. |
| Poplar Trees (Populus spp.) | Trichloroethylene, benzene | 65% | Suitable for larger areas due to size and growth rate. |
| Alfalfa (Medicago sativa) | Organic pollutants | 75% | Enhances soil quality while remediating. |
| Chinese Brake Fern (Pteris vittata) | Arsenic | 95% | High efficiency in arsenic removal from soils. |
| Cattails (Typha spp.) | Nutrients, some metals | 80% | Often used in CWs. |

**Table 2.** Physical and chemical properties of the site before and after phytoremediation

| Parameter | Before | After | % Change | Units |
|---|---|---|---|---|
| pH | 5.8 | 7.2 | +24% | pH Units |
| Heavy Metals (Pb) | 200 | 40 | -80% | mg/kg |
| Organic Pollutants (Benzene) | 150 | 30 | -80% | mg/L |
| Soil Conductivity | 25 | 15 | -40% | dS/m |
| Nitrate (NO3) | 300 | 100 | -67% | mg/L |
| Phosphate (PO4) | 50 | 20 | -60% | mg/L |
| Total Organic Carbon (TOC) | 10,000 | 7,000 | -30% | mg/kg |

Phyto-volatilization employs plants to absorb contaminants, which are then metabolized and released into the atmosphere as a less toxic volatile form [30]. This method is suitable for contaminants that can be converted into gaseous forms, such as selenium, mercury, and arsenic, as well as organic compounds like trichloroethylene and benzene. Phyto-volatilization not only cleans the soil and water but also assists in the partial cleaning of the air, making it a holistic environmental remediation technique [31]. Phyto-stabilization uses the roots of plants to stabilize contaminants in the soil, preventing them from leaching into groundwater or being carried away by surface runoff. The ideal plants for this method have a well-developed root system capable of adsorbing, absorbing, and accumulating contaminants, which then convert within the rhizosphere to less mobile or less soluble forms [32]. Phyto-stabilization is particularly useful for sites where contaminants are present in the upper soil layers and can be effectively immobilized by the plant root system. This method helps in reducing the spread of contaminants and stabilizing the soil against erosion.

**Table 3.** Summary of phytoremediation studies using various plants

| Name of Industrial Wastewater | Plant/Plant Part Used for Phytoremediation | Concluding Remarks |
|---|---|---|
| Mines wastewater, Sukinda, Orissa | Eichhornia crassipes | Removed 99.5% of Cr (VI) and other elements within 15 days. |
| Sugar industry wastewater | Eichhornia crassipes, Pistia stratiotes, Nelumbo lutea, Marsilea quadrifolia | Effectively reduced color, temperature, pH, TDS, DO, BOD, COD, chlorides, and sulphates. |
| Coke oven plant wastewater | Lemna minor | Reduced chloride by 30%, sulphate by 16%, and TDS by 14%. |
| Food and beverages industry wastewater, Ghana | Vetiveriazizanioides | Best contaminant removal observed in biogas effluent, followed by Pinora effluent. Oil Palm industry showed lowest removal efficiency. |
| Ceramic industry wastewater | Water hyacinth (Eichhornia crassipes) | Achieved up to 70% heavy metal removal, in the order of Fe > Zn > Cd > Cu > Cr > B. |
| Textile, metallurgical, and pharmaceutical wastewater | Eichhornia crassipes | Significantly reduced COD and efficiently removed Cd but was least efficient for Cu and Fe. |
| Dairy industry effluent | Typha sp., Eichhornia sp., Salvinia sp., Pistia sp., Azolla sp., Lemna sp. | Typha sp.-based CW proved promising for dairy effluent treatment. |
| Industrial effluent with high concentration of heavy metals | Euphorbia prostrata | Showed potential for both phyto-remediation and biosorption of heavy metals. |
| Swine industry wastewater containing Zn and Cu | Typha domingensis Pers., Eleocharis cellulosa | Efficiently phytoremediated Zn & Cu. |
| Paper mill effluent, JK Paper mill, Rayagada, Orissa | Eichhornia crassipes, Hydrilla verticillata, Jussiaea repens, Lemna minor, Pistia stratiotes, Trapa natans | Significantly reduced pH, conductivity, TDS, TSS, chlorine, sulfur, BOD, COD, and heavy metals. Hg was reduced by 66.5% by Lemna minor. |

Table 3 highlights the major industrial wastewater sources, phytoremediation plants, and treatment efficacy. *Eichhornia crassipes* removed 99.5% of Cr (VI) and other elements from mine wastewater in Sukinda, Orissa in 15 days [33]. *Eichhornia crassipes*, *Pistia stratiotes*, *Nelumbo lutea*, and *Marsilea quadrifolia* were shown to considerably lower color, temperature, pH, TDS, DO, BOD, COD, chlorides, and sulphates in sugar industrial effluent [34]. A research [35] found that treating coke oven plant effluent with *Lemna minor* reduced chloride, sulphate, and TDS by 30%, 16%, and 14%, respectively. In Ghana's food and beverage sector, Vetiveriazizanioides removed the most contaminants in biogas effluent, whereas oil palm effluent had the lowest removal efficiency [36]. Ceramic industry wastewater treated with *Eichhornia crassipes* removed up to 70% heavy metals, with Fe > Zn > Cd > Cu > Cr > B [37]. *Eichhornia crassipes* successfully decreased COD and removed Cd from textile, metallurgical, and pharmaceutical wastewater, but exhibited low Cu and Fe removal efficiency [38]. *Typha sp.*-based built wetlands have shown promise in treating dairy sector effluents [39]. *Euphorbia prostrata* has been shown to be capable of phytoremediation and heavy metal biosorption in industrial effluents with high concentrations. Swine industrial effluent including Zn and Cu was effectively treated using *Typha domingensis Pers.* and *Eleocharis cellulose* [41]. Finally, paper mill effluent in Rayagada, Orissa, treated with *Eichhornia crassipes*, *Hydrilla verticillata*, *Jussiaea repens*, *Lemna minor*, *Pistia stratiotes*, and *Trapa natans* significantly reduced pH, conductivity, TDS, TSS, chlorine, sulfur, BOD, COD, and heavy metals, with *Lemna minor* reducing Hg by 66.5% [42]. The study supports phytoremediation efficiency in cleaning industrial wastewater by removing heavy metals and organic pollutants using plant-based processes.

## 4. Advantages and Disadvantages of Phytoremediation

Phytoremediation, a bioremediation process, offers significant advantages for environmental cleanup due to its ability to directly reclaim contaminated sites, in situ. This method allows for the treatment of pollutants on-site without the need to remove and transport contaminated materials, thereby reducing further environmental disturbance and potential hazards. This advantage is underlined by its eco-friendly approach, as it avoids the production of secondary pollutants that are often a by-product of more traditional chemical extraction [43]. Another key benefit of phytoremediation is its contribution to soil health. The extensive root systems of the plants not only help to stabilize the soil and prevent erosion but also enhance the soil's structure and fertility by improving aeration and increasing productivity. These plants also create a conducive environment for the growth of beneficial microorganisms, which play a critical role in maintaining or restoring the natural ecosystem [44]. Furthermore, phytoremediation is cost-effective compared to conventional remediation techniques. It generally requires less specialized equipment and fewer resources, making it a financially viable option for large-scale environmental remediation projects. Additionally, its natural and less invasive approach tends to garner greater acceptance from local communities, who may be directly affected by such remediation activities [45]. However, despite this advantage, phytoremediation is not without its limitations. A significant drawback is the limited depth of treatment, particularly when aquatic plants with shallow root systems are employed. These systems may not effectively address contaminants that have penetrated deeper into the soil. Additionally, the rate of cleanup using phytoremediation is generally slow and can extend over a decade, which may not be feasible for projects requiring rapid remediation [46]. The effectiveness of phytoremediation can also vary seasonally. It tends to decrease during the winter months when plant growth slows or stops entirely. Adverse weather conditions, along with potential plant diseases and pest infestations, can further compromise the effectiveness of the remediation process, posing challenges to its reliability and consistency in performance [47].

Table 4 details the specific conditions under which each phytoremediation experiment was conducted by a recent study [48], including variations in plant species and contaminant types, which are critical for understanding the differential effectiveness noted across treatments. The table outlines the numerous plant species utilized in phytoremediation, the pollutants they target, concentration levels, treatment periods, control group presence, and the phytoremediation method used. Sunflowers (*Helianthus annuus*) were utilized to phytoextract lead (Pb) at 50 mg/L for 90 days, demonstrating their potential to absorb and accumulate heavy metals from polluted environments. Similarly, Indian Mustard (*Brassica juncea*) efficiently phytoextracted cadmium (Cd) at 20 mg/L for 90 days, making it a promising option for cadmium remediation in soils. Water Hyacinths (*Eichhornia crassipes*) used phytovolatilization to remove arsenic (As) at 10 mg/L in 60 days, relying on their capacity to absorb and release pollutants in volatile forms. Poplar trees (*Populus spp.*) were critical in the phytodegradation of trichloroethylene at 100 mg/L over 120 days, converting the pollutant into less toxic chemicals. Finally, Cattails (*Typha spp.*) used rhizofiltration to remove nutrients (N and P) at a total concentration of 150 mg/L during a 120-day period, improving water quality by absorbing surplus nutrients. Each study includes a control group to guarantee that the efficacy of phytoremediation is appropriately determined. This data demonstrates the wide range of applications for phytoremediation techniques in the treatment of heavy metals, organic contaminants, and nutrient pollution in various environmental contexts.

**Table 4.** Experimental design and treatment conditions [38]

| Plant Species | Contaminant | Concentration | Duration | Control Group | Treatment Method |
|---|---|---|---|---|---|
| Sunflowers (Helianthus annuus) | Lead (Pb) | 50 mg/L | 90 days | Yes | Phytoextraction |
| Indian Mustard (Brassica juncea) | Cadmium (Cd) | 20 mg/L | 90 days | Yes | Phytoextraction |
| Water Hyacinths (Eichhornia crassipes) | Arsenic (As) | 10 mg/L | 60 days | Yes | Phytovolatilization |
| Poplar Trees (Populus spp.) | Trichloroethylene | 100 mg/L | 120 days | Yes | Phytodegradation |
| Cattails (Typha spp.) | Nutrients (N, P) | 150 mg/L (total) | 120 days | Yes | Rhizofiltration |

**Table 5.** Performance data of phytoremediation over time [40]

| Time Interval | Contaminant | Initial Concentration (mg/L) | Concentration After 30 Days (mg/L) | Concentration After 60 Days (mg/L) | Concentration After 90 Days (mg/L) | Total Reduction (%) |
|---|---|---|---|---|---|---|
| 1 | Pb | 200 | 160 | 120 | 80 | 60% |
| 2 | Cd | 50 | 42 | 35 | 25 | 50% |
| 3 | As | 100 | 85 | 70 | 50 | 50% |
| 4 | Benzene | 300 | 240 | 180 | 120 | 60% |
| 5 | PO2-4 | 150 | 130 | 110 | 90 | 40% |

The data presented in Table 5 demonstrate the progressive decrease in pollutant concentrations over time, highlighting the time-dependent efficacy of phytoremediation strategies in reducing environmental contaminants [49]. The table depicts the reduction in pollutant concentrations over a 90-day treatment period, demonstrating the efficacy of phytoremediation and other treatment approaches. Lead (Pb) levels were first measured at 200 mg/L and gradually dropped to 160 mg/L at 30 days, 120 mg/L at 60 days, and 80 mg/L at 90 days, for a total reduction of 60%. Cadmium (Cd) and arsenic (As) began at 50 mg/L and 100 mg/L, respectively, and decreased by 50% each until reaching final concentrations of 25 mg/L and 50 mg/L after 90 days. Benzene, a prevalent organic pollutant, reduced from 300 mg/L to

120 mg/L over the treatment period, representing a 60% reduction. Similarly, phosphate ($PO_4^{3-}$), a primary cause to eutrophication, was lowered from 150 mg/L to 90 mg/L, representing a 40% drop. These data highlight the cleanup process's increasing efficiency, since contamination levels consistently fall over time.

## Conclusion

This review has systematically explored the application of phytoremediation as an innovative and sustainable method for water treatment, presenting a green alternative to conventional remediation techniques. Through the detailed examination of various phytoremediation strategies, including phytoextraction, phytostabilization, phytodegradation, rhizofiltration, and phytovolatilization, this paper has highlighted the diverse mechanisms through which plants can effectively mitigate and manage water pollution. Phytoremediation offers a cost-effective solution with the potential for wide-scale deployment in contaminated water bodies. The use of plants not only reduces the financial burden associated with mechanical and chemical treatments but also contributes to ecological restoration, enhancing both biodiversity and the aesthetic value of the landscape. Moreover, the non-invasive nature of this technology promotes community acceptance and facilitates its integration into existing environmental management frameworks. However, the review also acknowledges the limitations of phytoremediation, including the time required to achieve significant contaminant reduction and the variability in effectiveness due to climatic and soil conditions. These challenges underscore the necessity for ongoing research to optimize plant selection and genetic enhancement to improve both the efficiency and range of contaminants that can be addressed. Future perspectives in phytoremediation should focus on combining traditional approaches with advanced biotechnological innovations to develop hybrid systems that can address a broader spectrum of contaminants and site conditions. Furthermore, integrating phytoremediation with other green technologies such as CWs and biochar application could amplify its effectiveness and sustainability.

## References

[1] Alkio, M., Tabuchi, T. M., Wang, X., Colon-Carmona, A. (2005). Stress responses to polycyclic aromatic hydrocarbons in Arabidopsis include growth inhibition and hypersensitive response-like symptoms. *Journal of Experimental Botany*, 56: 2983-2994.

[2] Burken, J. G., Schnoor, J. L. (1999). Distribution and volatilization of organic compounds following uptake by hybrid poplar trees. *International Journal of Phytoremediation*, 1 (II): 139-151.

[3] Burns, R. G., DeForest, J. L., Marxsen, J., Sinsabaugh, R. L., Stromberger, M. E., Wallenstein, M. D., Weintraub, M. N., Zoppini, A. (2013). Soil enzymes in a changing environment: current knowledge and future directions. *Soil Biology and Biochemistry*, 58: 216-234.

[4] Coleman, J., Blake-Kalff, M., Davies, T. (1997). Detoxification of xenobiotics by plants: chemical modification and vacuolar compartmentation. *Trends in Plant Science*, 2 (IV): 144-151.

[5] Cunningham, S. D., Berti, W. R. (1993). Remediation of contaminated soils with green plants: an overview. *In Vitro, Cell & Developmental Biology*, 29 (III): 207-212.

[6] Dordio, A., Carvalho, A. J. P. (2013). Organic xenobiotics removal in CWs, with emphasis on the importance of the support matrix. *Journal of Hazardous Materials*, 252-253: 272-292.

[7] Edwards, R., Dixon, D. D. (2004). Metabolism of natural and xenobiotic substrates by the plant glutathione S-transferase superfamily. In H. Sandermann (Ed.), *Molecular Ecotoxicology of Plants, Ecological Studies,* 170 (pp. 17-50). Springer Verlag, Heidelberg.

[8] Frick, C. M., Farrell, R. E., Germida, J. J. (1999). *Assessment of phytoremediation as an in-situ technique for cleaning oil-contaminated site*s.

[9] Fu, J., Mai, B., Sheng, G., Zhang, G., Wang, X., Peng, P., Xiao, X., Cheng, R. R., Peng, X., Wang, Z., Tang, U. W. (2003). Persistent organic pollutants in the environment of the Pearl River Delta, China: an overview. *Chemosphere*, 52: 1411-1422.

[10] Gajewska, E., Skłodowska, M., Słaba, M., Mazur, J. (2006). Effect of nickel on antioxidative enzyme activities, proline and chlorophyll contents in wheat shoots. Biologia Plantarum, 50: 653-659.

[11] Mustafa HM, Hayder G (2021). Recent studies on applications of aquatic weed plants in phytoremediation of wastewater: a review article. *Ain Shams Eng J.* 12(1):355–65.

[12] Al-Baldawi IA, Abdullah SRS, Almansoory AF, Hasan HA, Anuar N. (2020). Role of *Salvinia molesta* in biodecolorization of methyl orange dye from water. *Sci Rep.* 10(1):1–9.

[13] Ali S, Abbas Z, Rizwan M, Zaheer I, Yavaş İ, Ünay A, (2020). Application of floating aquatic plants in phytoremediation of heavy metals polluted water: a review. *Sustainability.* 12(5):1927.

[14] Naseer A, Jamshaid A, Hamid A, Muhammad N, Ghauri M, Iqbal J, (2019). Lignin and lignin-based materials for the removal of heavy metals from wastewater-an overview. *Z für Phys Chem.* 233(3):315–45.

[15] Calderón OAR, Abdeldayem OM, Pugazhendhi A, Rene ER (2020). Current updates and perspectives of biosorption technology: an alternative for the removal of heavy metals from wastewater. *Curr Pollut Rep.* 6(1):8–27.

[16] Alalwan HA, Kadhom MA, Alminshid AH. (2020). Removal of heavy metals from wastewater using agricultural byproducts. *J Water Supply Res Technol-Aqua.* 69(2):99–112.

[17] Nharingo T, Moyo M. (2016). Application of *Opuntia ficus-indica* in bioremediation of wastewaters. A critical review. *J Environ Manag.* 166:55–72.

[18] Khalifa ME, Abou El-Reash YG, Ahmed MI, Rizk FW. (2020). Effect of media variation on the removal efficiency of pollutants from domestic wastewater in CW systems. *Ecol Eng.* 143:105668.

[19] Yu T, Meng L, Zhao Q. B., Shi Y, Hu H. Y., Lu Y. (2017) Effects of chemical cleaning on RO membrane inorganic, organic and microbial foulant removal in a full-scale plant for municipal wastewater reclamation. *Water Res.*113:1–10.

[20] Rehman K, Imran A, Amin I, Afzal M. (2018). Inoculation with bacteria in floating treatment wetlands positively modulates the phytoremediation of oil field wastewater. *J Hazard Mater.* 349:242–51.

[21] Deng Z, Cao L. (2017). Fungal endophytes and their interactions with plants in phytoremediation: a review. *Chemosphere.* 168:1100–6.

[22] Gupta D, Chatterjee S, Datta S, Voronina A, Walther C. (2016). Radionuclides: accumulation and transport in plants. *Rev Environ Contamination Toxicol.* 241:139–60.

[23] Chander PD, Fai CM, Kin CM, editors. Removal of pesticides using aquatic plants in water resources: a review. *IOP Conference Series: Earth and Environmental Science.* IOP Publishing; 2018.

[24] Olette R, Couderchet M, Biagianti S, Eullaffroy P. (2008). Toxicity and removal of pesticides by selected aquatic plants. *Chemosphere.* 70(8):1414–21.

[25] Gerhardt, K. E., Huang, X. D., Glick, B. R., Greenberg, B. M. (2009). Phytoremediation and rhizoremediation of organic soil contaminants: Potential and challenges. *Plant Science*, 176: 20-30.

[26] Gonçalves, J. F., Becker, A. G., Cargnelutti, D., Tabaldi, L. A., Pereira, L. B., Battisti, V., Spanevello, R. M., Morsch, V. M., Schetinger, M. R. C. (2007). Cadmium toxicity causes oxidative stress and induces response of the antioxidant system in cucumber seedlings. *Brazilian Journal of Plant Physiology*, 19: 223-232.

[27] Karadede, H., Unlü, E. (2000). Concentrations of some heavy metals in water, sediment and fish species from the Atatürk Dam Lake (Euphrates), Turkey. *Chemosphere,* 41: 1371–1376.

[28] Komives, T., Gullner, G. (2005). Phase I xenobiotic metabolic systems in plants. *Zeitschrift für Naturforschung,* 60: 179-185.

[29] Lu, Y., Li, X. R., He, M. Z., Wang, Z. N., Tan, H. J. (2010). Nickel effects on growth and antioxidative enzymes activities in desert plant Zygophyllumxanthoxylon (Bunge) Maxim. *Sciences in Cold and Arid Regions*, 2: 436-444.
[30] Macek, T., Macková, M., Káš, J. (2000). Exploitation of plants for the removal of organics in environmental remediation. *Biotechnology Advances*, 18: 23-24.
[31] Małecka, A., Tomaszewska, B. (2005). Reactive oxygen species in plant cells and antioxidative defence systems. *PostępyBiologiiKomór*ki, 32: 311-325.
[32] Marecik, R., Króliczak, P., Cyplik, P. (2006). Phytoremediation – an alternative method for environmental cleanup. *Biotechnologia,* 3: 88-97.
[33] Mielcarek, A., Krzemieniewski, M. (2013). Research on the use of selected macrophytes in the process of methane fermentation. *RocznikOchronaŚrodowiska,* 15: 2611-2624.
[34] Monferrán, M. V., Pignata, M. L., Wunderlin, D. A. (2012). Enhanced phytoextraction of chromium by the aquatic macrophyte Potamogeton pusillus in presence of copper. *Environmental Pollution,* 161: 15-22.
[35] Abdullah, S. R. S., Al-Baldawi, I. A., Almansoory, A. F., Purwanti, I. F., Al-Sbani, N. H., Sharuddin, S. S. N. (2020). Plant-assisted remediation of hydrocarbons in water and soil: Application, mechanisms, challenges and opportunities. *Chemosphere,* 247: 125932.
[36] Ahmad, J., Abdullah, S. R. S., Hassan, H. A., Rahman, R. A. A., Idris, M. (2017). Screening of tropical native aquatic plants for polishing pulp and paper mill final effluent. *Malaysian Journal of Analytical Sciences*, 21: 105–112.
[37] Al-Ajalin, F. A. H., Idris, M., Abdullah, S. R. S., Kurniawan, S. B., Imron, M. F. (2020a). Effect of wastewater depth to the performance of short-term batching-experiments horizontal flow CW system in treating domestic wastewater. *Environmental Technology and Innovation*, 20: 101106.
[38] Al-Ajalin, F. A. H., Idris, M., Abdullah, S. R. S., Kurniawan, S. B., Imron, M. F. (2020b). Evaluation of short-term pilot reed bed performance for real domestic wastewater treatment. *Environmental Technology and Innovation*, 20: 101110.
[39] Al-Ajalin, F. A. H., Idris, M., Abdullah, S. R. S., Kurniawan, S. B., Imron, M. F. (2020c). Design of a reed bed system for treatment of domestic wastewater using native plants. *Journal of Ecological Engineering,* 21: 22–28.
[40] Al-Baldawi, I. A., Abdullah, S. R. S., Anuar, N., Suja, F., Mushrifah, I. (2015). Phytodegradation of total petroleum hydrocarbon (TPH) in diesel-contaminated water using Scirpusgrossus. *Ecological Engineering,* 74: 463–473.
[41] Al Falahi, O. A., Abdullah, S. R. S., Hasan, H. A., Othman, A. R., Ewadh, H. M., Kurniawan, S. B., Imron, M. F. (2022). Occurrence of pharmaceuticals and personal care products in domestic wastewater, available treatment technologies, and potential treatment using CW: A review. *Process Safety and Environmental Protection,* 168: 1067–1088.
[42] Almansoory, A. F., Idris, M., Abdullah, S. R. S., Anuar, N., Kurniawan, S. B. (2021). Response and capability of Scirpusmucronatus (L.) in phytotreating petrol-contaminated soil. *Chemosphere,* 269: 128760.
[43] Alshekhli, A. F., Hasan, H. A., Muhamad, M. H., Sheikh Abdullah, S. R. (2020). Development of Adsorbent from Phytoremediation Plant Waste for Methylene Blue Removal. *Journal of Ecological Engineering,* 21: 207–215.
[44] Armstrong, J., Armstrong, W., Beckett, P. M. (1992). Phragmites australis: Venturi- and humidity-induced pressure flows enhance rhizome aeration and rhizosphere oxidation. *New Phytologist,* 21: 5559.
[45] Ayyappan, B., Srinivasan, V., Veeramani, P., Sangeetha, S. R., Tennyson, S. (2020). Bioremediation of wastewater using aquatic macrophytes and microalgae: Mechanisms and effectiveness. *Chemosphere,* 254: 126791.
[46] Bennicelli, R. P., Stępniewska, Z., Banach, A., Szajnocha, K., Ostrowski, J. (2004). The ability of Azolla caroliniana to remove heavy metals (Hg(II), Cr(III), Cr(VI)) from municipal wastewater. *Chemosphere,* 55: 141-146.
[47] Bhat, S. A., Singh, J., Vig, A. P. (2017). Phytoremediation potential of aquatic macrophyte, Azolla. *Ambio*, 46: 486-497.

[48] Brix, H. (1997). Do macrophytes play a role in constructed treatment wetlands? *Water Science and Technology*, 35: 11-17.

[49] Carranza-Álvarez, C., Alonso-Castro, A. J., Alfaro-De la Torre, M. C., García-De la Cruz, R. F. (2019). Removal of heavy metals by the wetland plant Phragmites australis from wastewater in CWs. *International Journal of Phytoremediation*, 21: 121-129.

# Chapter 6

# Microbes and Biofilms for Sustainable Wastewater Treatment

**Dembede Steve**[1]
**Ephraim Muriithi Kiarii**[1,2,*]
**and Jacinta Muthini Kavuluko**[1]
[1]Department of Science, School of Science, Health and Engineering Daystar University, Nairobi, Kenya
[2]Department of Physical sciences, University of Embu, Embu, Kenya

## Abstract

The growth spurt of industries and anthropogenic activities has produced harmful residues and effluents in the form of wastewater laden with numerous pollutants. Industrial wastewater, also known as sewage, is a major concern and a threat to water and soil quality and to human health. Industrial effluents usually are rich in toxic metals, inorganic compounds, and organic pollutants, so the proper treatment methods must be employed to reduce their environmental impact. Physicochemical treatment techniques are predominantly employed for wastewater management, but these processes have several downsides, namely high operational costs and the formation of secondary pollutants like sludge.

In response to these challenges, biological treatment and bioremediation have emerged as sustainable and cost-effective alternatives. These use a variety of microorganisms that can degrade the pollutants, produce very little sludge, and are environmentally friendly. This chapter discusses the principles and mechanisms of biological wastewater treatment and bioremediation and their advantages over conventional methods. It also discusses the problems encountered when applying these biological processes and the prospects for improving their efficiency. The chapter also assesses the environmental and sustainability implications of implementing bioremediation methods in wastewater treatment systems.

**Keywords:** microbes, biofilms, bioremediation, wastewater, fungi, pollutants, sustainability

* Corresponding Author's Email: kiarii.ephraim@embuni.ac.ke or emuriithi@daystar.ac.ke

In: Advancements in Wastewater Treatment
Editors: Neha Agarwal, Maulin P. Shahl, Vijendra Singh Solankil et al.
ISBN: 979-8-90134-007-3

## 1. Introduction

The recent development of industries and other anthropogenic activities has resulted in the generation of the harmful residues and effluents [1]. The most impactful threats of the industrial effluents are from the industrial wastewater that is commonly known as sewage. It is released into the environment resulting into major water and soil pollution and also serious health concerns to living beings. The various arrays of contaminants in the wastewater are toxic metals, inorganic and organic pollutants [2, 3]. Various techniques including biotechnological processes, are employed to treat industrial and municipal waste worldwide, helping to safeguard the environment from pollution [4, 5]. Among these, the physicochemical method is the most commonly used for wastewater treatment. However, it has certain limitations, such as high operational costs and the production of significant amounts of secondary pollutants, particularly sludge [6, 7]. Biological treatment and bioremediation methods serve as alternative approaches to wastewater treatment due to their cost-effectiveness, ease of operation, utilization of diverse microorganisms, minimal sludge production, and environmental friendliness [8]. The biological method of wastewater treatment has several benefits over other methods. With this method, the organic waste is broken down by microorganisms and transformed in to nontoxic components which serves as a source of food [9, 10].

Different microorganisms that are utilized in the biological treatment processes are grouped into various classes on the basis of their cell structure, energy and carbon sources and respiration methods. According to the cell structure the microorganisms are grouped into Prokaryotes and Eukaryotes. Prokaryotes are simple, small and uni-cellular structures that are less than 5 microns in the size, with the primitive nucleus that is made up of only chromosomes [11]. Eukaryotes are larger in structures with the multiple cells that bound with a membrane. The nucleus is also bound with the nuclear membrane and they are large in number [12]. The microorganisms that are employed in wastewater treatment are bacteria (prokaryote), fungi, algae, protozoa, crustaceans, nematodes and rotifers which are collectively known as eukaryotes [13]. The development of the microorganisms requires the sources of energy and carbon to sustain proper functioning and reproduction and for the synthesis of the new cellular materials [14]. Additionally, the inorganic nutrients required by the microorganisms includephosphorus, nitrogen, potassium, magnesium, calcium and sulfur. These nutrients are utilized as a source of energy by the microorganisms during cell growth and also act as electron acceptors during reactions that yield high energy. Among these, biofilms structured communities of microorganisms offer significant advantages in wastewater treatment due to their resilience, adaptability, and high degradation capacity. The use of microbes and biofilms in wastewater treatment represents a sustainable and environmentally friendly alternative to conventional physicochemical methods. Despite their benefits, microbial and biofilm-based wastewater treatment methods face many challenges such as biofilm clogging, high maintenance requirements, and process optimization. In view of this, current chapter explores the role of microorganisms in biological wastewater treatment in detail, highlighting their challenges, future prospects, and their impact on the environment and sustainability.

## 2. Microbes for Wastewater Treatment

The microorganisms utilized in wastewater treatment can be categorized into four groups based on the basis of energy sources they utilize:

- *Phototrophic organism*- utilize sunlight as energy sources
- *Chemotrophic organisms*- depend on the oxidation/reduction chemical reactions
- *Organotrophic organisms*- obtain their energy from the oxidation reaction of the organic compounds
- *Lithotrophic organisms*- utilize the oxidation of the inorganic compounds as the energy sources

The carbon source is an alternative method of grouping microorganisms. On the basis of carbon source these microorganisms can be grouped as:

- *Autotrophic*- the organisms use carbon dioxide as source of carbon
- *Heterotrophic*- derives carbon from organic matter

Another classification is based on the methods of respiration utilized by the microorganisms. On the basis of respiration these can be categorized as:

- *Facultative organisms*- the group utilizes the free oxygen or nitrite as electron acceptor
- *Anaerobic organisms*- this group of organisms utilize carbon dioxide or sulphate as electron acceptor and they do not obtain energy through aerobic respiration
- *Aerobic organisms*- utilize free oxygen during respiration.

Microorganisms play a vital role in wastewater treatment by breaking down organic and inorganic pollutants, removing nutrients, and detoxifying harmful substances. Various fungi, bacteria, protozoa, and algae contribute to the efficiency of biological treatment processes.

### 2.1. Fungi

Fungi usually obtain their nutrients by breakdown of dead organic matter [15], thus classified as saprophytes. The fungi are mostly free-living species that include molds, mushrooms and yeasts. The mode of respiration of the fungi is strictly aerobic and they have the ability to tolerate low nitrogen and low $p^H$ in the environment. In the biological wastewater treatment process (especially in the activated sludge process) the filamentous fungus has a higher rate of proliferation due to low nitrogen and low $p^H$ in the treatment tanks [16, 17]. The rapid expansion of fungi leads to settling difficulties in activated sludge. A substantial and varied fungal community plays a key role in the composting process of organic waste. These fungi are capable of breaking down cellulose, surviving in nutrient-poor environments, and thriving in conditions with low moisture and acidic $p^H$ levels. Under aerobic conditions; yeasts usually degrade the organic matter into water and carbon dioxide while in the absence of free oxygen yeast degrades the sugars to ethanol.

## 2.2. Protozoa

The protozoa are unicellular, free living, solitary organisms while others exists in the colonies [18]. Protozoa participating in the activated sludge process are classified into five groups: flagellates, amoebae, crawling ciliates, free-swimming ciliates, and stalked ciliates. Among these, ciliated protozoa are the most significant in the activated sludge process. They feature short, hair-like structures called cilia, which they use both for collecting food and for movement. These protozoa have the following benefits in the activated sludge process:

- Improve the settleability to the floc by adding their weight
- Feed on the dispersed cells in the wastewater therefore cleansing the wastewater
- Recycle the nutrients especially phosphorus and nitrogen through excretion
- Generate and release secretions that form a coating, aiding in the removal of fine solids from the bulk solution and attaching them to the surface of particles.

## 2.3. Bacteria

The biological treatment plants contain the most important organism which belongs to bacteria are eubacteria and archaebacteria. The entry of these organisms into the wastewater treatment plants is through fecal waste, infiltration and inflow from water and soil [19]. Eubacteria are primitive, simple life forms and comprise a vast number of organisms in terms of both total biomass and species diversity [20, 21]. The archaebacteria consists of thermacidophiles, halophiles and methanogens [22]. The methanogens (methane forming bacteria) are the most relevant group in the biological treatment plants since they breakdown the organic waste in to methane [23, 24]. The halophiles bacteria require a salt concentration of approximately 3.5% in their environment for the optimum growth, thus they are found in the salty water. For the proper enzymatic activity and maintenance of the cell wall, these marine organisms require an elevated concentration of the potassium ions and sodium ions. The halophilic bacteria and other photosynthetic bacteria have a buoyancy device in their cell structure known as gas vacuoles. The thermoacidophilic bacteria do not have to perform any role in the wastewater treatment plants. The habitat of these organisms is a hot acidic condition such as the volcanic vents. Although the bacteria can be categorized into various groups, but for the biological treatment process they are categorized based on the basis of the roles they perform and on the basis of energy and carbon sources.

Bacteria primarily derive their carbon from two sources: organic and inorganic compounds. Bacteria that obtain carbon from organic compounds are referred to as heterotrophs or organotrophs. In contrast, bacteria that acquire carbon from carbon dioxide or other inorganic compounds, such as carbonate or bicarbonate, are known as autotrophs. The energy sources for the bacteria are always through chemical oxidation reaction and sunlight. The bacterium that utilizes the chemical oxidation reaction as energy sources is called chemotroph. Chemotrophs are classified into two types: heterotrophs, which break down organic compounds, and lithotrophs, which use reduced inorganic compounds such as hydrogen sulphate, ionized ammonia, and nitrites as energy sources. The most significant lithotrophs in the biological treatment plants are iron bacteria, nitrifying bacteria and sulfur bacteria. The most significant lithotrophs are nitrifying bacteria, which play a key role in the biological removal of nitrogen from organic waste. The bacteria that use sunlight as the energy sources are the phototrophs. Bacteria perform both the positive and the negative roles in the biological treatment plants based upon the operational conditions of the plant. The negative role of the bacteria contributes to the increased operational costs, permit violations and the inefficient treatment of wastewater. A significant group of bacteria used in wastewater are summarized in Table 1.

**Table 1.** Significant groups of wastewater bacteria

| | |
|---|---|
| Acetongenic bacteria [25] | Hydrolytic bacteria [26] |
| Coliforms [27] | Methane-forming bacteria [28] |
| Cyanobacteria [29] | Nitrifying bacteria [30] |
| Denitrifying bacteria [31] | Nocardioform [32] |
| Fecal coliforms [33] | Pathogenic bacteria [34] |
| Fermentative (acid-forming) bacteria [35] | Poly-P bacteria [36] |
| Filamentous bacteria [37] | Saprophytic bacteria [38] |
| Floc-forming bacteria [39] | Sheathed bacteria [40] |
| Gliding bacteria [41] | Spirochetes [42] |
| Gram negative aerobic cocci and rods [43] | Sulfur-oxidizing bacteria [44] |
| Gram negative facultative anaerobic rods [45] | Sulfur-reducing bacteria [46] |

## 3. Wastewater Nutrients and Their Removal

The main nutrients found in the wastewater are phosphorus and nitrogen. These nutrients pose adverse impacts to the environment when the treated wastewater that is discharged into water bodies contain high concentrations of phosphorus and nitrogen [47]. The impact of high nutrient levels in the environment results in algal growth in lakes and reservoirs [48] and eutrophication in a marine environment [49]. The algal growth can create shade over coral skeletons, making them more vulnerable to breakage during storms. Algal growth in water bodies depends on three key nutrients: phosphorus, nitrogen, and carbon. The availability of these nutrients influences algal growth, with wastewater serving as a nutrient source. In freshwater, phosphorus is the primary limiting nutrient, while in marine environments, nitrogen is the limiting factor. The limiting nutrient for algal synthesis can be determined through a chemical equation as:

$$106\,CO_2 + 16NO_3{}^{3-} + HPO_4 + 122H_20 + 18H^+ \xrightarrow[\text{Trace elements}]{\text{Sunlight}} C_30_{106}H_{26110}N_{16}P(algae) + 1380_2$$

The four key elements that are involved in algal bloom are phosphorus, nitrogen, carbon dioxide and sunlight [13, 19, 50]. The only elements that can be controlled to a certain extent are phosphorus and nitrogen.

### 3.1. Biological Removal of Nitrogen

In wastewater nitrogen exist in the various forms such as organic nitrogen compounds, inorganic nitrogen compounds, nitrite and nitrate. The organic nitrogen in the wastewater accounts for the highest proportion i.e., approximately 60% while the inorganic nitrogen accounts for approximately 40% [51]. The biological removal of nitrogen occurs through the processes of nitrification and denitrification, facilitated by nitrifying and denitrifying bacteria, respectively.

#### *3.1.1. Nitrification*

Nitrification is the first stage in the biological oxidation process, during which ionized ammonia is oxidized to nitrite, and subsequently, nitrite is further oxidized to nitrate [52, 53]. The bacteria that oxidize ammonia to nitrite are Nitrosomonas and Nitrosospira, while the bacteria that convert nitrite to nitrate are Nitrobacter and Nitrospira. These groups of the bacteria belong to the categories of the nitrifying bacteria. Nitrifying bacteria obtain their energy for cellular activity such as reproduction by oxidizing ionized ammonia and nitrite. The nitrification process results in the increase in the sludge as a result of the reproduction in the activated sludge process. The bicarbonate alkalinity in the wastewater is the source of the carbon that is required by the nitrifying bacteria. Nitrifying bacteria are classified as chemolithoautotrophs. They derive their cellular energy through the oxidation of minerals such as nitrogen and utilize inorganic carbon for cellular synthesis. For autotrophic nitrifying bacteria, carbon dioxide serves as the primary source of inorganic carbon, which is absorbed in the form of bicarbonate ($HCO_3^-$). The efficiency of nitrifying bacteria in oxidizing ionized ammonia and nitrite can be attributed to nitrifying enzyme systems which is very unique and cytomembrane. Cyromembranes are infoldings of the cellular membrane that increase the surface area, enhancing the efficiency of nitrification.

These bacteria are free-living organisms, typically found in water and soil. They enter wastewater treatment plants through infiltration and inflow. Nitrifying bacteria, being strict aerobes, primarily reside in the upper layers of soil where oxygen is abundant.

#### *3.1.2. Denitrification*

This is the second step that follows the nitrification process in the biological removal of the nitrogen [14, 47, 54]. Denitrification process involves the use of nitrite ($NO_2^-$) or nitrate ($NO_3^-$) by facultative anaerobic denitrifying bacteria for the degradation of soluble wastewater. Denitrification requires a facultative anaerobic condition, which is established by the following conditions:

- A sufficient and active population of denitrifying or facultative anaerobic bacteria
- The availability of nitrite or nitrate
- The absence of free oxygen
- The presence of soluble carbonaceous Biochemical Oxygen Demand (cBOD).

Denitrifying bacteria possess the ability to utilize free molecular oxygen, nitrite, or nitrate to degrade soluble cBOD, which serves as a source of energy and carbon for their activity and growth. These bacteria enter activated sludge processes via fecal waste or through inflow and infiltration, as they are naturally present in water and soil environments. The denitrifying bacteria belong mostly in the following genera, *Pseudomonas, Bacillus and Alcaligens*. Therefore, in the activated sludge process it is important to have both the anaerobic and aerobic zones/ sections in the process of treatment for efficient removal of nitrogen in the wastewater. However; the retention time of the sludge for nitrification and denitrification is different. For example the time required for the organic removal is around 2-5 days, for nitrification is 10-20 days and for denitrification is 1-5 days.

### 3.2. Biological Removal of Phosphorus

The discharge of treated wastewater that contains high concentrations of phosphorus and nitrogen in to the water bodies stimulates the growth of algal species. The minimum amount of the phosphorus into the environment that leads to eutrophication in the lakes and other water bodies is 0.02 mg per litre. The biological removal of phosphorus involves two groups of bacteria: fermentative bacteria and Poly-P bacteria. Fermentative bacteria are facultative anaerobes or strict anaerobes, while Poly-P bacteria are strict aerobes. The treatment process utilizes both the anaerobic and aerobic tanks due to the specific bacteria involved and the breakdown of phosphorus-containing compounds, which helps reduce phosphorus in the waste material. The poly-P- bacteria and the fermentative bacteria do enter the treatment plants for the removal of the phosphorus through the inflow and infiltration from water and soil. This is because these bacteria are naturally present in the environment, while another source of their introduction into the treatment plant is the fecal route, as some bacteria reside in the alimentary canal.

## 4. Biofilm Mediated Wastewater Treatment

The industrial, municipal and domestic wastewater is composed of various contaminants which can be reclaimed through physical, chemical or the biological methods [17, 55]. The biological wastewater treatment is considered as the current technique of wastewater treatment which is seen as an alternative for the old conventional methods hence playing an important role in the improvement of water quality [56]. The organic pollutants in wastewater contribute to various issues, including a decrease in dissolved oxygen, toxin accumulation, and eutrophication [57, 58]. Therefore, it is necessary to remove these organic pollutants that find its way in wastewater. Bioremediation, an environmentally friendly technique for removing harmful contaminants, has gained popularity due to its safety and sustainability. It is preferred over conventional treatment plants, which face limitations such as high operational costs, lower efficiency, and complex startup requirements.

Bioremediations has been utilized to mineralize, eliminate, neutralize and degrade various organic and inorganic pollutants in the wastewater by using microbes such as fungi, bacteria and algae [8, 59, 60]. The biofilm- mediated wastewater remediation technique is considered to be an economical and environmentally friendly alternative [61-63]. Biofilms are communities of microbial cells, either single or mixed species, that adhere to a living or inert surface in an aquatic environment. Unlike planktonic cells, which are free-floating, bacteria within biofilms possess enhanced capabilities to compete for nutrients, exhibit greater resistance to pollutants, and create a protective environment for the cells. The survival rate of these bacteria is high in the extreme environmental conditions. Biofilm communities play a crucial role in metabolizing and absorbing heavy metals and organic pollutants due to the regulated expression of genes mediated by quorum sensing. They have been utilized in wastewater treatment for years because of their remarkable efficiency in pollutant removal. The biodeterioration of materials like plastics, concrete, and metals is expedited by the metabolic activities of microbes living within biofilms.

### 4.1. Biofilms

A biofilms is made up of well- structured microorganism groups that is immobilized in a matrix synthesized by microorganisms known as extracellular polymeric substances (EPS) [64]. The primary role of EPS is to protect microorganisms from predatory protozoa, contaminants, and various abiotic and biotic environmental challenges. EPS is composed of polysaccharides, lipids, proteins, humic compounds, and nucleic acids. Water, which makes up approximately 97% of biofilms, is essential for nutrient flow within the biofilm matrix. Biofilms typically have a thickness of 10–30 mm, with the EPS matrix layer accounting for about 0.2–1.0 mm. EPS constitutes approximately 65–95% of the biofilm, while bacteria make up the remaining portion.

### 4.2. Biofilm Formation and Development

The formation and development of biofilm involves a process that occurs in three phases i.e., adsorption, consolidation and colonization [65, 66]. The adsorption occurs when the macromolecules such as polysaccharides, proteins, humic acid and nucleic acids and the micro molecules such as lipids, fatty acids are adsorbed on the surface forming thin films. These materials when adsorbed on the surfaces have various effects such as:

- Alteration in the physiochemical features of surface.
- Enhancing and suppressing the toxic metal ions release from surface
- Supply nutrients and elements needed for biofilm
- Biofilm sloughing triggering
- Detoxification of the abundant solution via adsorption of hindering substances

The consolidation phase is the second stage of development, during which bacteria adhere to a surface and produce a slimy, adhesive substance known as EPS. EPS plays a crucial role in the maturation of biofilms by forming a protective, gel-like biofilm matrix. Colonization is the third phase in the development of the biofilm where the mature biofilms established on the surfaces propagates actively and eventually extend over the surface. The propagations mechanism of mature biofilms is more advanced and the microorganisms in the mature biofilms are attached within the EPS layers. The factors that affect the formation of biofilm are categorized into:

1. Biological factors.
   - Extracellular polymeric substances (EPS)
   - Extracellular DNA (eDNA)
   - Quorum sensing and gene regulation
2. Physical factors
   - Surface topology
   - Turbulence, hydrodynamics and velocity
3. Environmental factors.
4. Effect of PH
   - Effect of nutrients
   - Effect of temperature
   - Effect of oxygen availability
   - Divalent cations

## 4.3. Wastewater Treatment Using Biofilms

Different variety of pollutants such as organic, inorganic and micro pollutants can be treated in wastewater using biofilms.

### *4.3.1. Treatment of Organic Pollutants*

Bacterial biofilm community exhibit enhanced resistance to organic pollutants due to their high microbial biomass and immobile nature. The presence of extracellular enzymes in biofilms facilitates the sorption and metabolism of organic contaminants. The degradation of organic waste by biofilms occurs through the following mechanisms:

- The organic compounds act as substrates; the biofilms decompose it effectively.
- The enzymes involved in the breakdown process can be retained within the biofilm's EPS at the producing cell, enhancing the likelihood that both the enzyme and its products remain accessible for use by the original cells.
- The higher biomass of biofilms, compared to suspended cells, provides protection against chemical damage.
- The physiochemical structure of biofilm matrix maximizes sorption while permitting the coexistence of aerobic and anaerobic bacteria because the biofilm is more effective at breaking down the toxicants.

### *4.3.2. Treatment of Inorganic Pollutants*

Generally, the inorganic pollutants include heavy metals and commercial colors that contaminate the water bodies and threaten the health of human beings. Therefore, the bacterial biofilms are responsible for elimination of pollutants of inorganic wastes in the wastewater. The polysaccharides that make up the EPS have the potential of heavy metals chelation from the contaminated locations [66, 67]. A significant number of bacterial biofilms are utilized for the removal of heavy metal ions and xenobiotic chemicals, leveraging their metabolic capabilities to degrade or transform these pollutants. Indigenous microorganisms can mitigate the effects of toxic metals in wastewater through various adaptive mechanisms, including biotransformation, metal efflux from cells, and heavy metal chelation.

#### *4.3.3. Removal of Micro-Pollutants*

The micropollutants from pharmaceuticals, cosmetic and electronic industries are discharged into the environment, creating severe water pollution. The microbes found in the biofilms are good for bioremediation of micropollutants since they have the potential to degrade and absorb a broad range of grievous contaminants through various catabolic ways. The microbial cells that make up the biofilm are more adapted to the stress than other planktonic cells in many bioprocesses.

The house hold items such as paints and cosmetics substances contain titanium dioxide nanoparticles which result in the discharge of the nanoparticles in the wastewater from the residential areas. This substance is known to be hazard to the environment since the titanium is activated photo- catalytically. Many studies have been done on the utilization of biofilms in the removal of the toxic nanoparticles because the remediation of the nanoparticles is possible due to availability of surfactants and humic compounds in the wastewater.

The EPS matrix regulates contaminant adhesion and adsorption by influencing the physical trapping and diffusion of chemicals. In addition to diffusion transport, biofilm bacteria are affected by osmotic pressure, temperature, and nutrient availability [68]. Organic materials and dissolved substances in wastewater provide a platform for the adsorption of micropollutants.

### 4.4. Types of Bioreactors Based on Biofilms

Biofilm-based reactors have emerged as a promising technology in wastewater treatment, offering significant potential due to their various advantages. These include high efficiency, reduced sludge production, shorter hydraulic retention times, the presence of extracellular polymeric substances, high concentrations of active biomass, and microbial diversity. Different types of biofilm-based bioreactors are discussed below:

#### *4.4.1. Membrane Biofilm Reactors (MBR)*

MBR is been used over a long time and is a promising technology in pollutant recovery and removal in wastewater. The MBR system combines biological degradation using a biofilm process with physical separation via a membrane unit, replacing traditional secondary settlers. This technology offers several advantages over other wastewater reclamation methods for industrial and municipal use. It provides more efficient removal of organic and inorganic compounds, reduces sludge production, and offers greater flexibility in handling influent variations while maintaining high operational efficiency.

MBR systems have two main configuration types: the immersed (or submerged) configuration and the side-stream (or external) configuration. In MBR, the biofilm process can be implemented by incorporating media in fixed beds, moving beds, or aerated membranes within bioreactors, providing structural support to enhance biofilm growth. Currently, materials used to support biofilm growth include RBC media, cord media, plastic media, and sponge, all of which are commercially applied in these systems. This technology has demonstrated exceptional efficiency in wastewater reclamation, effectively removing various contaminants.

For instance, it can break down up to 90% of organic waste and is also highly effective in eliminating both organic and inorganic nitrogen.

However, its widespread use is restricted due to membrane fouling and blockage that results in hike in plant expenses. In membrane applications, such as MBR systems, membrane fouling is a frequent occurrence. It remains unclear which specific fraction is primarily responsible for this issue. Potential factors include colloidal matter, certain organic compounds like biopolymers or EPS, suspended particles, and various physical characteristics. Numerous studies have addressed these drawbacks by identifying less expensive membrane materials and controlling the membrane fouling mechanisms [69, 70]. In order to make this system a more practical and dependable substitute for activated sludge processes and other traditional technologies, it also sought to optimize energy consumption.

#### *4.4.2. Trickling Filter (TF)*

The Trickling Filter (TF) technology has been widely used for wastewater treatment over an extended period. It is a three-phase biofilm reactor that incorporates influent recirculation, a pump station, and a clarification unit. The effectiveness of TFs in achieving treatment goals, particularly for nitrification and carbon oxidation, has been well-documented. In this system, airborne oxygen is utilized by the biofilm on the filter media to oxidize carbon. When solid separation is included in the treatment process, TFs can be used for both carbon oxidation and a combined process of carbon oxidation and nitrification. Combining oxidation and nitrification in TFs has shown positive results, especially when ammonia concentrations are below 3 mg $NH_4^+$ N/L and BOD concentrations are under 10 mg/L.

The choice of filter media is a critical aspect of TF design. Traditionally, stones and gravel were commonly used as filter media, but these materials restricted airflow through the filter, limiting the oxygen availability necessary for biofilm growth. This constraint reduced the reactor's capacity to process a high volume of wastewater and decreased the surface area for microbial attachment, which ultimately restricted its ability to handle biochemical oxygen demand (BOD) loading. In systems with high organic loads, excessive microbial growth could cause blockages in the void spaces, further limiting the filter's effectiveness. Conversely, stone-bed trickling filters performed better under low organic loading conditions.

To overcome these limitations, alternative materials such as plastic rings, zeolite, and sponges have been introduced. These materials enhance oxygen transfer and allow better control over biofilm thickness compared to traditional rock media. Trickling filters using plastic modules with a specific surface area between 89 and 102 mm are suitable for both nitrification and carbon oxidation. However, a well-known drawback of this technology is that TFs are not highly volume-efficient systems, which can limit their scalability for large treatment volumes.

#### *4.4.3. Moving- Bed Biofilm Reactors (MBBR)*

MBBR is a wastewater biofilm-based technology that is currently utilized in many countries in the world. The technology was developed in 1980s and early 1990s in Norway for the treatment of both municipal and varied kinds of industrial wastewater. This technology operates on the principle of combining the activated sludge and biofilter processes. It is used in anoxic, aerobic, and anaerobic treatment processes. Biofilm growth occurs on carriers that move freely within the tank. In aerobic conditions, movement is driven by air agitation, while in anaerobic and anoxic processes; carrier motion is facilitated by mechanical means. Biofilms mainly grow in the plastic carriers that are protected from outside abrasions. Moving beds are preferred over fixed beds because they reduce clogging and make full use of the bioreactor's volume. The shearing forces generated in the reactor create turbulence, ensuring efficient substance diffusion within the biofilm and maintaining a minimal biofilm thickness. This approach significantly reduces the requirement for biomass recirculation, which is a major challenge in both fixed-bed biofilm and activated sludge systems, as only the excess biomass is removed from the solution. The carrier materials used in MBBR enhance the surface area for microbial growth and attachment. Moreover, the reactor's optimized mixing conditions facilitate the release of biogas and the uniform distribution of volatile acids within the aqueous solution. MBBR has demonstrated reliability and compactness as an effective reactor for wastewater treatment, efficiently supporting carbon oxidation as well as nitrification and denitrification, whether used independently or as part of combination systems. Both laboratory and large-scale operational results were satisfactory. However, a known drawback of these systems is that, to optimize reactor performance, the carrier must be removed.

#### *4.4.4. Fluidized- Bed Biofilm Reactors (FBBR)*

In FBBRs, the media particles are typically arranged along a size gradient that increases from the top to the bottom of the bioreactor. Compared to fixed biofilm reactors, FBBRs are known for their superior mass transfer capabilities. They have demonstrated high efficiency in treating municipal wastewater, particularly for tertiary denitrification, and are widely used to eliminate oxidized pollutants. The fluidization of media particles offers several advantages, including an increased surface area for interaction between microbial cells and wastewater, which significantly improves mass transfer and enhances treatment efficiency. However, to maintain optimal performance, a low sludge bed expansion is recommended, as it reduces flow velocity, minimizes energy consumption, and increases the concentration of effective biomass.

One of the critical aspects of FBBRs is their high microbial diversity, which results in a significant number of attached microbial cells. This characteristic not only improves biodegradation rates but also enables the system to recover quickly from operational disturbances. However, the high biomass concentration within the reactor also raises the volumetric oxygen demand. The fundamental principle of the FBBR process is to pump wastewater upward through a densely packed particle bed at a speed sufficient to fluidize or move the particles. As the wastewater ascends through the biological bed, the dense population of microorganisms on the particle surfaces enhances the biodegradation of waste pollutants, leading to effective and efficient wastewater treatment.

### 4.5. Challenges in Biofilm- Mediated Removal of Pollutants

The efficiency of biofilm-mediated treatment is influenced by the nature of pollutants and the thickness of the biofilm. Positively charged micro-contaminants, with a charge percentage exceeding 90%, have demonstrated significant sorption ability in activated sludge compared to negatively charged contaminants at a $p^H$ of 7.5. The $p^H$ level plays a crucial role in determining the charge and sorption capacity of micro-contaminants. For instance, the protonation of the amino group in macrolides results in strong ionic interactions between positively charged macrolides and negatively charged cell surfaces.

The biofilm's capacity to degrade phenol is directly related to its thickness, indicating that managing biofilm thickness can be an effective bioremediation strategy. The evaluation of biofilms mainly focuses on determining their thickness and cell diversity, as these factors directly impact treatment efficiency. Additionally, the porosity of the biofilm matrix significantly affects both its structure and functional performance. Studies have shown that thinner biofilms generally have fewer pores and higher cell density compared to thicker biofilms [71], highlighting the importance of controlling biofilm architecture for optimal pollutant removal.

## 5. Sustainability Challenges and Future Trends

### 5.1. Management Strategy for Increasing Demand of Water

The population in urban areas has been increasing due to high population growth and migration in recent years, leading to a greater demand for water [67, 68]. To address this, government agencies are implementing management strategies to mitigate the impact of rising water demand. These strategies focus on reducing water loss and managing demand more effectively through both structural and non-structural measures [69]. The non- structural measures employed includes:

- Restrictions on water use
- Education and awareness campaigns
- The creative and innovative presentation of the water bills
- New policies concerning water tariff structures

The structural measures utilized involves

- Reuses and recycling
- Active or reactive leak detection and repair
- Retrofitting of water appliances

### 5.2. Adjustment in Storm Water Management

In urban environments, contaminated water primarily runs off from streets, lawns, and other surfaces, eventually making its way into natural water bodies. The extensive presence of impervious surfaces such as pavements and rooftops prevents rainwater from naturally infiltrating the ground. Consequently, rainwater rapidly flows into sewer systems, storm drains, and drainage ditches, causing various issues including downstream flooding, increased turbidity, streambank erosion, habitat loss, combined sewer overflows, damage to infrastructure, and contamination of rivers, streams, and coastal waters.

Runoff from urban catchment areas often contains moderate to high levels of pollutants, including nutrients like nitrogen and phosphorus, heavy metals, suspended solids, fecal coliforms, salinity, and total dissolved solids [70]. To mitigate these impacts, various stormwater management strategies have been developed. One effective approach is stormwater harvesting, which involves capturing and storing excess rainwater to reduce flooding risks.

Additionally, biofiltration or bioretention systems are designed to store and treat runoff while gradually releasing it at a controlled rate, thereby minimizing its impact on downstream ecosystems [71]. Wetland treatment systems are also widely used, providing dual benefits of runoff regulation and wildlife habitat preservation in urban settings.

## 5.3. Minimization of the Nutrient Pollution

An increase in the nutrients pollution in the underground and surface water is a result of the intensification in the agricultural processes that are being practiced [71, 72]. The nutrient pollution can be reduced by a number of methods that include:

a. Animal density reduction
b. Fertilizer application in the ground or in irrigation water
c. Stream water runoff management
d. Erosion barriers creation
e. Implementations of the buffer strips along the streams

## 5.4. Adaptation of Water Reclamation and Reuse Options

Water reclamation involves treating wastewater to a level that meets specified quality standards, allowing it to be reused for various beneficial applications. This process ensures that treated water becomes suitable for reuse, thereby conserving freshwater resources and reducing environmental impacts [73]. Common applications of reclaimed water include groundwater recharge using treated industrial or municipal wastewater, irrigation for agriculture and golf courses, and car washing. These practices not only help in reducing the demand for potable water but also serve as indirect methods of potable water reuse, thereby enhancing water sustainability.

## 5.5. Utilization of Wastewater to Generate Electricity

Wastewater treatment is an energy-intensive process, especially during aeration. One effective way to reduce these energy costs is by harnessing renewable energy from the organic matter present in wastewater. This approach not only ensures efficient treatment but also contributes to energy sustainability. Anaerobic treatment is a prominent method for generating energy from wastewater, as it facilitates the production of biogas a mixture containing 60-70% methane. This biogas can be utilized to power engines for electricity generation or to heat anaerobic digesters, thereby creating a self-sustaining system that lowers operational expenses while producing renewable energy.

## Conclusion

The industrial development and other human activities have resulted in to hazardous effluents and residues in different environmental matrices. These contaminants are generally known as sewage; pose serious threats into the environment leading to major water and soil pollution and a greatest threat to human health. Biological sewage treatment techniques offer several advantages over conventional physical and chemical methods, including cost-effectiveness, ease of operation, the use of diverse microorganisms, reduced sludge production, and environmental sustainability. Microorganisms are ideal for sewage treatment due to their small size, metabolic versatility, adaptability, and ability to break down a wide range of natural substances. As environmental concerns continue to grow, microbial technology will receive increasing attention, with ongoing advancements and improvements to better address sewage treatment challenges. The focus remains on determining the most effective approach to tackling wastewater treatment issues.

Innovative ideas and technologies have the potential to revolutionize our relationship with water while ensuring its availability and quality for future generations. By investing in these advancements, we can promote environmental conservation and water sustainability. To achieve this, we must embrace key principles that maximize the opportunities presented by emerging science and technology. Our responsibility to protect this vital resource is as old as water itself, and we must uphold it with unwavering dedication.

## References

[1] Aghalari, Zahra, Dahms, Hans-Uwe, Sillanpää, Mika, Sosa-Hernandez, Juan Eduardo, & Parra-Saldívar, Roberto. (2020). Effectiveness of wastewater treatment systems in removing microbial agents: a systematic review. *Globalization and health,* 16, 1-11. https://link.springer.com/article/10.1186/s12992-020-0546-y.

[2] Ahmed, SF, Mofijur, M, Nuzhat, Samiha, Chowdhury, Anika Tasnim, Rafa, Nazifa, Uddin, Md Alhaz, . . . Chia, Wen Yi. (2021). Recent developments in physical, biological, chemical, and hybrid treatment techniques for removing emerging contaminants from wastewater. *Journal of hazardous materials*, 416, 125912. https://www.sciencedirect.com/science/article/pii/S0304389421008761.

[3] Ajiboye, Timothy O, Oyewo, Opeyemi A, & Onwudiwe, Damian C. (2021). Simultaneous removal of organics and heavy metals from industrial wastewater: A review. *Chemosphere,* 262, 128379. https://www.sciencedirect.com/science/article/pii/S0045653520325741.

[4] Akhtar, Naseem, Syakir Ishak, Muhammad Izzuddin, Bhawani, Showkat Ahmad, & Umar, Khalid. (2021). Various natural and anthropogenic factors responsible for water quality degradation: A review. *Water*, 13(19), 2660. https://www.mdpi.com/2073-4441/13/19/2660.

[5] Al-Ani, Sama, Guo, Hui, Fyfe, Sheila, Long, Zebo, Donnaz, Sylvain, & Kim, Younggy. (2024). Deep learning-based image analysis for filamentous and floc-forming bacteria in wastewater treatment. *Journal of Water Process Engineering,* 65, 105772. https://www.sciencedirect.com/science/article/pii/S2214714424010043.

[6] Amahmid, O, Asmama, S, & Bouhoum, K. (2019). Occurrence and fate of pathogenic parasites in an overland flow and percolation wastewater treatment system under arid climate. *International journal of environmental science and technology,* 16(7), 3291-3296. https://link.springer.com/article/10.1007/s13762-018-1964-8.

[7] Asghar, Zeeshan, Shah, Rehman Ali, Khan, Muhammad Waris Saeed, & Gondal, Muhammad Asif. (2024). Exploring the Bio-Mechanisms of Rod-Shaped Bacteria Moving on Rigid Substrate Coated with

Non-Newtonian Williamson Slime. *Chinese Journal of Physics*. https://www.sciencedirect.com/science/article/pii/S0577907324004519.

[8] Ayilara, Modupe S, & Babalola, Olubukola O. (2023). Bioremediation of environmental wastes: the role of microorganisms. *Frontiers in Agronomy,* 5, 1183691. https://www.frontiersin.org/articles/10.3389/fagro.2023.1183691/full.

[9] Borah, Preetismita, Kumar, Manish, & Devi, Pooja. (2020). *Types of inorganic pollutants: metals/metalloids, acids, and organic forms Inorganic pollutants in water* (pp. 17-31): Elsevier. https://www.sciencedirect.com/science/article/pii/B9780128189658000020.

[10] Cáceres, Rafaela, Malińska, Krystyna, & Marfà, Oriol. (2018). Nitrification within composting: A review. *Waste Management,* 72, 119-137. https://www.sciencedirect.com/science/article/pii/S0956053X1730795X.

[11] Carlton, Jeremy G, Jones, Hannah, & Eggert, Ulrike S. (2020). Membrane and organelle dynamics during cell division. *Nature Reviews Molecular Cell Biology*, 21(3), 151-166. https://www.nature.com/articles/s41580-019-0208-1.

[12] Chavan, Shraddha, Yadav, Bhoomika, Atmakuri, Anusha, Tyagi, Rajeshwar Dayal, Wong, Jonathan WC, & Drogui, Patrick. (2022). Bioconversion of organic wastes into value-added products: A review. *Bioresource Technology*, 344, 126398. https://www.sciencedirect.com/science/article/pii/S0960852421017405.

[13] Chazal, Natalie, Carr, Megan, Leight, Andrew K, Saia, Sheila M, & Nelson, Natalie G. (2024). Short-term forecasting of fecal coliforms in shellfish growing waters. *Marine Pollution Bulletin*, 200, 116053. https://www.sciencedirect.com/science/article/pii/S0025326X24000304.

[14] Cheng, Wei-Min, Liu, Jin-Di, Feng, Yue, Hu, Xiang-Ming, Zhao, Yan-Yun, & Liu, Yu. (2024). Study on the cooperation mechanism of urea-hydrolysis bacteria and biosurfactant bacteria for dust suppression. *Chemical Engineering Journal,* 480, 148008. https://www.sciencedirect.com/science/article/pii/S1385894723067402.

[15] Cole Jr, John R. (1990). Spirochetes. *Diagnostic Procedure in Veterinary Bacteriology and Mycology*, 41-60. https://www.sciencedirect.com/science/article/pii/B9780121617752500098.

[16] del Carmen Muñoz-Marín, María, López-Lozano, Antonio, Moreno-Cabezuelo, José Ángel, Díez, Jesús, & García-Fernández, José Manuel. (2024). Mixotrophy in cyanobacteria. *Current Opinion in Microbiology*, 78, 102432. https://www.sciencedirect.com/science/article/pii/S1369527424000080.

[17] Di Martino, Patrick. (2018). Extracellular polymeric substances, a key element in understanding biofilm phenotype. AIMS microbiology, 4(2), 274. https://www.ncbi.nlm.nih.gov/pmc/articles/PMC6604936/.

[18] Dighton, John. (2018). *Fungi in ecosystem processes*: CRC press. https://www.taylorfrancis.com/books/mono/10.1201/9781315371528/fungi-ecosystem-processes-john-dighton.

[19] Dos Santos, Stéphanie, Adams, EA, Neville, G, Wada, Y, De Sherbinin, A, Bernhardt, E Mullin, & Adamo, SB. (2017). Urban growth and water access in sub-Saharan Africa: Progress, challenges, and emerging research directions. *Science of The Total Environment*, 607, 497-508. https://www.sciencedirect.com/science/article/pii/S0048969717315759.

[20] Grandclément, Camille, Seyssiecq, Isabelle, Piram, Anne, Wong-Wah-Chung, Pascal, Vanot, Guillaume, Tiliacos, Nicolas, . . . Doumenq, Pierre. (2017). From the conventional biological wastewater treatment to hybrid processes, the evaluation of organic micropollutant removal: a review. *Water research*, 111, 297-317. https://www.sciencedirect.com/science/article/pii/S0043135417300052.

[21] Guo, Honghong, Li, Na, Xue, Shuhong, Zhangsun, Xuanzi, Huang, Tinglin, Zhang, Haihan, & Ma, Ben. (2024). Enhancing nitrogen removal in source water with intermittent aeration: Improved performance of iron-reducing denitrifying bacteria. *Chemical Engineering Journal*, 498, 154960. https://www.sciencedirect.com/science/article/pii/S1385894724064519.

[22] El-nshar, Esraa M, Kenawy, El-Refaie, Azaam, Mohamed M, & El-Khalafy, Sahar H. (2024). Investigation of the Biological Activity of Modified Branched Poly (P-Hydroxystyrene). *Sinai International Scientific Journal,* 1(2), 18-34. https://sisj.journals.ekb.eg/article_389245_e516ef099dda762a802cfb21dfa14ec9.pdf.

[23] Fisher-Jeffes, Lloyd, Carden, Kirsty, Armitage, Neil P, & Winter, Kevin. (2017). Stormwater harvesting: Improving water security in South Africa's urban areas. *South African Journal of Science*,

113(1-2), 1-4. https://www.scielo.org.za/scielo.php?pid=S0038235320170001000l5&script=sci_arttext.

[24] Goenka, Sheetal, Wanswett, Wanshisha, Jain, Manisha, Loomba, Poonam, Sharma, Abha, & Tyagi, Shivani. (2024). Prevalence and clinical significance of saprophytic bacteria in bloodstream infections among cancer patients: Colonizers in BSI from cancer patients. *Journal of Current Oncology and Medical Sciences*, 4(3), 875-882. http://submission.journalofcoms.com/index.php/JCOMS/article/view/238.

[25] Gupta, Munishwar Nath, Khare, Sunil Kumar, & Sinha, Rajeshwari. (2021). *Introduction to the Microbial World Interfaces Between Nanomaterials and Microbes* (pp. 1-16): CRC Press .https://www.taylorfrancis.com/chapters/edit/10.1201/9780429321269l/introduction-microbial-world-munishwar-nath-gupta-sunil-kumar-khare-rajeshwari-sinha.

[26] Hao, Zhen-Le, Ali, Amjad, Ren, Yi, Su, Jun-Feng, & Wang, Zhao. (2022). A mechanistic review on aerobic denitrification for nitrogen removal in water treatment. *Science of The Total Environment*, 847, 157452. https://www.sciencedirect.com/science/article/pii/S0048969722045508.

[27] Hayat, Kashif, Menhas, Saiqa, Bundschuh, Jochen, & Chaudhary, Hassan Javed. (2017). Microbial biotechnology as an emerging industrial wastewater treatment process for arsenic mitigation: a critical review. *Journal of Cleaner Production*, 151, 427-438. https://www.sciencedirect.com/science/article/pii/S0959652617305188.

[28] He, Chunyang, Liu, Zhifeng, Wu, Jianguo, Pan, Xinhao, Fang, Zihang, Li, Jingwei, & Bryan, Brett A. (2021). Future global urban water scarcity and potential solutions. *Nature communications*, 12(1), 4667. https://www.nature.com/articles/s41467-021-25026-3.

[29] Karmakar, Subhankar, & Musthafa, OM. (2020). *Lakes and reservoirs: Pollution Managing Water Resources and Hydrological Systems* (pp. 65-80): CRC Press. https://www.taylorfrancis.com/chapters/edit/10.1201/9781003045045-7/lakes-reservoirs-pollution-subhankar-karmakar-musthafa.

[30] Lan, Jiaxin, Liu, Pengfei, Hu, Xi, & Zhu, Shanshan. (2024). Harmful Algal Blooms in Eutrophic Marine Environments: Causes, Monitoring, and Treatment. *Water*, 16(17), 2525. https://www.mdpi.com/2073-4441/16/17/2525.

[31] Maclsaac, Erland A, & Stockner, John G. (2018). Enumeration of phototrophic picoplankton by autofluorescence microscopy *Handbook of methods in aquatic microbial ecology* (pp. 187-197): CRC Press. https://www.taylorfrancis.com/chapters/edit/10.1201/9780203752746-24/enumeration-phototrophic-picoplankton-autofluorescence-microscopy-erland-maclsaac-john-stockner.

[32] Malik, Anju, & Garg, Vinod Kumar. (2024). *Bioremediation for Sustainable Environmental Cleanup*: CRC Press. https://api.taylorfrancis.com/content/books/mono/download?identifierName=doi&identifierValue=10.1201/9781003277941&type=googlepdf.

[33] Martikainen, Pertti J. (2022). Heterotrophic nitrification–An eternal mystery in the nitrogen cycle. *Soil Biology and Biochemistry,* 168, 108611. https://www.sciencedirect.com/science/article/pii/S0038071722000682.

[34] Mazza, Marco G. (2016). The physics of biofilms—an introduction. Journal of Physics D: Applied Physics, 49(20), 203001. https://iopscience.iop.org/article/10.1088/0022-3727/49/20/203001/meta.

[35] Meshjel, Maysoon Hassan. (2024). Diversity of free-living protozoa. *Wasit Journal for Pure Sciences,* 3(2). https://www.iasj.net/iasj/download/2e8fb5e0025b15bd.

[36] Middelkoop, Stephanie JM, de Joode, Anoek AE, van Pelt, L Joost, Kampinga, Greetje A, Ter Maaten, Jan C, & Stegeman, Coen A. (2024). Clinical usefulness of urine Gram stain for diagnosing urinary tract infections at the emergency department. *Infectious Diseases,* 1-9. https://www.tandfonline.com/doi/abs/10.1080/23744235.2024.2389478.

[37] Mishra, Sandhya, Chowdhary, Pankaj, & Bharagava, Ram Naresh. (2019). Conventional methods for the removal of industrial pollutants, their merits and demerits. *Emerging and eco-friendly approaches for waste management,* 1-31. https://link.springer.com/chapter/10.1007/978-981-10-8669-4_1.

[38] Mubaraq, A, Sembiring, M, Widiastuti, E, Fachrial, E, Utomo, B, Turjaman, M, . . . Siregar, ES. (2024). Diversity and distribution of nitrifying bacteria play an important role in the nitrogen cycle in mangrove sediments. *Global Journal of Environmental Science & Management* (GJESM), 10(4). https://www.researchgate.net/profile/Mohammad-Basyuni/publication/381942891_Diversity_and_distribution_of_nitrifying_bacteria_play_an_important_role_in_the_nitrogen_cycle_in_mangrove_sed

iments/links/66b0156b299c327096af0319/Diversity-and-distribution-of-nitrifying-bacteria-play-an-important-role-in-the-nitrogen-cycle-in-mangrove-sediments.pdf.

[39] Mukherjee, Ayan, & Ghosh, Prabal Ranjan. (2023). *Cellular and Molecular Physiology Textbook of Veterinary Physiology* (pp. 21-35): Springer. https://link.springer.com/chapter/10.1007/978-981-19-9410-4_2.

[40] Mukherjee, Diptangshu, Selvi, Vetrivel Angu, Ganguly, Jhuma, & Masto, Reginald Ebhin. (2024). New insights into the coal-associated methane architect: the ancient archaebacteria. *Archives of Microbiology*, 206(5), 234. https://link.springer.com/article/10.1007/s00203-024-03961-1.

[41] Mulder, Eppe Gerke, & Deinema, Maria H. (1981). The sheathed bacteria The Prokaryotes: *a Handbook on Habitats, Isolation, and Identification of Bacteria* (pp. 425-440): Springer. https://link.springer.com/chapter/10.1007/978-3-662-13187-9_27.

[42] Müller, Volker, & Frerichs, Janin. (2013). *Acetogenic bacteria.* eLS. https://onlinelibrary.wiley.com/doi/abs/10.1002/9780470015902.a0020086.pub2.

[43] Ngatia, Lucy, Grace III, Johnny M, Moriasi, Daniel, & Taylor, Robert. (2019). Nitrogen and phosphorus eutrophication in marine ecosystems. *Monitoring of marine pollution*, 1,1-17. https://books.google.com/books?hl=en&lr=&id=mTj8DwAAQBAJ&oi=fnd&pg=PA77&dq=Ngatia,+Lucy,+Grace+III,+Johnny+M,+Moriasi,+Daniel,+%26+Taylor,+Robert.+(2019).+Nitrogen+and+phosphorus+eutrophication+in+marine+ecosystems.+Monitoring+of+marine+pollution,+1,+1-17.&ots=On1ItJtZt&sig=YPma0UTV60zZEDD_pB3aTP6_Sh8.

[44] Nguyen, Thu Thuy, Ngo, Huu Hao, Guo, Wenshan, Wang, Xiaochang C, Ren, Nanqi, Li, Guibai, Liang, Heng. (2019). Implementation of a specific urban water management-Sponge City. *Science of The Total Environment,* 652, 147-162. https://www.sciencedirect.com/science/article/pii/S0048969718340518.

[45] Nielsen, Chad, Rahman, Asif, Rehman, Asad Ur, Walsh, Marie K, & Miller, Charles D. (2017). Food waste conversion to microbial polyhydroxyalkanoates. *Microbial biotechnology*, 10(6), 1338-1352. https://ami-journals.onlinelibrary.wiley.com/doi/abs/10.1111/1751-7915.12776.

[46] Ojha, Nishita, Karn, Rama, Abbas, Sadiqa, & Bhugra, Sonal. (2021). *Bioremediation of industrial wastewater: a review*. Paper presented at the IOP Conference Series: Earth and Environmental Science. https://iopscience.iop.org/article/10.1088/1755-1315/796/1/012012/meta.

[47] Okorogbona, Alfred OM, Denner, Freddie DN, Managa, Lavhelesani R, Khosa, Tsunduka B, Maduwa, Khathutshelo, Adebola, Patrick O, . . . Macevele, Stanford. (2018). Water quality impacts on agricultural productivity and environment. *Sustainable Agriculture Reviews* 27, 1-35. https://link.springer.com/chapter/10.1007/978-3-319-75190-0_1.

[48] Oren, Aharon. (2019). Extremophiles and their natural niches on Earth. *Handbook of astrobiology*,635À660. https://api.taylorfrancis.com/content/chapters/edit/download?identifierName=doi&identifierValue=10.1201/b22230-42&type=chapterpdf.

[49] Osmane, Ahmed, Zidan, Khadija, Benaddi, Rabia, Sbahi, Sofyan, Ouazzani, Naaila, Belmouden, Moustapha, & Mandi, Laila. (2024). Assessment of the effectiveness of a full-scale trickling filter for the treatment of municipal sewage in an arid environment: Multiple linear regression model prediction of fecal coliform removal. *Journal of Water Process Engineering.* 64,105684. https://www.sciencedirect.com/science/article/pii/S2214714424009164.

[50] Pande, Veni, Pandey, Satish Chandra, Sati, Diksha, Pande, Veena, & Samant, Mukesh. (2020). Bioremediation: an emerging effective approach towards environment restoration. *Environmental Sustainability*, 3, 91-103. https://link.springer.com/article/10.1007/S42398-020-00099-W.

[51] Pokorna, Dana, & Zabranska, Jana. (2015). Sulfur-oxidizing bacteria in environmental technology. *Biotechnology Advances,* 33(6), 1246-1259. https://www.sciencedirect.com/science/article/pii/S0734975015000324.

[52] Pugazhendhi, Abinaya Sindu, Wei, Fei, Hughes, Megan, & Coathup, Melanie. (2022). Bacterial adhesion, virulence, and biofilm formation *Musculoskeletal Infection* (pp. 19-64): Springer. https://link.springer.com/chapter/10.1007/978-3-030-83251-3_2.

[53] Pujic, Petar, Valour, Florent, Lebeaux, David, & Rodriguez-Nava, Veronica. (2024). *Nocardia and actinomyces Molecular medical microbiology* (pp. 775-809): Elsevier. https://www.sciencedirect.com/science/article/pii/B9780128186190001325.

[54] Rotiroti, Marco, Sacchi, Elisa, Caschetto, Mariachiara, Zanotti, Chiara, Fumagalli, Letizia, Biasibetti, Michela, . . . Leoni, Barbara. (2023). Groundwater and surface water nitrate pollution in an intensively irrigated system: Sources, dynamics and adaptation to climate change. *Journal of Hydrology*, 623, 129868. https://www.sciencedirect.com/science/article/pii/S0022169423008107.

[55] Rout, Prangya Ranjan, Shahid, Muhammad Kashif, Dash, Rajesh Roshan, Bhunia, Puspendu, Liu, Dezhao, Varjani, Sunita, . . . Surampalli, Rao Y. (2021). Nutrient removal from domestic wastewater: A comprehensive review on conventional and advanced technologies. *Journal of Environmental Management*, 296, 113246. https://www.sciencedirect.com/science/article/pii/S0301479721013086.

[56] Saeed, Muhammad Usama, Hussain, Nazim, Sumrin, Aleena, Shahbaz, Areej, Noor, Saman, Bilal, Muhammad, . . . Iqbal, Hafiz MN. (2022). Microbial bioremediation strategies with wastewater treatment potentialities–A review. *Science of The Total Environment*, 818, 151754. https://www.sciencedirect.com/science/article/pii/S0048969721068303.

[57] Salgot, Miquel, & Folch, Montserrat. (2018). Wastewater treatment and water reuse. *Current Opinion in Environmental Science & Health*, 2, 64-74. https://www.sciencedirect.com/science/article/pii/S2468584417300314.

[58] Saralov, AI. (2019). Adaptivity of archaeal and bacterial extremophiles. *Microbiology*, 88, 379-401. https://link.springer.com/article/10.1134/S0026261719040106.

[59] Saravanan, A, Kumar, P Senthil, Jeevanantham, S, Karishma, S, Tajsabreen, B, Yaashikaa, PR, & Reshma, B. (2021). Effective water/wastewater treatment methodologies for toxic pollutants removal: Processes and applications towards sustainable development. *Chemosphere*, 280, 130595. https://www.sciencedirect.com/science/article/pii/S0045653521010663.

[60] Schauder, Rolf, & Müller, Eva. (1993). Polysulfide as a possible substrate for sulfur-reducing bacteria. *Archives of Microbiology*, 160, 377-382. https://link.springer.com/article/10.1007/BF00252224.

[61] Sharaf, Ahmed, Guo, Bing, & Liu, Yang. (2019). Impact of the filamentous fungi overgrowth on the aerobic granular sludge process. *Bioresource Technology Reports*, 7, 100272. https://www.sciencedirect.com/science/article/pii/S2589014X19301628.

[62] Sharma, Anjney, Jamali, Hena, Vaishnav, Anukool, Giri, Balendu Shekhar, & Srivastava, Alok Kumar. (2020). Microbial biofilm: An advanced eco-friendly approach for bioremediation New and future developments in microbial biotechnology and bioengineering: *Microbial biofilms* (pp. 205-219): Elsevier. https://www.sciencedirect.com/science/article/pii/B9780444642790000153.

[63] Shi, Hong-Xin, Liu, Shao-Yang, Guo, Jin-Song, Fang, Fang, Chen, You-Peng, & Yan, Peng. (2024). Potential role of AgNPs within wastewater in deteriorating sludge floc structure and settleability during activated sludge process: Filamentous bacteria and quorum sensing. *Journal of Environmental Management*, 349, 119536. https://www.sciencedirect.com/science/article/pii/S0301479723023241.

[64] Singh, Gurudatta, Singh, Anubhuti, Singh, Priyanka, Shukla, Reetika, Tripathi, Shashank, & Mishra, Virendra Kumar. (2021). The fate of organic pollutants and their microbial degradation in water bodies. *Pollutants and water management: resources, strategies and scarcity*, 210-240. https://onlinelibrary.wiley.com/doi/abs/10.1002/9781119693635.ch9.

[65] Singh, Jyoti, Yadav, Priyanka, Pal, Ashok Kumar, & Mishra, Vishal. (2020). Water pollutants: Origin and status. *Sensors in water pollutants monitoring*: *Role of material*, 5-20. https://link.springer.com/chapter/10.1007/978-981-15-0671-0_2.

[66] Teng, Jade LL, Chan, Elaine, Li, Tsz Tuen, Kwan, Tsz Ying, Chan, Ka Fai, Li, Wing Ho, . . . Woo, Patrick CY. (2024). Antibiotic Susceptibility of Aerobic and Facultative Anaerobic Gram-Negative Rods in Hong Kong and Implications on Usefulness of Ceftazidime-Avibactam and Ceftolozane-Tazobactam. *Antibiotics*, 13(9), 802. https://www.mdpi.com/2079-6382/13/9/802.

[67] Tóth, András József, Fózer, Dániel, Mizsey, Péter, Varbanov, Petar Sabev, & Klemeš, Jiří Jaromír. (2023). Physicochemical methods for process wastewater treatment: powerful tools for circular economy in the chemical industry. Reviews in Chemical Engineering, 39(7), 1123-1151. https://www.degruyter.com/document/doi/10.1515/revce-2021-0094/html.

[68] Wang, Li, Liu, Xiang, Li, Zhenwen, Wan, Chunli, & Zhang, Yi. (2023). Filamentous aerobic granular sludge: A critical review on its cause, impact, control and reuse. Journal of Environmental Chemical Engineering, 11(3), 110039. https://www.sciencedirect.com/science/article/pii/S2213343723007789.

[69] Wang, Xin, Xia, Kang, Yang, Xiaojing, & Tang, Chao. (2019). Growth strategy of microbes on mixed carbon sources. *Nature communications*, 10(1), 1279. https://www.nature.com/articles/s41467-019-09261-3.

[70] Weissbrodt, David G, Laureni, Michele, van Loosdrecht, Mark CM, Comeau, Yves, Chen, G, Ekama, GA, . . . Brdjanovic, D. (2020). *Basic microbiology and metabolism. Biological wastewater treatment: principles, modelling and design,* 2nd edn. IWA Publishing, London, 11-76. https://iwaponline.com/ebooks/book/chapter-pdf/1262064/9781789060362_0011.pdf.

[71] Wu, Yang, Zhu, Zizeng, Sun, Pengliang, Zhao, Zhengzheng, Zheng, Xiong, Long, Min, & Chen, Yinguang. (2024). Enhanced short-chain fatty acids production from food waste with magnetic biochar via anaerobic fermentation: Linking interfacial, extracellular, and intracellular electron transfer. *Chemical Engineering Journal*, 488, 150853. https://www.sciencedirect.com/science/article/pii/S1385894724023404.

[72] Zheng, Fang, Wang, Jie, Xiao, Rui, Chai, Wenbo, Xing, Defeng, & Lu, Huijie. (2021). Dissolved organic nitrogen in wastewater treatment processes: Transformation, biosynthesis and ecological impacts. *Environmental Pollution*, 273, 116436. https://www.sciencedirect.com/science/article/pii/S0269749121000142.

[73] Żuchowska, Karolina, & Filipiak, Wojciech. (2024). Modern approaches for detection of volatile organic compounds in metabolic studies focusing on pathogenic bacteria: Current state of the art. *Journal of Pharmaceutical Analysis*, 14(4), 100898. https://www.sciencedirect.com/science/article/pii/S209517792300271X.

## Chapter 7

# Removal of Micro-Pollutants from Wastewater: Challenges and Advancements in Treatment Technologies

**Apoorav Sharma**
**and Lovleen Marwaha***

Department of Zoology, School of Bioengineering and Biosciences, Lovely Professional University, Phagwara, Punjab, India

## Abstract

The global contamination of freshwater systems by a wide range of industrial and natural compounds has become one of the most pressing environmental issues these days. Wastewater is often laden with various pollutants, including detergents, personal care products (PCPs), pharmaceuticals and their byproducts, plastics, flame retardants, and other emerging contaminants. These pollutants, commonly referred to as Micropollutants (MPs), are defined as synthetic or naturally occurring compounds that are introduced into aquatic ecosystems at low concentrations from both; point and non-point sources. Their accumulation across different environmental compartments poses serious risks to all forms of life, making them a significant public health concern. The increasing presence of MPs, such as polyaromatic hydrocarbons, antibiotics, pesticides, contraceptives, and chemicals found in PCPs, in water bodies necessitates urgent intervention. Given their detrimental impacts on human health, it is crucial that wastewater containing these pollutants undergo thorough treatment prior to discharge. While conventional physicochemical methods have been extensively explored for the degradation of MPs, these approaches often fall short in achieving comprehensive removal and can sometimes generate secondary pollution. There is a growing need to develop innovative strategies for mitigating the impact of these hazardous substances. Traditional methods, while widely used, tend to be energy-intensive and not always effective in completely eliminating contaminants. Advanced treatment techniques such as activated carbon adsorption, advanced oxidation processes, ozonation, nanofiltration, membrane bioreactors, and algal-based technologies have shown promise in achieving more consistent and effective MPs removal. However, many of these methods are still limited in their ability to remove a broad spectrum of toxic substances efficiently from wastewater. This chapter underscores the urgent need for continued scientific and technological advancements to improve wastewater treatment and address the issue of MP removal in line with international environmental and public health goals.

---

* Corresponding Author's Email: lovleenmarwaha@gmail.com

In: Advancements in Wastewater Treatment
Editors: Neha Agarwal, Maulin P. Shahl, Vijendra Singh Solankil et al.
ISBN: 979-8-90134-007-3

**Keywords:** pollutants, wastewater, oxidation, ozonation, phytoremediation

## 1. Introduction

The escalating global consumption of chemical substances has resulted in the contamination of surface and groundwater bodies. This widespread pollution poses significant ecological risks and potential health hazards, particularly due to the presence of MPs. MPs, present in aquatic environments at extremely low concentrations pose significant risks to ecosystems (Figure 1). These contaminants encompass a wide range of substances, including pesticides, flame retardants, per- and polyfluoroalkyl substances (PFAS), pharmaceuticals, surfactants, and PCPs. Recent studies have identified these chemicals in aquatic environments [1].

MPs are primarily introduced into the environment through anthropogenic activities such as agriculture, urban runoff, municipal and industrial wastewater discharges, sludge disposal, and accidental spills (Figure 2). Once released, these contaminants undergo various environmental processes, including phase partitioning and both biotic and abiotic degradation, which influence their removal and bioavailability in ecosystems [2, 3]. The fate and transport of MPs in the environment are influenced by their physicochemical properties (e.g., polarity, solubility, vapor pressure) and the specific environmental conditions (e.g., groundwater, surface water, sediments, wastewater, or drinking water). These contaminants often undergo transformation processes that can generate metabolites with different environmental behaviors and toxicological profiles as compared to the parent compounds [4]. MPs pose a range of adverse effects on aquatic organisms, including acute and chronic toxicity, endocrine disruption, bioaccumulation, and bio magnification [5]. Pesticides, known for their toxicity and persistence, and emerging contaminants like pharmaceuticals, steroid hormones, PFAS, surfactants, and PCPs are particularly concerning due to their increasing presence in the environment.

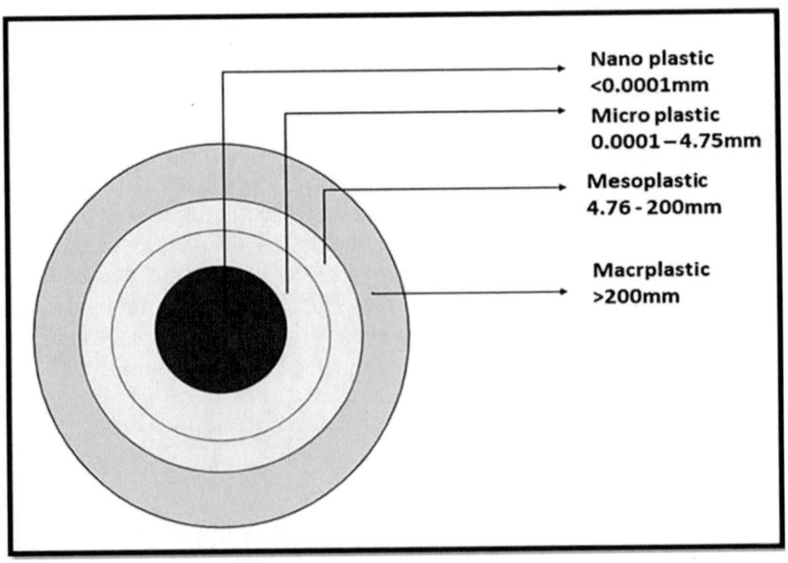


**Figure 1**. Different types of plastics on the basis of their size.

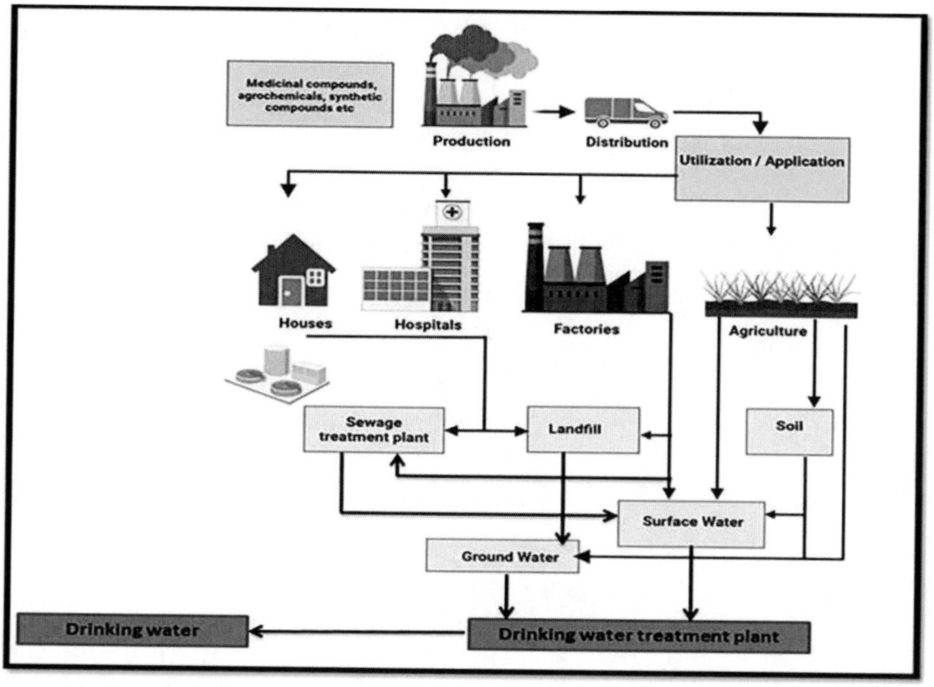


**Figure 2.** Entry of MPs into the water ecosystem.

## 2. Emerging Micro-Pollutants

### 2.1. Pesticides

Pesticides are the chemical compounds employed to control or eradicate unwanted organisms, such as insects, rodents, fungi, and weeds. These biologically active substances, often categorized as biocides; including herbicides, insecticides, and fungicides are classified based on their target pests. Prior to the 1940s, pesticides primarily consisted of inorganic compounds or natural plant-derived substances. The widespread use of synthetic pesticides began with the discovery of DDT in 1938 and accelerated significantly during and after World War II. Subsequent decades witnessed a rapid expansion in the production and application of synthetic pesticides, including classes like chlorinated hydrocarbons, organophosphates, urea, and triazines [6]. While synthetic pesticides have contributed to global economic growth and public health, but their extensive use has led to substantial environmental contamination and adverse impacts on ecosystems and human health. Insecticides, particularly organochlorines and organophosphates, are commonly used in pre- and post-harvest applications. Molecular structures of some common types of pesticides are shown in Figure 3. Both classes inhibit acetylcholinesterase, a crucial enzyme in the nervous system. Organochlorine insecticides, characterized by their halogenated, lipophilic nature and limited water solubility, exhibit high environmental persistence and tend to bioaccumulate in organisms.

Hexachlorocyclohexanes, chlorinated cyclodienes, and dichlorodiphenyltrichloroethanes (e.g., DDT) represent significant classes of insecticides that pose risks to both the environment and human health. Organophosphorous insecticides, synthetic organic esters of phosphorus-containing acids, exhibit higher water solubility and lower environmental persistence as compared to organochlorines [7]. Common examples of organophosphates include chlorpyrifos and methyl parathion. Herbicides, employed to control weed growth, can exert their effects through direct contact or systemic uptake by plants. They can be classified based on their application timing (pre-emergence or post-emergence) and their target spectrum (selective or broad-spectrum). Triazines and substituted urea are particularly noteworthy herbicides due to their widespread use, persistence, and toxicological implications [8, 9].

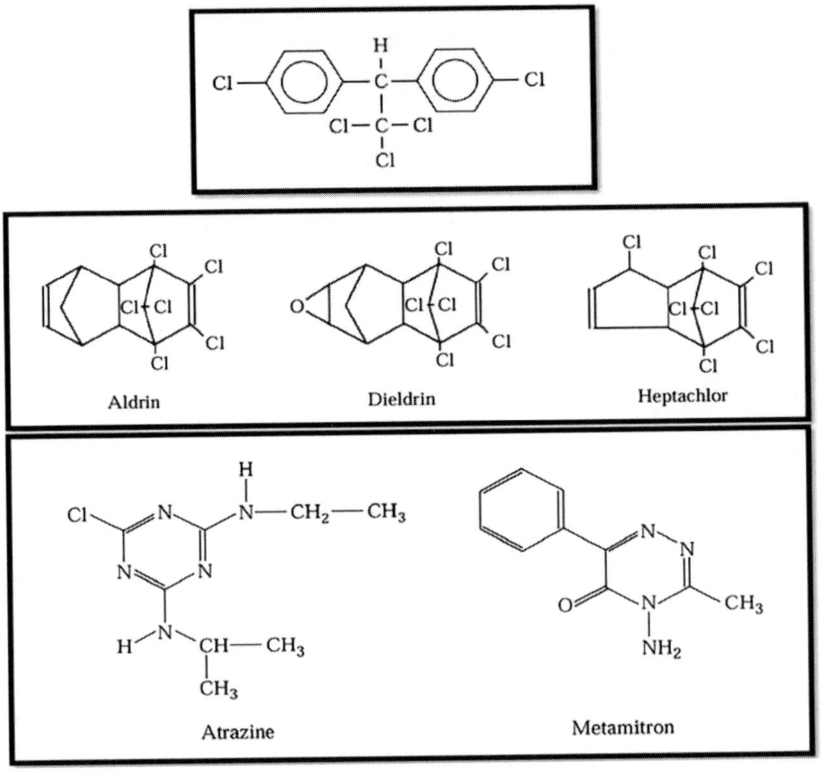


**Figure 3.** Chemical structures of organochlorine insecticide DDT, aldrin, dieldrin, heptachlor, atrazine and metamitron.

## 2.2. Pharmaceuticals

Pharmaceutically Active Compounds (PhACs) are designed to perform specific biological functions. Following human and animal use, over 4,000 of these compounds are released into the environment. Antibiotics, lipid regulators, anticonvulsants, and Non-Steroidal Anti-Inflammatory Drugs (NSAIDs) are commonly detected in aquatic environments and are recognized as potential environmental contaminants [10]. Chemical structures of Carbamazepine, Diclofenac, Ibuprofen, Gemfibrozil, Trimethoprim and Erythromycin are shown in Figure 4. Municipal wastewater, supplemented by hospital wastewater, industrial effluents, and landfill leachates, is a primary source of PhAC introduction into the environment. Pharmaceuticals are often found in environmental mixtures, and studies indicate that their combined and synergistic effects can induce toxicity to non-target organisms [11].

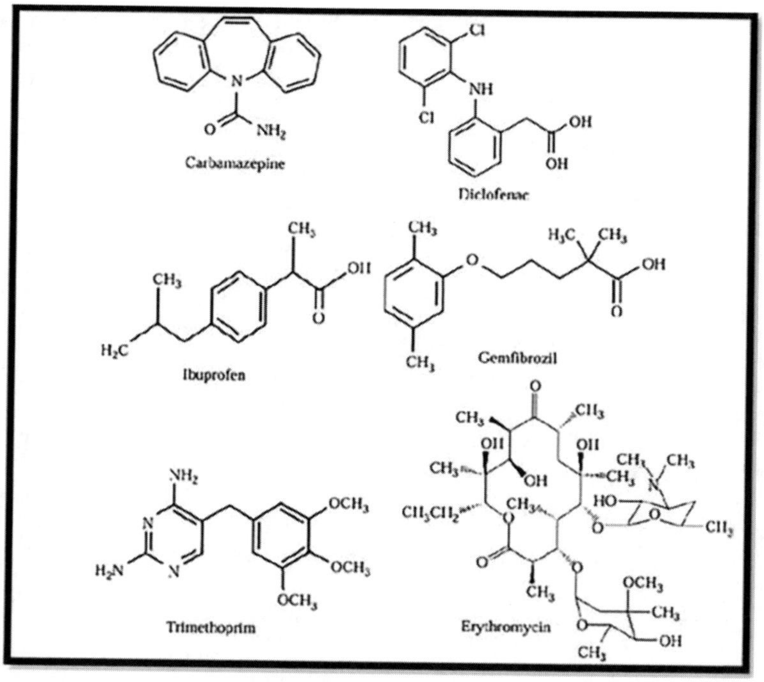


**Figure 4.** Chemical structures of Carbamazepine, Diclofenac, Ibuprofen, Gemfibrozil, Trimethoprim and Erythromycin.

## 2.3. Steroid Hormones

Steroid hormones, which regulate the endocrine and immune systems, encompass major classes such as estrogens, androgens, progestagens, and corticosteroids. Estrone (E1), 17β-estradiol (E2), estriol (E3), and ethinylestradiol (EE2) (as shown in Figure 5) are significant contributors to estrogenic activity in treated wastewater and surface water [12]. These hormones enter the environment through wastewater discharges, runoff, sewage sludge application, and aquaculture. Human excretion of steroid hormones varies based on gender, pregnancy, and menopause. Pregnant women, in particular, exhibit significantly higher excretion rates of estrogens, such as E1 [13]. Steroid hormones are primarily eliminated as sulfate or glucuronide conjugates, although free hormones can also be present in feces. While glucuronides can be hydrolyzed to release free hormones, sulfates are more resistant to biotransformation. Steroid hormones, especially estrogens, are potent endocrine disruptors, capable of feminizing male fish and inducing vitellogenesis, even at low concentrations. Other aquatic organisms, including amphibians, reptiles, and invertebrates, are also susceptible to the effects of these hormones [14].

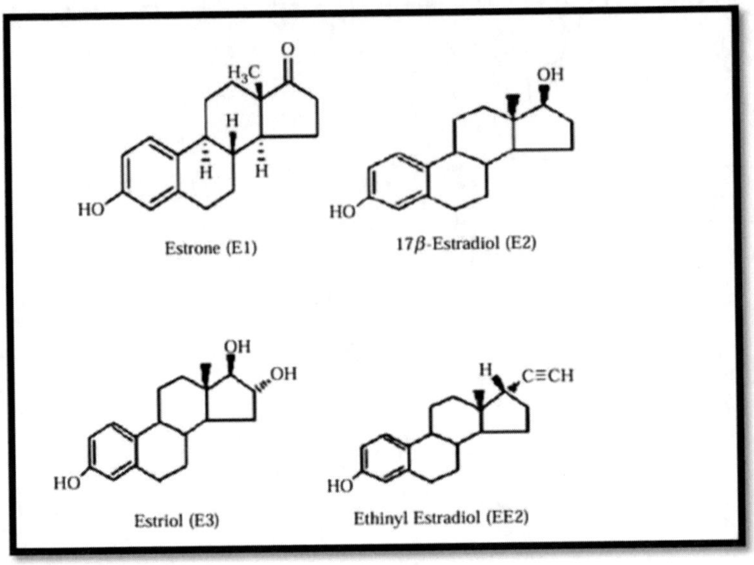


**Figure 5.** Chemical structure of major classes of steroid hormones.

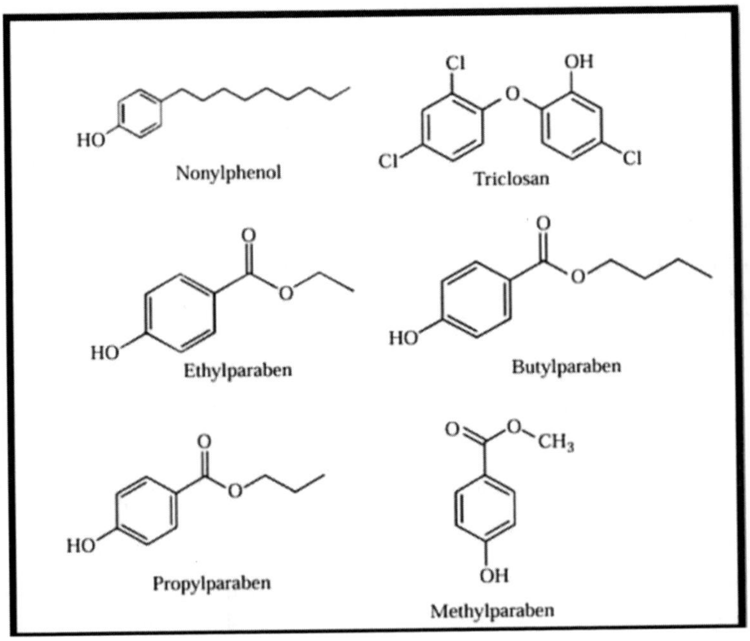


**Figure 6.** Chemical structure of surfactants and PCPs.

## 2.4. Surfactants and PCPs

Surfactants are widely used in detergents, textiles, and paper industries [7]. They are categorized into anionic (e.g., linear alkylbenzene sulfonates), cationic (e.g., quaternary ammonium compounds), and non-ionic (e.g., alkylphenol ethoxylates) types. PCPs, such as skincare items, soaps, shampoos, and dental products, contain significant amounts of synthetic chemicals, including antimicrobial agents and sunscreen agents. Molecular structure of surfactants and PCPs are shown in Figure 6. These chemicals enter aquatic ecosystems through routine use [15]. Triclosan (TCS) and parabens have attracted considerable research attention due to their widespread use and potential toxicological effects. Parabens, in particular, are found in over 22,000 cosmetic products [16].

## 2.5. Per Fluorinated Compounds

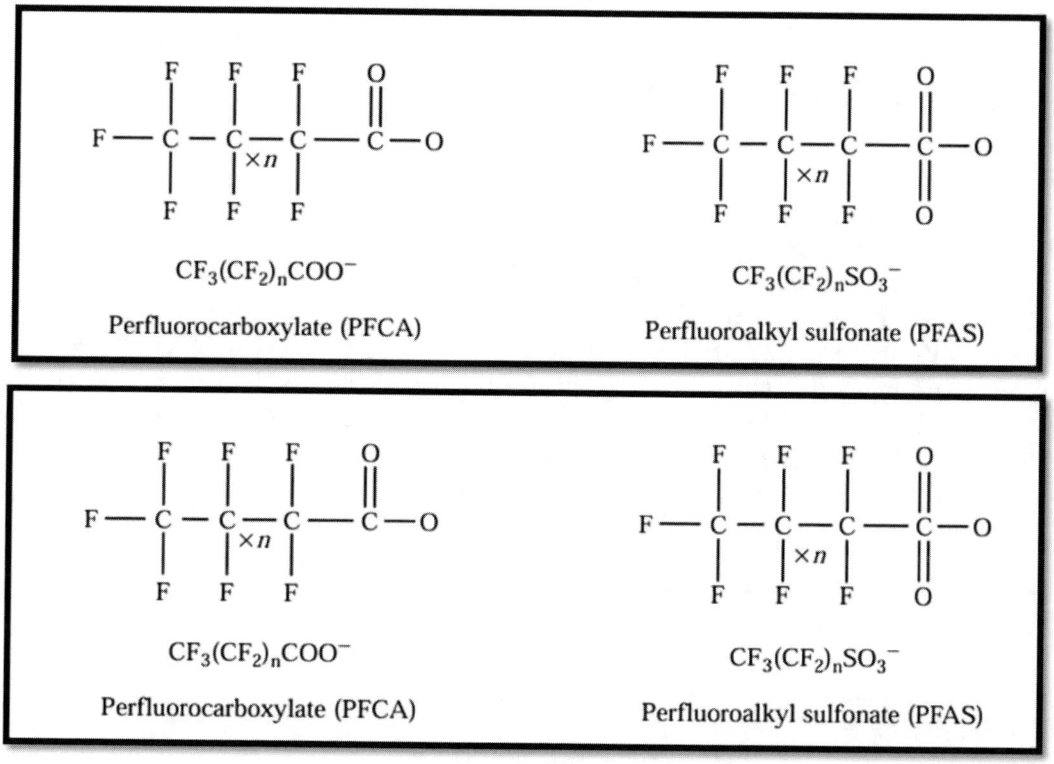


**Figure 7.** Chemical structure of per fluorinated compounds.

Perfluorinated compounds (PFCs), renowned for their thermal and chemical stability, surface activity, and low surface free energy, have been manufactured since the 1950s for diverse industrial and consumer applications. The strong carbon-fluorine (C-F) bonds in these molecules render them resistant to oxidation, reduction, and acid-base reactions [17]. PFAS and perfluorinated carboxylates (PFCAs) as shown in Figure 7, characterized by a perfluorinated carbon chain linked to a sulfonic or carboxylic group, respectively, are among the most extensively studied PFCs. Common environmental pollutants within this category include perfluorononanoic acid (PFNA), perfluorodecanoic acid (PFDA), perfluoroundecanoic acid (PFUnA), perfluorododecanoic acid (PFDoA), and perfluorooctanesulfonamide (PFOSA) [18]. These compounds have been detected globally in drinking water, surface water, groundwater, and wastewater, even in remote regions, highlighting their widespread environmental contamination. Extensive research on perfluorooctanesulfonate (PFOS) and perfluorooctanoic acid (PFOA) has raised concerns about their potential toxicity, including hepatotoxicity, immunotoxicity [19, 20], developmental toxicity, endocrine disruption, and carcinogenicity. Animal studies have shown that these chemicals primarily accumulate in the serum, liver, and kidneys. The elimination half-lives of PFOA and PFOS vary significantly across species and genders, ranging from several hours to 30 days. Human blood and tissue samples from individuals exposed to PFCs both occupationally and non-occupationally have revealed the presence of these compounds [21]. For the general population, food consumption and drinking water are the primary exposure routes. Recent studies indicate that PFOA concentrations in human blood can range from 2 to 8 mg/mL in populations exposed to ambient levels in developed countries [20].

## 3. Mechanisms of MP Removal

### 3.1. Biological Treatment of MPs

Biological treatment techniques, including activated sludge processes, bioreactors, and advanced membrane-based systems (as shown in Figure 8), can leverage the microorganisms for effective pollutant removal. Numerous bacterial and fungal species possess the ability to degrade MPs through microbial degradation, utilizing them as growth substrates. The efficiency of microbial growth and contaminant degradation is influenced by several operational factors, including temperature, pH, and the physicochemical properties of the pollutants [22]. *Pseudomonas* species are well-known for their capacity to oxidize various MPs. For instance, *Pseudomonas putida* can co-metabolize atrazine (ATZ) and oxidize diclofenac (DCF).

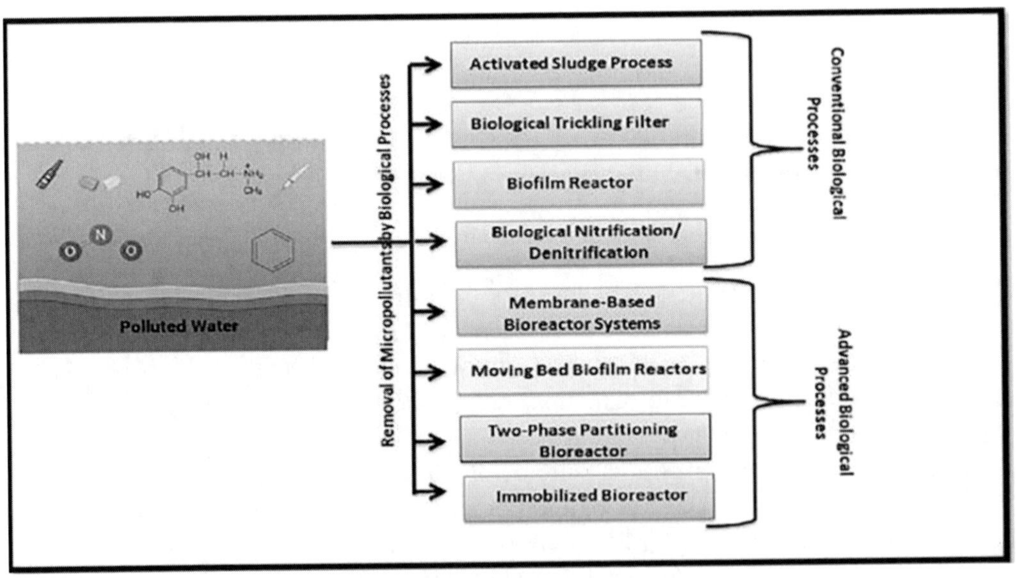


**Figure 8.** Conventional and advanced methods of biological processes for MP removal.

Additionally, strains of *Sphingomonas* and *Streptomyces* have been shown to degrade pesticides like 3-(3,4-dichlorophenyl)-1,1-dimethylurea (DIU). Fungi, particularly white rot fungi, secrete extracellular enzymes such as lignin peroxidase and laccase, which exhibit low substrate specificity, enabling them to degrade complex organic contaminants like endocrine-disrupting chemicals (EDCs) [3]. While most research focuses on aerobic degradation processes, recent studies have highlighted the anaerobic degradation of MPs like galaxolide and bisphenol, mediated by enzymes such as acetate kinase [23].

### *3.1.1. Biological Treatments Removal of MPs from Wastewater*

#### *3.1.1.1. Activated Sludge Process*

The Activated Sludge Process (ASP) is a conventional biological wastewater treatment process that utilizes microbial activity to remove nutrients and oxidize organic matter as shown in Figure 9. In this process, wastewater is mixed with a microbial culture in an aerated reactor, and activated sludge is recirculated to maintain a continuous treatment process [24]. Microbial degradation and sorption onto microbial flocs are the primary mechanisms of pollutant removal, with hydrophobic chemicals adsorbing to flocs due to their surface properties. Co-metabolism by microbial consortia enhances the removal of diverse contaminants. Hybrid ASP systems, combining suspended and attached biomass, have shown promise in effectively removing MPs such as diclofenac, achieving removal efficiencies of up to 25%. The biodegradation of bis(2-ethylhexyl) phthalate in sequencing batch reactors has also been demonstrated [5]. However, ASP has limitations, including challenges in meeting stringent discharge standards for MPs and the requirement for significant space for installation.

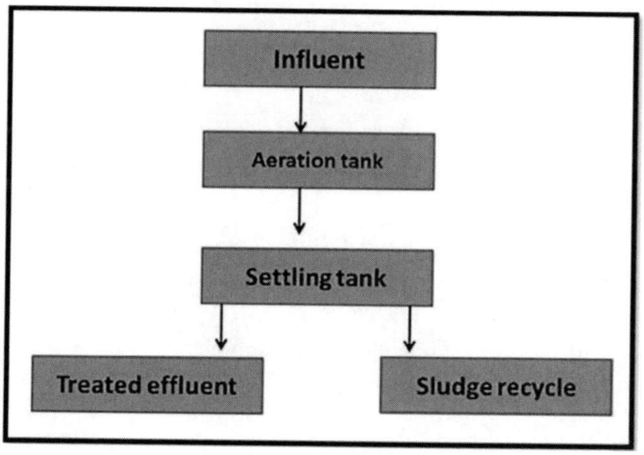


**Figure 9.** Flow Chart of Activated Sludge Process.

#### *3.1.1.2. Biological Trickling Filter and Biofilm Reactor*

Biological trickling filters, which utilize biomass immobilized on a support medium, are a common technology for MP treatment (Figure 10). In these systems, wastewater percolates through the substrate, where the immobilized biomass degrades the contaminants [25]. In spite of being historically used in wastewater treatment plants (WWTPs) to remove chemical oxygen demand (COD) and biochemical oxygen demand (BOD), their effectiveness in removing MPs is limited. Research by Kasprzyk-Hordern et al. demonstrated approximately 50% removal efficiency for over 50 pharmaceuticals and endocrine-disrupting chemicals (EDCs) over a five-month period using a combination of trickling filters and activated sludge processes [23].

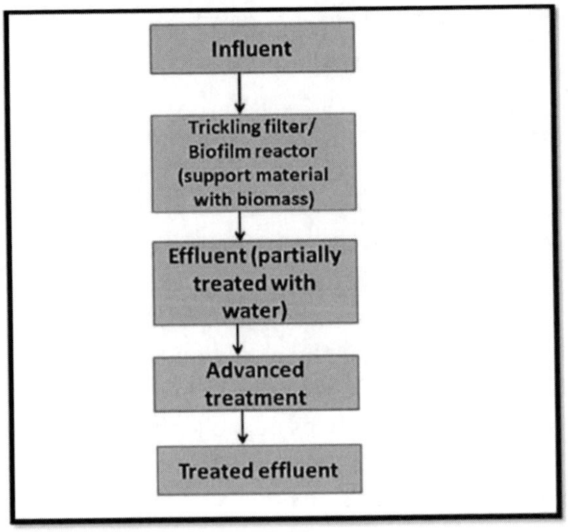


**Figure 10.** Flow chart of Biological Trickling Filter and Biofilm Reactor process.

### *3.1.1.3. Biological Nitrification and Denitrification*

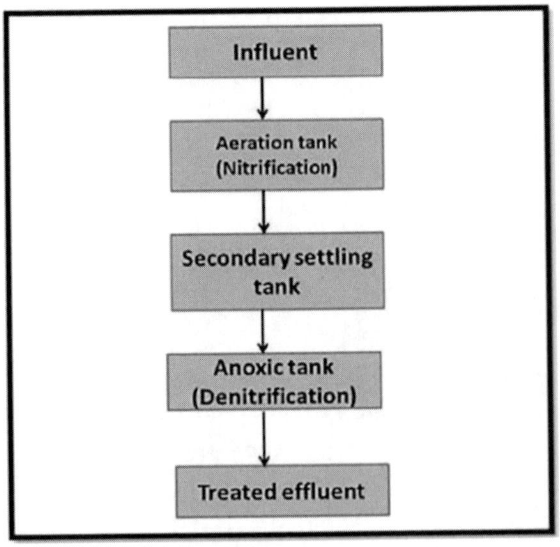


**Figure 11.** Flow chart of Biological Nitrification and Denitrification process.

Biological nitrification and denitrification are effective bioprocesses for removing MPs from wastewater as shown in Figure 11. Nitrification involves the biological oxidation of ammonium to nitrite and nitrate, while denitrification converts nitrate or nitrite to nitrogen gas. Research has demonstrated significant removal efficiencies for these processes [26]. Ahmed et al. reported that denitrification can remove up to 100% of endocrine-disrupting chemicals (EDCs) at microgram-per-liter concentrations, including salicylic acid, bisphenol A, galaxolide, and

17β-estradiol [12]. Additionally, Suarez et al. found that nitrification was more effective in removing compounds like erythromycin, galaxolide, and ibuprofen during wastewater treatment [10].

#### *3.1.1.4. Membrane-Based Bioreactor Systems*

Membrane bioreactors (MBRs) are employed in two primary configurations: side-stream and submerged as shown in Figure 12. The side-stream configuration positions the membrane module externally to the bioreactor, requiring a recirculation pump, which increases energy consumption and costs. To address these limitations, the submerged MBR was developed, which integrates the membrane within the bioreactor, enabling simultaneous treatment and removal of MPs [27]. In MBRs, MP removal occurs through physical retention, air stripping, and biodegradation. Sludge retention time (SRT) and hydraulic retention time (HRT) are critical factors influencing MBR performance. While low SRT can effectively remove MPs like ibuprofen and acetaminophen, hydrophilic substances such as carbamazepine and diclofenac are less efficiently removed under the same conditions. Temperature and pH are additional parameters that significantly impact MBR performance [28].

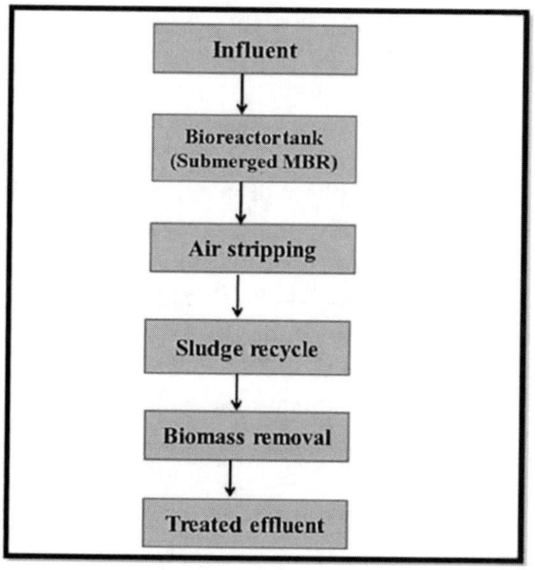


**Figure 12.** Flow chart for Membrane-Based Bioreactor process.

#### *3.1.1.5. Moving Bed Biofilm Reactors*

The Moving Bed Biofilm Reactor (MBBR) is a biofilm-based system that utilizes small biomass carriers to provide a large surface area for microbial growth is as shown in Figure 13. Introduced nearly three decades ago as an efficient alternative to the conventional Activated Sludge Process (ASP), MBBR offers several advantages, including a compact design, enhanced solid retention time for slow-growing organisms, and the ability to facilitate both aerobic and anaerobic growth within a single reactor [29]. MBBR has demonstrated significant effectiveness and cost-efficiency in removing MPs. Casas et al. observed a 20% increase in degradation efficiency in MBBR compared to MBR, successfully removing up to 79% of iohexol and 73% of diatrizoate from urban wastewater [21].

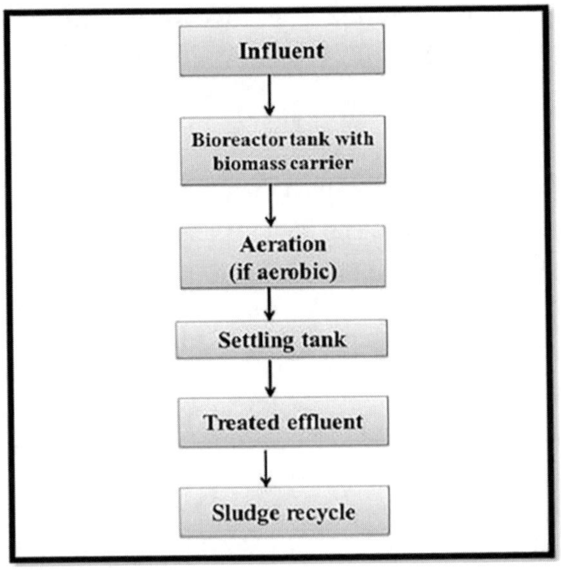


**Figure 13.** Flow chart for Moving Bed Biofilm Reactor process.

### *3.1.1.6. Two-Phase Partitioning Bioreactor*

Two-phase partitioning bioreactors (TPPBs) consist of two distinct phases: a biocompatible organic phase containing the target pollutant and an aqueous phase housing the microbe. High-speed agitation generates small organic phase droplets, facilitating efficient pollutant transfer to the aqueous phase for microbial consumption. As microbes degrade the substrate, the equilibrium shifts, leading to further pollutant partitioning [9]. TPPBs effectively degrade hydrophobic and toxic compounds such as PAH and phenols. The mechanism involves substrate absorption at the interface between the aqueous and organic phases, where agitation and phase volume significantly influence mass transfer rates. TPPBs have demonstrated significant effectiveness in degrading pollutants like pyrene, styrene, and 2,4-dimethylphenol, often achieving removal efficiencies exceeding 90%. However, TPPBs require high energy input due to the need for high agitation speeds, which may potentially damage microbial cells [30].

### *3.1.1.7. Immobilized Bioreactor*

There are two primary types of immobilization: active and passive. Both methods involve restricting microbial cell mobility within a specific area. In active immobilization, cells are trapped or adhered to support materials like alginate, agar, or chitosan using physical or chemical forces. In contrast, passive immobilization involves the natural formation of biofilms on solid supports [31]. This technique is commonly employed in biofilm-based systems such as biotrickling filters and moving bed biofilm reactors. Immobilized systems offer several advantages, including increased cell concentration, reusability of biomass, prevention of cell washout, enhanced genetic stability, and reduced cell separation costs. Studies have demonstrated the effective degradation of pollutants using immobilized microbes.

### 3.1.1.8. Enzyme-Based Method

Microorganisms exhibit a range of adaptive responses to stress, influencing their growth rate, energy metabolism, and macromolecular synthesis. These responses are fundamentally linked to enzyme activity, which is crucial for cellular functions, including the degradation of anthropogenic pollutants like microplastics. Microbial enzymes interact with the polymer structures of microplastics, facilitating their breakdown into monomers that can be utilized as carbon sources for microbial growth and energy metabolism [32]. Microplastic degradation involves several mechanisms, with two primary pathways being the most recognized. Surface modification by hydrolase enzymes, such as lipases, carboxylesterases, cutinases, and proteases, enhances the degradation of microplastic polymers. This process increases surface hydrophilicity, priming the material for further breakdown. These enzymes modify surface chemistry without compromising the polymer's structural integrity. Elemental spectroscopy chemical analysis (ESCA) has confirmed the interaction of enzymes with microplastic surfaces, indicating changes in C-O bonds, which suggests enzyme-mediated surface modifications. While only a limited number of cutinases can degrade the polymer interior, they play a crucial role in surface modification, increasing the material's susceptibility to biodegradation. Over the past decade, significant advancements have been made in identifying enzymes capable of degrading polymers into monomer units [30]. The following sections will delve into specific enzymes that target different types of microplastic polymers, considering their environmental prevalence.

### 3.1.1.9. Bioremediation

Bioremediation is a natural process that utilizes microorganisms like bacteria and fungi to degrade pollutants. Plastic waste, under environmental factors, breaks down into macro- and microplastics. Bioengineered bioremediation employs bacteria and fungi, often in conjunction with physicochemical processes, to specifically degrade microplastics. During degradation, the physical and chemical properties of polymers undergo significant alterations due to environmental factors and biological processes [33]. Primary degradation produces organic intermediates like aldehydes, alcohols, acids, and ketones through oligomer solubilization. While microplastic removal efficiencies typically exceed 20%, the specific efficiency depends on microbial species. Bioremediation offers advantages like cost-effectiveness, energy efficiency, and environmental friendliness [34]. However, challenges include identifying suitable microbial consortia and achieving complete microplastic degradation, which can be a slow process. Biodegradation is highly specific and produces minimal harmful byproducts. The biodegradation process involves four stages: biodeterioration, biofragmentation, assimilation, and mineralization. Studies have shown that *Arthrobacter sp.* and *Pseudomonas sp.* effectively degraded high-density polyethylene (HDPE), achieving weight losses of 12.23% and 15.18%, respectively, after 30 days [35]. Similarly, *Aspergillus tubingensis* degraded HDPE, with weight loss increasing in the presence of mineral oil. For low-density polyethylene (LDPE) remediation, *Penicillium pinophilum* exhibited mineralization rates of 0.64% in an ethanol medium and 0.37% without ethanol [35].

### 3.1.1.10. Phytoremediation

Phytoremediation is a sustainable, plant-based strategy that utilizes plants and their associated rhizosphere microorganisms to extract, stabilize, transfer, and degrade MPs from wastewater.

In this process, plants grow in contaminated water, absorbing pollutants along with essential nutrients and water through their root systems [36]. The pollutants are then transported throughout the plant, primarily accumulating in the roots, which can be harvested once saturated with contaminants. The success of phytoremediation hinges on selecting rapidly growing plant species with extensive root systems and high transpiration rates, as these characteristics enhance pollutant uptake and water flow. The involvement of plant growth-promoting rhizobacteria (PGPR) significantly improves the remediation process. These bacteria produce phytohormones, biosurfactants, and organic acids, which promote root development, mitigate phytotoxicity, and increase contaminant bioavailability [37]. Additionally, they protect plants from pathogens, alleviate abiotic stress, and facilitate the biosorption and bioaccumulation of toxic substances, enhancing MP removal efficiency.

#### *3.1.1.11. Anaerobic Digestion*

Anaerobic digestion is a critical process in wastewater sludge treatment, focusing on MP removal and organic matter stabilization. The process begins with sludge thickening, which can be achieved through air flotation or gravity settling. Thickened sludge then undergoes anaerobic digestion in a controlled environment, either at mesophilic or thermophilic temperatures [38]. Anaerobic digestion involves a series of biochemical reactions: hydrolysis, acidogenesis, acetogenesis, and methanogenesis. These stages facilitate the breakdown of complex organic compounds and MPs by microbial consortia, converting them into simpler compounds and producing methane-rich biogas as a renewable energy source [39].

**Table 1.** Advantages and disadvantages of different biological methods of removal of MPs from wastewater

| Biological processes for MP removal | Advantages | Disadvantages |
|---|---|---|
| Biological Trickling Filter | Need for complex water and air distribution system | Bed clogging Long startup period |
| Biofilm Reactor | High removal within less time | Selection of microbial consortia |
| Nitrification/Denitrification | Natural bio-process | Specific |
| Two-Phase Partitioning Bioreactor | High removal for hydrophobic compounds | Requirement of high agitation speed |
| Immobilized Bioreactor | Reuse of biomass Resistance to toxic substrate | Scaling up is difficult |
| Enzyme-based Method | Highly specific and eco-friendly | High production costs, and slower degradation rates |
| Bioremediation | Cost-effective, energy-efficient, environmentally friendly, and produces minimal harmful by-products | Slow method, with challenges in identifying suitable microbial consortia |
| Phytoremediation | Sustainable, cost-effective, and enhances pollutant removal through natural plant and microbial interactions, with minimal environmental disruption. | Slow, requires large land areas, and may be limited by plant tolerance to specific pollutants and environmental conditions. |
| Anaerobic Digestion | Effectively stabilizes organic matter, reduces sludge volume, removes MPs, and generates methane-rich biogas as a renewable energy source. | Careful control of conditions can be costly to set up and may have reduced efficiency with certain types of MPs. |

The final stage of sludge treatment is dewatering, typically achieved through centrifugation. This process efficiently separates water from biosolids, significantly reducing sludge volume and enhancing pollutant removal [40]. By optimizing these processes, wastewater treatment facilities can minimize the discharge of residual MPs into the environment.

## 3.2. Physiochemical Treatments

### *3.2.1. Adsorption*

The science of adsorption, dating back to the 18th century, has significant applications in modern environmental science, including the removal of microplastics from wastewater. Adsorption is a surface-based process where microplastic particles adhere to the surface of an adsorbent material, reducing their concentration in water through physical and chemical interactions. Early chemists like Felice Fontana, Karl Wilhelm Scheele, and Johann Tobias Lovitz pioneered adsorption techniques [41]. Materials like activated carbon, with their high surface area and affinity for organic compounds, are commonly used as adsorbents. When introduced to contaminated water, activated carbon traps microplastic particles through various forces, including Van der Waals forces, hydrogen bonding, and electrostatic interactions. The porous structure of activated carbon maximizes contact with contaminants, enhancing adsorption efficiency. Historically, charcoal was used to purify water by removing organic compounds and odors. Modern adsorbents, such as activated carbon and biochar, effectively bind microplastic particles, enabling their removal through filtration. Adsorption targets microplastic contaminants at the molecular level, interacting with their polymer structure and surface charge. This physiochemical treatment offers several advantages: low cost, high efficiency, and scalability issues for large-scale wastewater treatment [42].

### *3.2.2. Advanced Oxidation Process*

Advanced oxidation processes (AOPs), introduced in the 1980s for drinking water treatment, rely on the in-situ generation of highly reactive hydroxyl radicals (•OH) for effective water purification. The application of AOPs has expanded to include the production of additional powerful oxidizing agents, such as sulfate radicals ($SO_4^{\bullet-}$), enhancing their effectiveness in advanced wastewater treatment [43]. Unlike conventional oxidants like chlorine and ozone, which serve as both disinfection and contaminant removal purposes, AOPs are specifically designed to degrade and mineralize complex organic and inorganic contaminants. These processes transform toxic substances into less harmful or inert byproducts, providing an effective solution for persistent pollutants that traditional treatments often fail to address. Leveraging the high oxidation potential of reactive radicals, AOPs offer a powerful tool for water and wastewater treatment, particularly in the removal of MPs and emerging contaminants.

#### *3.2.2.1. Hydroxyl Radical Based AOP's*

AOPs leverage the high reactivity of •OH to effectively eliminate MPs from wastewater. Hydroxyl radicals, powerful oxidizing agents in water treatment, possess an oxidation potential ranging from 2.8 V (at pH 0) to 1.95 V (at pH 14) relative to the saturated calomel electrode (SCE). Their strong reactivity and non-selective nature enable rapid reactions with a diverse range of pollutants at rate constants between $10^8$ and $10^{10}$ $M^{-1}$ $s^{-1}$. Hydroxyl radicals degrade

MPs through four primary reaction mechanisms: radical addition, hydrogen abstraction, electron transfer, and radical combination. These reactions generate carbon-centered radicals, which, in the presence of oxygen, can form organic peroxyl radicals [44]. This cascade of radical reactions produces secondary reactive species, including $H_2O_2$ and superoxide anions ($O_2^{\bullet -}$), further promoting the degradation and mineralization of organic contaminants into harmless byproducts like water and carbon dioxide. Given the short lifespan of hydroxyl radicals, they are generated in situ during treatment. Various AOP methods, such as the activation of $H_2O_2$ and ozone ($O_3$), irradiation techniques (e.g., ultraviolet light or ultrasound), and catalytic systems (e.g., $Fe^{2+}$), facilitate the continuous production of hydroxyl radicals. This ensures the thorough destruction of MPs, enhancing the efficiency and environmental sustainability of advanced wastewater treatment [45].

*3.2.2.2. Ozone Based AOPs*

Ozone-based AOPs leverage the strong oxidative properties of ozone ($O_3$), which has an oxidation potential of 2.07 V relative to the saturated calomel electrode (SCE). Ozone acts as a selective oxidant, reacting primarily with ionized forms of organic pollutants at rate constants between 1.0 and $10^3$ $M^{-1}$ $s^{-1}$. Under certain conditions, ozone can generate highly reactive •OH, initiating non-selective oxidation processes that effectively degrade complex MPs [46]. Several mechanisms contribute to •OH production from ozone, including direct reactions in water and the presence of other oxidizing agents or irradiation. For example, in the peroxone system ($O_3/H_2O_2$), hydroperoxide ions ($HO_2^-$) catalyze ozone decomposition, enhancing •OH formation. Similarly, in the ozone/UV system, UV photolysis of $O_3$ generates $H_2O_2$, providing additional oxidant support. In summary, •OH can be generated through three primary mechanisms: direct ozonation, peroxone reactions, and $H_2O_2$ photolysis. These pathways drive a powerful oxidation process, leading to efficient MP degradation through both direct ozone oxidation and indirect •OH-mediated processes. The combination of ozone, $H_2O_2$, and UV light in AOPs significantly improves pollutant degradation, making it a highly effective method for advanced wastewater treatment.

### *3.2.3. Coagulation and Flocculation*

Coagulation and flocculation are widely used in wastewater treatment as an integrated physico-chemical process to remove nano- and microplastics (NMPs). Coagulation involves adding coagulants to destabilize suspended particles, while flocculation aggregates these destabilized particles into larger flocs for efficient separation and removal [47]. This approach is particularly effective for treating wastewater containing NMPs, as traditional methods often struggle to remove such small particles.

*3.2.3.1. Process of Coagulation and Flocculation*

Coagulation and flocculation are the processes that facilitate the aggregation and subsequent sedimentation or flotation of particle pollutants, including NMPs. Coagulation neutralizes the surface charges of suspended particles, reducing electrostatic repulsion and promoting particle interaction. Coagulants such as aluminum sulfate, ferric chloride, magnetic magnesium oxide, polyacrylamide (PAM), tannic acid, and protein amyloid fibrils are used to enhance coagulation efficiency. Following coagulation, flocculation consolidates smaller aggregates into larger flocs with the aid of flocculants like PAM. This process is crucial for NMP removal, as these small particles are effectively captured within larger flocs, enabling their removal through sedimentation or flotation. By combining chemical destabilization and physical aggregation, coagulation and flocculation effectively remove micro-scale pollutants, improving water quality and safety.

### *3.2.4. Membrane Filtration*

Membrane separation is a process that uses a selective, semi-permeable barrier to separate substances based on size, charge, and chemical affinity. In wastewater treatment, semi-permeable membranes allow water molecules to pass through while retaining pollutants through molecular sieving and diffusion [48]. This process often relies on a pressure gradient to drive pollutant separation. Various membrane technologies have been developed, each with specific membrane characteristics and separation mechanisms. Microfiltration, ultrafiltration, nanofiltration, and reverse osmosis (RO) are common membrane technologies, each designed to target specific contaminants based on their molecular properties.

#### *3.2.4.1. Nanofiltration*

Nanofiltration (NF) membrane technology, based on the solution-diffusion process, selectively removes MPs from wastewater by allowing monovalent ions to pass through a semi-permeable barrier while rejecting particles smaller than 0.002 μm. This technology is particularly effective for water-soluble plastics due to its high selectivity and charge-based repulsion properties. Nanofiltration membranes, such as titanium/polyethyleneimine hybrids, offer improved stability and can be tailored for specific pollutant rejection by adjusting parameters like solute rejection, permeability, and recovery rates. Nanofiltration selectively removes ions, favoring monovalent anions over multivalent ions [49]. The filtration process is influenced by factors like membrane hydrophilicity, charge, particle size, and hydrophobicity. Nanofiltration typically involves pretreatment, membrane filtration, and post-treatment stages. It often serves as a pre-treatment for RO, reducing fouling and operational costs by lowering scaling and pressure requirements. Most NF membranes are polymeric, offering good thermal and chemical stability and enabling production in flat-sheet or hollow-fiber configurations. Ceramic membranes, while chemically stable and offering a narrow pore size distribution are less commonly used commercially due to their higher cost compared to polymeric alternatives. NF membranes offer advantages such as low energy consumption, cost-effectiveness, and high permeability, making them suitable for various applications. However, challenges like fouling and the trade-off between flow and selectivity need to be addressed. Phase inversion is a common method for manufacturing NF membranes, enabling the production of flat-sheet and hollow-fiber membranes, expanding the applicability of nanofiltration in advanced wastewater treatment.

#### *3.2.4.2. Reverse Osmosis (RO)*

RO is an advanced water treatment technology that uses pressure to force water through a semi-permeable membrane, separating it from dissolved contaminants. RO is widely used in wastewater treatment, particularly in desalination. Operating at pressures between 7000 and 15,000 kPa, RO employs the solution-diffusion mechanism to filter solutes with molecular weights below 200 g/mol. RO membranes, often made from polyamide, cellulose acetate, or polysulfone, are sensitive to pH fluctuations and solute concentrations, requiring careful operational management to prevent fouling [50]. Pre-treatment is crucial for optimal RO performance, as it removes suspended particles that can lead to membrane fouling. While spiral-wound configurations are common, hollow-fiber configurations are also used. Polymeric membranes, especially cellulose acetate, function best in low pH ranges (4-6) and are chlorine-resistant. Some configurations offer higher flow rates, reducing fouling and energy consumption. Due to their limited pore size and susceptibility to fouling, RO systems are often integrated with ultrafiltration to enhance MP removal. This combined approach leverages RO's solution-diffusion mechanism with ultrafiltration's pre-filtration capabilities, providing an effective solution for removing MPs like microplastics from wastewater.

## 4. Challenges

The effective removal of MPs, such as microplastics, from wastewater remains a significant challenge due to their diverse physical and chemical properties, low concentrations, environmental persistence, and the high technological and economic demands of treatment processes. This chapter delves into the key challenges associated with microplastic removal in wastewater treatment systems, as highlighted by current scientific research.

### 4.1. Intricacy of MPs

MPs encompass a wide range of substances, including microplastics, pharmaceuticals, PCPs, and industrial chemicals. Each pollutant type exhibits unique physical and chemical properties, making it challenging to develop a single, universal removal method [51]. Microplastics, in particular, exhibit significant variability in size, polymer composition, morphology, density, and degradation potential, further complicating the development of standardized treatment approaches.

### 4.2. Minimal Concentrations

Microplastics and other MPs are often present at very low concentrations [10], typically in the nanogram to microgram per liter (ng/L to μg/L) range. This low concentration makes their detection, quantification, and removal challenging. Traditional wastewater treatment methods, designed to remove larger pollutants, are ineffective at capturing microplastics at such small scales. The need for advanced analytical techniques to detect and measure microplastics in wastewater, coupled with potential analytical variability, further complicates the issue.

### 4.3. Restricted Biodegradability

Many microplastics and synthetic chemicals are resistant to degradation by conventional biological treatment methods used in WWTPs. The polymeric nature of microplastics imparts exceptional stability and resistance to biodegradation, enabling them to survive these processes and persist in the environment. This resistance necessitates advanced physical or chemical treatment methods, which are better suited to address persistent contaminants like microplastics.

### 4.4. Operational Conditions and Environmental Fluctuations

The effectiveness of microplastic removal is significantly influenced by operational parameters in treatment facilities, such as pH, temperature, sludge retention time (SRT), and redox potential. Variations in these parameters can impact MP removal efficiency, leading to inconsistent results. Additionally, wastewater composition can fluctuate seasonally and regionally, requiring continuous optimization of treatment processes to ensure effective microplastic capture and removal. Effective management of these variables necessitates advanced monitoring and adaptive treatment strategies to maintain high-quality pollutant removal.

### 4.5. Elevated Expenses of Sophisticated Treatment Technologies

Advanced technologies like ozonation, UV-based AOPs, and membrane filtration, while effective in removing microplastics, often incur high costs due to specialized equipment, energy consumption, and maintenance requirements. The significant economic burden associated with large-scale implementation of these technologies may hinder their widespread adoption. Ongoing research focuses on developing cost-effective alternatives and improving the economic feasibility of advanced microplastic removal techniques.

### 4.6. Rigorous Regulatory Standards and Oversight Requirements

Regulatory agencies have imposed stricter regulations on the allowable concentrations of MPs and other MPs in treated wastewater due to growing concerns about environmental and human health. To comply with these standards, WWTPs must implement continuous monitoring and regularly update their treatment processes. The requirement for rigorous monitoring, coupled with the challenges of identifying and quantifying microplastics, adds operational complexity and increases the costs and labor involved in wastewater treatment.

### 4.7. Environmental and Health Consequences of MPs

Microplastics, even at low concentrations, can have significant environmental and health impacts [52]. They can disrupt food webs by accumulating in marine organisms, posing risks to both aquatic life and humans who consume seafood. Additionally, MPs can act as vectors for harmful substances and diseases, further exacerbating their potential harm. Therefore, there is an urgent need for effective and sustainable removal techniques to protect both ecological and human health. While minimizing microplastic discharge through advanced removal technologies can mitigate these risks, developing efficient and environmentally friendly methods remains a challenge.

### 4.8. Technological Obstacles

Advanced treatment methods, such as MBRs, AOPs, and activated carbon adsorption, are crucial for effective MP removal. However, these technologies present challenges, including high operational costs, maintenance requirements, and the need for infrastructure upgrades. Implementing these advanced techniques on a large scale poses significant economic and logistical challenges for many wastewater treatment facilities.

## Conclusion

The removal of MPs from wastewater remains a significant challenge in maintaining sustainable and safe water supplies. These pollutants, originating from various sources including industry, pharmaceuticals, agriculture, and households, pose serious risks to aquatic ecosystems and human health due to their persistence, bioaccumulation, and potential toxicity at low concentrations. Conventional wastewater treatment methods are often insufficient to completely eliminate MPs, necessitating the development and integration of advanced treatment technologies. While methods such as membrane filtration, adsorption, advanced oxidation processes, and biological treatment have shown promise in improving MP removal, each has limitations in terms of effectiveness, operational costs, scalability, and potential environmental impacts. Challenges like partial removal, generation of harmful byproducts, energy requirements, and high costs hinder the widespread adoption of these technologies. To address the complex nature of wastewater composition and the diversity of MPs, a tailored treatment approach that balances sustainability and efficiency is essential. Future research should focus on developing hybrid treatment systems and refining existing techniques to achieve high removal efficiency with minimal environmental impact. Overcoming these challenges and safeguarding water quality for future generations requires interdisciplinary collaboration, supportive policies, and technological innovation. This chapter underscores the urgent need for continued scientific and technological advancements to improve wastewater treatment and address the issue of MP removal in line with international environmental and public health goals.

## References


[1] Ghosh, S., Sinha, J. K., Ghosh, S., Vashisth, K., Han, S., & Bhaskar, R. (2023). Microplastics as an emerging threat to the global environment and human health. *Sustainability*, *15*(14), 10821 https://doi.org/10.3390/su151410821.

[2] Becker, A. M., Gerstmann, S. and Frank, H. (2008) Perfluoroctane surfactants in wastewaters, the major source of river pollution. Chemosphere 72, 115–121 https://doi.org/10.1016/j.chemosphere.2008.01.009.

[3] Thacharodi, A., Hassan, S., Meenatchi, R., Bhat, M. A., Hussain, N., Arockiaraj, J., ... & Pugazhendhi, A. (2024). Mitigating microplastic pollution: A critical review on the effects, remediation, and utilization strategies of microplastics. *Journal of Environmental Management, 351*, 119988 https://doi.org/10.1016/j.jenvman.2023.119988.

[4] Marwaha, L., Saini, M. K., Singh, D. J., & Kapahi, N. First Report on the Structural and Functional Aspects of Androconial Gland in the Greater Wax Moth (*Galleria Mellonella Linnaeus*) and the Lesser Wax Moth (*Achoria Grisella Fabricus*). *Available at SSRN 4720970.*

[5] Suarez, S., Lema, J. M., & Omil, F. (2010). Removal of pharmaceutical and PCPs (PPCPs) under nitrifying and denitrifying conditions. *Water research, 44*(10), 3214-3224 https://doi.org/10.1016/j.watres.2010.02.040.

[6] Jiang, J. Q., Zhou, Z., & Sharma, V. K. (2013). Occurrence, transportation, monitoring and treatment of emerging micro-pollutants in wastewater—A review from global views. *Microchemical Journal, 110*, 292-300 https://doi.org/10.1016/j.microc.2013.04.014.

[7] Tran, H. T., Lesage, G., Lin, C., Nguyen, T. B., Bui, X. T., Nguyen, M. K., ... & Nguyen, D. D. (2022). Activated sludge processes and recent advances. *Current developments in biotechnology and bioengineering*, 49-79 https://doi.org/10.1016/B978-0-323-99874-1.00021-X.

[8] Monroy, O., Famá, G., Meraz, M., Montoya, L., & Macarie, H. (2000). Anaerobic digestion for wastewater treatment in Mexico: state of the technology. *Water Research, 34*(6), 1803-1816 https://doi.org/10.1016/S0043-1354(99)00301-2.

[9] Muñoz, R., Daugulis, A. J., Hernández, M., & Quijano, G. (2012). Recent advances in two-phase partitioning bioreactors for the treatment of volatile organic compounds. *Biotechnology Advances, 30*(6), 1707-1720 https://doi.org/10.1016/j.biotechadv.2012.08.009.

[10] Kapahi, N., & Marwaha, L. (2022). Polyethylene degradation by larvae of wax moth. *Journal of Entomological Research, 46*(3), 620-625 https://doi.org/10.5958/0974-4576.2022.00107.4.

[11] Moosavi, S., Lai, C. W., Akbarzadeh, O., & Johan, M. R. (2021). Recycled activated carbon-based materials for the removal of organic pollutants from wastewater. *Waste Recycling Technologies for Nanomaterials Manufacturing*, 513-539 https://doi.org/10.1007/978-3-030-68031-2_18.

[12] Ahmed, M. B., Zhou, J. L., Ngo, H. H., Guo, W., Thomaidis, N. S., & Xu, J. (2017). Progress in the biological and chemical treatment technologies for emerging contaminant removal from wastewater: a critical review. *Journal of hazardous materials, 323*, 274-298 https://doi.org/10.1016/j.jhazmat.2016.04.045.

[13] Lau, C., Butenhoff, J. L., & Rogers, J. M. (2004). The developmental toxicity of perfluoroalkyl acids and their derivatives. *Toxicology and applied pharmacology, 198*(2), 231-241 https://doi.org/10.1016/j.taap.2003.11.031.

[14] Chen, J., Liu, Y. S., Deng, W. J., & Ying, G. G. (2019). Removal of steroid hormones and biocides from rural wastewater by an integrated constructed wetland. *Science of the Total Environment, 660*, 358-365 https://doi.org/10.1016/j.scitotenv.2019.01.049.

[15] Khasawneh, O. F. S., & Palaniandy, P. (2021). Occurrence and removal of pharmaceuticals in wastewater treatment plants. *Process Safety and Environmental Protection, 150*, 532-556 https://doi.org/10.1016/j.psep.2021.04.045.

[16] Freeling, F., Alygizakis, N. A., von der Ohe, P. C., Slobodnik, J., Oswald, P., Aalizadeh, R., ... & Scheurer, M. (2019). Occurrence and potential environmental risk of surfactants and their transformation products discharged by wastewater treatment plants. *Science of the Total Environment, 681*, 475-487 https://doi.org/10.1016/j.scitotenv.2019.04.445.

[17] Ilieva, Z., Hamza, R. A., & Suehring, R. (2024). The significance of fluorinated compound chain length, treatment technology, and influent composition on per-and polyfluoroalkyl substances removal in worldwide wastewater treatment plants. *Integrated Environmental Assessment and Management, 20*(1), 59-69 https://doi.org/10.1002/ieam.4778 .

[18] Ahrens, L., Felizeter, S., Sturm, R., Xie, Z., & Ebinghaus, R. (2009). Polyfluorinated compounds in wastewater treatment plant effluents and surface waters along the River Elbe, Germany. *Marine pollution bulletin, 58*(9), 1326-1333 https://doi.org/10.1016/j.marpolbul.2009.04.028.

[19] Kulkarni, D., & Jaspal, D. (2023). Surfactants in wastewater: Development, current status and associated challenges. *Materials Today: Proceedings* https://doi.org/10.1016/j.matpr.2023.11.022.

[20] Lakshmi, K. S., Sailaja, V. H., & Reddy, M. A. (2017). Phytoremediation-a promising technique in wastewater treatment. *International Journal of Scientific Research and Management (IJSRM)*, *5*(06), 5480-5489 https://doi.org/10.18535/ijsrm/v5i6.20.

[21] Casas, M. E., & Bester, K. (2015). Can those organic micro-pollutants that are recalcitrant in activated sludge treatment be removed from wastewater by biofilm reactors (slow sand filters)?. *Science of the Total Environment*, *506*, 315-322 https://doi.org/10.1016/j.scitotenv.2014.10.113.

[22] Kanaujiya, D. K., Paul, T., Sinharoy, A., & Pakshirajan, K. (2019). Biological treatment processes for the removal of organic MPs from wastewater: a review. *Current pollution reports*, *5*, 112-128 https://doi.org/10.1007/s40726-019-00110-x.

[23] Benner, J., Helbling, D. E., Kohler, H. P. E., Wittebol, J., Kaiser, E., Prasse, C., ... & Boon, N. (2013). Is biological treatment a viable alternative for MP removal in drinking water treatment processes?. *Water research*, *47*(16), 5955-5976 https://doi.org/10.1016/j.watres.2013.07.015.

[24] Ahansazan, B., Afrashteh, H., Ahansazan, N., & Ahansazan, Z. (2014). Activated sludge process overview. *International Journal of Environmental Science and Development*, *5*(1), 81 https://doi.org/10.7763/IJESD.2014.V5.455.

[25] Zhang, X., Chen, J., & Li, J. (2020). The removal of microplastics in the wastewater treatment process and their potential impact on anaerobic digestion due to pollutants association. *Chemosphere*, 251,126360. https://doi.org/10.1016/j.chemosphere.2020.126360.

[26] Lau, C., Anitole, K., Hodes, C., Lai, D., Pfahles-Hutchens, A., & Seed, J. (2007). Perfluoroalkyl acids: a review of monitoring and toxicological findings. *Toxicological sciences*, *99*(2), 366-394 https://doi.org/10.1093/toxsci/kfm128.

[27] Al-Asheh, S., Bagheri, M., & Aidan, A. (2021). Membrane bioreactor for wastewater treatment: A review. *Case Studies in Chemical and Environmental Engineering*, *4*, 100109 https://doi.org/10.1016/j.cscee.2021.100109.

[28] Liu, Z. H., Kanjo, Y., & Mizutani, S. (2009). Urinary excretion rates of natural estrogens and androgens from humans, and their occurrence and fate in the environment: a review. *Science of the Total Environment*, *407*(18), 4975-4985 https://doi.org/10.1016/j.scitotenv.2009.06.001.

[29] Quijano, G., Hernandez, M., Thalasso, F., Muñoz, R., & Villaverde, S. (2009). Two-phase partitioning bioreactors in environmental biotechnology. *Applied microbiology and biotechnology*, *84*, 829-846 https://doi.org/10.1007/s00253-009-2158-6.

[30] Melin, T., Jefferson, B., Bixio, D., Thoeye, C., De Wilde, W., De Koning, J., ... & Wintgens, T. (2006). Membrane bioreactor technology for wastewater treatment and reuse. *Desalination*, *187*(1-3), 271-282 https://doi.org/10.1016/j.desal.2005.04.086.

[31] Rashid, R., Shafiq, I., Akhter, P., Iqbal, M. J., & Hussain, M. (2021). A state-of-the-art review on wastewater treatment techniques: the effectiveness of adsorption method. *Environmental Science and Pollution Research*, *28*, 9050-9066 https://doi.org/10.1007/s11356-021-12395-x.

[32] Schwarzenbach, R. P., Escher, B. I., Fenner, K., Hofstetter, T. B., Johnson, C. A., Von Gunten, U., & Wehrli, B. (2006). The challenge of MPs in aquatic systems. *Science*, *313*(5790), 1072-1077 https://doi.org/10.1126/science.1127291.

[33] Sathasivam, M., Shanmugapriya, S., Yogeshwaran, V., & Priya, A. K. (2019). Industrial wastewater treatment using advanced oxidation process–A review. *Int. J. Eng. Adv. Technol*, *8*(3), 485-488.

[34] Yuliasni, R., Kurniawan, S. B., Marlena, B., Hidayat, M. R., Kadier, A., & Ma, P. C. (2023). Recent progress of phytoremediation-based technologies for industrial wastewater treatment. *Phycoremediation Processes in Industrial Wastewater Treatment*, 21-41 https://doi.org/10.12911/22998993/156621.

[35] Pistocchi, A., Andersen, H. R., Bertanza, G., Brander, A., Choubert, J. M., Cimbritz, M., ... & Thornberg, D. (2022). Treatment of MPs in wastewater: Balancing effectiveness, costs and implications. *Science of The Total Environment*, *850*, 157593.99 https://doi.org/10.1016/j.scitotenv.2022.157593.

[36] Kim, S. K., Im, J. K., Kang, Y. M., Jung, S. Y., Kho, Y. L., & Zoh, K. D. (2012). Wastewater treatment plants (WWTPs)-derived national discharge loads of perfluorinated compounds (PFCs). *Journal of hazardous materials*, *201*, 82-91 https://doi.org/10.1016/j.jhazmat.2011.11.036.

[37] Vethaak, A. D., & Legler, J. (2021). Microplastics and human health. *Science*, *371*(6530), 672-674 https://doi.org/10.1126/science.abe5041.

[38] Shah, A., & Shah, M. (2020). Characterisation and bioremediation of wastewater: a review exploring bioremediation as a sustainable technique for pharmaceutical wastewater. *Groundwater for Sustainable Development*, *11*, 100383 https://doi.org/10.1016/j.gsd.2020.100383.

[39] Maloney, A. J., Dong, C., Campbell, A. S., & Dinu, C. Z. (2015). Emerging enzyme-based technologies for wastewater treatment. *Green Polymer Chemistry: Biobased Materials and Biocatalysis*, 69-85 https://doi.org/10.1021/bk-2015-1192.ch005.

[40] Virkutyte, J., & Varma, R. (2010). *Treatment of MPs in water and wastewater*. IWA Publishing https://doi.org/10.2166/9781780401447.

[41] Nassiri Koopaei, N., & Abdollahi, M. (2017). Health risks associated with the pharmaceuticals in wastewater. *DARU Journal of Pharmaceutical sciences*, *25*(1), 9 https://doi.org/10.1186/s40199-017-0176-y.

[42] Ying, G. G. (2006) Fate, behavior and effects of surfactants and their degradation products in the environment. Environ. Int. 32, 417–431 https://doi.org/10.1016/j.envint.2005.07.004.

[43] Deng, Y., & Zhao, R. (2015). Advanced oxidation processes (AOPs) in wastewater treatment. *Current Pollution Reports*, *1*, 167-176 https://doi.org/10.1007/s40726-015-0015-z.

[44] Asghar, A., Raman, A. A. A., & Daud, W. M. A. W. (2015). Advanced oxidation processes for in-situ production of hydrogen peroxide/hydroxyl radical for textile wastewater treatment: a review. *Journal of cleaner production*, *87*, 826-838 https://doi.org/10.1016/j.jclepro.2014.09.010.

[45] Saidulu, D., Majumder, A., & Gupta, A. K. (2021). A systematic review of moving bed biofilm reactor, membrane bioreactor, and moving bed membrane bioreactor for wastewater treatment: Comparison of research trends, removal mechanisms, and performance. *Journal of Environmental Chemical Engineering*, *9*(5), 106112 https://doi.org/10.1016/j.jece.2021.106112.

[46] Saeed, M. U., Hussain, N., Sumrin, A., Shahbaz, A., Noor, S., Bilal, M., ... & Iqbal, H. M. (2022). Microbial bioremediation strategies with wastewater treatment potentialities–A review. *Science of the total environment*, *818*, 151754 https://doi.org/10.1016/j.scitotenv.2021.151754.

[47] Bratby, J. (2016). *Coagulation and flocculation in water and wastewater treatment*. IWA publishing https://doi.org/10.2166/9781780407500.

[48] Hube, S., Eskafi, M., Hrafnkelsdóttir, K. F., Bjarnadóttir, B., Bjarnadóttir, M. Á., Axelsdóttir, S., & Wu, B. (2020). Direct membrane filtration for wastewater treatment and resource recovery: A review. *Science of the total environment*, *710*, 136375 https://doi.org/10.1016/j.scitotenv.2019.136375.

[49] Abdel-Fatah, M. A. (2018). Nanofiltration systems and applications in wastewater treatment. *Ain Shams Engineering Journal*, *9*(4), 3077-3092 https://doi.org/10.1016/j.asej.2018.08.001.

[50] Colla, V., Branca, T. A., Rosito, F., Lucca, C., Vivas, B. P., & Delmiro, V. M. (2016). Sustainable reverse osmosis application for wastewater treatment in the steel industry. *Journal of cleaner production*, *130*, 103-115 https://doi.org/10.1016/j.jclepro.2015.09.025.

[51] Sarma, S. J., & Pakshirajan, K. (2011). Surfactant aided biodegradation of pyrene using immobilized cells of Mycobacterium frederiksbergense. *International Biodeterioration & Biodegradation*, *65*(1), 73-77 https://doi.org/10.1016/j.ibiod.2010.09.004.

[52] Kapahi, N., Marwaha, L., Kaur, M., & Bawa, G. (2023). Comparison study on low-density polyethylene film biodegradation potential of *Achoria grisella* and *Galleria mellonella* larvae. *Journal of Environmental Biology*, *44*(5), 682-690 http://doi.org/10.22438/jeb/44/5/MRN-5126.

# Chapter 8

# Emerging Technologies: A Revolution in Wastewater Treatment

**Unik Arora[1,*]**
**Ashish Tomar[1]**
**and Harveer Singh Cheema[2]**
[1]Department of Chemistry, Meerut College, Meerut (U.P.), India
[2]Department of Botany, Meerut College, Meerut (U.P.), India

## Abstract

Rapid industrialization and population growth have significantly increased the discharge of industrial effluents, making wastewater treatment a critical aspect of water resource management. Urbanization and agricultural expansion will further exacerbate water pollution, emphasizing the need for advanced treatment methods to mitigate environmental and health impacts. Traditional approaches such as chemical precipitation, ion exchange, and thermal treatment, often fall short due to inefficiencies, high-energy demands, and the generation of toxic-by products. Although hybrid methods combining chemical and biological processes show promise, the persistent challenge of secondary sludge production remains an unresolved issue. Recent advancements in nanotechnology offer transformative solutions in wastewater treatment, leveraging materials such as carbon nanotubes, graphene derivatives, and quantum dots. These nanomaterial (NMs) excel in removing heavy metals, dyes, and pharmaceuticals, providing enhanced adsorption, catalysis, and antimicrobial properties. Coupled with membrane-based technologies like nanofiltration and advanced polymer composites, these approaches offer sustainable, efficient, and scalable solutions for addressing global water challenges.

**Keywords:** wastewater treatment, nanotechnology, advanced nanomaterials, membrane filtration, water purification

* Corresponding Author's Email: unikmcm@gmail.com

In: Advancements in Wastewater Treatment
Editors: Neha Agarwal, Maulin P. Shahl, Vijendra Singh Solankil et al.
ISBN: 979-8-90134-007-3

## 1. Introduction

Water is a fundamental resource for sustaining life on earth, covering approximately 70% of the planet's surface, leading to its designation as the blue planet [1]. However, the majority of this vast resource is saline, with only a small proportion categorized as fresh water. Within this limited freshwater supply, a significant portion is sequestered in ice and glaciers, while a smaller share exists as groundwater, and a mere fraction is readily accessible for direct human use [2]. The chemical properties of water, particularly its function as a universal solvent, are integral to numerous biological processes, making it indispensable to all forms of life [3].

The exponential growth of population, urbanization, industrialization, and agricultural expansion has collectively contributed to the alarming degradation of water resources, thereby creating a critical global challenge [1, 4]. Currently, over one billion people lack access to safe drinking water, while approximately 2.6 billion struggle to obtain basic sanitation. The World Health Organization (WHO) has highlighted that water contamination is associated with substantial health risks and economic ramifications, with millions of deaths attributed to waterborne diseases annually [5]. Human activities lead to pollution that introduces toxic substances such as heavy metals, pesticides, and industrial waste into aquatic ecosystems. These contaminants originate from diverse sources and severely threaten freshwater sources and aquifers, comprising the safety of drinking water (Figure 1). It is estimated that nearly 80% of untreated municipal wastewater is directly discharged into the environment, further exacerbating the deterioration of water quality [6]. Additionally, the impact of climate change, irregular freshwater supplies, and increasing water demand underscores the urgent need to implement advanced treatment methodologies and explore alternative water sources [7].

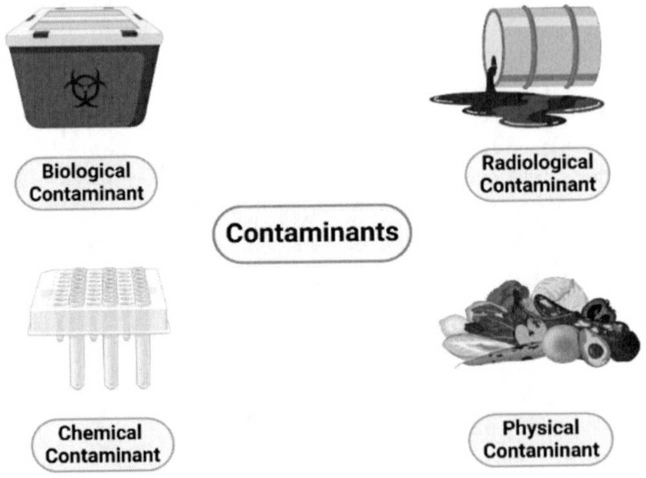


**Figure 1.** Different types of contaminants.

Various physical, chemical, and biological processes are commonly used in wastewater treatment. These methods include coagulation, flocculation, adsorption, ion exchange, and membrane filtration. While these traditional techniques can effectively reduce certain pollutants, they often face challenges in completely removing toxins, nitrogen, phosphorous, and heavy metals. Additionally, these methods can be energy-intensive and costly, and they may require chemical additives, leading to operational difficulties and environmental concerns [1,2,6]. Table 1 summarizes the conventional treatment methods and their limitations.

**Table 1.** Overview of the conventional treatment methods and their limitations [2, 6]

| Treatment method | Purpose | Limitations |
|---|---|---|
| Screening | Physical removal of debris | Ineffective for smaller particles |
| Flocculation/Coagulation | Promotes aggregation of particles for easier separation | Requires additives, pH adjustments, less accurate |
| Sedimentation | Allows suspended particles to settle | Limited by flow rate and particle density |
| Filtration | Removes sediments, algae, microorganisms, reduces turbidity | Filter clogging, ineffective for dissolved inorganics |
| Adsorption | Eliminates organic/inorganic contaminants, odors, and color | Filter replacement required, clogging, susceptible to mold |
| Disinfection | Kills or inactivates pathogens | High cost, by-product formation |
| Distillation | Separates contaminants by boiling | Ineffective for high-boiling contaminants, energy-intensive |
| Reverse osmosis | Removes dissolved solids and organics | Energy-intensive, does not remove volatile organics, acidic output water |
| Carbon filtration | Removes certain chemicals and organics | Frequent filter change needed, prone to clogging |
| Ultrafiltration | Filters suspended solids and macromolecules | Susceptible to particulate plugging, inefficient for dissolved inorganics |
| Microfiltration | Filters larger particles and microorganisms | Ineffective for viruses, regular cleaning required |
| Nanofiltration | Partially removes dissolved solids and organic molecules | Membrane fouling, high energy demand, requires pretreatment |
| Chemical precipitation | Removes heavy metals and phosphorus | Excess reagent use, disposal issues for resulting sludge |
| Biological treatment | Decomposes organic waste using microorganism | Time-consuming, costly, difficult microorganism control |

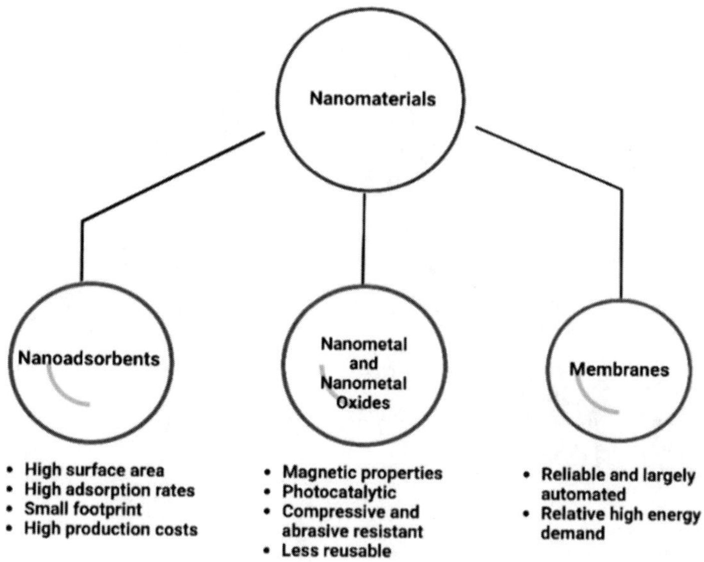


**Figure 2.** Types of NMs and their properties (adapted from [13]).

Nanotechnology represents a transformative approach to wastewater treatment, effectively addressing the limitations inherent in traditional methods by manipulating materials at the atomic level [8]. This innovative technique imparts distinctive properties to NMs, including an extensive surface area, enhanced reactivity, and specialized functionalization, rendering them particularly effective in removing a diverse array of contaminants [9]. NMs are typically defined as materials with at least one dimension measuring less than 100 nanometers [2,10]. NMs possess significant advantages, as they can adapt their structural functionalities to improve adsorption, catalysis, and the degradation of pollutants with minimal energy input [11]. Figure 2 illustrates the various types of NMs and highlights their properties that enhance wastewater treatment. This capability effectively removes organic compounds, biological toxins, and pathogens, leading to a cleaner and more sustainable water supply [12]. Consequently, nanotechnology emerges as a potent instrument for managing water resources in global scarcity. This chapter highlights the application of nanotechnology in wastewater treatment, focusing on recent advancements, key methods, and the potential environmental and safety implications of using NMs.

## 2. Overview of Wastewater Sources and Treatment Processes

Wastewater is generated from various sources, including domestic, agricultural, industrial, and commercial activities, each contributing distinct contaminants [1]. Domestic wastewater commonly contains organic materials, suspended solids, and pathogens, while industrial and agricultural discharges introduce a range of chemicals, heavy metals, pesticides, and nutrient-rich compounds [14]. The untreated release of wastewater poses significant risks to ecosystems, human health, and overall water quality, thereby underscoring the necessity of effective wastewater management [15].

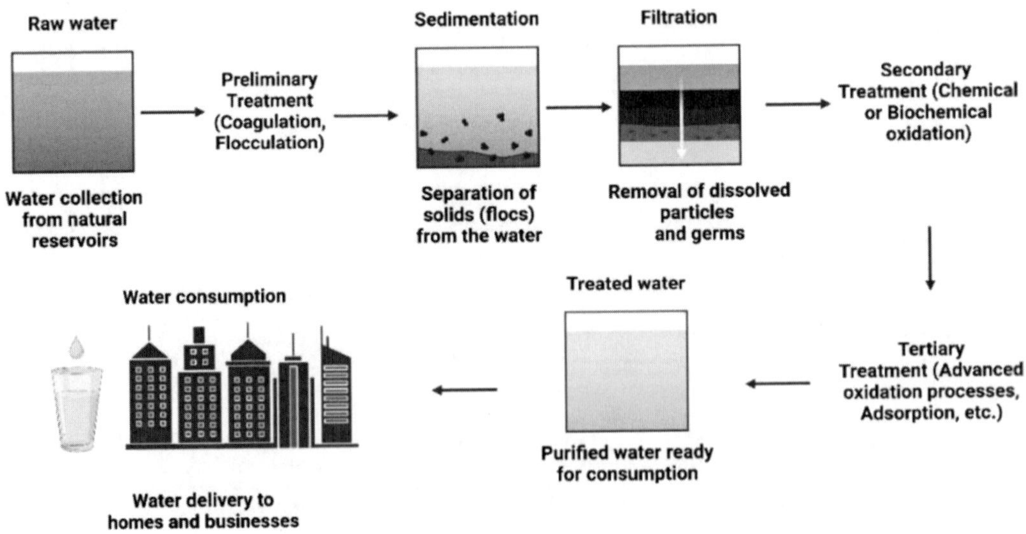


**Figure 3.** General steps for water treatment.

The wastewater treatment process encompasses four critical stages: preliminary, primary, secondary, and tertiary (Figure 3). The preliminary treatment stage focuses on removing large debris by employing screening techniques [1,14]. Following this, primary treatment involves sedimentation that allows suspended solids to settle, forming sludge that can be subsequently removed. This phase typically eliminates approximately 40 percent of the biological oxygen demand (BOD) and substantially reduces suspended solids [16]. In the secondary treatment stage, biological methods decompose organic pollutants further, such as activated sludge processes or bio-filters. This stage relies on oxidation and microbial activity, where bacteria and protozoa metabolize organic matter while diminishing nutrient concentrations, particularly nitrogen and phosphorous, which can contribute to eutrophication, if left unaddressed [17]. Tertiary treatment then targets the removal of any remaining contaminants. It incorporates disinfection processes, such as chlorination or ultraviolet (UV) radiation, to ensure that the treated water meets safety standards for discharge or reuse [14, 16].

## 3. NMS For Water Treatment

**Table 2.** Various water treatment processes utilizing NMs [6]

| Technique | Contaminants | Mechanism | Advantages | Drawbacks |
|---|---|---|---|---|
| Photocatalysis | Organic pollutants (e.g., pesticides, dyes) | Oxidative degradation using light | Complete pollutant breakdown, minimal secondary pollution | Slow reaction rate, UV dependency |
| Adsorption | Organic/inorganic pollutants, microbes | Binding contaminants on material surface | Cost-effective, reversible, simple operation | Limited selectivity, waste from spent adsorbents |
| Nanofiltration Membranes | Organic/inorganic pollutants, microbes | Filtration through nanoporous membranes | High efficiency, reduced chemical use, minimum solid waste | Membrane fouling, high operational cost |
| Antimicrobial applications | Pathogens and microbial contaminants | Direct cell disruption | Low-cost alternative to conventional disinfectants | Potential nanotoxicity, post-treatment NPs removal challenge |
| Catalytic degradation | Organic pollutants and pharmaceutical residues | Catalytic reactions (redox) | Effective degradation of non-biodegradable contaminants | High operation costs, catalyst recovery issues |
| Encapsulation (Dendrimers) | Heavy metals, organic pollutants | Encapsulation of contaminants | High pollutant binding capacity, renewable, minimum sludge | Expensive materials |
| Nanoclay adsorption | Heavy metals, anions, organic pollutants | Adsorption on nanostructured clay | Cost-effective, stable, recyclable, high sorption capacity | Sludge production |

NMs exhibit unique properties that render them highly effective for environmental applications, particularly in water pollution remediation. Their elevated surface area-to-volume ratio provides many active sites for adsorption and chemical reactions, facilitating the efficient removal of contaminants from aqueous environments [18]. Additionally, the chemical reactivity and stability of these materials can be significantly enhanced through functionalization with specific chemical groups, thus tailoring their affinity for targeted pollutants and supporting sustained applications over time [19]. NMs, such as titanium dioxide ($TiO_2$) and bimetallic nanoparticles (NPs), demonstrate photocatalytic properties that drive redox reactions, enabling the degradation of non-biodegradable pollutants when exposed to light [20,21]. In the context of water disinfection, the antimicrobial properties of NMs, including silver and chitosan NPs, disrupt microbial cell membranes, thus providing a broad-spectrum solution that avoids the generation of harmful byproducts [22]. Furthermore, specific NMs have been engineered for ease of recovery and reusability; for instance, magnetic NPs and dendrimers facilitate straightforward separation, minimizing waste production and reducing operational costs [23]. The environmental safety of engineered NMs is a critical consideration, with efforts to minimize toxic byproducts to ensure their suitability for environmental application and the safe disposal of treated sludge [24]. Table 2 outlines NM treatment techniques, targeted contaminants, mechanisms, advantages, and drawbacks in wastewater treatment. The interplay of these advantages, alongside the specific properties of NMs, positions them as a versatile and potent solution for wastewater treatment.

### 3.1. Carbon Based NMs

Traditional adsorbents, including activated carbon and various polymeric materials, are widely recognized for their user friendliness; however, they often demonstrate inefficiency in comprehensively removing biological and chemical contaminants from water [25,26]. These conventional materials frequently exhibit low adsorption capacities for heavy metals and are prone to biofouling, clogging, and potential bacteria leakage [25].

Nanotechnology has enabled the development of both metallic and carbon-based NMs characterized by high surface areas, tunable properties, and superior capabilities for contaminant removal. This section explores key carbon-based NMs, highlighting their potential as effective agents for advanced water purification processes.

#### *3.1.1. Carbon Nanotubes (CNTs)*

Carbon nanotubes (CNTs) have shown significant promise as adsorbents due to their high surface area and versatile structure, with single-walled (SWCNTs), double-walled (DWCNTs), and multi-walled (MWCNTs) variants each bringing unique capabilities to pollutant removal, which are shown in Figure 4 [27, 28].

The demand for pharmaceutical products has significantly increased in recent years, primarily due to rapid population growth and a rising prevalence of various diseases [29]. As medical needs continue to grow, the consumption of medications, especially antibiotics, has expanded considerably. However, this increase in pharmaceutical usage also leads to a more significant discharge of residuals into the environment, raising concerns about their impact on ecosystems and human health. A critical area of focus for addressing the issue of pharmaceutical contaminants in wastewater is the development of effective materials for their removal. Ncibi and Sillanpää explores the effectiveness of SWCNTs, DWCNTs, and MWCNTs in removing the antibiotics oxytetracycline and ciprofloxacin from water. SWCNTs showed the highest removal capacities, reaching 724 mg/g for ciprofloxacin and 554 mg/g for oxytertacycline. The Brouers-Sotolongo model best described the adsorption process [30]. Kariim et al. have prepared MWCNTs derived from activated carbon made from wood sawdust. These adsorbents are enhanced with nickel ferrites (Ni-Fe) to target pharmaceutical compounds such as metronidazole and levofloxacin [31].

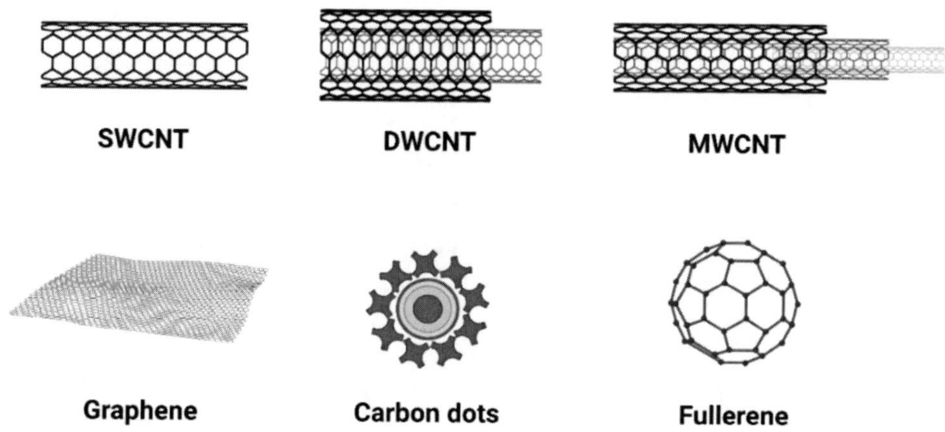


**Figure 4.** Carbon based NMs.

Dyes are a significant class of industrial pollutants widely used in various industries, including textiles, rubber, automotive, cosmetics, pharmaceuticals, printing, and photography to enhance product aesthetics [32]. When dye-laden wastewater enters natural water systems, it poses a treatment challenge due to synthetic dyes complex and stable structures [33]. These dyes are resilient and highly visible even at low concentrations [32]. CNTs have shown promising potential for removing dyes in wastewater treatment. Madrakian et al. developed magnetic-modified MWCNT to remove cationic dyes from water. Optimal dye removal occurred at pH 7.0, with high adsorption capacities observed for various dyes, as predicted by the Langmuir model [34]. The CNTs walls are generally unreactive, but their end tips are much more reactive. Functionalizing these ends has been shown to effectively enhance the reactivity of CNTs, allowing for adding various functional groups, including carboxylic, ester, alcohol, and ketone groups. This modification increases the chemical interaction potential of CNTs with target molecules, thereby expanding their applications in adsorption and catalysis [35,36]. Saber-Samandari et al. developed a magnetic adsorbent by incorporating carboxylic acid-functionalized MWCNT with iron oxide magnetic NPs based on gelatin. This magnetic biosorbent effectively removed both anionic and cationic dyes from water, achieving high removal efficiencies. The removal process followed a second-order kinetic model [37].

CNT-based adsorbents are widely used to remove heavy metals from water. Modifying these materials to enhance their functional properties can transform them into multifunctional carbon-based nanocomposites, offering improved environmental compatibility [38]. Alijani and Shariatinia successfully developed zero-valent iron-doped MWCNTs using chemical vapor deposition on natural hematite. The results showed that at an initial concentration of 10 mg/l, hematite alone removed 18% of arsenite and 26% of arsenate. In contrast, the oxidized nanohybrid enhanced the removal rates to 74% for arsenite and 77% for arsenate. Additionally, the raw nanohybrid exhibited the highest efficiency, removing 96% of arsenite and 98.5% of arsenate [39]. Ghasemi et al. synthesized an amino-functionalized silica coated MWCNTs for the removal of Pb(II) and Cd(II) from water. The adsorption efficiencies of amino-functionalized silica coated MWCNTs for Pb(II) and Cd(II) were in the range of 98-104% under the optimum conditions [40]. Huang et al. enhanced the adsorption of Cr(VI) in acidic solutions by modifying carboxylated MWCNTs with chitosan. This modification significantly improved the adsorption performance across varying temperatures. The modified nanocomposite also demonstrated excellent reusability across multiple Cr(VI) adsorption cycles with minimal loss in performance, effectively increasing adsorption capacity and reducing the time needed for Cr(VI) removal [41].

Despite the numerous benefits of CNTs, their application is predominantly restricted to industrial contexts due to their elevated costs [42]. Only limited amounts can be feasibly utilized for direct applications without jeopardizing economic viability. Consequently, further investigation is required to establish dependable and economical techniques for employing CNTs in water purification.

### *3.1.2. Graphene and its Derivatives*

Graphene is a two-dimensional material comprising a single layer of carbon atoms arranged in a hexagonal lattice [43]. Since its isolation as a material in 2004, graphene has emerged as a crucial material in both environmental and industrial applications due to its distinct structural and conductive properties. One notable area of use is in wastewater treatment, where derivatives of graphene, such as graphene oxides, quantum dots, nanoplatelets, and nanocomposites, play a significant role as effective components. Graphene based materials possess large surface areas and excellent conductivity, making them suitable for electrochemical treatment, photocatalysis, and adsorption [44]. These characteristics facilitate the efficient removal of contaminants from wastewater. Numerous methodologies exist for synthesizing graphene and graphene-based materials, encompassing mechanical exfoliation, electrochemical exfoliation, chemical vapor deposition, modified Hummers method, and epitaxial growth (Figure 5). In contemporary research, there is a notable emphasis on developing graphene-based nanocomposites, which incorporate carbon nanotubes and nanoclays [45].

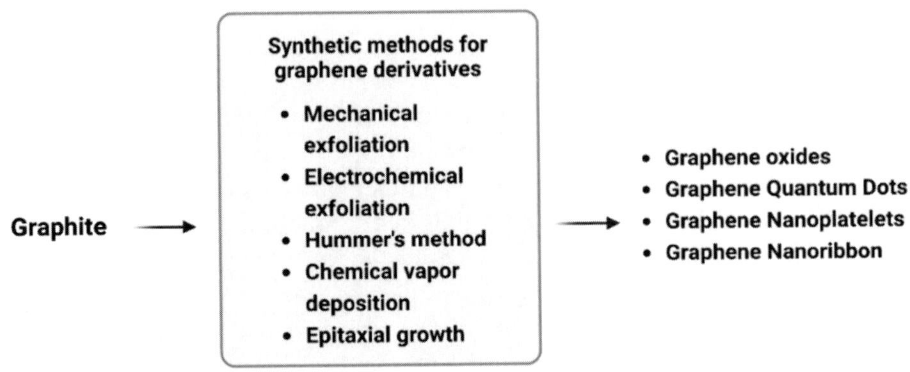


**Figure 5.** Different synthetic methods for graphene derivatives.

#### *3.1.2.1. Graphene Oxide*

Graphene oxide is a monolayer of graphite that incorporates various oxygen-containing functional groups, including hydroxyl, carbonyl, carboxyl, and epoxide groups [1]. Recently, Graphene oxide has garnered considerable interest within the scientific community, particularly when combined with magnetic particles, for its efficacy as an adsorbent in wastewater treatment applications [46]. Dubey et al. synthesized multi-layer graphene oxide and a graphene-based magnetic nanocomposite through a microwave-assisted modification of the Hummers method, aiming to facilitate the removal of toxic Pb(II) and Cd(II) from aqueous solutions. The adsorption experiments revealed a significant Pb(II) affinity for the graphene oxide sheets, thereby underscoring their efficacy in lead removal. Moreover, the graphene oxide-based magnetic nanocomposite demonstrated an enhanced Cd(II) adsorption capacity [47]. Kumar et al. synthesized hybrid materials by combining single-layer graphene oxide with manganese ferrite magnetic NPs to improve the removal of heavy metal ions from contaminated water. The experimental data on adsorption capacity exceeded previously reported values [48]. Hassan et al. investigated various graphene-based materials for the photodegradation of Bisphenol A, revealing that reduced graphene oxide supported on silica-zirconia exhibited the highest degradation efficiency compared to standalone graphene oxide, graphene, and silica-zirconia. Furthermore, the stability of reduced graphene oxide on silica-zirconia over multiple cycles highlights its potential as an effective and reusable method for removing Bisphenol A [49].

In addition to heavy metal removal, graphene oxide-based materials have also shown excellent potential for dye removal from wastewater. Bhattacharyya et al. developed a cross-linked bio-composite made from graphene oxide, potato starch, and glutaraldehyde, which was applied to adsorb the industrial dye methylene blue from water. This bio-composite displayed distinct auto-phase separability, allowing it to precipitate without mechanical assistance and enhancing its practicality as an adsorbent. Additionally, successful desorption was achieved over three cycles, confirming the reusability and economic viability of the material for wastewater treatment applications [50]. Rout and Jena employed reduced graphene oxide as an adsorbent for removing malachite green dye from water. The batch adsorption experiments conducted revealed a high removal efficiency of 96.3% and an impressive adsorption capacity of 279.85 mg/g. The Langmuir isotherm model best described the adsorption process, while kinetic analysis showed that the adsorption followed a pseudo-second-order model [51].

#### *3.1.2.2. Graphene Quantum Dots (GQDs)*

Graphene quantum dots, characterized by their nanoscale dimension (typically <30 nm) and few-layer graphene composition, exhibit exceptional physical and chemical stability due to their robust carbon framework [52]. Their crystalline structure, linked by edge chemical groups and their circular or elliptical morphology, facilitates a substantial surface area and efficient surface grafting due to $\pi$-$\pi$ conjugated network. Graphene quantum dots exhibit anisotropic properties due to their larger lateral dimensions compared to height and possess distinct size-dependent quantum behaviors linked to their $sp^2$-$sp^2$ carbon bonding [53]. Additionally, their non-toxic and environmentally friendly nature positions graphene quantum dots as potential candidates for diverse nanotechnology and material science applications.

Nagaraj et al. developed a novel ionic liquid-capped graphene quantum dot as an efficient adsorbent for removing Cr(VI) from water. The adsorbent demonstrated an adsorption capacity of 934.62 mg/g within a short period at pH 7. The study also indicated that lower pH levels further enhanced its adsorption efficiency. Additionally, the adsorption data aligned well with the Langmuir isotherm and pseudo-second-order kinetic models [54]. Sarkar et al. conducted an extensive study to develop and optimize a photocatalytic system incorporating liquid-exfoliated graphene quantum dots with titanium dioxide nanotubes, nanorods, and NPs. The findings indicate that the optimal photocatalyst formulation, comprising two wt.% graphene dots, exhibited a degradation efficiency of 96.7% for methylene blue under visible light irradiation over 120 minutes [55].

### *3.1.2.3. Graphene Nanoplatelets (GNPs)*

Graphene nanoplatelets (GNPs) are graphene derivatives characterized by their composition of stacked individual layers, with thicknesses varying from 0.34 nm to 100 nm. Their structural configuration positions them between single-layer graphene and multi-layer graphite in dimensionality [56]. GNPs exhibit an exceptional combination of electrical, thermal, and mechanical properties, often comparable to CNTs. Moreover, they present a cost-effective alternative to CNTs and nanofibers. Their lightweight nature, high aspect ratio, superior electrical and thermal conductivity, and mechanical strength render graphene nanoplatelets effective nanostructured fillers in material science [43]. When incorporated into polymer matrices through techniques such as solvent or melt compounding, GNPs can significantly enhance the properties of the resulting composites [57]. Recently, Zembet et al. developed composite nanofibers made of polyamide-6 embedded with GNPs using electrospinning techniques. The study showed that incorporating 0.5 wt.% GNPs significantly enhanced the mechanical properties of the nanofibers. Additionally, higher concentrations of GNPs resulted in reduced turbidity [58]. The electrical conductivity of GNPs, which exceeds that of graphene oxide, and their ease of dispersion in polymer systems make them versatile for applications across fields such as wastewater treatment and other industrial sectors [59].

In the field of environmental applications, GNPs have been widely investigated for their efficacy in antibiotic and heavy metal removal from aqueous systems. Al-Khateeb et al. employed GNPs to effectively remove aspirin, acetaminophen, and caffeine from aqueous solutions, with the adsorption process aligning closely with the pseudo-second-order kinetic model [60]. Similarly, Rosli et al. investigated the efficacy of GNPs as an adsorbent for removing antibiotic sulfamethoxazole and analgesic acetaminophen. GNPs achieved a maximum removal efficiency of 99% for both contaminants through an optimized batch adsorption process. The pseudo-second-order kinetic model best represented the adsorption kinetics, while the equilibrium data conformed to the Langmuir isotherm model [61].

For heavy metal removal, González-López et al. developed cellulose beads loaded with graphene nanoplatelets (cellulose-GNPs) using a physical gelation method. The adsorption equilibrium followed the Langmuir isotherm, showing that Cellulose-GNPs beads achieved significantly higher maximum adsorption capacities, with 98.1 mg/g for Cellulose alone and 139.6 mg/g for Cellulose-GNP beads [62].

### *3.1.2.4. Graphene Nanoribbons (GNRs)*

Graphene nanoribbons (GNRs) are narrow strips derived from graphene nanosheets [63]. These ribbons are composed of repetitive hexagonal carbon units at the nanoscale, with lengths ranging from 0.4 to 0.7 μm and width from 1 to 1000 nm. GNRs are planar, thin, segmented, and exhibit quasi-one-dimensional characteristics. They possess a finite bandgap due to quantum confinement, which enables the combination of graphene's distinct properties [64,65]. Gao et al. developed GNRs via atomic hydrogen-induced treatment of SWCNTs for cationic methylene blue and anionic orange II dyes. The GNRs showed enhanced adsorption capabilities due to their increased active surface area and uniform vertical arrangement, facilitating electrostatic and $\pi$–$\pi$ interactions with the dyes. The adsorption kinetics followed a pseudo-second-order model, while the adsorption process adhered to the Langmuir isotherm. The maximum adsorption capacities of GNRs for methylene blue and orange II dyes are reported as 280 mg/g and 265 mg/g [66]. Sadeghi et al. developed GONRs using MWCNTs and investigated their effectiveness in removing As(V) and Hg(II) ions from aqueous solutions through ultrasonic-assisted adsorption. The study demonstrated that GONRs exhibited a greater adsorption affinity for As(V) ions than Hg(II) ions. Notably, when the initial concentrations of As(V) and Hg(II) ions were increased to 300 ppm, the GONRs achieved equilibrium adsorption capacities of 155.61 mg/g for As(V) and 33.02 mg/g for Hg(II) ions, with a rapid separation time of only 12 minutes [67].

### *3.1.3. Carbon Dots (CDs)*

Carbon dots (CDs), nanoscale carbon particles (~10 nm), exhibit exceptional fluorescent properties due to quantum confinement effects and oxygenated surface functional groups [68]. These features enable tunable photoluminescence, efficient electron transfer, and photogenerated electron separation [69]. With high stability, biocompatibility, low toxicity, and excellent water dispersibility, CDs are promising for wastewater treatment, offering robust optimal properties, cost-effective synthesis, and chemical inertness [70].

Given their unique characteristics, researchers have extensively investigated CDs for the removal of heavy metal ions from wastewater. Yahaya Pudza at al. synthesized CDs from tapioca and demonstrated their effectiveness as adsorbents for removing Pb(II) ions from wastewater. The study achieved a removal efficiency of 80.6% in an aqueous environment. The adsorption phenomenon was analyzed using Langmuir and Freundlich isotherms, confirming the favorability of the process. The adsorption process followed pseudo-second-order kinetic model [71]. Perumal et al. prepared nitrogen-doped CDs from Red Malus floribunda and subsequently developed hybrid-spherical hydrogel particles (CGCDs) for the removal of heavy metal ions from aqueous environments. The CGCDs demonstrated exceptional efficiency in removing Hg(II), achieving nearly complete removal even in mixed-metal systems. Additionally, the CGCDs showed over 70% removal efficiency for all the targeted metal ions in multiple systems [72].

CDs have demonstrated significant potential in photocatalytic dye degradation, particularly when integrated with materials such as ZnO, $TiO_2$, and CdS to enhance catalytic efficiency [73]. The doping of heteroatoms, notably nitrogen, has been shown to improve their photoluminescence properties, thereby broadening their applicability in catalysis [74]. Velumani et al. synthesized carbon quantum dot-supported ZnO hollow spheres using a solvothermal method with polyethylene glycol-400 (PEG-400) as the solvent. The ZnO/C-dot nanocomposites showed unique photocatalytic efficiency, achieving methylene blue dye degradation of 96% under UV-VIS irradiation, significantly surpassing that of pure ZnO, which had an efficiency of only 63%. The enhanced results were attributed to improved light absorption in the visible range and reduced recombination of photoexcited electron-hole pairs. Further, the prepared composite demonstrated excellent reusability, maintaining stable photocatalytic performance over five cycles [75]. Nitrogen-doped CDs are particularly distinguished for their fluorescence characteristics and widespread use as probes. Further, the incorporation of CDs into single or few-layered nanocomposites enhances photocatalytic performance by increasing charge carrier mobility and surface area [73]. Beker at al. used advanced oxidation processes for degrading MB in wastewater, using iron and nitrogen-doped carbon dots (Fe, N-CDs) as Fenton-like catalysts. These catalysts were synthesized using a hydrothermal method and demonstrated near-complete dye removal of 20 mg/L MB solution under optimized conditions, including a temperature of $50^0$C, pH 8, hydrogen peroxide concentration of 147 mM, and a catalyst concentration of 0.5 mg/mL. The process achieved this efficiency within just 60 minutes, showcasing the potential of Fe, N-CDs as a sustainable and effective catalyst for wastewater treatment [76].

The presence of pharmaceutical contaminants in drinking water necessitates the development of sustainable removal methods. Recent research demonstrates that CDs effectively degrade and detect pollutants [77]. Cai et al. developed S-scheme heterostructured photocatalyst, $Cd_{0.5}Zn_{0.5}S$/carbon dots/$Bi_2WO_6$ (CZS/CDs/BWO), for effective tetracycline oxidation under visible light. This photocatalyst showed improved reactivity compared to pure CZS, BWO, and the CZS/BWO combination. Its enhanced performance results from the synergy between CDs and the S-scheme heterojunction [78]. Yin et al. synthesized a ternary photocatalyst CDs modified BiOI/g-$C_3N_4$ through a one-pot in situ co-precipitation method, using ethylene glycol as a chelating agent. The results indicated that under LED irradiation, the photocatalyst exhibited remarkable activity and stability for tetracycline mineralization, mainly when used in conjunction with persulfate (PS). Notably, even after five cycles, the tetracycline removal rate remained approximately 80%. Further, the experiments indicated that the system achieved an impressive 82.7% tetracycline removal, surpassing pure g-$C_3N_4$, BiOI, and BiOI/g-$C_3N_4$ photocatalyst by 6.7, 3.7, and 1.2 times [79]. Guo et al. developed a CDs-modified porous $ZnSnO_3$ cubic photocatalyst via a precipitation-calcination method for tetracycline degradation. With optimal CDs loading (3 wt.%), the photocatalyst exhibited enhanced degradation kinetics compared to pure $ZnSnO_3$. The photodegradation mechanism was elucidated using active species capture experiments and ESR techniques, while the degradation pathway was proposed based on intermediates identified through LC-MS analysis [80]. Although CDs have potential in wastewater treatment, challenges remain in enhancing their functionality, developing eco-friendly synthesis methods with high quantum yield, and understanding their physicochemical and photoluminescent properties. Future research should focus on these aspects, supported by theoretical studies, to optimize their use in environmental applications.

#### *3.1.4. Fullerenes*

Fullerene is a carbon allotrope of carbon atoms arranged in a fully or partially closed network of fused five- to seven-membered rings connected through single and double bonds. Fullerene has recently attracted considerable attention because of its diverse physical and chemical properties, including non-toxicity, solubility, superconductivity, and limiting effects. Its remarkable capabilities in photocatalytic processes have made it a primary emphasis for researchers exploring various applications, including wastewater treatment [81].

Building on these foundational properties, recent advancements in nanotechnology have focused on enhancing the performance of fullerene-based materials for targeted applications. Adam et al. developed a magnetic fullerene-based NP synthesized by combining fullerene, CNT, and $ZnFe_2O_4$. This NM demonstrated the effective removal of metal ions. The incorporation of $ZnFe_2O_4$ enhanced the adsorption capacity of fullerene while significantly improving the adsorbent's reusability [82]. Sudhaik et al. developed $C_{60}$/PCN nanocomposites by integrating fullerene ($C_{60}$) with phosphorous-doped graphitic carbon nitride (P-g-$C_3N_4$) using an ultrasonic-assisted method. These nanocomposites were evaluated for their photocatalytic efficiency in degrading the pesticide imidacloprid (IMI). At an optimal pH of 4, the 0.04 wt.% $C_{60}$/PCN nanocomposite achieved a removal efficiency of 91%, which increased to 95% with the addition of $H_2O_2$. The enhancement in photodegradation efficiency was attributed to $H_2O_2$ acting as an effective electron acceptor, thereby improving charge carrier separation and facilitating the degradation process [83].

### *3.1.5. Nano Membranes*

Nanomembranes play a critical role in water treatment due to their selective filtration capabilities, effectively removing contaminants such as salts, heavy metals, microorganisms, and pathogens. Nanomembranes are prepared using additive or subtractive techniques and top-down or bottom-up approaches. The top-down approach used physical methods like evaporation, sputtering, and vapor deposition or chemical methods such as chemical vapor deposition (CVD) and sol-gel techniques. The bottom-up approach relies on self-assembly, where structures organize autonomously. Techniques include Langmuir-Blodgett method, which uses surfactant molecules, layer-by-layer assembly for multilayer films, block copolymer polymerization, and sol-gel processes for forming solid networks. Dip-coating and drop-coating involve depositing materials through immersion or droplets on substrates.

Nanomembranes offer several advantages over traditional large-scale materials, primarily due to their nanoscale dimensions, which provide a large surface area. These materials exhibit unique magnetic, optical, and electrical properties that enhance their functionality. When integrated into membrane structures, NMs improve filtration efficiency and introduce diverse physical, chemical, and biological characteristics. Notable features include distinct electrical properties, enhanced adsorption capabilities, photocatalytic activity, antimicrobial effectiveness, and resistance to chlorine-induced degradation, making them highly effective for water purification applications [84]. Nanomembranes are broadly categorized into four primary types, which include inorganic, organic, hybrid, and synthetic biomembranes. This classification is based on the material composition used in their fabrication. These membranes are made from various materials, including polymers (polyamide and polysulfone), ceramics (alumina and silica), and metals. Nanomembranes offer several advantages, including high efficiency, compact design, ease of operation, and the ability to reduce water hardness. However, these membranes also face significant challenges, primarily fouling, which limits their efficiency, increases operational costs, and requires frequent maintenance and replacements. This issue is prevalent in pressure-driven processes like ultrafiltration and reverse osmosis. Among these types, nanofiltration stands out for effectively separating contaminants at lower pressures, selectively allowing monovalent salts to pass through while rejecting heavy metals and microorganisms. Nonetheless, addressing fouling and enhancing durability are essential for improving membrane performance [85].

Polymer-based nanocomposite membranes are gaining attention in sustainable water treatment. These membranes are formed by incorporating engineered NPs or nanotubes into a polymer matrix using advanced methods, which enhances their physicochemical properties and functional capabilities for better performance in water purification [86]. Rashidi et al. employed polyamide Nanomembranes to treat wastewater containing various fiber reactive dyes. Under optimal conditions within a near-neutral pH range, the chemical oxygen demand in the treated samples was completely eliminated, achieving a dye removal efficiency exceeding 90% [87].

## 4. Challenges and Future Perspectives

Nanotechnology presents a transformative opportunity in water treatment. However, its implementation is accompanied by several pressing challenges. One significant limitation is the lack of standardized NMs and regulatory frameworks, which complicates efforts by industries, policymakers, and stakeholders to ensure safe and effective applications [88]. Research into the health and environmental impacts of NMs frequently yields inconsistent and contradictory results, primarily due to the absence of universally accepted experimental protocols and measurement standards [89]. The environmental and health risks posed by NMs in water bodies are of particular concern. Substances such as Ag NPs, $TiO_2$, and CNTs have been shown to accumulate in aquatic organisms, potentially transferring harmful effects to humans through the food chain [90]. Furthermore, the high financial investment required for the large-scale implementation of nanotechnology, coupled with insufficient knowledge about behavior and fate of NPs, has raised concerns about their safety [88,89]. Minor changes in the chemical or structural properties of these materials can drastically alter their behavior, potentially leading to oxidative stress, DNA damage, cellular dysfunction, and adverse health outcomes such as cancer and apoptosis [91].

Despite these challenges, the future of nanotechnology in water treatment holds immense promise. Nanotechnology offers innovative, cost-effective solutions for water treatment, addressing pollutants such as recalcitrant compounds, biofilms, and viruses [89,92]. Materials like carbon nanotubes (CNTs) reduce energy consumption in filtration systems, while advanced nanoadsorbents, dendrimers, and polymers show significant promise. Nanosensors, capable of detecting contaminants at sub-ppm levels, represents a breakthrough in monitoring technologies [93]. Nanotechnology also plays a key role in preventing nuclear waste contamination of groundwater. Efforts to integrate green NMs, such as functionalized silica gels and ionic liquids, aim to make water treatment more sustainable. Techniques like hybrid nano-membranes and bio-nanotechnology hold significant potential, though validation under real-world conditions is essential. Cost barriers can be addressed through research focused on reuse and risk management, paving the way for efficient, multi-pollutant treatment systems with reduced operational costs [94].

## Conclusion

In summary, nanotechnology offers transformative solutions to the challenges of wastewater treatment by enhancing the removal of pollutants through advanced materials such as carbon nanotubes, graphene derivatives, and quantum dots. These materials provide superior adsorption, catalytic, and antimicrobial properties while addressing the limitations of conventional methods. Integrating NMs with membrane technologies has shown significant potential for sustainable and efficient water purification. However, challenges such as regulatory gaps, environmental risks, and scalability remain critical. Future research must prioritize eco-friendly designs and real-world applications to ensure the development of cost-effective and environmentally sustainable water treatment technologies.

## References

[1] Jain, K., Patel, A. S., Pardhi, V. P., Flora, S. J. S. 2021. Nanotechnology in Wastewater Management: A New Paradigm towards Wastewater Treatment. *Molecules 26* (6):1797. doi: https://doi.org/10.3390/molecules26061797.

[2] Bhat, S. A., Sher, F., Hameed, M., Bashir, O., et al. 2022. Sustainable Nanotechnology Based Wastewater Treatment Strategies: Achievements, Challenges and Future Perspectives. *Chemosphere 288*:132606. doi: https://doi.org/10.1016/j.chemosphere.2021.132606.

[3] Khan, S., Naushad, Mu., Al-Gheethi, A., Iqbal, J. 2021. Engineered NPs for Removal of Pollutants from Wastewater: Current Status and Future Prospects of Nanotechnology for Remediation Strategies. *Journal of Environmental Chemical Engineering 9* (5):106160. doi: https://doi.org/10.1016/j.jece.2021.106160.

[4] Olvera, R. C., Silva, S. L., Robles-Belmont, E., Lau, E. Z. 2017. Review of Nanotechnology Value Chain for Water Treatment Applications in Mexico. *Resource-Efficient Technologies 3* (1):1–11. doi: https://doi.org/10.1016/j.reffit.2017.01.008.

[5] Briggs, A. M., Cross, M. J., Hoy, D. G., Sànchez-Riera, L., et al. 2016. Musculoskeletal Health Conditions Represent a Global Threat to Healthy Aging: A Report for the 2015 World Health Organization World Report on Ageing and Health. *The Gerontologist 56*(Suppl_2):S243–S255. doi: https://doi.org/10.1093/geront/gnw002.

[6] Ajith, M. P., Aswathi, M., Priyadarshini, E., Rajamani, P. 2021. Recent Innovations of Nanotechnology in Water Treatment: A Comprehensive Review. *Bioresource Technology 342*:126000. doi: https://doi.org/10.1016/j.biortech.2021.126000.

[7] Ibrahim, R. K., Hayyan, M., AlSaadi, M. A., Hayyan, A., et al. 2016. Environmental Application of Nanotechnology: Air, Soil, and Water. *Environmental Science & Pollution Research 23* (14):13754–13788. doi: https://doi.org/10.1007/s11356-016-6457-z.

[8] Humplik, T., Lee, J., O'Hern, S. C., Fellman, B. A., et al. 2011. Nanostructured Materials for Water Desalination. *Nanotechnology 22*(29):292001. doi: https://doi.org/10.1088/0957-4484/22/29/292001.

[9] Khan, S., Anjum, R., Bilal, M. 2021. Revealing Chemical Speciation Behaviors in Aqueous Solutions for Uranium (VI) and Europium (III) Adsorption on Zeolite. *Environmental Technology & Innovation 22*:101503. doi: https://doi.org/10.1016/j.eti.2021.101503.

[10] Goh, P. S., Matsuura, T., Ismail, A. F., Hilal, N. 2016. Recent Trends in Membranes and Membrane Processes for Desalination. *Desalination 391*:43–60. doi: https://doi.org/10.1016/j.desal.2015.12.016.

[11] Leonel, A. G., Mansur, A. A. P., Mansur, H. S. 2021. Advanced Functional Nanostructures Based on Magnetic Iron Oxide NMs for Water Remediation: A Review. *Water Research 190*:116693. doi: https://doi.org/10.1016/j.watres.2020.116693.

[12] Kumar, S., Ahlawat, W., Bhanjana, G., Heydarifard, S., et al. 2014. Nanotechnology-Based Water Treatment Strategies. *Journal of Nanoscience and Nanotechnology 14* (2):1838–1858. doi: https://doi.org/10.1166/jnn.2014.9050.

[13] Khanna, N., Singh, S., Chatterji, T. 2022. Potential Application of Nanotechnology in Wastewater Management: A Paradigm Shift. *Materials Letters 315*:131975. doi: https://doi.org/10.1016/j.matlet.2022.131975.

[14] Koul, B., Yadav, D., Singh, S., Kumar, M., et al. 2022. Insights into the Domestic Wastewater Treatment (DWWT) Regimes: A Review. *Water 14* (21):3542. doi: https://doi.org/10.3390/w14213542.

[15] Ahmadi, F., Zinatizadeh, A. A., Asadi, A., McKay, T., et al. 2020. Simultaneous Carbon and Nutrients Removal and PHA Production in a Novel Single Air Lift Bioreactor Treating an Industrial Wastewater. *Environmental Technology & Innovation 18*:100776. DOI: https://doi.org/10.1016/j.eti.2020.100776.

[16] Kalfa, A., Shapira, B., Shopin, A., Cohen, I., et al. 2020. Deionization for Wastewater Treatment: Opportunities and Challenges. *Chemosphere 241*:125003. doi: https://doi.org/10.1016/j.chemosphere.2019.125003.

[17] Verlicchi, P., Al Aukidy, M., Zambello, E. 2012. Occurrence of Pharmaceutical Compounds in Urban Wastewater: Removal, Mass Load and Environmental Risk after a Secondary Treatment—A Review. *Science of The Total Environment 429*:123–155. doi: https://doi.org/10.1016/j.scitotenv.2012.04.028.

[18] Naskar, J., Boatemaa, M. A., Rumjit, N. P., Thomas, G., et al. 2022. Recent Advances of Nanotechnology in Mitigating Emerging Pollutants in Water and Wastewater: Status, Challenges, and Opportunities. *Water, Air, & Soil Pollut*ion *233*(5):156. doi: https://doi.org/10.1007/s11270-022-05611-y.

[19] Fernandes, S. P. S., Romero, V., Espiña, B., Salonen, L. M. 2019. Tailoring Covalent Organic Frameworks to Capture Water Contaminants. *Chemistry A European Journal 25*(26):6461–6473. doi: https://doi.org/10.1002/chem.201806025.

[20] Boudjemaa, A., Gómez-Ruiz, S. 2020. Titanium Oxide-Based NMs with Photocatalytic Applications in Environmental Chemistry. In Environmental Nanotechnology Volume 4; Dasgupta, N., Ranjan, S., Lichtfouse, E., Eds.; Environmental Chemistry for a Sustainable World; Springer International Publishing: Cham, Vol. 32:215–263. doi: https://doi.org/10.1007/978-3-030-26668-4_7.

[21] Kumari, H., Sonia, Suman, Ranga, R., et al. 2023. A Review on Photocatalysis Used For Wastewater Treatment: Dye Degradation. *Water, Air, & Soil Pollution 234*(6):349. doi: https://doi.org/10.1007/s11270-023-06359-9.

[22] Crisan, C. M., Mocan, T., Manolea, M., Lasca, L. I., Tăbăran, F.-A., Mocan, L. 2021. Review on Silver NPs as a Novel Class of Antibacterial Solutions. *Applied Sciences 11*(3):1120. doi: https://doi.org/10.3390/app11031120.

[23] Kainz, Q. M., Reiser, O. 2014. Polymer- and Dendrimer-Coated Magnetic NPs as Versatile Supports for Catalysts, Scavengers, and Reagents. *Accounts of Chemical Research 47*(2):667–677. doi: https://doi.org/10.1021/ar400236y.

[24] Zhao, J., Lin, M., Wang, Z., Cao, X., Xing, B. 2021. Engineered NMs in the Environment: Are They Safe? *Critical Reviews in Environmental Science and Technology 51*(14):1443–1478. doi: https://doi.org/10.1080/10643389.2020.1764279.

[25] Smith, S. C., Rodrigues, D. F. 2015. Carbon-Based NMs for Removal of Chemical and Biological Contaminants from Water: A Review of Mechanisms and Applications. *Carbon 91*:122–143. doi: https://doi.org/10.1016/j.carbon.2015.04.043.

[26] Zhang, S., Shao, T., Karanfil, T. 2011. The Effects of Dissolved Natural Organic Matter on the Adsorption of Synthetic Organic Chemicals by Activated Carbons and Carbon Nanotubes. *Water Research 45*(3):1378–1386. doi: https://doi.org/10.1016/j.watres.2010.10.023.

[27] Sayed, E. T., Alawadhi, H., Elsaid, K., Olabi, A. G., et al. 2020. A Carbon-Cloth Anode Electroplated with Iron Nanostructure for Microbial Fuel Cell Operated with Real Wastewater. *Sustainability 12*(16):6538. DOI: https://doi.org/10.3390/su12166538.

[28] Li, Y., Zhou, B., Zheng, G., Liu, X., et al. 2018. Continuously Prepared Highly Conductive and Stretchable SWNT/MWNT Synergistically Composited Electrospun Thermoplastic Polyurethane Yarns for Wearable Sensing. *Journal of Material Chemistry C 6*(9):2258–2269. doi: https://doi.org/10.1039/C7TC04959E.

[29] Salam, Md. A., Al-Amin, Md. Y., Salam, M. T., Pawar, J. S., et al. 2023. Antimicrobial Resistance: A Growing Serious Threat for Global Public Health. *Healthcare 11*(13):1946. doi: https://doi.org/10.3390/healthcare11131946.

[30] Ncibi, M. C., Sillanpää, M. 2015. Optimized Removal of Antibiotic Drugs from Aqueous Solutions Using Single, Double and Multi-Walled Carbon Nanotubes. *Journal of Hazardous Materials 298*:102–110. doi: https://doi.org/10.1016/j.jhazmat.2015.05.025.

[31] Kariim, I., Abdulkareem, A. S., Abubakre, O. K. 2020. Development and Characterization of MWCNTs from Activated Carbon as Adsorbent for Metronidazole and Levofloxacin Sorption from Pharmaceutical Wastewater: Kinetics, Isotherms and Thermodynamic Studies. *Scientific African 7*:e00242. doi: https://doi.org/10.1016/j.sciaf.2019.e00242.

[32] Mashkoor, F., Nasar, A., Inamuddin. 2020. Carbon Nanotube-Based Adsorbents for the Removal of Dyes from Waters: A Review. *Environmental Chemistry Letters 18*(3):605–629. doi: https://doi.org/10.1007/s10311-020-00970-6.

[33] Agarwal, S., Sadegh, H., Monajjemi, M., Hamdy, A. S., et al. 2016. Efficient Removal of Toxic Bromothymol Blue and Methylene Blue from Wastewater by Polyvinyl Alcohol. *Journal of Molecular Liquids 218*:191–197. doi: https://doi.org/10.1016/j.molliq.2016.02.060.

[34] Madrakian, T., Afkhami, A., Ahmadi, M., Bagheri, H. 2011. Removal of Some Cationic Dyes from Aqueous Solutions Using Magnetic-Modified Multi-Walled Carbon Nanotubes. *Journal of Hazardous Materials 196*:109–114. doi: https://doi.org/10.1016/j.jhazmat.2011.08.078.

[35] Fu, F., Wang, Q. 2011. Removal of Heavy Metal Ions from Wastewaters: A Review. *Journal of Environmental Management 92*(3):407–418. doi: https://doi.org/10.1016/j.jenvman.2010.11.011.

[36] Addo Ntim, S., Mitra, S. 2011. Removal of Trace Arsenic To Meet Drinking Water Standards Using Iron Oxide Coated Multiwall Carbon Nanotubes. *Journal of Chemical & Engineering Data, 56*(5):2077–2083.doi: https://doi.org/10.1021/je1010664.

[37] Saber-Samandari, S., Saber-Samandari, S., Joneidi-Yekta, H., Mohseni, M. 2017. Adsorption of Anionic and Cationic Dyes from Aqueous Solution Using Gelatin-Based Magnetic Nanocomposite Beads Comprising Carboxylic Acid Functionalized Carbon Nanotube. *Chemical Engineering Journal 308*:1133–1144. doi: https://doi.org/10.1016/j.cej.2016.10.017.

[38] Yuan, B., Zhang, S., Ren, D., Zhang, X. 2024. Research Progress on the Removal of Heavy Metals in Water and Soil by Modified Carbon Nanotubes: A Review. *Water, Air, & Soil Pollution 235*(7):418. doi: https://doi.org/10.1007/s11270-024-07238-7.

[39] Alijani, H., Shariatinia, Z. 2017. Effective Aqueous Arsenic Removal Using Zero Valent Iron Doped MWCNT Synthesized by in Situ CVD Method Using Natural α-$Fe_2O_3$ as a Precursor. *Chemosphere 171*:502–511. doi: https://doi.org/10.1016/j.chemosphere.2016.12.106.

[40] Ghasemi, E., Heydari, A., Sillanpää, M. 2019. Ultrasonic Assisted Adsorptive Removal of Toxic Heavy Metals from Environmental Samples Using Functionalized Silica-Coated Magnetic Multiwall Carbon Nanotubes (MagMWCNTs@$SiO_2$). *Engineering in Agriculture, Environment and Food 12*(4):435–442. doi: https://doi.org/10.1016/j.eaef.2019.07.002.

[41] Huang, Y., Lee, X., Macazo, F. C., Grattieri, M., et al. 2018. Fast and Efficient Removal of Chromium (VI) Anionic Species by a Reusable Chitosan-Modified Multi-Walled Carbon Nanotube Composite. *Chemical Engineering Journal 339*:259–267. doi: https://doi.org/10.1016/j.cej.2018.01.133.

[42] Madhura, L., Singh, S., Kanchi, S., Sabela, M., et al. 2019. Nanotechnology-Based Water Quality Management for Wastewater Treatment. *Environmental Chemistry Letters 17*(1):65–121. doi: https://doi.org/10.1007/s10311-018-0778-8.

[43] Saravanan, A., Kumar, P. S., Srinivasan, S., Jeevanantham, S., et al. 2022. Insights on Synthesis and Applications of Graphene-Based Materials in Wastewater Treatment: A Review. *Chemosphere 298*:134284. doi: https://doi.org/10.1016/j.chemosphere.2022.134284.

[44] Lee, J., Noh, S., Pham, N. D., Shim, J. H. 2019. Top-down Synthesis of S-Doped Graphene Nanosheets by Electrochemical Exfoliation of Graphite: Metal-Free Bifunctional Catalysts for Oxygen Reduction and Evolution Reactions. *Electrochimica Acta 313*:1–9. doi: https://doi.org/10.1016/j.electacta.2019.05.015.

[45] Young, R. J., Kinloch, I. A., Gong, L., Novoselov, K. S. 2012. The Mechanics of Graphene Nanocomposites: A Review. *Composites Science and Technology 72*(12):1459–1476. doi: https://doi.org/10.1016/j.compscitech.2012.05.005.

[46] Jayakaran, P., Nirmala, G. S., Govindarajan, L. 2019. Qualitative and Quantitative Analysis of Graphene-Based Adsorbents in Wastewater Treatment. *International Journal of Chemical Engineering 2019*:1–17. doi: https://doi.org/10.1155/2019/9872502.

[47] Dubey, A., Bhavsar, N., Pachchigar, V., Saini, M., et al. 2022. Microwave Assisted Ultrafast Synthesis of Graphene Oxide Based Magnetic Nano Composite for Environmental Remediation. *Ceramics International 48*(4):4821–4828. doi: https://doi.org/10.1016/j.ceramint.2021.11.018.

[48] Kumar, S., Nair, R. R., Pillai, P. B., Gupta, S. N., et al. 2014. Graphene Oxide–$MnFe_2O_4$ Magnetic Nanohybrids for Efficient Removal of Lead and Arsenic from Water. *ACS Applied Materials & Interfaces 6*(20):17426–17436. doi: https://doi.org/10.1021/am504826q.

[49] Hassan, N. S., Jalil, A. A., Azami, M. S., Rahman, A. F. A., et al. 2021. Photodegradation of Bisphenol A from Aqueous Solution over Reduced Graphene Oxide Supported on Tetragonal Silica-Zirconia Nanocatalysts: Optimization Using RSM. *Process Safety and Environmental Protection 156*:496–507. doi: https://doi.org/10.1016/j.psep.2021.10.032.

[50] Bhattacharyya, A., Banerjee, B., Ghorai, S., Rana, D., et al. 2018. Development of an Auto-Phase Separable and Reusable Graphene Oxide-Potato Starch Based Cross-Linked Bio-Composite Adsorbent for Removal of Methylene Blue Dye. *International Journal of Biological Macromolecules 116*:1037–1048. doi: https://doi.org/10.1016/j.ijbiomac.2018.05.069.

[51] Rout, D. R., Jena, H. M. 2021. Removal of Malachite Green Dye from Aqueous Solution Using Reduced Graphene Oxide as an Adsorbent. *Materials Today: Proceedings 47*:1173–1182. doi: https://doi.org/10.1016/j.matpr.2021.03.406.

[52] Baptista, F. R., Belhout, S. A., Giordani, S., Quinn, S. J. 2015. Recent Developments in Carbon NM Sensors. *Chemical Society Reviews 44*(13):4433–4453. DOI: https://doi.org/10.1039/C4CS00379A

[53] Iravani, S., Varma, R. S. 2020. Green Synthesis, Biomedical and Biotechnological Applications of Carbon and Graphene Quantum Dots. A Review. *Environmental Chemistry Letters 18*(3):703–727. doi: https://doi.org/10.1007/s10311-020-00984-0.

[54] Nagaraj, A., Munusamy, M. A., Al-Arfaj, A. A., Rajan, M. 2019. Functional Ionic Liquid-Capped Graphene Quantum Dots for Chromium Removal from Chromium Contaminated Water. *J. Chem. Eng. Data 64*(2):651–667. doi: https://doi.org/10.1021/acs.jced.8b00887.

[55] Sarkar, S., Raghavan, A., Giri, A., Ghosh, S. 2021. Graphene Quantum Dots Decorated $TiO_2$ Nanostructures: Sustainable Approach for Photocatalytic Remediation of an Industrial Pollutant. *ChemistrySelect 6*(40):10957–10964. doi: https://doi.org/10.1002/slct.202102242.

[56] Cataldi, P., Athanassiou, A., Bayer, I. S. 2018. Graphene Nanoplatelets-Based Advanced Materials and Recent Progress in Sustainable Applications. *Applied Sciences 8*(9):1438. doi: https://doi.org/10.3390/app8091438.

[57] Kausar, A., Anwar, Z., Muhammad, B. 2016. Recent Developments in Epoxy/Graphite, Epoxy/Graphene, and Epoxy/Graphene Nanoplatelet Composites: A Comparative Review. *Polymer-Plastics Technology and Engineering 55*(11):1192–1210. doi: https://doi.org/10.1080/03602559.2016.1163589.

[58] Zembat, A. A., Gezmis-Yavuz, E., Koseoglu-Imer, D. Y., Cansoy, C. E. 2024. The Use of Graphene Nanoplatelet-embedded PA -6 Nanofibres to Remove Turbidity from Water. The *Canadian Journal of Chemical Engineering* doi: https://doi.org/10.1002/cjce.25469.

[59] Jiménez-Suárez, A., Prolongo, S. G. 2020. Graphene Nanoplatelets. *Applied Sciences 10*(5):1753. doi: https://doi.org/10.3390/app10051753.

[60] Al-Khateeb, L. A., Almotiry, S., Salam, M. A. 2014. Adsorption of Pharmaceutical Pollutants onto Graphene Nanoplatelets. *Chemical Engineering Journal 248*:191–199. doi: https://doi.org/10.1016/j.cej.2014.03.023.

[61] Rosli, F. A., Ahmad, H., Jumbri, K., Abdullah, A. H., et al. 2021. Efficient Removal of Pharmaceuticals from Water Using Graphene Nanoplatelets as Adsorbent. *Royal Society Open Science 8*(1):201076. doi: https://doi.org/10.1098/rsos.201076.

[62] González-López, M. E., Laureano-Anzaldo, C. M., Pérez-Fonseca, A. A., Gómez, C., et al. 2021. Congo Red Adsorption with Cellulose-Graphene Nanoplatelets Beads by Differential Column Batch Reactor. *Journal of Environmental Chemical Engineering 9*(2):105029. doi: https://doi.org/10.1016/j.jece.2021.105029.

[63] Rajaji, U., Arumugam, R., Chen, S. M., Chen, T. W., et al. 2018. Graphene Nanoribbons in Electrochemical Sensors and Biosensors: A Review. *International Journal of Electrochemical Science 13*(7):6643–6654. doi: https://doi.org/10.20964/2018.07.51

[64] Casiraghi, C., Prezzi, D. 2017. Raman Spectroscopy of Graphene Nanoribbons: A Review. In *GraphITA*; Morandi, V., Ottaviano, L., Eds., Carbon Nanostructures; Springer International Publishing: Cham, 19–30. doi: https://doi.org/10.1007/978-3-319-58134-7_2.

[65] Shende, P., Augustine, S., Prabhakar, B. 2020. A Review on Graphene Nanoribbons for Advanced Biomedical Applications. *Carbon Letters 30*(5):465–475. doi: https://doi.org/10.1007/s42823-020-00125-1.

[66] Gao, Y., Liang, X., Han, S., Wu, L., et al. 2020. High-Efficiency Adsorption for Both Cationic and Anionic Dyes Using Graphene Nanoribbons Formed by Atomic-Hydrogen Induced Single-Walled Carbon Nanotube Carpets. *Carbon Letters 30*(2):123–132. doi: https://doi.org/10.1007/s42823-019-00089-x.

[67] Sadeghi, M. H., Tofighy, M. A., Mohammadi, T. 2020. One-Dimensional Graphene for Efficient Aqueous Heavy Metal Adsorption: Rapid Removal of Arsenic and Mercury Ions by Graphene Oxide Nanoribbons (GONRs). *Chemosphere 253*:126647. doi: https://doi.org/10.1016/j.chemosphere.2020.126647.

[68] Joshi, P. N., Mathias, A., Mishra, A. 2018. Synthesis of Ecofriendly Fluorescent Carbon Dots and Their Biomedical and Environmental Applications. *Materials Technology 33*(10):672–680. doi: https://doi.org/10.1080/10667857.2018.1492683.

[69] Liang, Z., Yang, J., Zhou, C., Mo, Q., et al. 2018. Carbon Quantum Dots Modified BiOBr Microspheres with Enhanced Visible Light Photocatalytic Performance. *Inorganic Chemistry Communications 90*:97–100. doi: https://doi.org/10.1016/j.inoche.2018.02.013.

[70] Saud, P. S., Pant, B., Alam, A.-M., Ghouri, Z. K., et al. 2015. Carbon Quantum Dots Anchored TiO2 Nanofibers: Effective Photocatalyst for Wastewater Treatment. *Ceramics International 41*(9):11953–11959. doi: https://doi.org/10.1016/j.ceramint.2015.06.007.

[71] Yahaya Pudza, M., Zainal Abidin, Z., Abdul Rashid, S., Md Yasin, F., et al. 2020. Eco-Friendly Sustainable Fluorescent Carbon Dots for the Adsorption of Heavy Metal Ions in Aqueous Environment. *NMs 10*(2):315. doi: https://doi.org/10.3390/nano10020315.

[72] Perumal, S., Atchudan, R., Thirukumaran, P., Yoon, D. H., et al. 2022. Simultaneous Removal of Heavy Metal Ions Using Carbon Dots-Doped Hydrogel Particles. *Chemosphere 286*:131760. doi: https://doi.org/10.1016/j.chemosphere.2021.131760.

[73] Basavaraj, N., Sekar, A., Yadav, R. 2021. Review on Green Carbon Dot-Based Materials for the Photocatalytic Degradation of Dyes: Fundamentals and Future Perspective. *Materials Advances 2*(23):7559–7582. doi: https://doi.org/10.1039/D1MA00773D.

[74] Bourlinos, A. B., Trivizas, G., Karakassides, M. A., Baikousi, M., et al. 2015. Green and Simple Route toward Boron Doped Carbon Dots with Significantly Enhanced Non-Linear Optical Properties. *Carbon 83*:173–179. doi: https://doi.org/10.1016/j.carbon.2014.11.032.

[75] Velumani, A., Sengodan, P., Arumugam, P., Rajendran, R., et al. 2020. Carbon Quantum Dots Supported ZnO Sphere Based Photocatalyst for Dye Degradation Application. *Current Applied Physics 20*(10):1176–1184. doi: https://doi.org/10.1016/j.cap.2020.07.016.

[76] Beker, S. A., Khudur, L. S., Cole, I., Ball, A. S. 2022. Catalytic Degradation of Methylene Blue Using Iron and Nitrogen-Containing Carbon Dots as Fenton-like Catalysts. *New Journal of Chemistry 46*(1):263–275. doi: https://doi.org/10.1039/D1NJ04761B.

[77] Korah, B. K., Chacko, A. R., Abraham, T., Mathew, B. 2022. Recent Progress and Future Perspectives of Carbon Dots in the Detection, Degradation, and Enhancement of Drugs. *Particle & Particle Systems Characterization 39*(2):2100264. doi: https://doi.org/10.1002/ppsc.202100264.

[78] Cai, M., Wang, C., Liu, Y., Yan, R., et al. 2022. Boosted Photocatalytic Antibiotic Degradation Performance of Cd0.5Zn0.5S/Carbon Dots/Bi2WO6 S-Scheme Heterojunction with Carbon Dots as the Electron Bridge. *Separation and Purification Technology 300*:121892. doi: https://doi.org/10.1016/j.seppur.2022.121892.

[79] Yin, C., Liu, Y., Lv, X., Lv, S., et al. 2022. Carbon Dots as Heterojunction Transport Mediators Effectively Enhance BiOI/g-$C_3N_4$ Synergistic Persulfate Degradation of Antibiotics. *Applied Surface Science 601*:154249. doi: https://doi.org/10.1016/j.apsusc.2022.154249.

[80] Guo, F., Huang, X., Chen, Z., Sun, H., et al. 2020. Investigation of Visible-Light-Driven Photocatalytic Tetracycline Degradation via Carbon Dots Modified Porous $ZnSnO_3$ Cubes: Mechanism and Degradation Pathway. *Separation and Purification Technology 253*:117518. doi: https://doi.org/10.1016/j.seppur.2020.117518.

[81] Srivastava, V., Sinha, S., Kanaujia, S., Bartaria, D., et al. 2024. Fullerene-Based Photocatalysis as an Eco-Compatible Approach to Photochemical Reactions. *Synthetic Communications* 55(1):1–18. doi: https://doi.org/10.1080/00397911.2024.2410947.

[82] Adam, A. M., Saad, H. A., Atta, A. A., Alsawat, M., et al. 2021. An Environmentally Friendly Method for Removing Hg(II), Pb(II), Cd(II) and Sn(II) Heavy Metals from Wastewater Using Novel Metal–Carbon-Based Composites. *Crystals 11*(8):882. doi: https://doi.org/10.3390/cryst11080882.

[83] Sudhaik, A., Raizada, P., Singh, P., Hosseini-Bandegharaei, A., et al. 2020. Highly Effective Degradation of Imidacloprid by $H_2O_2$/ Fullerene Decorated P-Doped g-$C_3N_4$ Photocatalyst. *Journal of Environmental Chemical Engineering 8*(6):104483. doi: https://doi.org/10.1016/j.jece.2020.104483.

[84] Tripathy, D. B., Gupta, A. 2023. Nanomembranes-Affiliated Water Remediation: Chronology, Properties, Classification, Challenges and Future Prospects. *Membranes 13*(8):713. doi: https://doi.org/10.3390/membranes13080713.

[85] Bishoge, O. K., Zhang, L., Suntu, S. L., Jin, H., et al. 2018. Remediation of Water and Wastewater by Using Engineered NMs: A Review. *Journal of Environmental Science and Health, Part A 53*(6):537–554. doi: https://doi.org/10.1080/10934529.2018.1424991.

[86] Puri, N., Gupta, A., Mishra, A. 2021. Recent Advances on Nano-Adsorbents and Nanomembranes for the Remediation of Water. *Journal of Cleaner Production 322*:129051. doi: https://doi.org/10.1016/j.jclepro.2021.129051.

[87] Rashidi, H. R., Sulaiman, N. M. N., Hashim, N. A., Hassan, C. R. C., et al. 2015. Reactive Dye Wastewater Treatment by Using Nano-Membrane Filtration. *Desalination and Water Treatment 55*(1):86–95. doi: https://doi.org/10.1080/19443994.2014.912964.

[88] Brame, J., Li, Q., Alvarez, P. J. J. 2011. Nanotechnology-Enabled Water Treatment and Reuse: Emerging Opportunities and Challenges for Developing Countries. *Trends in Food Science & Technology 22*(11):618–624. doi: https://doi.org/10.1016/j.tifs.2011.01.004.

[89] Gehrke, I., Geiser, A., Somborn-Schulz, A. 2015. Innovations in Nanotechnology for Water Treatment. *Nanotechnology, Science, and Applications* 8:1-17. doi: https://doi.org/10.2147/NSA.S43773.

[90] Asghari, S., Johari, S. A., Lee, J. H., Kim, Y. S., et al. 2012. Toxicity of Various Silver NPs Compared to Silver Ions in Daphnia Magna. *Journal of Nanobiotechnology 10*(1):14. doi: https://doi.org/10.1186/1477-3155-10-14.

[91] Fu, P. P., Xia, Q., Hwang, H.-M., Ray, P. C., et al. 2014. Mechanisms of Nanotoxicity: Generation of Reactive Oxygen Species. *Journal of Food and Drug Analysis 22*(1):64–75. doi: https://doi.org/10.1016/j.jfda.2014.01.005.

[92] Hlongwane, G. N., Sekoai, P. T., Meyyappan, M., Moothi, K. 2019. Simultaneous Removal of Pollutants from Water Using NPs: A Shift from Single Pollutant Control to Multiple Pollutant Control. *Science of The Total Environment 656*:808–833. doi: https://doi.org/10.1016/j.scitotenv.2018.11.257.

[93] Bethi, B., Sonawane, S. H., Bhanvase, B. A., Gumfekar, S. P. 2016. NMs-Based Advanced Oxidation Processes for Wastewater Treatment: A Review. *Chemical Engineering and Processing - Process Intensification 109*:178–189. doi: https://doi.org/10.1016/j.cep.2016.08.016.

[94] Anjum, M., Miandad, R., Waqas, M., Gehany, F., et al. 2019. Remediation of Wastewater Using Various Nano-Materials. *Arabian Journal of Chemistry 12*(8):4897–4919. doi: https://doi.org/10.1016/j.arabjc.2016.10.004.

## Chapter 9

# Biotechnological Approaches for Sustainable Wastewater Remediation

**A. H. D. Pushpa Latha***
Department of Botany, Dr. V. S. Krishna Government Degree College (A), Andhra Pradesh, India

## Abstract

Wastewater treatment is urgently needed due to its significant impact on both the environment and public health. Although, industrialization drives economic development, it also serves as a major source of wastewater generation. Hazardous wastes generated by chemical operations and processes are treated by the respective industries using physico-chemical and biological procedures to meet regulatory standards. Global research efforts are focused on addressing this issue through sustainable solutions. A number of wastewater treatment techniques such as mechanical, physical and chemical techniques are available but these techniques have significant drawbacks and sophisticated procedures. These challenges include high expenditures; toxic byproducts that would need additional purification due to secondary contaminants, and the insufficient removal of pollutants. Although biological treatment techniques (green methods) offer an environmentally sustainable option, they face several challenges. These methods are often more costly, time consuming, temperature-sensitive, and may release undesirable gases as a result of microbial activity. Hence, recent advancements in biotechnology have addressed many challenges of conventional wastewater treatment. These advancements use genetic engineering, synthetic biology, and biological systems to create more economical, environmentally friendly, and effective solutions. This chapter elucidates the recent developments in bio-remediation of wastewater in the contaminated areas using biotechnological methods of wastewater treatment.

**Keywords:** wastewater, biotechnological methods, bio-remediation, environmental sustainability, contaminants

* Corresponding Author's Email: lathahani2004@gmail.com

In: Advancements in Wastewater Treatment
Editors: Neha Agarwal, Maulin P. Shahl, Vijendra Singh Solankil et al.
ISBN: 979-8-90134-007-3

## 1. Introduction

Water stands as the quintessential elixir of life, indispensable to all forms of existence. Population explosion, industrialization, urbanization, and globalization have led to the overexploitation of water more than any other natural resource. Majority of countries around the world are facing a water crisis. Water pollution is one of the key factors limiting the availability of clean water. Consumption of contaminated water causes many diseases in humans and aquatic species. Wastewater recycling has become necessary to meet the increasing demand of water in today's world. However, its widespread implementation is often hindered due to high costs. Meanwhile, the direct discharge of untreated wastewater into water bodies continues to cause significant impacts on environment, agriculture and human beings [1]. Direct discharge of untreated industrial effluent, sewage sludge, and agricultural runoff into water bodies is one of the main causes of contamination in rivers. These pollutants are discharged into water bodies from various sources which are illustrated in Figure 1. Various categories of pollutants such as pharmaceuticals and personal care products (PCPs), dyes, pesticides, radionuclides, Poly- and perfluoroalkyl substances (PFASs), chemical and non-chemical contaminants, heavy metals etc. and their sources are also summarized in Table 1.

**Table 1.** Different categories of pollutants and their sources

| No | Type of Pollutant | Example | Source | Reference |
|---|---|---|---|---|
| 1 | Chemical contaminants | Nutrients-Ammonium, phosphate | Urine | [2] |
| 2 | Pharmaceuticals and personal care products | Antibiotics | Medicines | [3] |
| 3 | Poly- and perfluoroalkyl substances (PFASs) | Perfluorooctanoic acid (PFOA) and perfluorooctane sulfonate (PFOS) | Multiple consumer products | [4] |
| 4 | Biocides and pesticides | Carbendazim<br>Atrazine | Food processing industry | [5-7] |
| 5 | Heavy metals | Cadmium, Chromium, Copper, Iron, Lead, Mercury, Selenium or Zinc | Industrial wastes<br>Agricultural wastes | [8] |
| 6 | Dyes | Azo dyes | Textile industry | [9] |
| 7 | Radionuclides | Radioactive isotopes of elements including Strontium, Selenium or Uranium | Laundry wastewater from nuclear power plants regular wastewater treatment plants hospital wastewater | [10-12] |
| 8 | Non-Chemical contaminants<br>Micropalstics | Discharge of treated wastewater | Personal care products (cosmetics) | [13] |
| 9 | Nanoparticlces | Engineered nanomaterials include for instance silver (Ag), titanium dioxide ($TiO_2$), zinc oxide (ZnO), silicon dioxide ($SiO_2$) or carbon nanotubes | Personal care products, paints, food packaging | [14] |

These contaminants enter the soil through irrigation with different types of wastewaters such as sewage sludge, municipal and agricultural waste, industrial effluents and many other sources. Crops may absorb heavy metals present in wastewater. As a result of this contamination these metals may eventually make their way into the food chain, impacting ecosystems and human health. Prolonged use of untreated wastewater can lead to degradation of soil quality and agricultural productivity. A decrease in biodiversity may also be observed by direct application of wastewater for irrigation or farming, which retards the growth of native plants and grasses.

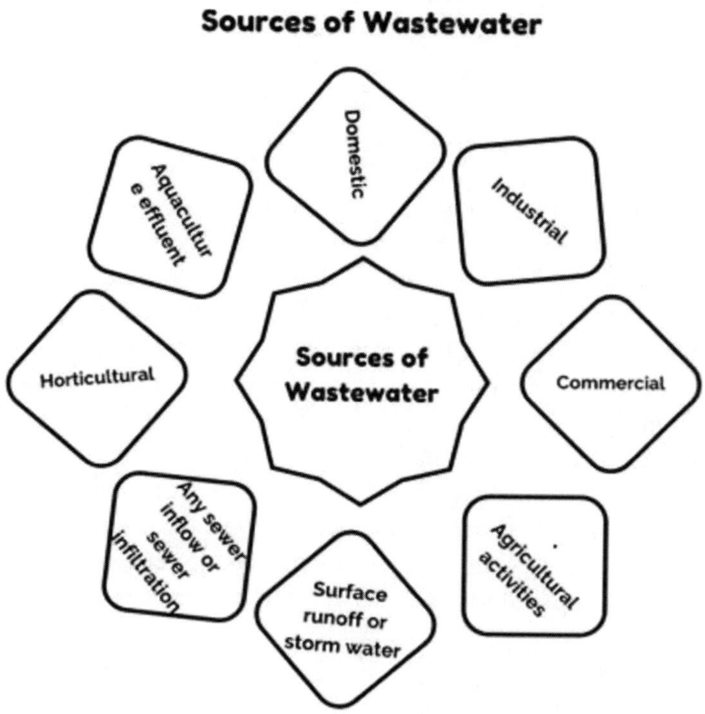


**Figure 1.** Different sources of wastewater.

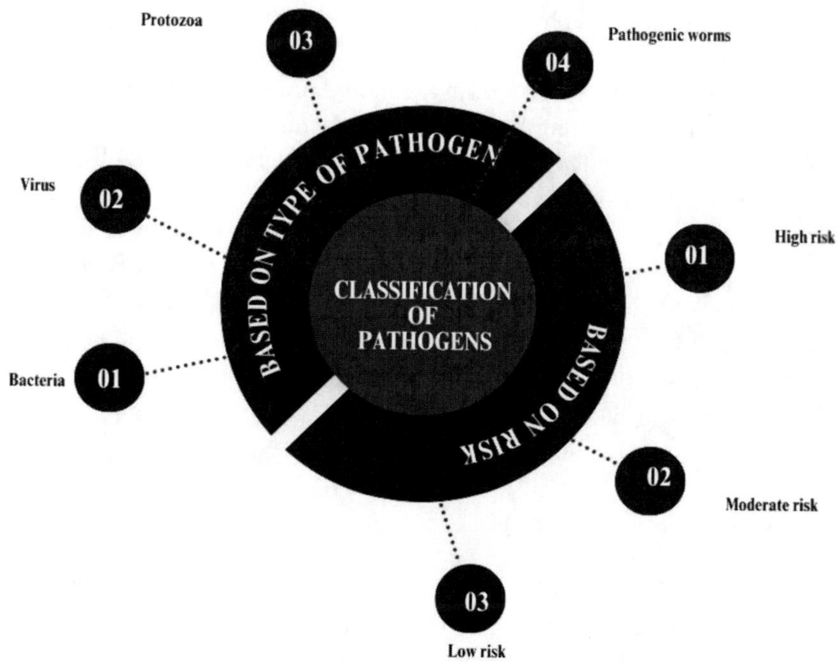


**Figure 2.** Classification of pathogens in wastewater.

Industrial wastewater discharges pose serious environmental problems, particularly by contributing to eutrophication, which increases the cost of wastewater treatment and encourages excessive algal blooms. These circumstances degrade recreational water quality, disturb aquatic ecosystems, lower oxygen levels, and jeopardize the human health and animals. Furthermore, these pollutants have the ability to contaminate aquatic resources, which increases ecological dangers and spreads pollution. To lessen these negative consequences and safeguard the environment and public health, wastewater must be properly monitored, treated and managed [15].

Wastewater frequently contains organic materials and pathogenic germs that can transmit diseases including typhoid, dysentery, diarrhea, and mal-absorption. Perfluorinated compounds (PFCs) that are commonly present in wastewater; have been connected to a number of serious health problems, such as pre-eclampsia, birth abnormalities, and increased sterility (Figure 2). Serious health issues like skin cancer, lung infections, and neurotoxicity can also result from prolonged exposure to the toxins present in wastewater. To reduce these threats and protect the environment and public health, wastewater management must be done effectively [16, 17].

The two main categories of pollutants found in untreated wastewater are chemical and microbiological. Microbial contamination can cause infectious diseases, while chemical contaminants can expose people to harmful concentrations. Both types pose serious risks to public health and ecosystems, underscoring the urgent need for efficient wastewater treatment and management techniques to reduce these dangers and protect water quality [18, 19].

## 2. Wastewater Treatment Methods

In traditional wastewater treatment, sediments, organic matter, and occasionally nutrients are eliminated by a combination of physical, chemical, and biological processes. Wastewater treatment typically progresses through four phases of treatment that include preliminary, primary, secondary, and tertiary/advanced, in ascending order of intensity. Primary Treatment involves screening and sedimenting suspended solids and floating debris. It lowers total suspended solids (TSS) by 50–60% and biological oxygen demand (BOD) by 20–30% [20]. In secondary treatment the dissolved organic matter is removed by biological processes. Waste is transformed by microbes into energy, water, and carbon dioxide. In order to separate biological sludge for energy recovery in biogas plants, this stage eliminates 85% of BOD and TSS. To be reused for irrigation, clear water must first be nitrified, filtered, and disinfected [21]. Tertiary Treatment involves the advanced methods like ultrasonication, UV treatment, and ozone exposure that remove bacteria, heavy metals, and contaminants. This process ensures 99% impurity removal, producing safe water for drinking or domestic use [22].

### 2.1. Physical Methods

Physical techniques like flotation and sedimentation are used in primary treatment to remove 65% of oil and grease, 50-70% of TSS, and 25-50% of BOD. However, it does not remove dissolved or colloidal materials. Additional treatment is required to further reduce organic matter and suspended solids. It primarily relies on biological processes involving controlled microorganisms to target the dissolved residues and colloidal pollutants. This step is necessary to meet higher water quality criteria that basic treatment alone cannot achieve [21].

#### *2.1.1. Mechanical Method*

Filtration and screening are mechanical techniques used in wastewater treatment. As one of the earliest techniques, screening was developed to remove large contaminants from wastewater, safeguarding downstream machinery, prevent damage, and ensure efficient plant operation. Additionally, it prevents large suspended debris from entering the main settling tanks. Although they are usually considered as biological rather than mechanical, filters are often used in aerobic attached-growth biological therapy. By facilitating microbial attachment to a medium and enabling efficient decomposition, filters aid in the removal of organic matter from wastewater. These techniques are for ensuring the overall efficiency of the treatment process [23].

#### *2.1.2. Chemical Method*

Chemical precipitation is the process of adding specific chemicals to wastewater to react with heavy metals to form insoluble precipitates. Filtration or sedimentation can subsequently be used to remove these precipitates. Following treatment, the water is decanted and made suitable for reuse or disposal. Alum, ferric chloride, ferric sulfate, ferrous sulfate, and lime are common chemical coagulants used in wastewater treatment for this purpose. By aiding in the removal of impurities, these chemicals ensure that the treated water satisfies the necessary quality standards [24].

Adsorption refers to the process by which soluble substances from a solution accumulate on the surface of a solid. The purpose of this technique is to remove any remaining dissolved organic debris and further purify the water. Due to its large surface area, activated carbon efficiently absorbs these compounds, reducing any residual pollutants and improving the overall quality of the water. Before treated water is released or utilized again, this procedure is essential [24].

### 2.2. Biological Methods

#### *2.2.1. Composting*

Composting is an economical and environmentally friendly method of converting organic waste into humus through microbial activity. Mesophilic and thermophilic microorganisms, such as bacteria, fungus, and actinomycetes, break down complex materials including lignin, protein, and cellulose in regulated aerobic process. In this process, *Aspergillus*, *Pseudomonas*, *Bacillus*, and *Penicillium* are important cellulolytic microorganisms that break down organic matter and turn it into compost that is rich in nutrients. This procedure improves the quality of soil in addition to reducing waste [25].

Plant diseases, including *Alternaria solani* and other pathogens, can be decreased by co-composting food waste with residual herbal materials. In addition to supporting landfill cover, erosion prevention, and land reclamation, composting improves soil fertility and serves as a natural pesticide. By adding vital nutrients to the soil, encouraging healthier plant development, and lowering the need for chemical fertilizers, it also improves the quality of soil [26].

### *2.2.2. Anaerobic Digestion*

Anaerobic digestion is a biological process that produces biogas (mostly methane) from organic waste by using microorganisms in an oxygen-free atmosphere. Hydrolysis is the rate-limiting phase of the process, followed by acidogenesis, acetogenesis, methanogenesis. White-rot fungi like as *Pleurotus ostreatus* and *Phanerochaete chrysosporium* produce lignin-degrading enzymes that improve the production of biogas when they are coupled with microorganisms that break down cellulose. This procedure supports waste management and sustainable resource recovery by turning wastewater into useful fertilizer in addition to producing renewable energy and lowering greenhouse gas emissions [27, 28].

### *2.2.3. Membrane Aerated Biofilm Reactor*

An advanced biofilm technology called the Membrane Aerated Biofilm Reactor (MABR) is intended to treat surface water that has been contaminated by industrial discharges, sewage, and agricultural runoff that contain pesticides and fertilizers, all of which can lead to eutrophication and health hazards. This technology delivers oxygen in a controlled way while promoting biofilm formation using porous hollow fiber membranes. Through a counter-diffusion mechanism, contaminants in the wastewater travel in the direction opposite to that of oxygen diffusion across the membrane into the biofilm, thereby, increasing the efficiency of contaminant removal. By effectively eliminating carbon and nitrogen molecules, MABR is very successful in treating pharmaceutical pollutants, dye-laden effluents, and wastewater from industries and municipalities [25-29].

### *2.2.4. Activated Sludge*

Activated sludge, composed of bacteria, fungi, and protozoa, treats wastewater by removing organic carbon, nitrogen, and phosphorus. In aeration tanks when organic matter is biodegraded by microorganisms, flocs are formed. These flocs can be subsequently separated by gravity-based settlement prior to the release of treated water [29].

### *2.2.5. Wastewater Stabilization Pond*

Wastewater stabilization ponds are an affordable treatment option, especially in places with limited resources. Wastewater is introduced in to ponds where diseases and pollutants are reduced by natural processes. Time, temperature, high pH (above 9), UV radiation, and dissolved oxygen are all important purifying factors. During therapy, these ponds also successfully get rid of helminths. This approach is sustainable since pollutants and pathogen levels are considerably reduced, the treated water may be safely reused for irrigation, meeting both agricultural and environmental demands [30].

## 2.3. Advanced Methods

The goals of advanced wastewater treatment methods are to improve environmental sustainability, cost effectiveness, and efficiency. By focusing on contaminants that are challenging to break down, these strategies supplement conventional biological and physical techniques. Treatment results are greatly enhanced by technologies like microbial fuel cells, artificial wetlands, and Upflow Anaerobic Sludge Blanket (UASB) reactors. For example,

UASB reactors minimize energy usage and sludge formation while achieving up to 85% elimination of COD. Microbial fuel cells provide bioenergy during treatment, whereas artificial wetlands filter pollutants using natural processes. Constant observation guarantees that treated water meets safety requirements, fostering environmental preservation and resource recovery [31]. Membrane filtration techniques like electrodialysis (ED), nanofiltration (NF), reverse osmosis (RO), and ultrafiltration (UF) are commonly employed to effectively remove heavy metals from wastewater, offering efficient solutions for treating toxic and persistent contaminants [32].

UF operates at low transmembrane pressures to remove dissolved and colloidal contaminants but cannot eliminate small metal ions like copper or nickel. Enhanced methods, such as micellar-enhanced ultrafiltration (MEUF) and polymer-enhanced ultrafiltration (PEUF) use surfactants or polymers to improve metal removal. Combining MEUF with electrolysis or activated carbon further enhances metal ion removal efficiency. RO; employing a semi-permeable membrane effectively removes a broad range of dissolved substances, including metals, making it invaluable for wastewater treatment and desalination. NF; positioned between UF and RO is gaining popularity due to its higher flux rates and lower energy demands, efficiently removing heavy metals like arsenic and chromium [33].

ED employs ion-exchange membranes and electric potential to extract ions from water, making it suitable for applications like metal recovery, wastewater treatment, and brackish water desalination. This technique is particularly effective for separating ionic contaminants. Additionally, adsorption using materials such as hydrogels or activated carbon offers an economical and environmentally friendly solution for heavy metal removal from wastewater. These adsorbents effectively capture and immobilize metal ions, making them versatile for various treatment processes. Combining ED with adsorption methods can further enhance treatment efficiency, providing a sustainable approach to addressing heavy metal pollution in industrial and municipal wastewater systems [33].

## 3. Microbial Bioremediation

Microbial bioremediation in wastewater treatment harnesses the metabolic abilities of microorganisms to degrade pollutants, remove nutrients, and detoxify harmful substances. This eco-friendly process is vital for maintaining water quality and preventing environmental contamination. By utilizing bacteria, fungi, and algae, wastewater bioremediation efficiently removes organic matter, heavy metals, and toxic compounds. Its cost-effectiveness, sustainability, and adaptability make it an essential component of modern water treatment systems and environmental management strategies. Various microorganisms are employed for the treatment of wastewater generated from different sources (Table 2).

### 3.1. Bacterial Bioremediation

Using the activities of microorganisms such as *Bacillus subtilis*, *Pseudomonas*, endophytes, and other microbial strains, bacterial remediation is a popular technique for treating industrial effluent. An efficacy of 70–80% is attained by bacterial bioremediation using a variety of microbial consortia. Aerobic techniques like membrane bioreactors, sequencing batch reactors (SBRs), and activated sludge are frequently used; SBRs are especially economical and effective. Because of their low energy consumption, anaerobic processes are extremely well-liked, particularly in the pharmaceutical sector, where upflow anaerobic stage reactors are extensively used.

Improved process control, operating efficiency, and pollutant targeting are provided by hybrid systems that combine anaerobic and aerobic methods. Furthermore, cutting-edge techniques like Fenton oxidation combined with bioremediation are showing promising ways to improve treatment efficiencies. These advanced methods, which integrate chemical and biological processes, offer a more sustainable and efficient approach to managing complex industrial wastewater, addressing a wide range of contaminants, and minimizing environmental impacts [34].

According to Suresh and Abraham (2018), the activated sludge method entails adding a flocculent suspension of microorganisms, such as bacteria that oxidize ammonia and nitrite, like *Nitrosococcus mobilis* and *Nitrospira* species, and then aerating the mixture [35]. Submerged diffusers are frequently used to stir the tank in this technique, which is divided into step aeration and standard aeration. To maximize treatment for various wastewater models, a number of parameters can be changed, including pH, aeration time, and the usage of oxidation ditches. Due to its adaptability, the activated sludge process is a flexible and successful biological wastewater treatment method that effectively removes nitrogen compounds and organic contaminants [35]. SBRs offer a space efficient alternative to the activated sludge process, as they require less area and eliminate the need for sludge return. According to a study; one of the key advantages of SBRs is the integration of aeration and sedimentation into a single unit, thereby streamlining the wastewater treatment process [36].

The lower energy consumption of anaerobic processes (roughly 1 kWh of fossil energy per kilogram of COD removed) gives them a major advantage over aerobic ones. In contrast, the kind of aeration system that is employed has a significant impact on the energy requirements of the activated sludge process. Depending on the microorganisms involved, anaerobic biodegradation operates as a fermentation process using a variety of processes, such as hydrolysis, acetogenesis, acidogenesis, and methanogenesis. The Anaerobic Contact Reactor (ACR), Fluidized Bed Reactor (FBR), UASB, and Anaerobic Film Reactor are common anaerobic reactor layouts for treating industrial wastewater. In addition to providing energy-efficient solutions and recovering biogas as a useful byproduct, these reactors efficiently treat high-strength industrial wastewater [37].

Pharmaceutical wastewater is particularly well suited for anaerobic wastewater treatment; with a UASB Reactor often preferred over anaerobic film or contact reactors. To enhance overall treatment efficiency, these reactors are frequently used as a pre-treatment step before the application of activated sludge processes [37].

### 3.2. Algal Bioremediation (Phycoremediation)

Heavy metals such as chromium can be effectively removed from wastewater using a process called phycoremediation, a process that utilizes cyanobacteria and algae. Species such as *Scenedesmus, Chlorella, Nodularia, Phormidium, Botryococcus, Spirulina, Oscillatoria, Desmodesmus, Arthrospira, Nostoc, Chlamydomonas,* and *Cyanothece* are particularly effective at absorbing nutrients and heavy metals from contaminated water. In addition to effectively removing pollutants, this biological technique generates biomass that may be converted into biofuels, offering a sustainable and environmentally friendly solution [38]. Living *Chlorella* species cells, for instance, help reduce Cr (VI) to Cr (III), thereby, greatly reducing the amount of residual Cr (VI) in water. The efficacy of *Chlorella* and *Nitrella* species in treating Chromium contaminated wastewater was studied that demonstrated that Cr (VI) concentrations were decreased by more than 50%, with only 10–21% remaining [38].

**Table 2.** Various biological organisms involved in degradation of pollutants in wastewater

| No | Biological organism category | Biological organism | Pollutant | Reference |
|---|---|---|---|---|
| 1 | Bacteria | Desulfovibrio, Desulfotomaculum, Desulfobacter, and Desulfococcus genera | sulfate | [42] |
| 2 | Bacteria | Aeromonas hydrophila, Bacillus subtilis, Bacillus cereus | Azo Dyes | [43] |
| 3 | Bacteria | Pseudomonas sp., Pseudomonas luteola, Proteus mirabilis | Azo Dyes | [44] |
| 4 | Bacteria | Acinetobacter Mycobacterium, Burkholderia, Gordonia, Brevibacterium, Dietzia, Aeromicrobium, Pseudomonas, Flavobacteria, Chrobacteria, Nocardia, Moraxella, Aeromonas, Modococci, Cyanobacteria, Streptomyces, Bacilli, Arthrobacter, | petroleum products | [45] |
| 5 | Bacteria | Bacillus and Staphylococcus Pseudomonas putida and Acinetobacter | pesticides | [46] |
| 6 | Bacteria | Rhizobium leguminosarum | Heavy metals- Cd | [47] |
| 7 | Bacteria | Desufuromonas sp | Heavy metals- Cr | [48] |
| 8 | Bacteria | Vibrio harveyi | Lead | [49] |
| 9 | Bacteria | Bacillus sp., Clostridium sp., Pseudomonas sp., and Escherichia sp | Hg | [50] |
| 10 | Algae | Scenedesmus sp | Organic waste | [51] |
| | | Chlamydomonas mexicana | Salicylic acid, paracetamol | [52] |
| | | Chlorella pyrenoidosa | Ciprofloxacin, Progesterone, paracetamol, diclofenac | [53] |
| | | Desmodesmussubspicatus | 17 α-Ethynylesteradiol | [54] |
| | | Chlorella vulgaris | Tetracycline, Organic waste | [55] |
| 11 | Cyanobacteria | Aphanocapsu sp. and Plectonema sp. | Crude oil | [56] |
| | | Anabaena sp., Osillatoria sp., Spirulina sp., Chrooous sp., Pseudospongioous sp., Sytonema sp., Dolihospermum | | [45 & 56] |
| 12 | Fungi | Pleurotus pulmonar ius, Stachybotrys sp., Cephalosporium aphidicola, Aspergillus parasitica, Verticillum terrestre, Candida sp., Acremonium sp., Glomus sp., Minimedusa sp., Talaromyces, Hydnobolites, Peziza | Heavy metals | [57] |
| 13 | | Saccharomyces cerevisiae, Galactomyces geotrichum, Trichosporon beigelii, and Candida krusei | dyes | [58] |

Seaweeds play a vital role in wastewater treatment due to their effectiveness in removing dyes, phenols, hydrocarbons, biphenyls, nutrients, and heavy metals. Their adaptability makes them particularly useful for addressing pharmaceutical waste contamination. Microalgal bioremediation is gaining attention for its ability to treat real wastewater samples while simultaneously generating biofuels. The residual algal biomass can also serve as a biosorption media in subsequent treatment stages. Algae are a great resource for environmental restoration due to their multiple functions [39].

Furthermore, it has been shown that the cyanobacterial isolate *Nostoc carneum* may biodegrade low-density polyethylene (PE) strips. It is thought that enzymes, which used about 3% of the carbon from PE, aided in the decomposition process. These results highlighted the substantial potential of cyanobacteria and algae in tackling the problems of plastic biodegradation and wastewater pollution [39].

### 3.3. Fungal Bioremediation

*Aspergillus niger, Aspergillus fumigatus*, and *Aspergillus niveus* are commonly used fungal strains that are utilized in bioremediation; however, their applications are limited due to the long duration of their spore generation cycles. Ascomycetes fungi, on the other hand, are well known for their capacity to efficiently extract COD from industrial wastewater, particularly that originating from distilleries [40].

An increasing number of studies are exploring the biochemical potential of filamentous fungal taxa, particularly basidiomycetes, for the enzymatic biotransformation of various pharmaceuticals. These fungi have the capacity to alter pharmaceutical compounds, offering a cutting-edge approach to bioremediation and waste management [41].

## 4. Biotechnological Approach-Genetic Engineering for Enhanced Bioremediation

### 4.1. Genetically Modified Organisms (GMOs) in Wastewater Treatment

Two important bioremediation mechanisms are bioaccumulation and biosorption. While biosorption removes heavy metals by binding metal ions to negatively charged microbial cell walls through electrostatic attraction, bioaccumulation includes pollutants being digested by living cells. GMOs, particularly transgenic plants and microbes have demonstrated enhanced capabilities in degrading pollutants and xenobiotics, making them vital tools for the remediation of contaminated soils. By efficiently treating soil, groundwater, wastewater, and activated sludge under controlled settings, these synthetic organisms have demonstrated great promise in environmental cleaning. The controlled use of these organisms offers a potential solution to various environmental pollution problems [59].

While GMOs are significantly more effective at bioaccumulation, wild bacteria such as *Aspergillus fumigatus*, *Bacillus cereus*, and *Pseudomonas veronii* exhibit only modest bioaccumulation capability. Genetically modified fungi such as *Aspergillus* and *Penicillium*, as well as yeasts like *Saccharomyces cerevisiae*, can absorb up to three times more heavy metals compared to their non-modified counterparts. GM bacteria, such as *Pseudomonas putida* and *E. coli*, are very efficient in bioremediation efforts because they exhibit improved metal resistance and quicker pollutant breakdown, even in the presence of severe pH and temperature circumstances [60]. While bacteria such as *Deinococcus geothermalis* are excellent in remediating radioactive waste, *Pseudomonas fluorescens* and related strains are designed to break down hydrocarbons like naphthalene and camphor. Due to their improved growth, tolerance, and pollutant removal capabilities, these modified organisms are very effective in environmental remediation. Bioremediation techniques such as bioaccumulation and biosorption offer effective solutions for controlling different kinds of environmental contaminants by using both living and dead cells to bind cationic heavy metals through electrostatic attraction [61].

While industrialization drives human progress, it also produces xenobiotic pollutants like 1,2,3-Trichloropropane (TCP), a persistent, poisonous, and carcinogenic chlorinated hydrocarbon. TCP is widely employed as an intermediate in the synthesis of polysulfone and hexafluoropropylene polymers, as well as a cleaning agent and in the manufacturing of paint. When TCP is improperly disposed of, it frequently contaminates groundwater and poses serious health risks. No naturally occurring microbial cultures have been found to breakdown TCP aerobically, despite the fact that it may be dechlorinated anaerobically [62, 63].

## 4.2. Engineering Microbes for Targeted Pollutant Removal

In contaminated soil and water, bacterial isolates have demonstrated an exceptional capacity to break down polycyclic aromatic hydrocarbons (PAHs). *Sphingobium* and *Sphingomonas* strains are especially noteworthy for their robust biodegradative potential and capacity to adapt to a wide range of ecological settings. Numerous aromatic substances, such as dioxins, biphenyl, naphthalene, carbazole, m-xylene, phenanthrene, chrysene, benzo[a]pyrene, and several methyl-, chloro-, hydroxyl-, and nitroaromatic compounds, can be effectively broken down by these bacteria. They are useful instruments for environmental cleanup and bioremediation due to their adaptability [64].

PAH bioremediation holds great promise, further research is needed in several areas, such as degradation byproducts, the evolution and transfer of genes linked to biodegradation, and underlying regulatory processes. Significant bioremediation potential was shown by the presence of oxygenase-encoding genes, with the β-ketoadipate and homogentisate pathways which are emerging as key metabolic pathways [65, 66].

The fusant strain F14, which was developed by merging the pyrene-degrading *Pseudomonas* strain GP3A and the phenanthrene-degrading *Sphingomonas* strain GY2B, represents a breakthrough in this field. Additionally, GMOsexhibit improved breakdown of PAHs. For example, *E. coli* producing salicylate oxygenase from *S. yanoikuyae* B1 and *Pseudomonas putida* which is designed to overexpress PHPhA depolymerase, both exhibit exceptional efficiency. These developments demonstrate the increasing capability of synthetic biology to successfully tackle PAH pollution [66].

## 5. Phytoremediation: Plants in Wastewater Treatment

Phytoremediation provides an economical and environmentally friendly way to purify contaminated soil or water using plants. Through processes such as phytoextraction, phytovolatilization, phytodegradation, rhizofiltration, and phytostabilization, this technique enables plants to efficiently absorb, break down, or stabilize contaminants [67].

### 5.1. Phytoextraction or phytoaccumulation

Phytoextraction is the process by which plants absorb metal pollutants from the soil and translocate them to above-ground biomass, such as leaves and shoots. This technique is most effective for large areas of shallow, low-to-moderately contaminated soil. Harvesting plants at peak growth and metal accumulation enables metal recovery and recycling. With over 400 plant species capable of absorbing metals like copper, zinc, and nickel, phytoextraction has proven to be an effective remediation technique [68].

### 5.2. Phytodegradation/Phyto-transformation

Phytoremediation is a natural remediation process in which organic contaminants are broken down into simpler molecules by the enzymatic activity of plants and their associated endophytic bacteria. Plant produced enzymes help degrade harmful pollutants, facilitating their uptake into plant tissues and promoting plant growth. The interaction among soil, water, bacteria, and plants is essential for effective remediation. By enhancing microbial activity, reducing runoff, promoting pollutant adsorption onto roots, and stimulating transpiration, plants play a central role in making this strategy a sustainable and efficient method for managing environmental contaminants [69].

### 5.3. Rhizofiltration/Phytofiltration

In contrast to phytoextraction, rhizofiltration (sometimes known as phytofiltration) uses plants that are cultivated in water to extract contaminants from aqueous streams. Before wastewater is introduced for treatment, plants are first grown hydroponically to establish vast root systems. Over time, pollutants accumulate in the roots or other plant parts, which are then harvested and properly disposed of. A more sophisticated version of this technique called blastofiltration can absorb more pollutants. Rhizofiltration effectively removes heavy metals like chromium (Cr), lead (Pb), and zinc (Zn), while its effectiveness varies depending on the particular metal being treated and the metabolic activities of the plant [70].

### 5.4. Phyto-Stimulation and Rhizodegradation

Microorganisms like bacteria, fungus, and yeast in the rhizosphere—the soil surrounding plant roots—play a crucial role in phyto-stimulation and rhizodegradation. Toxins, fuels, solvents, and other organic contaminants are broken down by these bacteria into simpler chemicals that provide nutrients and energy. Certain compounds released by plant roots nourish these microbes and encourage their growth and metabolism. In exchange, the microorganisms break down organic matter, turning it into nutrients that enhance plant growth and contributes to the overall process of environmental remediation [71].

### 5.5. Phytostabilization (Phytorestoration)

Using plants to stabilize pollutants in the soil and stop them from spreading by wind or water, a process known as Phytostabilization (or phytorestoration) involves using plants to stabilize pollutants in the soil and prevents their spread by wind or water, thereby reducing their bioavailability. By chemically attaching or adsorbing pollutants onto plant roots, this technique also regulates hydraulic flow, minimizing pollution seepage into groundwater. Plants used in phytostabilization must be capable to withstanding high metal concentrations, grow rapidly, establish themselves, and prevent contaminants from reaching edible aerial parts. This approach offers a sustainable and eco-friendly solution to address contamination, working particularly well in fine-textured, highly polluted soils. It promotes long-term environmental health by lowering the mobility of pollutants without requiring their removal from the site [72].

### 5.6. Phytovolatilization

In phytovolatilization, contaminants are absorbed from the soil by plant roots and transferred to aerial portions such as stems and leaves. Through metabolic processes in the leaves, these pollutants are converted into volatile chemicals, which are then expelled into the atmosphere through transpiration. Although this approach simplifies plant management and avoids the difficulties associated with pollutants disposal, it raises questions about the environmental safety of releasing volatile metals such as selenium and mercury. Photovoltalization involves both the transpiration of pollutants and their transfer from different plant parts to the leaves, where they are volatilized [73].

### 5.7. Role of Plants in Removing Contaminants

Phytoremediation is a low-cost, green technology that uses plants that accumulate metals to remove organic pollutants, radionuclides, pesticides, and hazardous metals from soil, water, and air. It fixes pollution at its source, restores habitats, and lowers remedial expenses. In industrial effluents, *Typha latifolia* and *Phragmites karka* lower COD, BOD, and solids, while plants like *Eichhornia crassipes* effectively remove heavy metals like Zn, Cr, Cu, Pb, and Cd. *Dahlia* also absorbs nickel and lead, whereas *Pteris vittata,* a pteridophyte; accumulates arsenic. Through the reduction of physical, chemical, and biological pollutants in industrial effluents, these plants also treat wastewater efficiently [74].

## 6. Genetic Modification of Plants for Enhanced Phytoremediation

Transgenic plants have emerged as a promising solution in the evolving strategies for soil decontamination, which have become the focus of latest remediation research. By introducing functional genes or particular promoters, these genetically modified plants can enhance heavy metal accumulation, detoxification, and translocation. For example, in the presence of elevated Cu and Zn levels, transgenic canola expressing the OsMyb4 gene under a stress-inducible promoter demonstrated enhanced survival. This alteration increases the plant's resistance to metal stress by inducing proline synthesis and phenylpropanoid pathways. These developments demonstrate how transgenic plants can be used to improve soil remediation by increasing their capacity to withstand and detoxify environmental pollutants [75].

Similarly, transgenic *Arabidopsis thaliana* exhibited improved resistance to cadmium (Cd) stress due to increased lignin formation caused by overexpressing lignin biosynthesis genes such as VsCCoAOMT. By encouraging lignin buildup, a crucial part of the plant's defensive systems against harmful substances, this genetic change increases the plant's resistance to heavy metal pollution. These developments show how genetically modified plants might enhance stress tolerance and facilitate more effective bioremediation of polluted soils [76].

To improve tolerance and accumulation of heavy metals, transgenic plants have been developed. For instance, tobacco was modified to increase the uptake of cadmium (Cd) by introducing heavy metal. Likewise, endophytic bacterial genes, such those of *Bacillus shackletonii*, have enhanced plant absorption of heavy metals. Polychlorinated biphenyls and heavy metals were successfully eliminated from soil by genetically engineered nettle (*Urtica dioica*). Genes like StGCS-GS were introduced into sugar beet to boost glutathione synthesis and increase tolerance to metals like Cd, Cu, and Zn [77]. PtABCC1 and other metal transport genes greatly enhanced the removal of mercury (Hg) in transgenic poplar and *Arabidopsis*. To increase the potential of phytoremediation, novel strategies, such as CRISPR technology, are now being investigated to target genes involved in metal binding, transport, and stress tolerance. These advancements demonstrate the potential of genetic engineering to reduce environmental pollution by improving metal uptake and purification [78].

## 7. Enzyme-Based Wastewater Treatment

### 7.1. Role of Enzymes in Breaking Down Complex Pollutants

Living cells from bacteria, fungi, and plants create enzymes, which function as reaction-specific biocatalysts are useful in the treatment of wastewater. While hydrolytic enzymes from bacteria and fungi are commonly employed, plant-derived enzymes such as peroxidases alsoplay an important role. For instance, lipase from *Candida rugosa* effectively breaks down the triglycerides in palm oil mill effluent, and laccase from *Trametes versicolor* breaks down chlorpyrifos. Wastewater communities' microbial enzymes are highly effective in biotransforming antibiotics. For pretreatment, lipases hydrolyze wastewater that contains oil and grease into glycerol and fatty acids. Cellulase, amylase, lipase, and protease are examples of hydrolytic enzymes that break down proteins, lipids, and carbohydrates into smaller molecules. The effectiveness of wastewater treatment is increased by enzymatic methods in conjunction with activated sludge or anaerobic fermentation [79, 80].

Antibiotics, hormones, and metabolic agents are among more than 200 pharmaceutical substances that are frequently detected in aquatic habitats as organic micropollutants. Only 20–90% of these residues are eliminated by conventional wastewater treatment, and persistent antibiotics like cephalosporins and tetracyclines pose ecotoxicological hazards. Surface water, groundwater, and drinking water have all been shown to have traces of these contaminants. Up to 95% of medications can be eliminated by enzymatic techniques employing oxidoreductases like laccase and chloroperoxidase. However, obstacles including complex degradation routes and toxic byproducts prevent broad use. To optimize enzymatic treatments for the safe and efficient removal of pharmaceutical pollutants, more research is required [81,82].

Despite their widespread usage, organophosphorus and carbamate insecticides are dangerous for both human health and the environment because they block cholinesterase and have other negative effects. Although highly hazardous pesticides such as carbofuran and triazophos are banned, they continue to have a substantial market; particularly in developing countries. Enzymatic methods have shown promise in the degradation of pesticides also. For instance, laccase immobilized on magnetic iron nanoparticles eliminated nearly 99% of the chlorpyrifos in 12 hours at pH $^{7}$ and 60°C. At a concentration of 26.7 μmol/L, immobilized chloroperoxidase (CPO) completely degraded isoproturon in less than 10 minutes, while enzymes including OpdA, OPAA, and Ophs were able to effectively eliminate contaminated organophosphates [83-85].

Synergistic enzyme systems enhance degradation efficiency. For instance, acetylcholinesterase converts cholinergic chloride to choline, which is further broken down by choline oxidase into harmless metabolites. However, some enzymatic transformation products can have adverse environmental impacts. Using multiple enzymes in combination offers a sustainable approach to pesticide degradation, promoting environmentally beneficial treatment outcomes [86-88].

Basically three different types of enzyme systems are used for wastewater treatment (Figure 3). Free enzyme systems use both refined and crude enzymes to eliminate contaminants. Sometimes, crude enzymes—which are frequently combined with natural mediators—perform better than purified ones.

However, due to the residual nutrients, they pose a risk of increasing contamination. Purified enzymes, although more effective and easier to control, are significantly more expensive. Their cost is increased by the use of sophisticated production methods like chromatography. Despite these distinctions, both types of enzymes are essential for the elimination of pollutants; the choice depends on factors such as cost, effectiveness, and the specific requirements of the environmental treatment process [81-89].

Cost and enzyme recovery challenges limit the feasibility of both approaches. However, immobilized enzymes help overcome these limitations by binding enzymes to solid matrices, thereby reduce costs and contamination while enhancing stability, reusability, and resilience to environmental factors. Enzyme activity may be influenced by the choice of matrix and immobilization methods, with trade-offs between chemical and physical immobilization.

Synthetic materials, organic matrices like cellulose, and inorganic materials like silica gel are popular supports, each with pros and cons [81-89].

Enzyme mediators are stable, light compounds that act as electron shuttles, expanding the range of substrates and improving the efficiency of pollution removal. Although ABTS and syringaldehyde (SA) are the two examples of synthetic and natural mediators respectively, both enhance enzyme performance, however; their toxicity and effectiveness vary. For instance, SA and ABTS improve the breakdown of antibiotics, but their high expense and possible adverse environmental impacts prohibit their widespread usage. The concentration of the mediator is important since improper dosing may decrease efficiency or increase effluent toxicity. Despite these challenges, the breakdown of complex pollutants like triclosan requires mediators, underscoring the need for affordable and ecologically friendly substitutes. The resistant pollutants that enzymatic processes can biodegrade include grease, lubricants, pharmaceuticals, POPs, insecticides, and industrial chemicals. However, this technology is primarily utilized as a pretreatment approach because enzymes primarily break down complex chemicals into simpler ones without completely eradicating them. To ensure comprehensive wastewater treatment, enzymatic processes must be used in conjunction with other methods. Current challenges include the development of multi-enzyme systems, monitoring of transformation products (TPs), improving enzyme efficiency, and expanding applications. Further research and real wastewater testing are required to advance this promising technique [89].

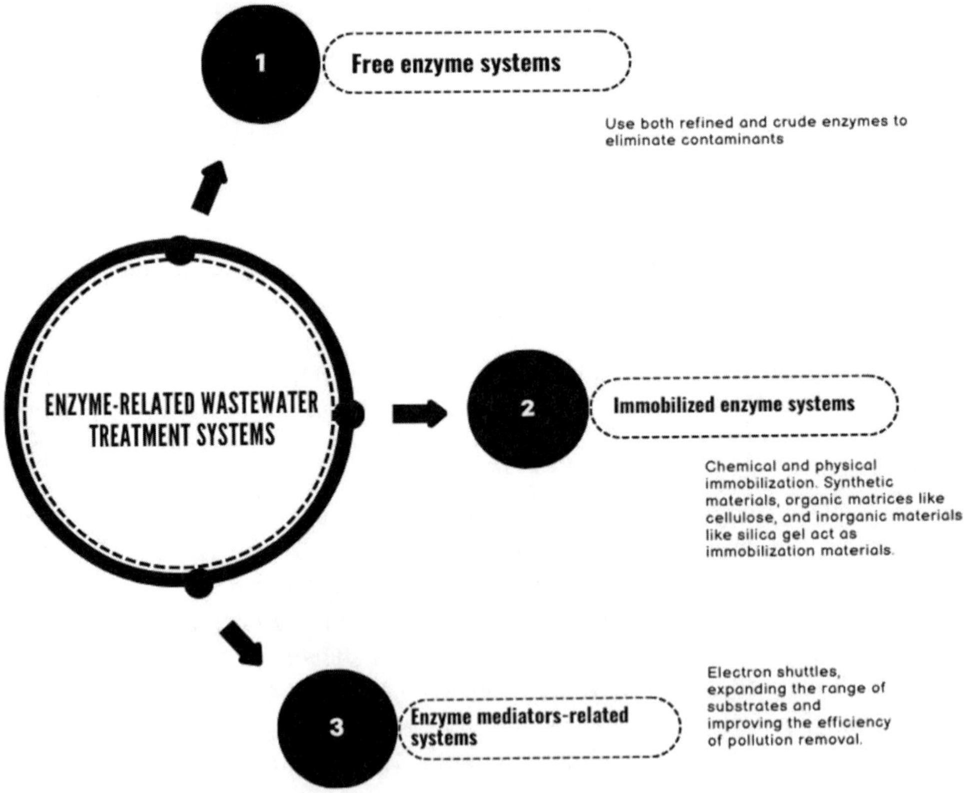


**Figure 3.** Types of Enzyme related Wastewater Treatment Systems.

## Conclusion

Biotechnological methods for sustainable wastewater treatment provide creative, effective, and environmentally responsible answers to the urgent problems of resource management and water contamination. Utilizing biological systems like microbes, enzymes, and plants, these techniques offer efficient substitutes for traditional chemical and physical therapies, often with lower energy consumption and environmental impact. Technologies such as bioreactors, artificial wetlands, and bioaugmentation have shown great promise in eliminating pathogens, heavy metals, and organic and inorganic pollutants from wastewater, improving public health and water quality.

Furthermore, by promoting resource recovery, biotechnological solutions are consistent with the circular economy's tenets. The creation of biogas, biofertilizers, and other useful byproducts from wastewater treatment, highlights the advantages of both resource development and pollution reduction. Due to their scalability and adaptability, these methods are especially appropriate for a variety of geographic and socioeconomic circumstances, from urban centers to rural locations with little infrastructure. Ongoing research and innovation are necessary to address issues like cost-effectiveness, operational optimization, and integration into current wastewater treatment frameworks. To sum up, biotechnological developments have great potential to manage wastewater sustainably, opening the door to a cleaner environment and a more sustainable future. To fully realize their potential, collaboration among scientists, policymakers, and industry leaders will be essential.

## References


[1] Patel, T. M., Mahour, P. K., Mahour, R. K., Lautre, H. K., & Shah, P. J. (2017). Physico-chemical analysis of ground water quality of Bhanvad. *International Journal of Economy, Energy and Environment, 2*(5), 87-89.

[2] Lofrano, G., & Brown, J., (2010). Wastewater management through the ages: A history of mankind. Science of The Total Environment 408, 5254-5264.

[3] Petrie, B., Barden, R., Kasprzyk-Hordern, B., (2015). A review on emerging contaminants in wastewaters and the environment: Current knowledge, understudied areas and recommendations for future monitoring. Water Research 72, 3-27.

[4] Richardson, S. D., & Kimura, S. Y., (2017). Emerging environmental contaminants: Challenges facing our next generation and potential engineering solutions. *Environmental Technology & Innovation* 8, 40-56.

[5] Merel, S., Benzing, S., Gleiser, C., Di Napoli-Davis, G., Zwiener, C., (2018). Occurrence and overlooked sources of the biocide carbendazim in wastewater and surface water. *Environmental Pollution* 239, 512-521.

[6] Campos-Mañas, M. C., Plaza-Bolaños, P., Martínez-Piernas, A. B., Sánchez-Pérez, J. A., Agüera, A., (2019). Determination of pesticide levels in wastewater from an agro-food industry: Target, suspect and transformation product analysis. *Chemosphere* 232, 152-163.

[7] Becouze-Lareure, C., Dembélé, A., Coquery, M., Cren-Olivé, C., Barillon, B., Bertrand-Krajewski, J. L. (2016). Source characterisation and loads of metals and pesticides in urban wet weather discharges. *Urban Water Journal* 13, 600-617.

[8] Westerhoff, P., Lee, S., Yang, Y., Gordon, G. W., Hristovski, K., Halden, R. U., Herckes, P., (2015). Characterization, Recovery Opportunities, and Valuation of Metals in Municipal Sludges from US Wastewater Treatment Plants Nationwide. *Environmental Science & Technology* 49, 9479-9488.

[9] Bilal, M., Asgher, M., Parra-Saldivar, R., Hu, H., Wang, W., Zhang, X., & Iqbal, H. M. (2017). Immobilized ligninolytic enzymes: an innovative and environmental responsive technology to tackle dye-based industrial pollutants–a review. *Science of the Total Environment*, *576*, 646-659.

[10] Park, Y., Lee, Y. C., Shin, W. S., Choi, S. J., (2010). Removal of cobalt, strontium and cesium from radioactive laundry wastewater by ammonium molybdophosphate-polyacrylonitrile (AMP-PAN). *Chemical Engineering Journal* 162, 685-695.

[11] Montaña, M., Camacho, A., Devesa, R., Vallés, I., Céspedes, R., Serrano, I., Blàzquez, S., Barjola, V., (2013). The presence of radionuclides in wastewater treatment plants in Spain and their effect on human health. *Journal of Cleaner Production* 60, 77-82.

[12] Verlicchi, P., Galletti, A., Petrovic, M., Barceló, D., (2010). Hospital effluents as a source of emerging pollutants: An overview of micropollutants and sustainable treatment options. Journal of Hydrology 389, 416-428.

[13] Mintenig, S. M., Int-Veen, I., Loeder, M. G. J., Primpke, S., Gerdts, G., (2017). Identification of microplastic in effluents of wastewater treatment plants using focal plane array-based micro-Fourier-transform infrared imaging. *Water Research* 108, 365-372.

[14] Zhang, Y., Leu, Y. R., Aitken, R. J., Riediker, M., (2015). Inventory of Engineered Nanoparticle-Containing Consumer Products Available in the Singapore Retail Market and Likelihood of Release into the Aquatic Environment. *International journal of environmental research and public health* 12, 8717-8743.

[15] Kesari, K. K., Soni, R., Jamal, Q. M. S., Tripathi, P., Lal, J. A., Jha, N. K., ... & Ruokolainen, J. (2021). Wastewater treatment and reuse: a review of its applications and health implications. *Water, Air, & Soil Pollution*, *232*, 1-28.

[16] Webster, G. M. (2010). Chemicals, Health and Pregnancy Study (CHirP). Vancouver, BC: University of British Columbia, Centre for Health and Environment Research (CHER) and School for Occupational and Environmental Hygiene (SOEH).

[17] Bonefeld-Jorgensen, E. C., Long, M., Bossi, R., Ayotte, P., Asmund, G., Kruger, T., et al. (2011). Perfluorinated compounds are related to breast cancer risk in Greenlandic Inuit: A case control study. *Environmental Health Perspectives*, 10, 88–95.

[18] Soni, R., Pal, A. K., Tripathi, P., Lal, J. A., Kesari, K., & Tripathi, V. (2020). An overview of nanoscale materials on the removal of wastewater contaminants. *Applied water science*, *10*(8), 189. Jia, S., Zhang, X. (2020). Biological HRPs in wastewater. In High-risk pollutants in wastewater. Eds. Ren, H. and Zhang, X. 41-67.

[19] Jia, S., Zhang, X. (2020). Biological HRPs in wastewater. In *High-risk pollutants in wastewater*. Eds. Ren, H. and Zhang, X. 41-67.

[20] Anjaneyulu, Y., Sreedhara Chary, N., & Samuel Suman Raj, D. (2005). Decolourization of industrial effluents–available methods and emerging technologies–a review. *Reviews in Environmental Science and Bio/Technology*, *4*, 245-273.

[21] Von-Sperling, M., & Chernicharo, C. A. L. (2005). Biological Wastewater treatment in warm climate regions (1st ed.p. 810). IWA Publishing. *Water Science,* 10, 189.

[22] Blume, T., & Neis, U. (2004). Improved wastewater disinfection by ultrasonic pre-treatment. *Ultrasonics Sonochemistry,* 11, 333–336.

[23] Rizvi, H., Ahmad, N., Abbas, F., Bukhari, I. H., Yasar, A., Ali, S., ... & Riaz, M. (2015). Start-up of UASB reactors treating municipal wastewater and effect of temperature/sludge age and hydraulic retention time (HRT) on its performance. *Arabian Journal of Chemistry*, *8*(6), 780-786.

[24] Raouf, M. E. A., Maysour, N. E., Farag, R. K., & Abdul-Raheim, A. M. (2019). Wastewater treatment methodologies, review article. *Int J Environ & Agri Sci*, *3*, 018.

[25] Ebah, E. E., Yange, I. A., Ohie, I. R., & Inya, O. J. (2022). Application of Genetically Modified Organisms in Waste Management– A Review. *Stamford Journal of Microbiology*, *12*(1), 15-20.

[26] Sánchez, A. (2022). Decentralized composting of food waste: a perspective on scientific knowledge. *Frontiers in Chemical Engineering*, *4*, 850308.

[27] Leow, C. W., Van Fan, Y., Chua, L. S., Muhamad, I. I., Klemes, J. J., & Lee, C. T. (2018). A review on application of microorganisms for organic waste management. *Chemical Engineering Transactions*, *63*, 85-90.

[28] Abdel-Hamid, A. M., Solbiati, J. O., & Cann, I. K. (2013). Insights into lignin degradation and its potential industrial applications. In *Advances in applied microbiology* (Vol. 82, pp. 1-28). Academic Press.

[29] Ahansazan, B., Afrashteh, H., Ahansazan, N., & Ahansazan, Z. (2014). Activated sludge process overview. *International Journal of Environmental Science and Development*, *5*(1), 81.

[30] Ramadan, H., & Ponce, V. M. (2003). Design and performance of waste stabilization ponds. *Carlsbad: Virgin GLOBE, Storm Water Surface Water Management Service*, *56*, 333-344.

[31] Pesoutova, R., Hlavinek, P., & Matysikova, J. (2011). Use of advanced oxidation processes for textile wastewater treatment-A review. *Food and Environmental Safety*, 10, 59–65.

[32] Charerntanyarak, L. (1999). Heavy metals removal by chemical coagulation and precipitation. *Water Science and Technology,* IWA Publishing 39(10-11): 135-138.

[33] Chakinala, A. G., Gogate, P. R., Burgess, A. E., & Bremner, D. H. (2009). Industrial wastewater treatment using hydrodynamic cavitation and heterogeneous advanced Fenton processing. *Chemical Engineering Journal*, *152*(2-3), 498-502.

[34] Shah, A., & Shah, M. (2020). Characterisation and bioremediation of wastewater: a review exploring bioremediation as a sustainable technique for pharmaceutical wastewater. *Groundwater for Sustainable Development*, *11*, 100383.

[35] Suresh, A., & Abraham, J., 2018. Bioremediation of hormones from wastewater. In: Hussain, C. M. (Ed.), *Handbook of Environmental Materials Management*, pp. 1–31.

[36] Ileri, R., Sengil, I. A., Kulac, S., Damar, Y. (2003). Treatment of mixed pharmaceutical industry and domestic wastewater by sequencing batch reactor. *J. Environ. Sci. Health* A 38 (10), 2101–2111.

[37] Shi, X., Lin, J., Zuo, J., Li, P., Li, X., Guo, X. (2017). Effects of free ammonia on volatile fatty acid accumulation and process performance in the anaerobic digestion of two typical bio-wastes. *J. Environ. Sci.* 55, 49–57.

[38] Dubey, S. K., Dubey, J., Mehra, S., Tiwari, P., Bishwas, A. (2013). Potential use of cyanobacterial species in bioremediation of industrial effluents. *Afr. J. Biotechnol.* 10, 1125–1132.

[39] Arumugam, N., Chelliapan, S., Kamyab, H., Thirugnana, S., Othman, N., Nasri, N. (2018). Treatment of wastewater using seaweed: a review. *Int. J. Environ. Res. Publ. Health* 15 (12), 2851.

[40] Mohammad, P., Azarmidokht, H., Fatollah, M., Mahboubeh, B. (2006). Application of response surface methodology for optimization of important parameters in decolorizing treated distillery wastewater using Aspergillus fumigatus UB2.60. *Int. Biodeterior. Biodegrad.* 57, 195–199.

[41] Olicon-hernandez, D. R., Gonzalez-l opez, J., Aranda, E., Varese, G. C., Wick, L. Y., Aranda, E. (2017). Overview on the biochemical potential of filamentous fungi to degrade pharmaceutical compounds.

[42] Franca, R. D. G., Vieira, A., Carvalho, G., Oehmen, A., Pinheiro, H. M., Crespo, M. T. B., & Lourenco, N. D. (2020). Oerskovia paurometabola can efficiently decolorize azo dye Acid Red 14 and remove its recalcitrant metabolite. *Ecotoxicology and Environmental Safety*, *191*, 110007.

[43] Khandare, R. V., & Govindwar, S. P. (2015). Microbial degradation mechanism of textile dye and its metabolic pathway for environmental safety. *Environmental waste management*, 399-439.

[44] Chang, J. S., Chou, C., Lin, Y. C., Lin, P. J., Ho, J. Y., & Hu, T. L. (2001). Kinetic characteristics of bacterial azo-dye decolorization by *Pseudomonas luteola*. *Water research*, *35*(12), 2841-2850.

[45] Sharma, M., Agarwal, S., Agarwal Malik, R., Kumar, G., Pal, D. B., Mandal, M., ... & Gupta, V. K. (2023). Recent advances in microbial engineering approaches for wastewater treatment: a review. *Bioengineered*, *14*(1), 2184518.

[46] Mónica, P., Darwin, R. O., Manjunatha, B., Zúñiga, J. J., Diego, R., Bryan, R. B., ... & Maddela, N. R. (2016). Evaluation of various pesticides-degrading pure bacterial cultures isolated from pesticide-contaminated soils in Ecuador. *African Journal of Biotechnology*, *15*(40), 2224-2233.

[47] Boonsong, K., & Chansiri, M. (2008). Domestic wastewater treatment using vetiver grass cultivated with floating platform technique. *AU Journal of Technology*, *12*(2), 73-80.

[48] Wu, Y. H., Zhou, P., Cheng, H., Wang, C. S., Wu, M., & Xu, X. W. (2015). Draft genome sequence of Microbacterium profundi Shh49T, an *Actinobacterium* isolated from deep-sea sediment of a polymetallic nodule environment. *Genome Announcements*, *3*(3), 10-1128.

[49] Guo, Y., Huang, T., Wen, G., & Cao, X. (2017). The simultaneous removal of ammonium and manganese from groundwater by iron-manganese co-oxide filter film: The role of chemical catalytic oxidation for ammonium removal. *Chemical Engineering Journal, 308*, 322-329.

[50] Pugazhendhi, A., Boovaragamoorthy, G. M., Ranganathan, K., Naushad, M., & Kaliannan, T. (2018). New insight into effective biosorption of lead from aqueous solution using Ralstonia solanacearum: characterization and mechanism studies. *Journal of cleaner production, 174*, 1234-1239.

[51] Sátiro, J., Cunha, A., Gomes, A. P., Simões, R., & Albuquerque, A. (2022). Optimization of microalgae–bacteria consortium in the treatment of paper pulp wastewater. *Applied Sciences, 12*(12), 5799.

[52] Maqbool, M., Bhatti, H. N., Sadaf, S., Zahid, M., & Shahid, M. (2019). A robust approach towards green synthesis of polyaniline-Scenedesmus biocomposite for wastewater treatment applications. *Materials Research Express, 6*(5), 055308.

[53] Fontalvo, N. P. M., Gamero, W. B. M., Ardila, H. A. M., Gonzalez, A. F. P., Ramos, C. G., & Muñoz, A. E. P. (2022). Removal of nitrogenous compounds from municipal wastewater using a bacterial consortium: an opportunity for more sustainable water treatments. *Water, Air, & Soil Pollution, 233*(8), 339.

[54] Du, H., Zeng, F., Song, T., Liu, K., Wang, K., & Liu, M. (2021). Water depletion of climax forests over humid karst terrain: Patterns, controlling factors and implications. *Agricultural Water Management, 244*, 106541.

[55] Su, Y., Mennerich, A., & Urban, B. (2012). Synergistic cooperation between wastewater-born algae and activated sludge for wastewater treatment: influence of algae and sludge inoculation ratios. *Bioresource technology, 105*, 67-73.

[56] Basutkar, M. R., Fatima, N., Chakra, P., Swami, R. S., Vishwaraj, S. S., & Shivannavar, C. T. (2019). Decolorisation of reactive red 120 azo dye by micrococcus sp. *Int J Adv Res Biol Sci, 6*(5), 118-125.

[57] Assress, H. A., Selvarajan, R., Nyoni, H., Ntushelo, K., Mamba, B. B., & Msagati, T. A. (2019). Diversity, co-occurrence and implications of fungal communities in wastewater treatment plants. *Scientific reports, 9*(1), 14056.

[58] Dönmez, G., & Aksu, Z. (2001). Bioaccumulation of copper (II) and nickel (II) by the non-adapted and adapted growing Candida sp. *Water Research, 35*(6), 1425-1434.

[59] Coelho, L. M., Rezende, H. C., Coelho, L. M., De Sousa, P. A., Melo, D. F., Coelho, N. M. (2015). Bioremediation of polluted waters using microorganisms. *Advances in bioremediation of wastewater and polluted soil, 10*(10.5772), 60770.

[60] Norton-Brandão, D., Scherrenberg, S. M., van Lier, J. B. (2013). Reclamation of used urban waters for irrigation purposes–a review of treatment technologies. *Journal of environmental management, 122*, 85-98.

[61] Samin, G., Pavlova, M., Arif, M. I., Postema, C. P., Damborsky, J., Janssen, D. B. (2014). A Pseudomonas putida strain genetically engineered for 1, 2, 3-trichloropropane bioremediation. *Applied and Environmental Microbiology, 80*(17), 5467-5476.

[62] Kathiresan, K. (2003). Polythene and plastics-degrading microbes from the mangrove soil. *Revista de biologia tropical, 51*(3-4), 629-633.

[63] Ibiene, A. A., Stanley, H. O., Immanuel, O. M. (2013). Biodegradation of polyethylene by Bacillus sp. indigenous to the Niger delta mangrove swamp. *Nigerian Journal of Biotechnology, 26*, 68-78.

[64] Aylward, F. O., McDonald, B. R., Adams, S. M., Valenzuela, A., Schmidt, R. A., Goodwin, L. A., ... Poulsen, M. (2013). Comparison of 26 sphingomonad genomes reveals diverse environmental adaptations and biodegradative capabilities. Appl. Environ. Microbiol. 79, 3724–3733.

[65] Zhao, Q., Yue, S., Bilal, M., Hu, H., Wang, W., & Zhang, X. (2017). Comparative genomic analysis of 26 Sphingomonas and Sphingobium strains: dissemination of bioremediation capabilities, biodegradation potential and horizontal gene transfer. *Science of the Total Environment, 609*, 1238-1247.

[66] Buyukgungor, H., & Gurel, L. (2009). The role of biotechnology on the treatment of wastes. *African Journal of Biotechnology, 8*(25).

[67] Saha, P., Banerjee, A., Sarkar, S. (2015). Phytoremediation potential of Duckweed (Lemna minor L.) on steel wastewater. *International journal of phytoremediation, 17*(6), 589-596.

[68] Kumar, P. N., Dushenkov, V., Motto, H., Raskin, I. (1995). Phytoextraction: the use of plants to remove heavy metals from soils. *Environmental science & technology*, *29*(5), 1232-1238.

[69] Lakshmi, K. S., Sailaja, V. H., Reddy, M. A. (2017). Phytoremediation-a promising technique in wastewater treatment. *International Journal of Scientific Research and Management (IJSRM)*, *5*(06), 5480-5489.

[70] Dushenkov, V., Kumar, P. N., Motto, H., Raskin, I. (1995). Rhizofiltration: the use of plants to remove heavy metals from aqueous streams. *Environmental science & technology*, *29*(5), 1239-1245.

[71] Garrison, A. W., Nzengung, V. A., Avants, J. K., Ellington, J. J., Jones, W. J., Rennels, D., Wolfe, N. L. (2000). Phytodegradation of p, p '-DDT and the enantiomers of o, p '-DDT. *Environmental Science & Technology*, *34*(9), 1663-1670.

[72] Vázquez, S., Agha, R., Granado, A., Sarro, M. J., Esteban, E., Peñalosa, J. M., Carpena, R. O. (2006). Use of white lupin plant for phytostabilization of Cd and As polluted acid soil. *Water, Air, and Soil Pollution*, *177*, 349-365.

[73] Özyiğit, İ. İ., & Doğan, İ. (2014). Plant-microbe interactions in phytoremediation. *Soil remediation and plants: Prospects and challenges*.

[74] Yapoga, S., Ossey, Y. B., & Kouame, V. (2013). Phytoremediation of Zinc, Cadmium, Copper and Chrome From Industrial Wastewater by Eichhornia Crassipes. *International Journal of Conservation Science*, *4*(1).

[75] Raldugina, G. N., Maree, M., Mattana, M., Shumkova, G., Mapelli, S., Kholodova, V. P., ... & Kuznetsov, V. V. (2018). Expression of rice OsMyb 4 transcription factor improves tolerance to copper or zinc in canola plants. *Biologia plantarum*, *62*, 511-520.

[76] Xia, Y., Liu, J., Wang, Y., Zhang, X., Shen, Z., Hu, Z. (2018). Ectopic expression of *Vicia sativa* Caffeoyl-CoA O-methyltransferase (VsCCoAOMT) increases the uptake and tolerance of cadmium in *Arabidopsis*. *Environmental and Experimental Botany*, *145*, 47-53.

[77] Wang, X., Zhi, J., Liu, X., Zhang, H., Liu, H., Xu, J. (2018). Transgenic tobacco plants expressing a P1B-ATPase gene from Populus tomentosa Carr.(PtoHMA5) demonstrate improved cadmium transport. *International journal of biological macromolecules*, *113*, 655-661.

[78] Viktorova, J., Jandova, Z., Madlenakova, M., Prouzova, P., Bartunek, V., Vrchotova, B., ... & Macek, T. (2016). Native phytoremediation potential of *Urtica dioica* for removal of PCBs and heavy metals can be improved by genetic manipulations using constitutive CaMV 35S promoter. *PLoS One*, *11*(12), e0167927.

[79] Cheng, D., Liu, Y., Ngo, H. H., Guo, W., Chang, S. W., Nguyen, D. D., ... & Liu, Y. (2020). A review on application of enzymatic bioprocesses in animal wastewater and manure treatment. *Bioresource technology*, *313*, 123683.

[80] Meng, Y., Li, S., Yuan, H., Zou, D., Liu, Y., Zhu, B., ... & Li, X. (2015). Evaluating biomethane production from anaerobic mono-and co-digestion of food waste and floatable oil (FO) skimmed from food waste. *Bioresource technology*, *185*, 7-13.

[81] Naghdi, M., Taheran, M., Brar, S. K., Kermanshahi-pour, A., Verma, M., & Surampalli, R. Y. (2018). Biotransformation of carbamazepine by laccase-mediator system: kinetics, by-products and toxicity assessment. *Process Biochemistry*, *67*, 147-154.

[82] Becker, D., Giustina, S. V. D., Rodriguez-Mozaz, S., Schoevaart, R., Barcelo, ́D., de Cazes, M., Belleville, M. P., Sanchez-Marcano, J., de Gunzburg, J., Couillerot, O., Volker, J., Oehlmann, J., Wagner, M. (2016). Removal of antibiotics in wastewater by enzymatic treatment with fungal laccase – Degradation of compounds does not always eliminate toxicity. *Bioresour. Technol*. 219, 500–509.

[83] Jiang, B., Zhang, N., Xing, Y., Lian, L., Chen, Y., Zhang, D., ... & Song, Y. (2019). Microbial degradation of organophosphorus pesticides: novel degraders, kinetics, functional genes, and genotoxicity assessment. *Environmental Science and Pollution Research*, *26*, 21668-21681.

[84] Das, A., & Singh, J., K. N, Y. (2017). Laccase immobilized magnetic iron nanoparticles: Fabrication and its performance evaluation in chlorpyrifos degradation. *Int. Biodeterior. Biodegrad*. 117, 183-189.

[85] Wu, L., Zhou, M., Liu, C., Chen, X., Chen, Y. (2021). Double-enzymes-mediated $Fe^{(2+)}/Fe^{(3+)}$ conversion as magnetic relaxation switch for pesticide residues sensing. *J. Hazard. Mater*. 403, 123619.

[86] Mohammadi, S., Kargari, A., Sanaeepur, H., Abbassian, K., Najafi, A., Mofarrah, E. (2015). Phenol removal from industrial wastewaters: a short review. *Cheminform*. 46 (12), 2215–2234.

[87] Moraes, B. S., Loro, V. L., Glusczak, L., Pretto, A., Menezes, C., Marchezan, E., de Oliveira Machado, S., 2007. Effects of four rice herbicides on some metabolic and toxicology parameters of teleost fish (*Leporinus obtusidens*). *Chemosphere.* 68 (8), 1597–1601.

[88] Mouni, L., Belkhiri, L., Bollinger, J. C., Bouzaza, A., Assadi, A., Tirri, A., Dahmoune, F., Madani, K., Remini, H. (2018). Removal of Methylene Blue from aqueous solutions by adsorption on Kaolin: Kinetic and equilibrium studies. *Appl. Clay Sci.* 153, 38–45.

[89] Cabana, H., Jiwan, J. L. H., Rozenberg, R., Elisashvili, V., Penninckx, M., Agathos, S. N., & Jones, J. P. (2007). Elimination of endocrine disrupting chemicals nonylphenol and bisphenol A and personal care product ingredient triclosan using enzyme preparation from the white rot fungus *Coriolopsis polyzona*. *Chemosphere*, *67*(4), 770-778.

# Chapter 10

# Tailored Solutions for Industrial Wastewater Challenges

**Yash Bisht**
**Ashruta M. Dhali**
**Nilufer Todiwala**
**Almas Ansari**
**Muazzama Ansari**
**and Punita Jain***
Department of Biotechnology, St. Xavier's College, Mumbai, India

## Abstract

The global expansion of industrialization has led to an increase in wastewater pollution. Industrial effluents across various sectors consist of toxic anthropogenic compounds that, when discharged into waterbodies, contribute to eutrophication and algal blooms, adversely impacting both natural ecosystems and public health. In recent years, multiple wastewater treatment strategies have been emerged. Conventional treatment processes involve various physicochemical methods like flocculation, adsorption, and coagulation for removal of suspended solids. Biological methods use moving bed biofilm reactors, up-flow anaerobic sludge blankets, and fluidized bed reactors to tackle the pollutants. Various polymeric and inorganic hybrid membrane systems, when combined with advanced oxidation processes, offer a promising 21st century solution for treating toxic and persistent effluents. This chapter focuses on highlighting various effluent treatment strategies employed by pharmaceutical, textile, distillery and pulp and paper industries, which are classified as "Red" by Central Pollution Control Board (CPCB) due to their high pollution index. The integration of Effluent Limitations Guidelines and Standards (ELG) database provides valuable information regarding pollutants, their characteristics and effective treatment options. For sustainable industrialization and environmental conservation, effluent treatment strategies must be tailored and targeted. Further collaboration among industries, regulatory bodies and researchers is vital to ensure the treatment and safe disposal of toxic effluents, leading towards a greener and healthier future.

* Corresponding Author's Email: punita.jain@xaviers.edu

In: Advancements in Wastewater Treatment
Editors: Neha Agarwal, Maulin P. Shahl, Vijendra Singh Solankil et al.
ISBN: 979-8-90134-007-3

**Keywords:** environmental conservation, eutrophication, effluent treatment, sustainable industrialization, regulatory collaboration

## 1. Introduction

Industrial effluents are by-products of industrial activities that pose a growing threat to environmental safety, a concern exacerbated by rapid industrialization. According to the Economic Survey 2023-24 (Ministry of Finance, India), the industrial and manufacturing sectors have expanded at annual rates of 9.5% and 5.2%, respectively, over the past decade [1]. This has contributed to a persistent water shortage, driven by increasing wastewater production from industries. The World Meteorological Organization (2023) reports that 3.6 billion people struggle to access sufficient water at least one month each year [2]. Worldwide, annual wastewater production is 380 billion $m^3$, with a projected 51% increase by 2050 [3]. Untreated wastewater, accounting for 48% of the total discharge [4] contributes to soil and water pollution, and leads to eutrophication and algal blooms, resulting in decreased dissolved oxygen and reduced solar penetration, thereby harming aquatic life [5]. According to Frost & Sullivan (2022), the Indian market for water and wastewater treatment is expected to grow at a 9.7% CAGR (compound annual growth rate), increasing from $1.13 billion in 2022 to $ 2.08 billion by 2050 [6]. However, the composition and concentration of industrial effluents vary across sectors, making their remediation a complex challenge. To address this, it's essential to shift from conventional wastewater technologies to advanced methods capable of treating a broader range of pollutants. Advanced physical treatment technologies such as nanofiltration, ultrafiltration, and reverse osmosis have been developed to meet this need. Photocatalytic processes and electro-coagulation techniques offer novel chemical treatment options for effluents [7]. Additionally, hybrid technologies like microbial fuel cells that incorporate microbes and enzymes have emerged as a biological treatment method [8]. When integrated with artificial intelligence and machine learning, these technologies can optimize processes more efficiently, transforming them into a sustainable solution for wastewater management [9]. As industrial effluent management becomes increasingly critical in the coming years, there will be a stronger focus on recycling and reusing effluents. This shift will help reduce the environmental burden and conserve resources. Details on these technologies and their applicability on an industrial scale are discussed further in this chapter.

## 2. Major Pollutants and Sources of Industrial Wastewater

Improper handling and disposal of toxic industrial pollutants in wastewater can harbor pathogenic microorganisms, posing significant health and environmental risks [10]. Table 1 shows major pollutants commonly found in different types of industrial wastewater, their impacts on human health and environment, and treatment methods used to mitigate these risks.

**Table 1.** Major pollutants of industrial wastewater and their treatment methods

| Industry Sector | Major Pollutants | Harmful effects | Treatment | References |
|---|---|---|---|---|
| Pharmaceutical | Heavy metals Antibiotics, plant and animal steroids, beta lactams, cytostatic agents | Mutagenic, genotoxic, ecotoxicological effects on humans, plants and animals | Advanced Oxidation Process (AOPs), Adsorption, Membrane Bioreactor | [11, 12] |
| Textile | Dyes, phenolic compounds, heavy metals, bleaching agents nitrogenous and phosphorous wastes | Carcinogenic, clastogenicity, microbial toxicity, cytotoxicity | Moving Bed Biofilm Reactor, Nanofiltration | [13, 14] |
| Distillery | Anthocyanins, caramel, tannins, xenobiotic compounds, phthalates | Endocrine disrupting compounds, reduces agricultural fertility, carcinogenic | Reverse Osmosis | [15] |
| Pulp and Paper | Resin, Tannins | Cause mutagenicity and genotoxicity | AOPs | [16] |
| Dairy | Proteins, phosphorous, nitrogen, chlorides, sulfates, oil, grease | Promotes eutrophication, promotes microbial growth | 3D printed membranes | [17] |

## 3. Characteristics of Industrial Effluents

Industrial wastewater varies in biological, physical and chemical characteristics due to factors such as substrate utilization, production process, and byproduct formation. Understanding its composition is crucial for ensuring effective treatment and safe discharge into the environment.

### 3.1. Physical Characteristics

The main physical characteristics of effluents include turbidity, color, odor and temperature. Industries such as food, textile, beverage, pulp and paper often produce high suspended solids during raw material processing, contributing to turbidity. The color in wastewater is caused by the decomposition of non-biodegradable dyes and oils. Foul odor can result from compounds such as sulfur, amines, mercaptans, indole and skatole. Industries involved in heating and cooling processes may discharge effluents at higher temperature, which can reduce dissolved oxygen levels and negatively affect aquatic flora and fauna [18].

### 3.2. Chemical Characteristics

The chemical characteristics of industrial effluents include $p^H$, TDS, TSS, Dissolved oxygen, biological oxygen demand (BOD) and chemical oxygen demand (COD). The $p^H$ level indicates the acidity or alkalinity of effluent, and industrial effluents can vary widely in $p^H$. Change in the $p^H$ of the water bodies can be harmful to the marine flora and fauna [19]. Total Dissolved Solids (TDS) encompass inorganic (fixed solids) and organic (volatile solids) substances in suspended, ionized, or micro-granular (colloidal) form. TDS particles are typically smaller than 2-micrometers and primarily consist of sodium, magnesium, potassium, calcium, iron, chlorides and sulphates [20]. Total Suspended Solids (TSS) are particles larger than 2 micrometers, including hydrophobic pollutants, heavy metals, phosphorus, and mercury. These solids infiltrate aquatic environments, obstructing sunlight and reducing photosynthetic activity [21].

Water quality can be measured by its dissolved oxygen content as aerobic organisms require dissolved oxygen for respiration. Discharge of industrial effluents can reduce dissolved oxygen levels (>4 ppm) negatively affecting the aquatic ecosystem. BOD measures the dissolved oxygen consumed by microbes during the decomposition for organic matter. High BOD levels indicate the presence of significant organic pollutants which can be detected by BOD sensors [22]. COD estimates the oxygen consumed through the chemical oxidation of organic matter. Together, the BOD/COD ratio helps assess the biodegradable content of the effluent [18].

### 3.3. Biological Characteristics

The effluent treatment of industrial wastewater can be significantly impacted by microorganisms such as bacteria, viruses, fungi, algae, and protozoan [18]. Recent studies have observed that concentration of phosphate and nitrate, in the range of 0.01 to 2.5 ug/ml and 0.01 to 15.89 ug/ml respectively, contribute to development of algal blooms [22]. Food processing and pharmaceutical industries release effluents rich in organic content, which can promote the proliferation of antibiotic-resistant bacteria causing a risk to human health [23, 24].

## 4. Regulatory Standards

According to the United Nations Sustainable Development Goals (SDGs), SDG 6.3 aims to achieve clean water and sanitation by 2030 [25]. To support this goal, regulatory standards have been established to monitor industry effluents discharge. These standards primarily focus on the pollutants like benzene, acrolein, acrylonitrile etc., as defined under the Clean Water Act (1972) in the U.S. [26]. In addition, industrial practices and treatment technologies are regulated by the Environmental Protection Agency (EPA). In the European Union, the Water Framework Directive (WFD) outlines regulations for water quality, while in India; Central Pollution Control Board (CPCB) enforces the industry-specific regulatory standards. The ELG (Effluent Limitation Guidelines and Standards) and the IWTT (Industrial Wastewater Treatment Technology) databases contain comprehensive information on these observations. Further the Environmental Impact Assessment (EIA) processes help mitigate the ecological impacts caused by industry effluents. The general standards for discharge of effluents discharge, as mandated by the CPCB are provided in the Table 2 below. This table outlines the key parameters used to assess industrial effluent discharge and their impact on environment and human health.

**Table 2.** General parameters, their impact and discharge standards of pollutants by CPCB

| S. No | Parameter | Impact | Standards | |
|---|---|---|---|---|
| | | | Inland surface water | Marine coastal areas |
| 1 | Suspended Solids (SS) | Obstructs sunlight, lowering photosynthetic activity | 100 mg/l | 100 mg/l |
| 2 | pH | Extreme levels can be deleterious for the aquatic flora and fauna | 5.5-9 | 5.5-9 |
| 3 | Temperature | High-temperature limits dissolved oxygen | Up to 5°C above receiving temperature | Up to 5°C above receiving temperature |
| 4 | Oil and grease | Oil film on water surface interferes with aquatic ecosystem, limiting oxygen supply | 10 mg/l | 20 mg/l |
| 5 | Ammonia nitrogen (N) | Decreases dissolved oxygen | 50 mg/l | 50 mg/l |
| 6 | Free ammonia (NH3) | Biocidal effects, reduced dissolved oxygen, results in nitrogenous BOD | 5 mg/l | 5 mg/l |
| 7 | BOD (3 days) | Indicative of biodegradable pollutants | 30 mg/l | 100 mg/l |
| 8 | COD | Indicative of biodegradable and non-biodegradable pollutants. | 250 mg/l | 250 mg/l |
| 9 | Dissolved Phosphates (P) | Decreases dissolved oxygen | 5 mg/l | – |
| 10 | Sulphide (S) | produces hydrogen sulphide | 2 mg/l | 5 mg/l |

## 5. Conventional Method for Wastewater Treatment

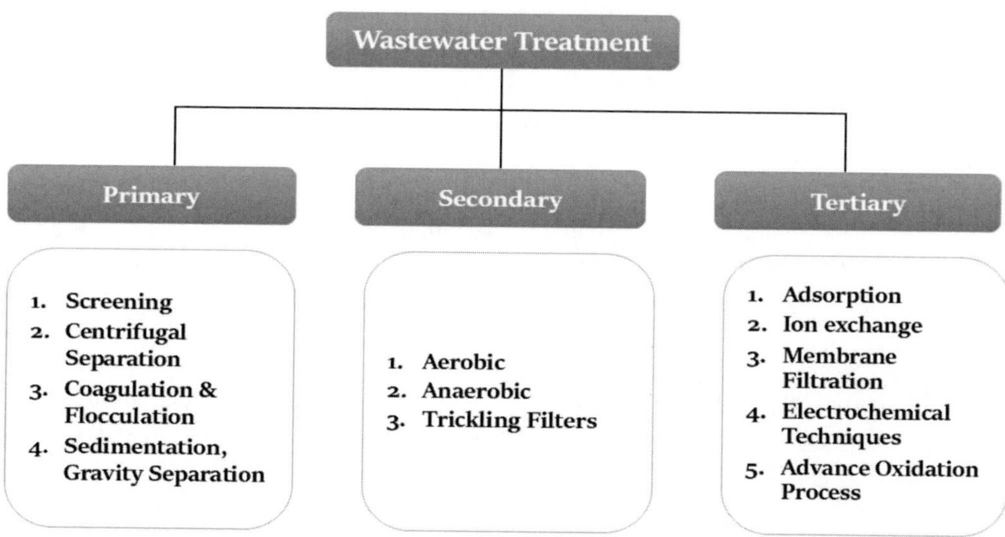


**Figure 1. Wastewater Treatment by conventional methods** (Flowchart illustrates three different stages for wastewater treatment. Primary treatment focuses on removing solids through screening, centrifugation, coagulation, flocculation and sedimentation. Secondary treatment stage focuses on breaking down organic matter using aerobic, anaerobic processes and trickling filters. Tertiary treatment enhances the quality of effluent by employing various advanced techniques.).

The wastewater treatment encompasses various stages and subsystems, are designed to achieve specific treatment goals for a given region. Figure 1 shows the classification of wastewater treatment into primary, secondary and tertiary stages [27]. The specifics of each treatment stage, including the method employed, are discussed in a later section.

## 5.1. Classification of Wastewater Treatment

Wastewater treatment can be classified into three main stages: primary treatment, secondary treatment, and tertiary treatment. The details of these treatment processes are discussed below.

### *5.1.1. Primary Treatment*

Primary treatment focuses on removing large solids and debris from effluent through processes such as screening and grit removal.

a. *Screening* - Screening, grit chambers, and skimming tanks form the initial stages of wastewater treatment that are designed to remove various debris, SS, and oily substances. In the screening process large floating objects, such as sticks, dead animals, and plastic bottles are intercepted by screens with uniform openings to prevent clogging or equipment damage. [28].
   *Limitations:* Ineffective at removing finer particles or dissolved impurities, necessitating additional treatment processes [28].
b. *Centrifugal separation* - Centrifugation separates solids from wastewater by spinning wastewater at a varying speed to separate solids based on density, aiding in liquid-solid separation, sludge thickening, and dewatering. This process reduces sludge discharge to align with disposal requirements [29].
   *Limitations*: High initial cost and need for frequent maintenance due to the use of high-power motors [29].
c. *Coagulation & flocculation* - These physicochemical processes are used to remove turbid particles and various organic contaminants from wastewater. Commonly used coagulants are hydrolytic salts of aluminum and iron [30].
   *Limitations:* Produces large amounts of chemical sludge, while aluminum-based coagulants elevate the residual aluminum levels in treated water, affecting water quality and posing health risks [30].
d. *Sedimentation* - After screening, grit chambers slow water flow, allowing dense inorganic solids to settle by gravity. This protects equipment from excessive wear, and manages debris such as silt, sand, and gravel from storm runoff [31].
   *Limitations:* Ineffective at removing lighter organic materials and requires regular maintenance [31].

### *5.1.2. Secondary Treatment*

Secondary treatment reduces organic, inorganic matter and SS using microorganisms, lowering the BOD to COD ratio to ~0.6. Common methods of secondary treatment include activated sludge and trickling filters. In activated sludge processes, microbial communities in aeration tanks degrade organic pollutants. In trickling filters, wastewater flows through a biofilm coated medium where biological oxidation occurs. Post treatment, sludge undergoes moisture reduction, anaerobic/aerobic digestion, and composting to enhance bio-fertilizer quality [32]. Secondary treatment includes the following processes.

a. *Aerobic Process* – Secondary treatment employs systems like active sludge, lagoons, oxidation ponds, and methods such as aerobic-digestion and surface aeration. These systems are effective in reducing BOD, COD, organic matter, nitrates, and phosphates with optimal BOD: P: N ratio for microbial growth being 100:5:1 [33].
   *Limitations:* These processes generate considerable bio-solids, which require additional management and disposal [33].
b. *Anaerobic Process* – The anaerobic process operates in an oxygen free environment, where bacteria degrade organic matter into compounds like nitrogen, ammonia and hydrogen sulfide with optimal BOD: P: N ratio for microbial growth being 100:2.5:0.5. Methane, released as byproduct, is used for energy generation [34].
   *Limitations:* Slow reaction rates, require careful monitoring and produces odorous byproducts like hydrogen sulfide, ammonia and methane [34].

### *5.1.3. Tertiary Treatment*

Tertiary treatment enhances water quality for reuse or discharge into sensitive ecosystems. It includes adsorption, filtration and disinfection, removing nitrogen, phosphorus, fine particles through sand, multimedia, and membrane filters. Pathogen removal is done specifically by UV radiation and chlorination [35]. Tertiary treatment includes the following processes:

- *Adsorption:* This removes organic pollutants by concentrating adsorbate molecules on the liquid's surface or solid adsorbent by a surface-based mechanism. Key factors include particle size, adsorbate concentration, temperature, molecular weight, $p^H$, and iodine value. Adsorption occurs through ion exchange, chemical and physical adsorption. Each type has distinct characteristics, such as porosity and surface structure [36].
  - *Limitations:* Limited adsorption capacity, selectivity, and ability of adsorbents to retain and isolate target substances from complex mixtures. Additionally, disposal of spent adsorbents and regulatory constraints complicate their application [36].
- *Ion exchange–* It involves substitution of undesirable ions with more acceptable ions through an exchange with resins. This process is used for softening water and removing nitrogen, heavy metals, and dissolved solids in wastewater treatment facilities [37].
  - *Limitations:* Resin beds require recharging with a salt brine solution, increasing operational costs and can lead to potential secondary pollution due to the regeneration process [37].
- *Membrane Filtration Techniques–* These techniques function by selectively filtering substances through semipermeable membranes. This chapter discusses several methods through case studies. Each method is characterized by its ability to remove particles of specific sizes, like microfiltration filters, ultrafiltration membranes, nanofiltration membranes and reverse osmosis having a diameter range from 0.1-10 μm, 0.001- 0.1 μm, 0.001 and- 0.01 μm and 0.1- 5000 nm respectively [38].

*Limitations:* High energy demand and membrane fouling require frequent maintenance [38].

- *Electrochemical Techniques-* These methods use redox reactions to remove pollutants from wastewater or generate reactive chemical species that break down contaminants, key techniques include electro dialysis and electrocoagulation [39].
  - *Limitations:* These processes are energy-intensive and generate concentrated waste streams, which require further disposal or treatment [39].
- *Advanced Oxidation Processes–* It relies on generating highly reactive hydroxyl radicals [•OH] to break down organic matter that is resistant to biodegradation. Commonly used AOPs are Ozone ($O_3$)/ hydrogen peroxide ($H_2O_2$), Ozone/UV and $H_2O_2$/ UV [60]. The $H_2O_2$/UV process is effective at removing non-degradable and toxic byproducts from pharmaceutical and other industries such as volatile organic compounds, antibiotics, pesticides, aromatic compounds, chlorine, Sulphur-based dyes and petroleum constituents [16].
  - *Limitations:* The Fenton process generates iron sludge as a byproduct, requiring disposal and increasing operational costs. Additionally, oxidative degradation process in the presence of $Cl^-$ and $NO_2^-$, forms chlorinated and nitrated byproducts, which are highly toxic as compared to the original pollutants [16].

All these drawbacks of each treatment must be weighed against its effectiveness in achieving high levels of pollutant removal from the wastewater.

## 6. Disadvantages of Conventional Wastewater Treatment Methods

Conventional methods have been used for decades due to their process simplicity, economic viability and established infrastructure. However, these methods face significant challenges such as difficulty in removing metal ions present at a low concentration, high energy and resource requirements, and the need for enhanced treatment management; leading to increased operational cost especially in areas with high wastewater or sludge volumes [40]. Additionally, conventional approaches for pollutant removal are increasingly insufficient to comply with the latest stringent waste management regulations [41]. To address these limitations, innovative tailored treatment methods are required to mitigate the environmental concerns posed by pollutants.

## 7. Tailored Wastewater Treatment Technologies

The wastewater from various industries contains toxins, pharmaceuticals, dyes, pesticides, and other hazardous compounds. Innovative wastewater treatment systems, such as hydrodynamic cavitation, electrocoagulation, 3D-printed membranes etc. are efficient, adaptable and cost-effective as shown in Figure 2. These technologies effectively reduce pollutants while addressing sustainability and regulatory issues [41].

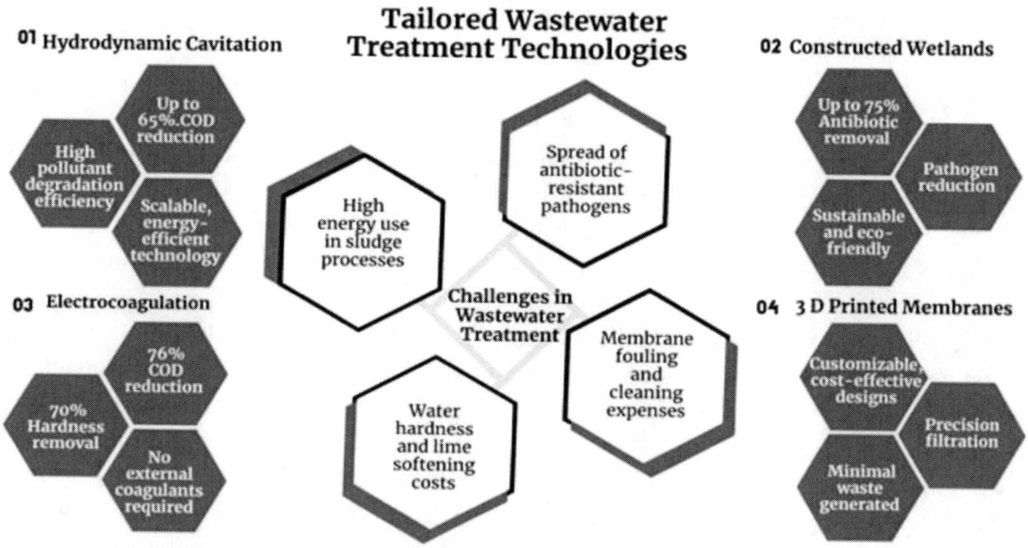


**Figure 2. Tailored Wastewater Treatment Technologies** (Figure shows advanced wastewater treatment technologies such as hydrodynamic cavitation, constructed wetlands, electrocoagulation, and 3D-printed membranes, highlighting their benefit and addressing key challenges).

## 7.1. Hydrodynamic Cavitation (HC) as a Tailored Solution for Treating High-Strength Wastewater

Traditional methods, like activated sludge process (ASP), utilizes microbial consortia to break down organic contaminants. This method accounts for 45-75% of overall energy expenditure due to high aeration rates required to maintain dissolved oxygen levels, thereby limiting the cost-effectiveness of the wastewater treatment process. Additionally, 25-50% COD of the influent is converted into primary and secondary sludge, resulting in a reduced sludge retention time (SRT) and increased sludge output [42]. This surplus sludge gets frequently pathogen-contaminated and demands expensive stabilization as well as disposal accounting for 60% of total treatment expenses. Due to these high costs, there has been growing interest in tailored technologies such as HC, which can effectively minimize excess activated sludge [43].\

### *7.1.1. Tailored Solutions/Advancements*

Cavitation is a viable alternative approach for disinfecting complex wastewater. During the cavitation process, high temperatures and critical pressure levels- approximately 4000-5000 atm- are generated as vaporous cavities form and collapse in the reactor [44]. This results in the degradation of the contaminants by chemical and mechanical method. HC occurs when liquid is forced through a constriction, causing rapid pressure drop and the formation of vaporous cavities. This highly energy-efficient technology offers significant scalability [45].

### *7.1.2. Advantages*

Hybrid HC technology consists of a powerful machinery configuration, faster treatment cycles, high efficiency and high scale-up capacities. According to several studies, this method has achieved 100% disinfection with 49% reduced COD levels within one hour. This hybrid technique when coupled with $H_2O_2$ shows enhances BOD and COD degradation achieving 100% microbial disinfection [46].

## 7.2. Constructed Wetlands (CWs) as Tailored Solution for Treating Antibiotic-Resistance in Wastewater

Wastewater discharge containing non-metabolized drugs and heavy metals ions, nurtures antibiotic-resistant (AR) pathogens, contributing to global health crisis. Traditional wastewater treatment plants (WWTPs) introduce antibiotic resistant bacteria's (ARBs) and antibiotic-resistant genes (ARGs) into the natural ecosystem by facilitating horizontal gene transfer [47].

### *7.2.1. Challenges*

Conventional WWTPs designs are ineffective in removing antibiotics during primary treatment, as they rely primarily on biodegradation and adsorption. Secondary treatments such as activated sludge and aerobic/anaerobic treatments enhance elimination, although results can be inconsistent due to impact of environmental conditions like pH and temperature. Tertiary treatments like chlorination are inadequate for removing specific antibiotics. Advanced processes like ozonation and photo catalysis are effective but costly. Overall, the efficiency of all treatments varies, affecting the removal of antibiotics, ARBs, and ARGs. Hence, traditional WWTPs face challenges in eradicating antibiotics from wastewater, raising significant concerns over potential environmental and public health risks [47].

### *7.2.2. Tailored Solutions/Advancements*

Constructed wetlands (CWs) offer an alternative approach to traditional systems, serving as non-traditional wastewater treatment system that provides a customized and efficient biological treatment option to remove a wide range of contaminants, including antibiotics and antibiotic-resistant bacteria/genes (ARB/ARG). CWs exhibit high removal efficiency for organic matters such as BOD, COD, SS and nitrogen. These flexible systems facilitate pollutant removal through nitrification, microbial denitrification, organic carbon degradation, and heavy metal filtration [48]. CWs utilize plants, microbes, and substrates like alum sludge cakes for efficient treatment of hospital and municipal wastewater including organic pollutants, inorganic contents and antibiotic-resistant bacteria [49].

### *7.2.3. Removal Efficiency of CWs for Antibiotics*

The efficacy of the various CW systems in removing antibiotics depends on the process parameters, site-specific factors and system configuration. According to Li et al. (2020), an integrated vertical flow CWs demonstrated 75% removal efficacy for tetracycline within a three-day hydraulic retention time [50]. Vertical subsurface flow CW (VSSF-CW) operates with intermittent input from the top, allowing wastewater to pass through a filter bed. This intermittent flow improves nitrification as compared to continuously saturated systems like Horizontal subsurface flow CW (HSSF-CW), enhancing microbial antibiotic degradation and $NH_3$-N removal [51]. VSSF-CWs were identified as the optimal configuration with an average antibiotic clearance rate of 70% and 81% reduction in macrolide content [50, 52].

**Table 3.** Efficiency of different types of CWs

| Type of CWs | Target ARG Categories | Removal Efficiency | Advantages | Limitations |
|---|---|---|---|---|
| Hybrid/integrated flow CWs [55, 56] | Sulfonamide, tetracycline beta-lactam and macrolide-lincosamide resistance gene. | >90% | High removal efficiency for pollutants. | Requires significant land area and increased operational complexity. |
| | Fecal indicator bacteria (E. coli and thermotolerant coliforms). | 99.5% | Removes fecal indicator bacteria. | Limited data on long term sustainability for ARGs removal. |
| Vertical flow CWs [57] | Sulfonamide, tetracycline, macrolide resistance genes, chloramphenicol resistance genes and 16S rRNA genes. | 63.9 and 84.0% | High oxygen and nutrient removal efficiency. | Reduces efficiency for pollutants in low hydraulic conductivity soils. |
| | Sulfonamide, tetracycline, genes (sul1, sul2, sul3, tetM and tetO). | 95.1% | Enhanced pathogen and ARG removal with proper design and optimization. | Seasonal variations can affect efficiency, may require pretreatment for heavily polluted water. |
| Horizontal flow CWs [58] | ARGs (sul1, sul2, ermB, tetW) and antibiotics like sulfonamides, macrolides, quinolones, tetracyclines, oxytetracycline. | 28% to 100% | Consistent removal of ARGs with lower operational cost. | Limited oxygen availability may hinder removal of specific ARGs. It may also accumulate antibiotics and ARGs in the sediment. |

#### *7.2.4. Removal Efficiency of CWs for ARBs and ARGs*

CWs are also effective in removing ARG and ARB's. Subsurface flow systems (SSFCW) are efficient in removing ARGs compared to surface flow systems (SFCWs). Hien et al. (2017) used CWs with *Phragmites australis* to study the removal of five tetracycline-resistant genes (tet C, M, O, Q, and W), achieving complete removal within a one-day retention period [53]. A comparative study by Pazda et al. (2020) showed that CWs coupled with activated sludge systems are more effective at minimizing sulfonamide- and tetracycline-resistant genes than conventional activated sludge alone [54]. Table 3 demonstrates different type of CWs and its wastewater treatment efficiency.

### 7.3. Electrocoagulation as a Tailored Solution to Remove Water Hardness

Untreated Wastewater contains dissolved minerals like calcium, magnesium, potassium, nitrate, bicarbonates, chloride sulfate which contribute to water hardness. Conventional lime softening reduces ~80% of water hardness by altering the ion equilibrium using $Ca(OH)_2$ [59]. While effective, this approach involves significant operating costs due to sludge disposal and pH adjustment. Lime softening can achieve 92-100% removal efficacy at pH 11. However, the availability and cost of $Ca(OH)_2$ can vary by region, and the process often requires additional steps to reduce hardness, complicating the overall treatment. Despite high chemical demand and rigorous maintenance practices, byproducts such as sludge are also formed [60].

#### *7.3.1. Tailored Solutions/Advancements*

Electrocoagulation (EC) is a versatile, robust, and economical technology that functions without the need for external coagulants. EC is one of the emerging electrochemical processes involving three main steps: (1) At anode, electrolytic oxidation occurs as the current passes through the electrolytic cell, releasing $M^+$, $Fe^{2+}$or $Al^{3+}$ ions (depending on the material chosen); (2) water molecules on the cathode undergoes hydrolysis; (3) pollutants attach to the coagulant and rise to the surface as flocs, where they are removed by sedimentation [61].

#### *7.3.2. Advantages*

EC is widely used to treat wastewater from industries such as textiles, automobiles, municipalities, petroleum, and paint. Notably, EC is highly effective for treating textile effluents. Tyagi et al. (2014) reported 76% removal efficiencies for COD and 95% for color when treating textile effluents using EC [62]. Lime softening achieves the highest hardness removal efficiency (70.7%) but combining it with soda ash is necessary to enhance the removal rate. EC also successfully reduces both TDS and groundwater hardness levels. Furthermore, it effectively treats wastewater with emerging pollutants like titanium dioxide nanoparticles (TNPs). These nanoparticles are used in food and medicines. They can damage DNA and red blood cells, posing serious health risks [63]. Researchers have developed an EC reactor using iron electrodes and HCl as an electrolyte, achieving over 95% removal of TNP [64].

### 7.4. 3D Printed Membranes for Wastewater Filtration

Membrane processes are the sustainable alternatives to the conventional processes and purification methods. They offer low footprint and consistent quality of treated water. It includes microfiltration, ultrafiltration, nanofiltration and reverse osmosis based on pore size. This membrane allows the components to pass through transmembrane pressure (TMP) based on its molecular weight cut-off (MWCO) which is measured in Daltons [65].

#### *7.4.1. Challenges*

A major drawback of membrane processes is reduced flow and productivity due to membrane fouling and concentration polarization. It occurs when certain species accumulate on membrane surface while others flow through it, reducing permeate flow. Fouling is caused by the interactions between foulants such as particulate materials, colloidal substances, biomolecules, and the membrane surface. This interaction can cause both chemical and physical degradation of the membrane material [66].

#### *7.4.2. Tailored Solutions/Advancements*

Tailored solutions aim to reduce the fouling and membrane surface interaction. It is achieved by altering the membrane's wetting properties or enhancing fluid turbulence at the membrane surface through surface structuring. Recent advancements, such as 3D orbiting, enable the development of complex membrane designs and structures, which are created using CAD software and then sliced into 2D layers for printing, allowing for the scalable, effective, and customizable membranes for wastewater treatment [67].

#### *7.4.3. Advantages*

This approach offers sustainability, cost-effectiveness, low energy use, minimal waste, mobility, precision and a reduced carbon footprint. Extrusion, selective laser sintering (SLS) and vat photo polymerization are the widely used techniques for fabrication of membranes. 3D printing offers more precise membrane designs providing better control parameters such as porosity and thickness compared to conventional membranes. Ceramic- based membranes with intricate structure has been widely used for water filtration and separation. Treatment tests on surface water have shown efficient turbidity and pollutants removal with minimal fouling [67].

## 8. Industry Specific Case Studies

### 8.1. Pharmaceutical Industry Effluents

The variability in effluents ranging from the combination of various manufacturing processes and the diverse types of chemicals used in drug production, leads to increased waste from the pharmaceutical industry. The issue arises from the widespread use of active pharmaceutical ingredients (APIs) and final dosage forms, which are released into the water bodies through households, agriculture, hospitals and industries [68].

Over 600 APIs are contaminating water bodies due to the use of pharmaceutical drugs. Despite their low concentrations (mg/l to μg/l), their impact is significant. NSAIDs like diclofenac have contributed to sharp decline in vulture population in South Asia. While; ibuprofen (IBU) and ketoprofen have shown harmful effects on aquatic life and plants [68]. Many antibiotics persist despite conventional treatments, leading to the rise of antibiotic-resistant bacteria [69].

### *8.1.1. Challenges and Treatment Methods for IBU*

With an annual consumption of 15,000 tons per year, IBU largely remains unmetabolized in the body and enters domestic wastewater at 0.22 μg/L. Despite its effectiveness in treating conditions like arthritis, its concentration ranges from 10 ng/L and 169 μg/L, posing risks to aquatic environments [11]. Even at low ibuprofen concentrations, IBU disrupts fish reproduction and causes algal toxicity in *Chlorella vulgaris* (Ec50 of 89.65mg/l) [70]. Additionally, its presence in surface and drinking water poses human health risks as these compounds can bio-accumulate in the food chain [71].

### *8.1.2. Treatment Techniques for IBU Removal*

Efforts to mitigate the impact of ibuprofen in wastewater include various treatment methods like AOPs, filtration, adsorption and active sludge treatment. Among these, adsorption is widely used due to its cost effectiveness, renderability and high efficiency [72]. Researchers are exploring a range of adsorbents, from activated carbon to graphene-based nanomaterials, to enhance ibuprofen removal.

#### *8.1.2.1. Adsorption*

Various adsorbents have been studied for their efficiency in removing the IBU from the water. Adoption efficiency is influenced by factors like pH, temperature, and adsorbent properties, with studies showing steam-activated biochar (SA-MBHB) achieves the highest IBU removal, outperforming activated carbon, graphene oxide, and metal organic frameworks while maintain the Langmuir isotherm and pseudo-second order kinetics (Table 4).

#### *8.1.2.2. Advanced Oxidation Processes (AOPs)*

These technologies are designed to improve wastewater treatment, particularly for the degradation of recalcitrant contaminant. Mostly used AOPs for IBU degradation is the electro-Fenton process. Other widely used AOPs for IBU removal are ozonation, UV disinfection, Fenton oxidation and the Photo- Fenton process [12]. Table 5 shows the different technologies used for IBU removal, highlighting their efficiencies, advantages, limitations and operational cost.

When comparing IBU removal techniques, adsorption proves to be the most effective method for the IBU removal with its success depending on selecting an appropriate adsorbent based on effluent size [11].

**Table 4.** Techniques for removal of ibuprofen (IBU) from wastewater

| Method | Adsorbent | Conditions | Efficiency | Observations |
|---|---|---|---|---|
| Activated Carbon Adsorption [73] | Activated Carbon | Temperature: 25° C<br>Model: Langmuir | High | – IBU efficiency influenced by pKa and log Kow.<br>– Higher log Kow and lower pKa improve absorption.<br>– Increasing pH reduces adsorption, temperature unaffected |
| Steam Activated Biochar (SA- MBHB) [74] | SA- MBHB | Adsorbent concentration: 0.1 g/L<br>pH: 2<br>Agitation: 200 rpm<br>Contact time: 120 min<br>Temperature: 20°C<br>Model: Langmuir, Pseudo second order | 99.26 % | – High removal rate at optimal conditions.<br>– Cost-efficient |
| Graphene Oxide Nanoplatelet (GONPs) [75] | GONPs | Adsorbent concentration: 0.1 g/L<br>pH: 6<br>Agitation: 180 rpm<br>Contact time: 60 min<br>Temperature: 35°C<br>Model: Langmuir, Pseudo second order | 98.17 % | – High efficiency achieved for IBU with 6 mg/L initial concentration. |
| Comparison of Adsorbent [76] | Activated Carbon, MOF (ZIF-8), Porous Carbon from MOF (PCPMC) | PPCMC Showed 3x higher performance than AC, 11x higher than ZIF-8 | Highest for PPMC | – PPCMC demonstrated the best adsorption performance due to structural properties |

**Table 5.** Comparison of technologies for ibuprofen removal from wastewater

| Technology | Removal Efficiency (%) | Advantage | Limitations | Conditions and Chemical Used |
|---|---|---|---|---|
| Coagulation- Flocculation [77] | 12 ± 4.8% | Simple, cost effective | Low efficiency; chemical residue | $FeCl_3/Al_2(SO_4)_3$ (25 mg/L, 50 mg/L) |
| Membrane Processes [77] | >99% | High efficiency; broad applicability | High operational cost; fouling risk | 9.5 ± 10.2 bar pressure, Filmtec TW30. |
| AOPs [12] | ~100% (Photo- Fenton) | Exceptional efficiency | High energy demand; complex setup | UV524, $H_2O_2$ (50 mg/L), 30 min |
| | ~99% (Ozonation) | Highly effective | Requires high temperature and pH | pH 9, 35°C, reduces 160 mg/L IBU to 1 mg/L in 20 min |
| | ~50% (Fenton oxidation) | Effective for diverse pollutants | Requires precise conditions | 0.87 mM $Fe_2$ and 0.32 Mm $H_2O_2$, 30°C, 3 pH for 2 h |
| Membrane Bioreactor [77] | ~ 100% | Consistent and high efficiency | Expensive installation | Full-scale HF, MA of 0.047 $m^2$<br>Diameter: 0.1 -0.2 µm |
| Steam Activated Biochar (SA- MBHB) [74] | 99.26 % | Environmentally friendly | Requires regeneration | 0.1 g/L adsorbent concentration at 20°C, contact time of 120 min and agitation of 200 rpm at pH 2 |
| Activated Sludge with High Nitrifying Activity in SBR [77] | 76% | Cost-effective | Long treatment duration | Biodegradation Time: 24 h |
| Microbial Biodegradation [77] | - | Eco-friendly and sustainable | Slow process | Variovorax sp. Ibu-1 strain cultured for 75 h with a concentration of 200 mg/L |

## 8.2. Dairy Industry Effluents

### *8.2.1. Environmental Impact of Dairy Effluents*

In India, four liters of wastewater are produced for every liter of milk processed [78]. Producing 1 $m^3$ of dairy products requires ~ 1 to 10 $m^3$ of water, with around 2% of the processed milk being discharged as wastewater [73]. Consequently, effluents from dairy processing plants contain large amounts of organic compounds such as fats, proteins, carbohydrates and lipids, inorganic compounds such as nitrogen and phosphorus, detergents, and minerals like sodium, potassium and calcium which contribute to high BOD and COD levels [17, 73]. The limitations of conventional treatment methods highlight the need for tailored approaches to effectively treat effluents from the dairy industry.

### *8.2.2. 3D-Printed Turbulence Promoters to Alleviate Membrane Fouling*

The issue of membrane fouling can be tackled by altering the membrane chemical properties. Patterned structures like indents can be used to generate fluid turbulence at the membrane surface ends [80]. Three-dimensional (3D) printing techniques can fabricate irregular patterns that promote the formation of vortices on membrane surface. Membrane fouling can be reduced through increased fluid turbulence, while features such as vortices or indents can enhance mass transfer efficiency. Additionally, modifying hydrodynamic conditions can help reduce particle deposition on the membrane [81].

Recent advancements in spacer and turbulence promoter designs have increased the efficiency of membrane filtration. Different researchers have shown improvements in efficiency with various approaches as summarized in Table 6. A comparison of efficiency of ultrafiltration models with various geometries of turbulence promoters for treating dairy effluent was studied by Szpisják et al. 2024. In conclusion, the turbulence promoters enhanced shear rates on the membrane surface, which enhanced flux performance under high pressure (0.3 MPa) and high-speed stirring (400 rpm). This demonstrated their effectiveness in mitigating membrane fouling during the filtration of dairy effluents [82].

### *8.2.3. Applications of Dairy Wastewater for Treating Coal Mine Drainage (CMD)*

The utilization of dairy wastewater as a substrate for treating coal mine drainage has both environmental remediation and resource recovery opportunities. CMD consists of high amounts of sulphate and thus, its treatment reliant on sulphate-reducing bacteria (SRB). SRB anaerobically degrades organic compounds by utilizing sulfate as a terminal electron acceptor. In this process, the lactose from the dairy effluent serves as a substrate for SRB in the treatment of CMD using CWs. In a recent study, Tripathi et al. 2024 designed a lab scale horizontal subsurface flow-constructed wetland (HSSF-CW) and the study combined dairy wastewater and CMD in a 3:1 ratio. The metals such as Fe, Zn, Cr, etc., were found to be below the recommended discharge limit as per USEPA. However, sulphate removal was inadequate, likely due to the inhibitory effect of sulphate on SRB. To improve treatment efficiency, sludge and coal fly ash from CMD can be used for phosphate removal from dairy wastewater and may also aid in the biodegradation of organic matter in CMD [88]. However, further studies are needed to assess whether treatment methods can maintain efficacy on a larger scale.

**Table 6.** Advancements in 3D-printed turbulence promoters for reduced membrane fouling

| Study | Approach | Key Findings | Performance Improvements |
|---|---|---|---|
| [83] | Built-in spacers with mini module | Enhanced flux through bovine serum albumin (BSA) filtration | Improved filtration efficiency |
| [84] | 3-D-printed micro spacer | Reduced concentration polarization and fouling: improved mixing | Enhanced membrane filtration performance |
| [85] | 3-D-printed honeycomb-shaped spacer | 16 % increase in permeate flux compared to standard spacers | Higher flux performance in nanofiltration |
| [86] | 3-D-printed feed spacers with triply periodic minimal surfaces (TPMS) | Flux enhancements of 15.5% in RO, 38% in UF using sodium alginate solution | Significant flux improvements in reverse osmosis and ultrafiltration |
| [87] | Polylactic acid (PLA) spacers. | Reduced fouling by 10% while maintaining tensile strength, flow of water as well as counter solute flow. | Less fouling, maintained mechanical and flow properties |
| [82] | 3-D-printed turbulence promoter for ultrafiltration | Promoters reduced flux decline ratio and fouling coefficient | Enhanced flux, reduced fouling |

## 8.3. Textile Industry Effluents

The textile industry is a significant contributor to wastewater effluent. On an average, 0.2 to 0.35 $m^3$ of textile effluent is produced per kg of the product [89]. The wastewater discharged from this industry consists of hazardous pollutants, including dyes, sodium hydroxide, acids, starch, and other toxic contaminants as shown in Figure 3 below:

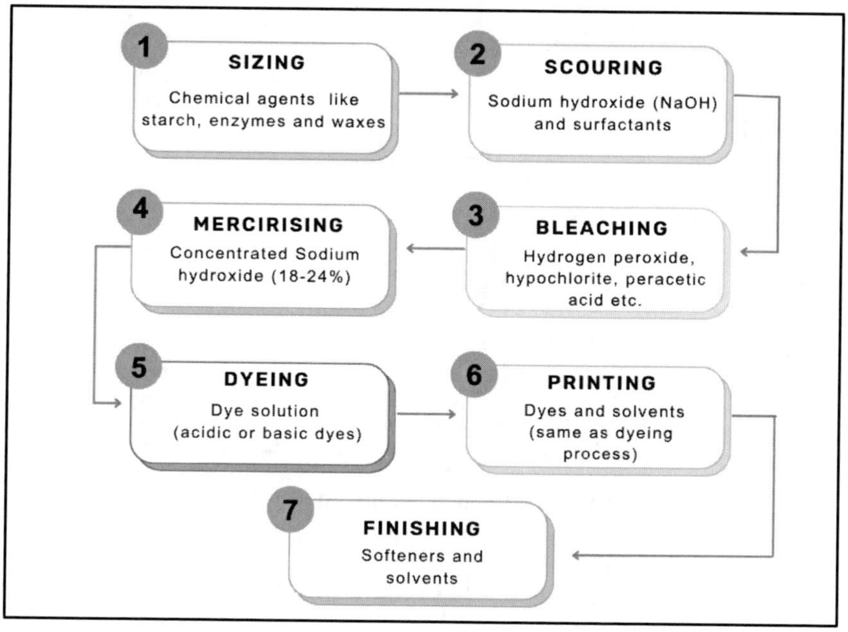


**Figure 3.** Major pollutants discharged at each step of wet processing [14].

### *8.3.1. Treatment Techniques for Textile Effluents*

The treatment methods for textile effluents include physical methods like ultrafiltration, nanofiltration, loose nanofiltration, and reverse osmosis [90]. Chemical treatment methods include UV/ $H_2O_2$, acoustics and hydrodynamic cavitation. Physicochemical treatment methods include membrane filtration, coagulation, and electrocoagulation. Biological treatment methods include fungal, algae, and microbial fuel cell technology along with Moving bed biofilm reactor (MBBR) [91].

### *8.3.2. Moving Bed Biofilm Reactor (MBBR) Technology*

In MBBR, the carriers provide a large surface area that supports the bacterial growth. As wastewater flows through the reactor, bacteria adhere to the carrier's surface and form a biofilm [92]. The bacteria in the biofilm metabolize the pollutants in the wastewater converting them into nontoxic by-products. The key steps involved in MBBR process are depicted in Figure 4 below.

According to Moga et al. 2018, the efficiency of MBBR can be increased by optimizing technical parameters such as, $p^H$, temperature and aeration speed [94]. In addition to MBBR, nanofiltration technology offers effective solution for separating solutes from the textile effluents based on their molecular weight ranging from 200-1500 Da [95]. This technology is efficient in the water softening process as it removes calcium and magnesium ions during filtration process [96]. With lower operational cost and better energy efficiency compared to other membrane-based filtration technologies, nanofiltration is an ideal choice for wastewater treatment.

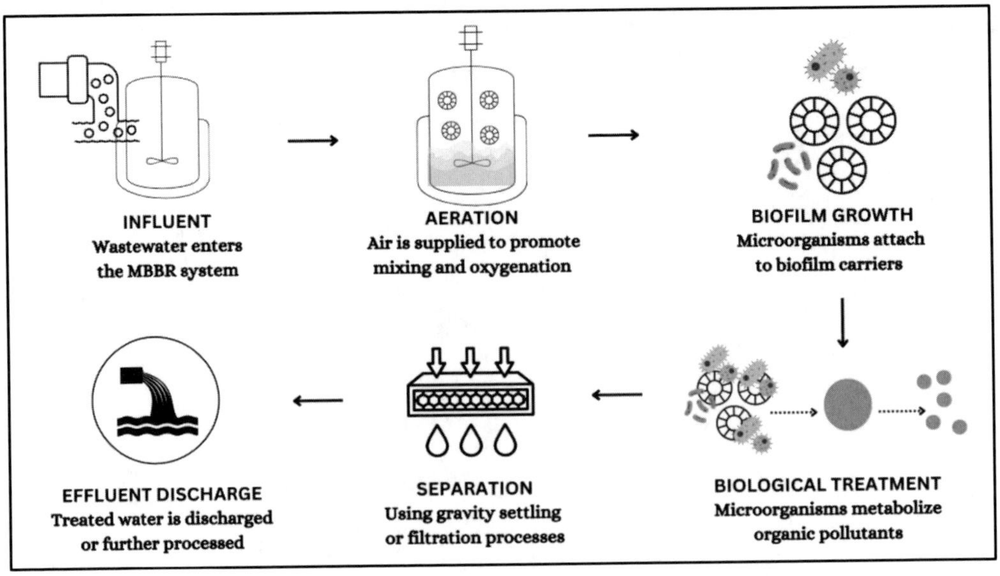


**Figure 4.** Key steps involved in MBBR for treating textile effluents [93].

## 8.4. Distillery Industry Effluents and Its Environmental Impact

Various industries such as cosmetics, chemicals, pharmaceuticals, food and beverages, and perfumery rely on products from distillery industry- including ethanol, methanol, glycerol, polyphenols, rectified spirit and denatured alcohol- as solvents. The effluent produced by distilleries is characterized by low $p^H$, high COD/BOD ratio, and a distinctive dark brown color due to melanoidin and other organic compounds like lignin, caramel, polyphenols, tannins and anthocyanins. Compounds like skatole and indole contribute to effluent's strong odor. The recalcitrant nature of this discharge has a detrimental effect on the aquatic ecosystems and the environment. Distilleries produce approximately 12-15 times the volume of effluent per liter of alcohol produced. Due to its high pollution index, CPCB has categorized the distillery industry in the "red" category [97].

### *8.4.1. Treatment Techniques for Distillery Effluents: Conventional and Emerging Approaches*

Given the significant environmental challenges posed by distillery effluents, various treatment techniques have been developed to mitigate these impacts. While conventional methods have proven effective to some extent, emerging techniques aim to address the limitations of these approaches by offering innovative solutions for more sustainable distillery effluent management, as outlined in Table 7.

**Table 7.** Conventional and emerging treatment techniques for distillery effluents

| Technique | Type | Efficiency | Advantages | Limitations |
|---|---|---|---|---|
| Primary Treatment: (Sedimentation, filtration, coagulation and flocculation) [98] | Conventional | Removes TSS; may reduce turbidity | Cost-effective, easy implementation. | Sludge generation, insufficient pollutant removal |
| Aerobic Secondary Treatment [99] | Conventional<br>Source: Coriolus versicolour, Oscillatoria boryana and Pseudomonas fluorescens. | Decolorizes melanoidin up to 90% | Effective for treating effluent with organic pollutants | Energy-intensive; requires continuous aeration and monitoring. |
| | Source: Hyacinth (Eichhornia crassipes) and duckweed (Lemna gibba) | bioaccumulate heavy metals like chromium, copper, lead and zinc | Low cost and bioaccumulates high concentrations of heavy metals | Slow process, plant growth can be affected by environmental conditions |
| Anaerobic Secondary Treatment [15] | Conventional<br>Sequencing batch reactors | 98% COD removal | treats high-strength effluents, is cost-effective, and produces biogas<br>High biomethane production | High capital cost |
| | Up-flow Anaerobic Sludge Blanket (UASB) | ~ 90% COD removal | | |
| Tertiary Treatment (Reverse Osmosis) [99] | Conventional | 99.9% reduction in TDS, COD, and potassium | Highly effective at removing persistent pollutants. | High TDS levels lead to concentration polarization; high operational costs, membrane fouling. |
| Modified PES Membranes with aluminum silicate particles (ASPs) [100] | Emerging | Up to 92% BOD and color reduction | Enhanced pollutant removal efficiency; low fouling tendency. | Cost and scalability of membranes need further optimization. |
| Ozone Electrocoagulation [101] | Emerging | 100% COD and color removal within 4 hours | Rapid and highly effective for color and COD removal. | High energy demand for ozone generation. |
| Forward Osmosis and anaerobic membrane bioreactors (FO-AnMBRs) [102] | Emerging | High water recovery efficiency; significant COD and color removal | Energy-efficient | Sensitive to high salt concentrations; potential membrane fouling. |
| Enzymatic Treatments (Immobilized Keratinase and Manganese peroxidase) [103] | Emerging<br>Keratinase immobilized on modified bagasse | ~65% melanoidin decolorization ~69% COD reduction | High specificity and biodegradable by-products produced | Enzyme stability and cost, require further research for large-scale applications. |
| | Manganese peroxidase produced from Candida glabrata | ~70% decolorization of synthetic melanoidins | | |

These emerging solutions such as modified membranes, AOPs, forward osmosis, and enzymatic treatments, are promising techniques to enhance efficiency and promote environmental sustainability. However, ongoing research and optimization are essential for large scale implementation for their cost effectiveness and scalability.

## 8.5. Pulp and Paper Industry Effluents

The effluents from pulp and paper industry contain large amount of lignocellulosic biomass like pulp and paper sludge, black liquor, and discarded sulfite liquor. This wastewater is rich in recalcitrant organic waste, resulting in high levels of BOD, TSS, COD, which negatively impacts the aquatic ecosystems. The waste also contains solid particles of various shapes, sizes, densities, and content such as phenols, chlorophenols, dioxins, and furans. The characteristics of these particles influence their behaviors in the liquid phase and their removal efficiency [16].

### *8.5.1. Environmental Impact of Pulp and Paper Effluents*

When dark-colored, turbid, and foul-smelling effluents are deposited into water bodies, they can damage the ecosystem [16]. Suspended particles from the pulping processes like sulfur compounds and nitrogen oxides, chlorinated and organic compounds, dark color and high turbidity contribute to water opacity and blanketing, resulting in release of hydrogen sulfide and reducing photosynthetic activity in aquatic plants [104]. Therefore, proper treatment of these effluents is essential before their release into the aquatic systems.

### *8.5.2. Treating Pulp and Paper Effluents*

Conventional wastewater treatment systems primarily rely on primary clarification and activated sludge processes. In recent years, Physico-chemical treatments have become increasingly popular for primary clarifiers, enhancing treatment efficiency and cost-effectiveness. Wastewater is biologically treated before secondary clarification, where microorganisms break down organic debris. Secondary clarification separates fine particulates and biological sludge using settling tanks; settled sludge is collected and treated, while collected water is released or sent to tertiary treatment. To address rising pollution levels and presence of physiologically resistant chemicals, advanced oxidation, biological treatments, 3FM, and membrane filtration are required as these pose challenges for conventional treatment [105].

### *8.5.3. Advanced Oxidation Processes (AOPs) for Treating Physiologically Resistant Chemicals*

Tertiary integrated processing technologies, such as, AOPs, are promising methods for managing organic pollutants in effluents from the pulp and paper industry. AOPs include methods like ozone, catalysts, UV radiation, and the Fenton reaction. The Fenton process involves electron transfer between $H_2O_2$ and metal ions generating high-oxidation potential hydroxyl radicals. These radicals are effective in degrading challenging compounds like lignin. This process can reduce the toxicity and color of effluents, making it more biodegradable. AOPs are particularly effective in addressing lignin-related species that are resistant to biodegradation and in reducing the concentration of harmful and refractory compounds in wastewater [16].

## Conclusion

Industrial effluent management is crucial for achieving sustainable industrialization, especially in light of the challenges associated with resource conservation and environmental protection. The rapid growth and diversification of industries have led to threats to biodiversity and human health, primarily due to the rising toxicity and complexity of industrial effluents. This chapter focused on effluent treatment challenges and innovations in the pharmaceutical, textile, distillery, and pulp and paper industries, each of which contributes distinct pollutants to the wastewater stream. These tailored technologies are becoming increasingly essential to overcome the limitations of traditional methods of wastewater treatment. Advanced technologies such as hydrodynamic cavitation, electrocoagulation, CWs, and membrane processes offer efficient, cost-effective, and sustainable solutions, addressing emerging contaminants and antibiotic-resistant bacteria. Customizing treatment based on wastewater characteristics and integrating innovations such as 3D printing and surface structuring can enhance pollutant removal while reducing energy consumption, waste generation and overall costs. As the demand for cleaner and safer water rises, optimizing wastewater treatment is crucial for saving environment and public health. Success requires a holistic approach that integrates policy enforcement, research, and stakeholder collaboration. Regulatory frameworks like the Clean Water Act and Effluent Limitation Guidelines (ELG), and global initiatives like United Nations Sustainable Development Goals (SDGs) ensure compliance and drive innovations in effluent management.

## References

[1] 9.5 Percent growth in the industrial sector. (n.d.). https://pib.gov.in/PressReleaseIframePage.aspx?PRID=2034939.

[2] State of Global Water Resources 2023. (2024, December 4). *World Meteorological Organization.* https://wmo.int/publication-series/state-of-global-water-resources-2023.

[3] Qadir, M., Drechsel, P., Cisneros, B. J., Kim, Y., Pramanik, A., Mehta, P., & Olaniyan, O. (2020). Global and regional potential of wastewater as a water, nutrient and energy source. *Natural Resources Forum, 44*(1), 40–51. https://doi.org/10.1111/1477-8947.12187.

[4] Jones, E. R., Van Vliet, M. T. H., Qadir, M., & Bierkens, M. F. P. (2021). Country-level and gridded estimates of wastewater production, collection, treatment and reuse. *Earth System Science Data*, 13(2), 237–254. https://doi.org/10.5194/essd-13-237-2021.

[5] Jia, H., & Yuan, Q. (2016). Removal of nitrogen from wastewater using microalgae and microalgae–bacteria consortia. *Cogent Environmental Science*, *2*(1), 1275089. https://doi.org/10.1080/23311843.2016.1275089.

[6] Frost and Sullivan. (2022). Indian WWW Treatment Market to Experience a Boom as the Country witnesses increase in Private Investments. *Indian Water and Wastewater Treatment Market to Experience a Boom.* https://www.frost.com/news/press-releases/energy-environment-press-releases/indian-www-treatment-market-to-experience-a-boom-as-the-country-witnesses-increase-in-private-investments/.

[7] Bilińska, L., & Gmurek, M. (2021). Novel trends in AOPs for textile wastewater treatment. Enhanced dye by-products removal by catalytic and synergistic actions. *Water Resources and Industry*, *26*, 100160. https://doi.org/10.1016/j.wri.2021.100160.

[8] Khandaker, S., Das, S., Hossain, M. T., Islam, A., Miah, M. R., & Awual, M. R. (2021). Sustainable approach for wastewater treatment using microbial fuel cells and green energy generation – A comprehensive review. *Journal of Molecular Liquids*, *344*, 117795. https://doi.org/10.1016/j.molliq.2021.117795.

[9] Alam, G., Ihsanullah, I., Naushad, M., & Sillanpää, M. (2021). Applications of artificial intelligence in water treatment for optimization and automation of adsorption processes: Recent advances and prospects. *Chemical Engineering Journal*, *427*, 130011. https://doi.org/10.1016/j.cej.2021.130011.

[10] Nahiun, K. M., Sarker, B., Keya, K. N., Mahir, F. I., Shahida, S., & Khan, R. A. (2021). A review on the methods of industrial wastewater treatment. *Scientific Review*, *73*, 20–31. https://doi.org/10.32861/sr.73.20.31.

[11] Davarnejad, R., Soofi, B., Farghadani, F., & Behfar, R. (2018). Ibuprofen removal from a medicinal effluent: A review on the various techniques for medicinal effluents treatment. *Environmental Technology & Innovation*, *11*, 308–320. https://doi.org/10.1016/j.eti.2018.06.011.

[12] Cuerda-Correa, E. M., Alexandre-Franco, M. F., & Fernández-González, C. (2019). Advanced Oxidation Processes for the Removal of Antibiotics from Water. An Overview. *Water*, *12*(1), 102. https://doi.org/10.3390/w12010102.

[13] Islam, M. T., Mamun, M. a. A., Halim, A. F. M. F., Peila, R., & Ramirez, D. O. S. (2024). Current trends in textile wastewater treatment—bibliometric review. *Environmental Science and Pollution Research*, *31*(13), 19166–19184. https://doi.org/10.1007/s11356-024-32454-3.

[14] Adane, T., Adugna, A. T., & Alemayehu, E. (2021). Textile industry effluent treatment techniques. *Journal of Chemistry*, *2021*, 1–14. https://doi.org/10.1155/2021/5314404.

[15] Fito, J., Tefera, N., & Van Hulle, S. W. H. (2019). Sugarcane biorefineries wastewater: bioremediation technologies for environmental sustainability. *Chemical and Biological Technologies in Agriculture*, *6*(1). https://doi.org/10.1186/s40538-019-0144-5.

[16] Ponnusami, A. B., Sinha, S., Ashokan, H., Paul, M. V., Hariharan, S. P., Arun, J., Gopinath, K., Le, Q. H., & Pugazhendhi, A. (2023). Advanced oxidation process (AOP) combined biological process for wastewater treatment: A review on advancements, feasibility and practicability of combined techniques. *Environmental Research*, *237*, 116944. https://doi.org/10.1016/j.envres.2023.116944.

[17] Raghunath, B. V., Punnagaiarasi, A., Rajarajan, G., Irshad, A., Elango, A., & Kumar, G. M. (2016b). Impact of Dairy Effluent on Environment—A Review. In *Environmental science and engineering* (pp. 239–249). https://doi.org/10.1007/978-3-319-27228-3_22.

[18] Rene, E. R., Ge, J., Kumar, G., Singh, R. P., & Varjani, S. (2020). Resource recovery from wastewater, solid waste, and waste gas: engineering and management aspects. *Environmental Science and Pollution Research*, 27(15), 17435–17437. https://doi.org/10.1007/s11356-020-08802-4.

[19] Saravanakumar, K., De Silva, S., Santosh, S. S., Sathiyaseelan, A., Ganeshalingam, A., Jamla, M., Sankaranarayanan, A., Veeraraghavan, V. P., MubarakAli, D., Lee, J., Thiripuranathar, G., & Wang, M. (2022). Impact of industrial effluents on the environment and human health and their remediation using MOFs-based hybrid membrane filtration techniques. *Chemosphere*, *307*, 135593. https://doi.org/10.1016/j.chemosphere.2022.135593.

[20] Hanif, M. A., Nadeem, F., Tariq, R., & Rashid, U. (2022). Energy and global environment. In *Elsevier eBooks* (pp. 673–753). https://doi.org/10.1016/b978-0-12-818150-8.00003-4.

[21] Patel, N., Ruparelia, J., & Barve, J. (2020). Prediction of total suspended solids present in effluent of primary clarifier of industrial common effluent treatment plant: Mechanistic and fuzzy approach. *Journal of Water Process Engineering*, 34, 101146. https://doi.org/10.1016/j.jwpe.2020.101146.

[22] Sidabutar, T., Srimariana, E. S., & Wouthuyzen, S. (2020). The potential role of eutrophication, tidal and climatic on the rise of algal bloom phenomenon in Jakarta Bay. *IOP Conference Series Earth and Environmental Science*, *429*(1), 012021. https://doi.org/10.1088/1755-1315/429/1/012021.

[23] Yashwant, C. P., Rajendran, V., Krishnamoorthy, S., Nagarathinam, B., Rawson, A., Anandharaj, A., & Sivanandham, V. (2022). Antibiotic resistance profiling and valorization of food waste streams to

starter culture biomass and exopolysaccharides through fed-batch fermentations. *Food Science and Biotechnology, 32*(6), 863–874. https://doi.org/10.1007/s10068-022-01222-9.

[24] Samreen, N., Ahmad, I., Malak, H. A., & Abulreesh, H. H. (2021). Environmental antimicrobial resistance and its drivers: a potential threat to public health. *Journal of Global Antimicrobial Resistance, 27*, 101–111. https://doi.org/10.1016/j.jgar.2021.08.001.

[25] UN DESA, 2024, The Sustainable Development Goals Report 2024 – June 2024, New York, USA: UN DESA. https://unstats.un.org/sdgs/report/2024/.

[26] *Summary of the Clean Water Act | US EPA*. (2024, June 12). US EPA. https://www.epa.gov/laws-regulations/summary-clean-water-act.

[27] Riffat, R., & Husnain, T. (2022). *Fundamentals of wastewater treatment and engineering*. https://doi.org/10.1201/9781003134374.

[28] Kato, S., & Kansha, Y. (2024). Comprehensive review of industrial wastewater treatment techniques. *Environmental Science and Pollution Research, 31*(39), 51064–51097. https://doi.org/10.1007/s11356-024-34584-0.

[29] Altowayti, W. a. H., Shahir, S., Othman, N., Eisa, T. a. E., Yafooz, W. M. S., Al-Dhaqm, A., Soon, C. Y., Yahya, I. B., Rahim, N. a. N. B. C., Abaker, M., & Ali, A. (2022). The Role of Conventional Methods and Artificial intelligence in the wastewater Treatment: A Comprehensive review. *Processes, 10*(9), 1832. https://doi.org/10.3390/pr10091832 .

[30] Iwuozor, K. O. (2019). Prospects and Challenges of Using Coagulation-Flocculation method in the treatment of Effluents. *Advanced Journal of Chemistry-Section A*, 105–127. https://doi.org/10.29088/sami/ajca.2019.2.105127

[31] Ambulkar, Archis, Nathanson, & A, J. (2025, January 17). *Wastewater treatment | Process, History, Importance, Systems, & Technologies*. Encyclopedia Britannica. https://www.britannica.com/technology/wastewater-treatment.

[32] Malakar, S., Saha, P. D., Baskaran, D., & Rajamanickam, R. (2017). Comparative study of biofiltration process for treatment of VOCs emission from petroleum refinery wastewater—A review. Environmental Technology & Innovation, 8, 441–461. https://doi.org/10.1016/j.eti.2017.09.007.

[33] Nguyen, P. Y., Carvalho, G., Reis, M. A. M., & Oehmen, A. (2021). A review of the biotransformations of priority pharmaceuticals in biological wastewater treatment processes. *Water research, 188*, 116446. https://doi.org/10.1016/j.watres.2020.116446.

[34] Rout, P. R., Zhang, T. C., Bhunia, P., & Surampalli, R. Y. (2021). Treatment technologies for emerging contaminants in wastewater treatment plants: A review. *The Science of the total environment, 753*, 141990. https://doi.org/10.1016/j.scitotenv.2020.141990.

[35] Bairagi, S., & Ali, S. W. (2020). Conventional and advanced technologies for wastewater treatment. *Environmental Nanotechnology for Water Purification*, 33–56. https://doi.org/10.1002/9781119641353.ch2.

[36] Satyam, S., & Patra, S. (2024). Innovations and challenges in adsorption-based wastewater remediation: A comprehensive review. *Heliyon, 10*(9), e29573. https://doi.org/10.1016/j.heliyon.2024.e29573.

[37] Ahmer, M. F., & Uddin, M. K. (2024). Structure properties and industrial applications of anion exchange resins for the removal of electroactive nitrate ions from contaminated water. *RSC advances, 14*(45), 33629–33648. https://doi.org/10.1039/d4ra03871a.

[38] Nqombolo, A., Mpupa, A., Moutloali, R. M., & Nomngongo, P. N. (2018). Wastewater treatment using membrane technology. In *InTech eBooks*. https://doi.org/10.5772/intechopen.76624.

[39] Feng, Y., Yang, L., Liu, J., & Logan, B. E. (2016). Electrochemical technologies for wastewater treatment and resource reclamation. *Environmental Science Water Research & Technology, 2*(5), 800–831. https://doi.org/10.1039/c5ew00289c.

[40] Patil, P. M., Ingavale, R. R., Matkar, A. R., Hwang, S., Gurav, R., Dhanavade, M. J. (2023). An Innovative and Effective Industrial Wastewater Treatments: A Brief History and Present Scenario. In: Shah, M. P. (eds) *Advanced and Innovative Approaches of Environmental Biotechnology in Industrial Wastewater Treatment*. Springer, Singapore. https://doi.org/10.1007/978-981-99-2598-8_10.

[41] Corpus, R. M. B., Bayani, M. S., Aguilar, J. L., Aguilar, J. C. B., & Aguilar, H. B. (2024). Emerging pollutants in wastewater: Challenges and advancements in treatment technology. *IOP Conference*

*Series Earth and Environmental Science*, *1372*(1), 012037. https://doi.org/10.1088/1755-1315/1372/1/012037.

[42] Leite, W. R. M., Gottardo, M., Pavan, P., Filho, P. B., & Bolzonella, D. (2015). Performance and energy aspects of single and two phase thermophilic anaerobic digestion of waste activated sludge. *Renewable Energy*, *86*, 1324–1331. https://doi.org/10.1016/j.renene.2015.09.069.

[43] Wu, Q. L., Guo, W. Q., Zheng, H. S., Luo, H. C., Feng, X. C., Yin, R. L., & Ren, N. Q. (2016). Enhancement of volatile fatty acid production by co-fermentation of food waste and excess sludge without pH control: The mechanism and microbial community analyses. *Bioresource technology*, *216*, 653–660. https://doi.org/10.1016/j.biortech.2016.06.006.

[44] Wang, B., Su, H., & Zhang, B. (2021). Hydrodynamic cavitation as a promising route for wastewater treatment – A review. *Chemical Engineering Journal*, *412*, 128685. https://doi.org/10.1016/j.cej.2021.128685.

[45] Bis, M. (2021). Current research trends in the application of hydrodynamic cavitation in environmental engineering. In *Routledge eBooks* (pp. 167–178). https://doi.org/10.1201/9781003171669-17.

[46] Patil, Y., Sonawane, S. H., Shyam, P., Sun, X., & Manickam, S. (2023). Hybrid hydrodynamic cavitation (HC) technique for the treatment and disinfection of lake water. *Ultrasonics Sonochemistry, 97,* 106454. https://doi.org/10.1016/j.ultsonch.2023.106454.

[47] Hazra, M., & Durso, L. M. (2022). Performance Efficiency of Conventional Treatment Plants and Constructed Wetlands towards Reduction of Antibiotic Resistance. *Antibiotics*, *11*(1), 114. https://doi.org/10.3390/antibiotics11010114.

[48] Zhao, L., Fu, G., Pang, W., Tang, J., Guo, Z., & Hu, Z. (2022). Biochar immobilized bacteria enhances nitrogen removal capability of tidal flow constructed wetlands. *The Science of the total environment*, *836*, 155728. https://doi.org/10.1016/j.scitotenv.2022.155728.

[49] Huang, X., Ye, G., Yi, N., Lu, L., Zhang, L., Yang, L., Xiao, L., & Liu, J. (2019). Effect of plant physiological characteristics on the removal of conventional and emerging pollutants from aquaculture wastewater by constructed wetlands. *Ecological Engineering*, *135*, 45–53. https://doi.org/10.1016/j.ecoleng.2019.05.017.

[50] Li, X., Zhu, W., Meng, G., Zhang, C., & Guo, R. (2020). Efficiency and kinetics of conventional pollutants and tetracyclines removal in integrated vertical-flow constructed wetlands enhanced by aeration. *Journal of Environmental Management, 273,* 111120. https://doi.org/10.1016/j.jenvman.2020.111120.

[51] Kahl, S., Nivala, J., Van Afferden, M., Müller, R. A., & Reemtsma, T. (2017). Effect of design and operational conditions on the performance of subsurface flow treatment wetlands: Emerging organic contaminants as indicators. *Water Research*, *125*, 490–500. https://doi.org/10.1016/j.watres.2017.09.004.

[52] Fernandes, J. P., Almeida, C. M. R., Pereira, A. C., Ribeiro, I. L., Reis, I., Carvalho, P., Basto, M. C. P., & Mucha, A. P. (2015). Microbial community dynamics associated with veterinary antibiotics removal in constructed wetlands microcosms. *Bioresource Technology, 182,* 26–33. https://doi.org/10.1016/j.biortech.2015.01.096.

[53] Hien, P. T., Tadashi, T., & Kazuhiro, M. (2017). Removal of Tetracycline and Tetracycline Resistance Genes from Municipal Wastewater in Microcosm Fill-and-Drain Constructed Wetlands. *Japanese Journal of Water Treatment Biology, 53(1), 11–21.* https://doi.org/10.2521/jswtb.53.11.

[54] Pazda, M., Rybicka, M., Stolte, S., Bielawski, K. P., Stepnowski, P., Kumirska, J., Wolecki, D., & Mulkiewicz, E. (2020). Identification of Selected Antibiotic Resistance Genes in Two Different Wastewater Treatment Plant Systems in Poland: A Preliminary Study. *Molecules, 25(12),* 2851. https://doi.org/10.3390/molecules25122851.

[55] Yi, X., Tran, N. H., Yin, T., He, Y., & Gin, K. Y. (2017). Removal of selected PPCPs, EDCs, and antibiotic resistance genes in landfill leachate by a full-scale constructed wetlands system. *Water Research, 121,* 46–60. https://doi.org/10.1016/j.watres.2017.05.008.

[56] Šereš, M., Innemanová, P., Hnátková, T., Rozkošný, M., Stefanakis, A., Semerád, J., & Cajthaml, T. (2021). Evaluation of Hybrid Constructed Wetland Performance and Reuse of Treated Wastewater in Agricultural Irrigation. *Water, 13(9),* 1165. https://doi.org/10.3390/w13091165.

[57] Du, L., Zhao, Y., Wang, C., Zhang, H., Chen, Q., Zhang, X., Zhang, L., Wu, J., Wu, Z., & Zhou, Q. (2020). Removal performance of antibiotics and antibiotic resistance genes in swine wastewater by integrated vertical-flow constructed wetlands with zeolite substrate. *The Science of the Total Environment, 721,* 137765. https://doi.org/10.1016/j.scitotenv.2020.137765.

[58] N. A. Sabri, H. Schmitt, B. M. van der Zaan, H. W. Gerritsen, H. H. M. Rijnaarts, A. A. M. Langenhoff, Performance of full scale constructed wetlands in removing antibiotics and antibiotic resistance genes, *Science of The Total Environment, Volume 786, 2021,* 147368, ISSN 0048-9697, https://doi.org/10.1016/j.scitotenv.2021.147368.

[59] S. E. H. Comstock, T. H. Boyer, Combined magnetic ion exchange and cation exchange for removal of DOC and hardness, Chem. Eng. J. 241 (Apr. 2014) 366–375, https://doi.org/10.1016/j.cej.2013.10.073.

[60] Qu, D., Wang, J., Wang, L., Hou, D., Luan, Z., & Wang, B. (2009). Integration of accelerated precipitation softening with membrane distillation for high-recovery desalination of primary reverse osmosis concentrate. *Separation and Purification Technology, 67(1),* 21–25. https://doi.org/10.1016/j.seppur.2009.02.021.

[61] Jonathan I. Mendez-Ruiz, Angie N. Medina-Toala, Leonardo Gutierrez, Priscila E. Valverde-Armas, Comparative evaluation of an advanced electrocoagulation treatment system versus a conventional lime softening treatment for removing Ca2+, SO42−, and Mn in groundwater, *Case Studies in Chemical and Environmental Engineering, Volume 8, 2023,* 100448, ISSN 2666-0164, https://doi.org/10.1016/j.cscee.2023.100448.

[62] Tyagi, N., Mathur, S., & Kumar, D. (2014). Electrocoagulation process for textile wastewater treatment in continuous upflow reactor. Journal of Scientific & Industrial Research, 73(3), 195–198. http://nopr.niscair.res.in/bitstream/123456789/27380/1/JSIR%2073%283%29%20195-198.pdf.

[63] Yangyang Sun, Yan Wang, Nan Xue, Cong Yu, Yingjie Meng, Baoyu Gao, Qilin Li, The effect of DOM on floc formation and membrane fouling in coagulation/ultrafiltration process for treating TiO2 nanoparticles in various aquatic media, *Chemical Engineering Journal, Volume 316, 2017,* Pages 429-437, ISSN 1385-8947, https://doi.org/10.1016/j.cej.2017.01.126.

[64] Ghanbari, F., Zirrahi, F., Olfati, D., Gohari, F., & Hassani, A. (2020). TiO2 nanoparticles removal by electrocoagulation using iron electrodes: Catalytic activity of electrochemical sludge for the degradation of emerging pollutant. *Journal of Molecular Liquids, 310,* 113217. https://doi.org/10.1016/j.molliq.2020.113217.

[65] Sutrisna, P. D., Kurnia, K. A., Siagian, U. W., Ismadji, S., & Wenten, I. G. (2022). Membrane fouling and fouling mitigation in oil–water separation: A review. *Journal of Environmental Chemical Engineering,* 10(3), 107532. https://doi.org/10.1016/j.jece.2022.107532.

[66] Othman, N. H., Alias, N. H., Fuzil, N. S., Marpani, F., Shahruddin, M. Z., Chew, C. M., Ng, K. M. D., Lau, W. J., & Ismail, A. F. (2021). A review on the use of membrane technology systems in developing countries. *Membranes*, *12*(1), 30. https://doi.org/10.3390/membranes12010030.

[67] Barman, S. R., Gavit, P., Chowdhury, S., Chatterjee, K., & Nain, A. (2023b). 3D-Printed materials for wastewater treatment. *JACS Au*, *3*(11), 2930–2947. https://doi.org/10.1021/jacsau.3c00409.

[68] Zhang, K., Yuan, G., Werdich, A. A., Zhao, Y., 2020. Ibuprofen and diclofenac impair the cardiovascular development of zebrafish (Danio rerio) at low concentrations. *Environ. Pollut*. 258, 113613. https://doi.org/10.1016/j.envpol.2019.113613.

[69] Phoon, B. L., Ong, C. C., Mohamed Saheed, M. S., Show, P. L., Chang, J. S., Ling, T. C., Lam, S. S., Juan, J. C., 2020. Conventional and emerging technologies for removal of antibiotics from wastewater. *J. Hazard Mater.* 400, 122961. https://doi.org/ 10.1016/j.jhazmat.2020.122961.

[70] Ding, T., Yang, M., Zhang, J., Yang, B., Lin, K., Li, J., & Gan, J. (2017). Toxicity, degradation and metabolic fate of ibuprofen on freshwater diatom Navicula sp. *Journal of Hazardous Materials*, *330*, 127–134. https://doi.org/10.1016/j.jhazmat.2017.02.004.

[71] Blasco, J., & Trombini, C. (2023). Ibuprofen and diclofenac in the marine environment - a critical review of their occurrence and potential risk for invertebrate species. *Water Emerging Contaminants & Nanoplastics*, *2*(3), 14. https://doi.org/10.20517/wecn.2023.06.

[72] Osman, A. I., Ayati, A., Farghali, M., Krivoshapkin, P., Tanhaei, B., Karimi-Maleh, H., Krivoshapkina, E., Taheri, P., Tracey, C., Al-Fatesh, A., Ihara, I., Rooney, D. W., & Sillanpaä, M. (2023). Advanced

adsorbents for ibuprofen removal from aquatic environments: a review. *Environmental Chemistry Letters*, *22*(1), 373–418. https://doi.org/10.1007/s10311-023-01647-6.

[73] Baccar, R., Sarrà, M., Bouzid, J., Feki, M., Blánquez, P., 2012. Removal of pharmaceutical compounds by activated carbon prepared from agricultural byproduct. *Chem. Eng. J.* 211, 310–317. https://doi.org/10.1016/j.cej.2012.09.099.

[74] Baccar, R., Sarrà, M., Bouzid, J., Feki, M., & Blánquez, P. (2012). Removal of pharmaceutical compounds by activated carbon prepared from agricultural by-product. *Chemical Engineering Journal*, *211–212*, 310–317. https://doi.org/10.1016/j.cej.2012.09.099.

[75] Banerjee, P., Das, P., Zaman, A., Das, P., 2016. Application of graphene oxide nanoplatelets for adsorption of Ibuprofen from aqueous solutions: Evaluation of process kinetics and thermodynamics. *Process Saf. Environ. Prot.* 101, 45–53.https://doi.org/10.1016/j.psep.2016.01.021.

[76] Bhadra, B. N., Ahmed, I., Kim, S., & Jhung, S. H. (2016). Adsorptive removal of ibuprofen and diclofenac from water using metal-organic framework-derived porous carbon. *Chemical Engineering Journal*, *314*, 50–58. https://doi.org/10.1016/j.cej.2016.12.127.

[77] Chopra, S., & Kumar, D. (2020). Ibuprofen as an emerging organic contaminant in environment, distribution and remediation. *Heliyon*, *6*(6), e04087. https://doi.org/10.1016/j.heliyon.2020.e04087.

[78] Chaudhry, S., & Garg, S. (2018). Industrial wastewater pollution and advanced treatment techniques. In *Advances in environmental engineering and green technologies book series* (pp. 74–97). https://doi.org/10.4018/978-1-5225-5754-8.ch006.

[79] B, R. K., Bhuvaneshwari, S., Majeed, F., M, M. M., Jose, E., & Mohan, A. (2022). Different treatment methodologies and reactors employed for dairy effluent treatment - A review. *Journal of Water Process Engineering*, *46*, 102622. https://doi.org/10.1016/j.jwpe.2022.102622.

[80] Al-Shimmery, A, Mazinani, S, Ji, J, Chew, YMJ & Mattia, D 2019, '3D printed composite membranes with enhanced anti-fouling behaviour,' *Journal of Membrane Science*, vol. 574, pp. 76-85. https://doi.org/10.1016/j.memsci.2018.12.058.

[81] Armbruster, S., Cheong, O., Lölsberg, J., Popovic, S., Yüce, S., Wessling, M. Fouling Mitigation in Tubular Membranes by 3D-Printed Turbulence Promoters. *J. Memb. Sci.* 2018, 554, 156–163. https://doi.org/10.1016/j.memsci.2018.02.015.

[82] Szpisják-Gulyás, N., László, Z., Kertész, S., & Hodúr, C. (2024b). Effects of a 3D-Printed turbulence promoter on membrane fouling during the ultrafiltration of dairy wastewater. *Dairy*, 5(4), 842–862. https://doi.org/10.3390/dairy5040061.

[83] Baitalow, K., Wypysek, D., Leuthold, M., Weisshaar, S., Lölsberg, J., & Wessling, M. (2021). A mini-module with built-in spacers for high-throughput ultrafiltration. *Journal of Membrane Science*, *637*, 119602. https://doi.org/10.1016/j.memsci.2021.119602.

[84] Van Dang, B., Charlton, A. J., Li, Q., Kim, Y. C., Taylor, R. A., Le-Clech, P., & Barber, T. (2020b). Can 3D-printed spacers improve filtration at the microscale? *Separation and Purification Technology*, *256*, 117776. https://doi.org/10.1016/j.seppur.2020.117776.

[85] Park, S., Jeong, Y. D., Lee, J. H., Kim, J., Jeong, K., & Cho, K. H. (2020). 3D printed honeycomb-shaped feed channel spacer for membrane fouling mitigation in nanofiltration. *Journal of Membrane Science*, *620*, 118665. https://doi.org/10.1016/j.memsci.2020.118665.

[86] Sreedhar, N., Thomas, N., Al-Ketan, O., Rowshan, R., Hernandez, H., Al-Rub, R. K. A., & Arafat, H. A. (2017). 3D printed feed spacers based on triply periodic minimal surfaces for flux enhancement and biofouling mitigation in RO and UF. *Desalination*, *425*, 12–21. https://doi.org/10.1016/j.desal.2017.10.010.

[87] Yanar, N., Son, M., Yang, E., Kim, Y., Park, H., Nam, S., & Choi, H. (2018). Investigation of the performance behavior of a forward osmosis membrane system using various feed spacer materials fabricated by 3D printing technique. *Chemosphere*, *202*, 708–715. https://doi.org/10.1016/j.chemosphere.2018.03.147.

[88] Tripathi, P., & Chakraborty, S. (2024). Application of dairy wastewater as substrate for bioremediation of coal mine drainage in planted horizontal flow constructed wetland. *International journal of phytoremediation,* 1–11. Advance online publication. https://doi.org/10.1080/15226514.2024.2417375.

[89] Raj, S., Singh, H., & Bhattacharya, J. (2022). Treatment of textile industry wastewater based on coagulation-flocculation aided sedimentation followed by adsorption: Process studies in an industrial ecology concept. *The Science of the Total Environment*, *857*, 159464. https://doi.org/10.1016/j.scitotenv.2022.159464.

[90] Verma, A., Kumar, V., Varjani, S., & Pant, D. (2022). Current status and research trends of textile wastewater treatments—A bibliometric-based study. *Frontiers in Environmental Science*, 10. https://doi.org/10.3389/fenvs.2022.1042256.

[91] Arabgol, R., Vanrolleghem, P. A., & Delatolla, R. (2022). Influence of MBBR carrier geometrical properties and biofilm thickness restraint on biofilm properties, effluent particle size distribution, settling velocity distribution, and settling behaviour. *Journal of environmental sciences* (China), 122, 138–149. https://doi.org/10.1016/j.jes.2021.09.029.

[92] Deena, S. R., Kumar, G., Vickram, A., Singhania, R. R., Dong, C., Rohini, K., Anbarasu, K., Thanigaivel, S., & Ponnusamy, V. K. (2022). Efficiency of various biofilm carriers and microbial interactions with substrate in moving bed-biofilm reactor for environmental wastewater treatment. *Bioresource Technology*, *359*, 127421. https://doi.org/10.1016/j.biortech.2022.127421 .

[93] Yang, X., López-Grimau, V., Vilaseca, M., & Crespi, M. (2020). Treatment of Textile Wastewater by CAS, MBR, and MBBR: A Comparative Study from Technical, Economic, and Environmental Perspectives. *Water*, *12*(5), 1306. https://doi.org/10.3390/w12051306.

[94] Corina, M. I., Ioan, A., Gabriel, P., Nicolae, C., & Radu, P. (2018). The potential of biofilms from moving bed bioreactors to increase the efficiency of textile industry wastewater treatment. *Industria Textila*, *69*(05), 412–418. https://doi.org/10.35530/it.069.05.1500.

[95] Emek, A. A., Yılmazer, C. E., Çınar, S., Kiraz, O., Paşaoğlu, M., Taşdemir, R. Ş., Özgün, Ö. K., Koyuncu, İ., Çitil, E., Orhon, A. K., Şıltu, E., Güçver, S., Turan, Z., Dikmen, B., Toröz, İ., Tanik, A., Yetiş, Ü., & Kınacı, C. (2021). Removal of endocrine-disrupting chemicals from textile industry effluents by nanofiltration. *Desalination and Water Treatment*, *211*, 411–421. https://doi.org/10.5004/dwt.2021.26542.

[96] Alhussaini, M. A., Souza-Chaves, B. M., Felix, V., & Achilli, A. (2024). Comparative analysis of reverse osmosis and nanofiltration for the removal of dissolved contaminants in water reuse applications. *Desalination,* 117822. https://doi.org/10.1016/j.desal.2024.117822.

[97] Shinde, P. A., Ukarde, T. M., Pandey, P. H., & Pawar, H. S. (2020). Distillery spent wash: An emerging chemical pool for next generation sustainable distilleries. *Journal of Water Process Engineering*, 36, 101353. https://doi.org/10.1016/j.jwpe.2020.101353.

[98] Jorge, N., Teixeira, A. R., Lucas, M. S., & Peres, J. A. (2022). Combined organic coagulants and photocatalytic processes for winery wastewater treatment. *Journal of Environmental Management*, *326*, 116819. https://doi.org/10.1016/j.jenvman.2022.116819

[99] Kharayat, Y. (2012). Distillery wastewater: bioremediation approaches. *Journal of Integrative Environmental Sciences,* 9(2), 69–91. https://doi.org/10.1080/1943815X.2012.688056.

[100] Velu, S., Arthanareeswaran, G., Lade, H., 2017. Removal of organic and inorganic substances from industry wastewaters using modified aluminosilicate-based polyethersulfone ultrafiltration membranes. *Environ. Prog. Sustain. Energy* 36, 1612–1620. https://doi.org/10.1002/ep.12614.

[101] Aziz, A. R. A., Asaithambi, P., Daud, W. M. A. B. W., 2016. Combination of electrocoagulation with advanced oxidation processes for the treatment of distillery industrial effluent. *Process Saf. Environ.* 99, 227–235. https://doi.org/10.1016/j.psep.2015.11.010. https://doi.org/10.1016/j.heliyon.2023.e14906.

[102] Schneider, C., Onoro, A. E., H´elix-Nielsen, C., Fotidis, I. A., 2020. Forward-osmosis anaerobic-membrane bioreactors for brewery wastewater remediation. *Sep. Purif. Technol.* 117786. https://doi.org/10.1016/j.seppur.2020.117786.

[103] Ratna, S., Rastogi, S., & Kumar, R. (2021). Current trends for distillery wastewater management and its emerging applications for sustainable environment. *Journal of Environmental Management*, 290, 112544. https://doi.org/10.1016/j.jenvman.2021.112544.

[104] Gavrilescu, Dan & Puitel, Adrian & Dutuc, Gheorghe & Craciun, Grigore. (2012). Environmental impact of pulp and paper mills. *Environmental Engineering and Management Journal*. 11. 81-86. https://doi.org/10.30638/eemj.2012.012.

[105] Liang, X., Xu, Y., Yin, L., Wang, R., Li, P., Wang, J., & Liu, K. (2023). Sustainable utilization of pulp and paper wastewater. *Water,* 15(23), 4135. https://doi.org/10.3390/w15234135.

# Chapter 11

# MXene-Integrated Carbon-Based Materials: Structural Engineering and Wastewater Remediation

**Gajendra Kumar Inwati[1,*]**
**Yamini Patel[1]**
**Deepak K. K. Kori[2]**
**Archana Chaudhary[3]**
**Pratibha Sharma[1]**
**Urmila Raghuvanshi[4]**
**and Shivraj Gangadhar Wanale[5]**

[1]School of Chemical Sciences, Devi Ahilya Vishwavidyalaya, Indore, M.P., India
[2]Department of Chemistry, Indian Institute of Technology, Indore, India
[3]Centre for Translational Research, Office of Research and Development, Indian Institute of Technology, Indore, India
[4]Department of Chemistry, Shri G S Institute of Technology and Science, Indore, M.P., India
[5]Department of Chemistry, Netaji Subhashchandra Bose Art's, Commerce, and Science College, Nanded, Maharashtra, India

## Abstract

Carbon containing hybrid materials have been explored for multiple applications particularly 2D materials such as graphene, carbon nitrides, MXenes (aligned carbides and nitrides of transition metals) and others, owing to their notable biological, chemical and physical properties. The structural, surface and compositional configurations of such advanced materials have been designed and used by researchers and scientists in recent years due to the versatile performance of these materials in the field of science and technology. Among the various classes of these 2D materials, MXene are the most promising candidates for multipurpose uses including biomedical, pharmaceutical, energy storage and conversion, and environmental applications such as wastewater treatment, and bioremediation. The heavy metal removal, dye degradation and antibacterial functions has been thoroughly studied, prompting a comprehensive review of relevant research articles and data to provide detailed insights into their fundamental properties. Moreover, MXenes

[*] Corresponding Author's Email: gajendrainwati@gmail.com

In: Advancements in Wastewater Treatment
Editors: Neha Agarwal, Maulin P. Shahl, Vijendra Singh Solankil et al.
ISBN: 979-8-90134-007-3

have been widely synthesized and structurally modified to enhance its catalytic properties, particularly in environmental science and research. This chapter thoroughly describes the essentials of MXene-based metamaterials or composites and their applications, such as industrial wastewater treatment.

**Keywords:** carbon-composites, nano-adsorbents, MXenes, structural modifications, environmental uses, antibacterial, 2D materials

## 1. Introduction

The industrialization and chemical production at a larger scale are major contributors to the discharge of organic and inorganic hazardous materials, causing serious health issues in humans and biodiversity [1]. The continuous progress in advanced techniques and new materials requires multiple ionic species for their stable construction and chemical stability and hence scientists and engineers from multiple backgrounds utilize large quantities of chemicals and metal ions, leading to the formation of by-products or secondary compounds. The production of these by-products is harmful due to their hazardous nature and long lifespan in the environment. The discharged electronic, hydrocarbon and multi-metallic components disrupt the equilibrium of chemical components in nature, posing environmental challenges by negatively impacting soil, air and water. Therefore, many researchers are working to resolve worldwide problems by designing and investigating structural modifications of hybrid materials such as metal-oxides, carbon-based nanostructures like CNTs, graphene, fullerenes, carbon nitrides [2] and MXenes [3, 4]. Structural and surface engineering of hybrid nanomaterials has been a focal point for researchers and material scientists in recent years due to the unique physical and chemical properties of these materials. In this regard, MXenes, a class of 2D structures have been found to be stronger layered materials than graphene, CNTs, $MoS_2$ and others [5, 6]. MXenes have outstanding metallic conductivity, hydrophilicity, filtration capabilities, exceptional adaptability, adsorption, solvent extraction and ion intercalation capabilities. As a result, they are gaining a lot of interest and have significant promise for use in environmental sciences to clean wastewater, gases, and remove heavy metals [7-10]. The transition metal-based carbon nitrides are oriented in a way to exhibit exceptional physicochemical and catalytic properties, enabling enhanced recyclability and durability. The chemically modified structure of these transition metal ions, C, and N known as MXenes with the $M_{n+1}AX_n$ formula, where M stands for a class of specific transition metal elements such as Ti, Sc, V, Cr, Mn, Zr, while A denotes the C/N element contribution in the structure, and X represents halogen elements such as Cl or I [11, 12]. This ternary complex is strongly recommended by several scientists for improving the functionality and processing efficiency of MXenes. MXenes show superior intrinsic qualities in terms of conductance, thinner interfaces between the 2D sheets, hydrophilic nature, photocatalytic properties, and chemical binding with components, making them excellent 2D materials for decorating 2D structures and metal ions to construct hybrid MXenes catalysts [13]. However, pure MXenes are modified to enhance reaction rates and chemical stability by functionalizing their surfaces and band alignments with additional 2D structures or suitable metal ions. To date, 2D materials such as graphene, CNTs and $MoS_2$ have been well integrated with MXenes and possess extraordinary performance in various optoelectronic systems for environmental remediation. Various transition, rare earth, alkali metals, lanthanides and actinides ions/atoms have been incorporated into bare 2D carbon-based compounds like MXenes to improve catalytic efficiency. The multifunctional applications of such hybrids have been developed and explored across various sectors, including environmental, pharmaceutical, energy harvesting and optoelectronic devices due to the durability, electron transport properties and chemical stability of both pure and doped MXenes [4,14]. Hence, the modified structures of 2D materials are widely used in environmental tasks such as cleaning industrial and pharmaceutical wastewater. However, ultrafiltration and nanofiltration techniques have been developed by integrating polymeric compounds with MXenes to filter toxic metals like Pb, As, Pd, and other divalent salts, while organic compounds such as dyes are degraded using photocatalysts under light irradiation [11, 15]. The catalytic reaction was enhanced by the doping and incorporation of suitable chemical compositions. In this chapter, we have addressed the systematic structural design and catalytic actions of hybrid MXenes (carbon-based materials) for industrial wastewater treatment, focusing on the removal of heavy metals, organic dyes, and bacteria.

## 2. Morphological Analysis of the Modified C-Based MXenes

Before using MXenes, it is an essential task to analyze prepared samples for their composition, shape, size, and surface properties. These parameters regulate the catalytic properties of the samples for different applications, necessitating structural and morphological investigations of bare and hybrid MXenes using advanced instruments [14, 16]. So, to identify the structural arrays, surface groups and crystalline nature, several techniques are employed such as SEM, TEM, EDX, XRD, Raman, XPS, SIMS, etc. These characterizations are essential for determining the morphological and spectral features of modified MXenes [17,18].

### 2.1. Morphological Confirmation

The morphological studies like shape, sizes and distribution patterns, were precisely analysed using SEM and TEM techniques, which confirm the size control and dispersion tendency. For example, Ag, Au and Pd-doped MXenes were studied using TEM techniques by Elumalai Satheeshkumar and co-workers [3]. They fabricated Ag, Au and Pd-MXenes hybrid materials, where the bare MXenes of approximately 2–3 μm sizes and thicknesses greater than 1 nm were prepared that was confirmed by HRTEM. The elemental confirmation was carried out by EDX studies which revealed the presence of only Ti and C content in the bare MXenes sample. However, the plasmonic metals like Ag were dispersed in the range of 10-70 nm while the Au NPs were found to be 40-50 nm on MXenes surfaces with a homogeneous distribution. Further, a sheet-like structure was found in the case of Pd NPs possibly due to different reduction rates compared to Ag and Au NPs. Therefore, the shapes and sizes were accurately examined using TEM and EDX techniques in the reported study. Likewise, N and S doped MXenes were analyzed using SEM and TEM to examine their porous structures with an interspacing of 1.02 nm (for N-$Ti_3C_2T_x$) and to observe the morphology of the doped nanoparticles (NPs) [11]. Also, the Pt, Ru Pd, and Au dispersion behaviour on MXenes was studied with the help of TEM. Moreover, the experimental conditions had a critical impact on the morphology of the prepared samples and thus, variations in size and shape were identified using these techniques. Of course, newly designed materials or catalysts need to be verified based on their surface, structure, and compositional ratios to provide reliable data for assessing catalytic performance. The spectral, and diffraction techniques are highly recommended and used for the investigation of hybrid materials, particularly at micro- and nanoscales. Scientific communities and researchers consistently characterize these nanomaterials before their application in specific tasks.

## 3. MXenes as a Catalyst for the Treatment of Industrial Wastewater

Recently, effluents from chemical industries have increased significantly due to the widespread use of hazardous chemicals, electronic components, and unsaturated organic compounds, particularly dyes, colors, and paint components [19]. The modern age of architecture, medical advances and material has led to large-scale chemical use in laboratories, academic institutions, and industrial sectors, resulting in the uncontrolled discharge of toxic elements into water bodies. Consequently, removing these hazardous chemicals from wastewater poses a significant challenge for researchers and scientists aiming to ensure a healthy environment and protect human life. Several forms of MXenes, such as $Cr_2TiAlC_2$, $Cr_2AlC$, $Mn_2AlC$, $Cr_2GaC$, $V_2CTx$, $Ti_3C_2$-$SO_3H$, MXene/$Fe_3O_4$, $Ti_3C_2Tx$/PVA, $Ti_3C_2Tx$, and poly (diallyldimethylammonium chloride) (PDAC) membranes, among others, have been designed and utilized for environmental applications [11, 19]. The filtration techniques have been employed for a long time, and thus, micro- and nanofiltration technologies have been developed and implemented for industrial wastewater treatment using metamaterials at various concentrations. The decomposition of azo-group compounds and other transition-metal-based hazardous molecules has been efficiently achieved through environmental research and science.

### 3.1. Designing and Exploration of Hybrid C-based MXenes for Wastewater Treatment

To treat effluents, researchers have made multiple attempts to design new materials and study their catalytic properties. For this, the layered stacking route was mainly proposed and applied to produce stable multi-layered membrane-like films with fixed thickness and a defined number of layers. The stacking of multi-layer sheets on oppositely charged species facilitates the formation of efficient and stable membranes. However, the various types of C-based materials such as graphene, CNTs, carbon nitrides, and MXenes have demonstrated significant potential for dye elimination and heavy metal removal in environmental science applications. Similarly, experimental conditions such as pH, are critical for conducting these experiments, prompting extensive research in this area. The catalytic activities of such 2D layered structures vary with the doping of specific elements or polymeric organic/inorganic compounds, leading to improved dye and salt rejection properties.

In this regard, Huang and co-workers [20] engineered MXenes by doping $Ni^{2+}$, resulting in magnetic behaviour, and then deposited them onto the PES polymeric substrate. The resulting material was known as MXene/Ni nanosheets. The HA rejection rate was higher than the that of the PES substrate alone, achieving approximately 99.8% HA removal. The layered structures provided a large surface area with numerous active sites, demonstrating enhanced catalytic properties of Ni-loaded MXenes. Approximately 2.5 times greater reactivity was observed for HA removal using these hybrid structures compared to bare samples. Therefore, the membrane design and structural analysis of such 2D materials have recently shown promise, particularly for dye removal and salt degradation in industrial wastewater or effluents. Research on C-based MXene materials is highly recommended, as they are effectively used to remove wastewater contaminants, such as heavy metals, dye molecules, and bacteria, particularly in industrial wastewater. The pure and hybrid MXenes were found to be highly suitable for treating industrial wastewater due to their high chemical stability, reactivity, and durability. For example, polyethyleneimine (PEI) based MXenes materials were prepared by Xing *et al.*, where both the components were found to be neutral charge, and were used to remove dye molecules. The best combination a 1:1 ratio of PEI and MXenes, was highly effective, removing approximately 99.9% of red dye, 99.3% of Acid Blue 90, and 99.82% of Congo dye. The mechanically stable and strong hybrid MXenes showed over 99% removal efficiency for the studied dyes, indicating their potential as excellent adsorbents for industrial wastewater contaminants [21].

### 3.2. Case Studies: MXenes Membranes as Nano Filtration Systems

These research studies focus on nanofiltration using hybrid MXene membranes, incorporating carbon-based elemental combinations to optimize catalytic and physicochemical properties. Several studies have reported on hybrid MXene membranes used to remove industrial dyes, with Table 1 showing the percentage of dye removal using specific carbon-based MXene catalysts. However, PEI-MXene composite was coated onto a polydopamine substrate to form a membrane used for dye molecule filtration. Primarily, Reactive Blue, Congo Red, Methylene Blue (MB), and Coomassie Blue dyes were removed using this hybrid MXene membrane, with a water permeability of approximately 38.2 L $m^{-2}$ $h^{-1}$ $bar^{-1}$. Moreover, the studies were conducted effectively in acidic and alkaline media to remove dye molecules commonly found in wastewater solutions [22]. Polysulfone (PSF) was mixed with a fixed amount of MXenes by Dashtbozorg et al., to improve the permeable capacity of the polysulfone whereas pure PSF exhibited low hydrophilicity. To enhance hydrophilicity, MXenes, which contain –OH functional groups, were incorporated. These active sites enhance the hydrophilicity of the combined PSF-MXene materials, achieving excellent dye elimination efficiency of over 90% for dyes such as Methylene Blue, as well as some divalent salts [22]. Therefore, the chemical inertness and large surface area of MXenes provide reliable catalytic action for adsorbing dye components and heavy metals from industrial wastewater.

Dye components and divalent salts were effectively removed due to the filtration gaps between the layered structures of MXenes. The cost-effective monovalent, neutral ions transition metal- or alkali metal-based oxide species are highly effective as intercalated molecules in 2D nanostructures. The compact structure of the MXenes may exhibit limited filtration capacity for salts, dyes, and heavy metals. For instance, hyperbranched polyethyleneimine (HPEI)-decorated AgNP-$Ti_3C_2T_x$ materials were designed by Li et al. [23] *to improve the permeability for dyes and salts. They engineered the surfaces of MXenes by incorporating Ag NPs which reduce the density of layered structures and facilitate greater transport of filtrates. HPEI acts as a cross-linker to produce highly branched 2D MXene materials for targeted applications. It* was observed that a 99.61% rejection rate for Rhodamine B dye was achieved, with a water permeability of 20.07 L $m^{-2}$ $h^{-1}$ $bar^{-1}$.The $MgCl_2$, $MgSO_4$, $Na_2SO_4$ and NaCl solutions were found to be highly permeable in wastewater treatment studies.

**Table 1.** Hybrid MXenes membranes for the removal of dyes as nanofiltration

| S. No. | MXenes as Membranes | Feed concentration | Rejection (%) | References |
|---|---|---|---|---|
| 1 | $Ti_3C_2Tx$ /AgNP- hyperbranched polyethyleneimine | RhB (10 ppm) | 99.61 | [23] |
| 2 | $Ti_3C_2Tx$-polyimide (P84) | Congo red (100 ppm) | 78.5 | [24] |
| 3 | $Ti_3C_2Tx$-polyamide | $Na_2SO_4$ (1000 ppm) | 99.9 | [25] |
| 4 | $Ti_3C_2Tx$- polyethersulfone | Congo red (100 ppm) | 92.3 | [26] |
| 5 | References$Ti_3C_2Tx$– mixed cellulose ester | Methylene blue (15 ppm) | 94.63 | [27] |
| 6 | $Ti_3C_2Tx$ @$TiO_2$ - polyether sulfone | BSA (1000 ppm) | 95 | [28] |
| 7 | $Ti_3C_2Tx$- aramid nanofibers | Congo Red | 96.1 | [29] |

## 4. Inorganic Metal and Metal-Oxides-Based MXenes Composites as Catalysts

Several categories of metal- and metal-oxide-based MXene composites have been designed and explored for wastewater treatment from industrial and laboratory sources. The transition metal ions and alkali metal ions are highly incorporated into C-based materials such as 2D MXenes and their hybrid frames. Metal oxides, particularly traditional semiconductors, are also used to form heterojunctions, engineering the bandgap structures of carbon-based compounds to enhance charge carrier capacity for photocatalytic activities [13, 27]. The transition elements such as Ti, Cr, Ni, co, Zn etc. and their oxide form are valued for multifunctional applications, including biomedical, energy harvesting, optical sensors, and superconductors due to the electronic transitions at fixed energy ranges and chemical stability [30-32]. Combinations with substrate materials, such as dielectric media, polymeric organic chains, and carbon-based materials like graphene, carbon nitrides, and MXenes, are designed and produced on a large scale for various applications. Consequently, many academic and scientific communities have focused on addressing these metal- and semiconductor-based research challenges to promote environmental cleanup [33, 34]. In this section, we highlight key case studies related to this theme in a clear and fundamental manner.

### 4.1. Noble Metal (Plasmon) Doped MXenes for the Removal of Organic Molecules

The plasmonic NPs such as Ag, Pd, Pt and Cu were incorporated to enhance the adsorption capacity of 2D carbon-based materials. Due to their inertness and plasmonic properties, these metals are highly valued in designing MXenes, particularly for improving restacking and stability. The uniform sheets and large surface area of MXenes enable significant adsorption of organic and inorganic substances from wastewater, with metals intercalated into the layered structures. The varying sizes and shapes of the metal NPs also positively influence the catalysis of adsorption characteristics. Conceptually, plasmonic metals exhibit intense surface plasmon resonance (SPR) when interacting with specific electromagnetic energies, leading to localized electron fluctuations on their surfaces [35-37]. This enhanced light interaction is primarily used to induce spectral signals and catalytic actions in applications such as environmental remediation, third-order optical responses, and other multifunctional purposes [38-40]. The fabrication of metal-based 2D materials has been explored for various applications due to their unique catalytic properties. The plasmon metals such as Ag, Au, Pt, Pd, and Cu, have been extensively studied at the nanoscale, exhibiting exceptional electronic and surface properties compared to bulk materials and transition metals, which typically lack plasmonic effects. For example, an advanced poly (vinyl alcohol)/poly (acrylic acid):$Fe_3O_4$-MXene: Ag nanomaterial-based film was prepared using the electro-spinning method [41]. The structured films were then processed using the temperature-induced self-reduction approaches. The resulting MXene flakes exhibited uniform dispersion, stability, and compatibility with self-reduced Ag NPs. The catalytic reduction activities enhanced by the addition of Ag metals, which facilitate exceptional separation between MXene sheets, creating larger open surface sites. These plasmonic nanomaterials are widely recommended for environmental applications, including antibacterial treatment, metal removal, and dye elimination, due to their high reactivity and physicochemical activity [41].

## 4.2. Metal-Oxides Based MXenes Composites for Wastewater Cleaning

The band gap engineering concept is employed to induce photocatalysis for the degradation of industrial wastewater. The heterojunction formation between the meta-oxides such as $TiO_2$, CdS, CuO, $Fe_3O_3$, ZnO etc., were used to create the interfaces on MXenes which boost the charge transfer rate for better catalytic functions [27, 42]. These composites are designed by optimizing surface, electronic, and elemental ratios to enhance the catalytic capabilities of the catalysts, particularly through heterojunctions that exhibit exceptional physicochemical and electronic properties for catalyzing chemical reactions in fields such as water purification, energy storage, biomedicine, and optoelectronic devices. The combined moieties of the metallic and organic components provide improved performance, and thus, dynamic hybrid materials are introduced in this section for wastewater treatment and environmental applications. For example, $BiVO_4$-MXene, $C_3N_4$-MXene, CdS, $TiO_2$ [43], and g-$C_3N_4$ M-MXene-$Z_XC_{1-X}S$, are designed for these targeted environmental applications. Enhanced free charge transmission and separation can be achieved by combining 2D MXenes with selected semiconductors. The objective is to adjust the conduction band (CB) and valence band (VB) potentials. Such nano semiconducting 0D-2D, 1D-2D, and 2D-2D heterojunctions and their best combinations provide adopted charge carrier properties for catalysis [44, 45]. The band gap adjustments of a mixed MXenes photocatalyst could be further tuned via the mixing of multiple metal oxides. Zhuang and co-workers [43] fabricated hybrid M-MXene: $TiO_2$ materials under the electrostatic-based self-assembled route for the chemical reactive performances by using them as photocatalysts. The etching of carbon-based $Ti_3AlC_2$ with 50 wt.% hydrofluoric acid at 25°C was performed, and the MXene/TiO2 exhibited superior photocatalytic activity for $H_2$ production due to enhanced charge transfer, reactive (101) facets, and a larger light absorption area [46]. Similarly, the surface functionalized MXenes: $TiO_2$ materials were also carried out for environmental aspects including bioremediations. For example, Siwanat et al. designed a microwave hydrothermal assisted M-MXenes: $TiO_2$ compounds dipped in NaCl and HCl solutions for the cleaning of enrofloxacin from wastewater. The approach involves incorporating Na ions from NaCl into M-MXene/$TiO_2$ nanocomposites to improve adsorption rates and photocatalytic degradation of antibiotics in wastewater. Additionally, a range of antibiotic rejections were achieved using M-MXene/$TiO_2$, likely due to active functional groups on MXenes, such as –OH hydroxyl groups [47].

## 5. Future Challenges and Opportunities

The wastewater, discharged from chemical industries and laboratories is hazardous due to its high toxicity and uncontrolled harmful compositions. The unsaturated dye components, divalent salts, toxic metal ions and bacterial colonies adversely affect environmental resources, including water, air, and soil. This has become a critical task for scientific communities and researchers [48, 49]. Therefore, the controlled and precise production of azo-complexes, closures, and other compounds is regulated by pollution control boards on international platforms. The use of pesticides and chemical complexes has been restricted by specific regulations to minimize chemical discharge into water. However, the pharmaceutical, and agricultural sectors heavily utilize these materials, which contain high levels of radical and ionic species that are harmful to human health and atmospheric conditions. Consequently, several researchers are focusing on the design and development of eco-friendly materials and technologies. The morphological, structural and elemental modifications of such materials are being investigated and utilized to reduce toxic and hazardous chemicals in the environment. The metal and metal-oxides based 2D materials are thoroughly described in this chapter as effective catalysts.

## Conclusion

The increasing challenges in treating industrial wastewater are a serious concern for researchers, scientists, and the general public, as water demand and utilization increase rapidly, day by day. The industrial waste includes high levels of organic dye molecules, heavy metals, bacteria, and other toxic chemicals. In this, chapter, we have presented a fundamental approach to treating wastewater discharged from industries using hybrid 2D MXene, a carbon-based nanocomposite layered structure. The structural features, compositional alterations and surface engineering are discussed to enhance catalytic performance in environmental science applications. In particular, the removal of various dyes and divalent salts is explained, supported by relevant case studies. These findings will be highly useful for researchers and scholars addressing wastewater treatment challenges, such as the elimination of Rhodamine B, Congo Red, Methylene Blue dyes, divalent salts, bacteria, and fungi using 2D carbon-based nanostructures or composites.

## References


[1] C. K. Ong, S. Dennison, S. Fearn, K. Hellgardt, and G. H. Kelsall, "Behaviour of titanium-based Fe2O3 photo-anodes in photo-electrochemical reactors for water splitting," *Electrochim. Acta*, vol. 125, pp. 266–274, 2014, doi: 10.1016/j.electacta.2014.01.086.

[2] J. Safaei et al., "Graphitic carbon nitride (g-C3N4) electrodes for energy conversion and storage: A review on photoelectrochemical water splitting, solar cells and supercapacitors," *J. Mater. Chem. A*, vol. 6, no. 45, pp. 22346–22380, 2018, doi: 10.1039/c8ta08001a.

[3] E. Satheeshkumar, T. Makaryan, A. Melikyan, H. Minassian, Y. Gogotsi, and M. Yoshimura, "One-step Solution Processing of Ag, Au and Pd@MXene Hybrids for SERS," *Sci. Rep.*, vol. 6, no. July, pp. 1–9, 2016, doi: 10.1038/srep32049.

[4] T. Teranishi, M. Eguchi, M. Kanehara, and S. Gwo, "Controlled localized surface plasmon resonance wavelength for conductive NPs over the ultraviolet to near-infrared region," *J. Mater. Chem.*, vol. 21, no. 28, pp. 10238–10242, 2011, doi: 10.1039/c0jm04545d.

[5] N. A. Abd Malek et al., "Ultra-thin MoS2 nanosheet for electron transport layer of perovskite solar cells," *Opt. Mater. (Amst).*, vol. 104, no. April, p. 109933, 2020, doi: 10.1016/j.optmat.2020.109933.

[6] S. Lu, M. Chen, Y. Wang, R. Li, J. Lin, and X. Zhang, "Highly efficient MoS2/rGO electrocatalysts for triiodide reduction as Pt-free counter electrode for dye-sensitized solar cells," *Sol. Energy*, vol. 220, no. March, pp. 788–795, 2021, doi: 10.1016/j.solener.2021.03.086.

[7] H. K. Makadia and S. J. Siegel, "Poly Lactic-co-Glycolic Acid (PLGA) as biodegradable controlled drug delivery carrier," *Polymers (Basel).*, vol. 3, no. 3, pp. 1377–1397, 2011, doi: 10.3390/polym3031377.

[8] B. R. da Cunha, L. P. Fonseca, and C. R. C. Calado, "Antibiotic discovery: Where have we come from, where do we go?," *Antibiotics*, vol. 8, no. 2, pp. 1–21, 2019, doi: 10.3390/antibiotics8020045.

[9] Q. T. Bu et al., "Rational construction of genome-reduced and high-efficient industrial Streptomyces chassis based on multiple comparative genomic approaches," *Microb. Cell Fact.*, vol. 18, no. 1, pp. 1–17, 2019, doi: 10.1186/s12934-019-1055-7.

[10] N. M. Julkapli, S. Bagheri, S. Bee, and A. Hamid, "Recent Advances in Heterogeneous Photocatalytic Decolorization of Synthetic Dyes," vol. 2014, 2014.

[11] A. Dey et al., "Doped MXenes—A new paradigm in 2D systems: Synthesis, properties and applications," *Prog. Mater. Sci.*, vol. 139, no. January, p. 101166, 2023, doi: 10.1016/j.pmatsci.2023.101166.

[12] M. Q. Zhao et al., "Flexible MXene/carbon nanotube composite paper with high volumetric capacitance," *Adv. Mater.*, vol. 27, no. 2, pp. 339–345, 2015, doi: 10.1002/adma.201404140.

[13] H. An et al., "Construction of a novel hybrid plasmonic metal-2D MXene catalyst for plasmon-driven photoreduction of nitroaromatics," *Appl. Surf. Sci.*, vol. 613, no. September 2022, 2023, doi: 10.1016/j.apsusc.2022.156055.

[14] Y. Gogotsi and M. W. Barsoum, "*Two-Dimensional Transition Metal Carbides*," no. 2, pp. 1322–1331, 2012.

[15] H. Zhang et al., "Physical deposition improved SERS stability of morphology controlled periodic micro/nanostructured arrays based on colloidal templates," *Small*, vol. 11, no. 7, pp. 844–853, 2015, doi: 10.1002/smll.201402630.

[16] F. Dixit et al., "Application of MXenes for water treatment and energy-efficient desalination: A review," *J. Hazard. Mater.*, vol. 423, no. PB, p. 127050, 2022, doi: 10.1016/j.jhazmat.2021.127050.

[17] V. Mauchamp et al., "Enhanced and tunable surface plasmons in two-dimensional $Ti_3C_2$ stacks: Electronic structure versus boundary effects," *Phys. Rev. B - Condens. Matter Mater. Phys.*, vol. 89, no. 23, pp. 1–6, 2014, doi: 10.1103/PhysRevB.89.235428.

[18] D. Magne, V. Mauchamp, S. Célérier, P. Chartier, and T. Cabioc'h, "Spectroscopic evidence in the visible-ultraviolet energy range of surface functionalization sites in the multilayer $Ti_3C_2$ MXene," *Phys. Rev. B - Condens. Matter Mater. Phys.*, vol. 91, no. 20, pp. 1–5, 2015, doi: 10.1103/PhysRevB.91.201409.

[19] R. M. Ronchi, J. T. Arantes, and S. F. Santos, "Synthesis , structure , properties and applications of MXenes : Current status and perspectives," *Ceram. Int.*, vol. 45, no. 15, pp. 18167–18188, 2019, doi: 10.1016/j.ceramint.2019.06.114.

[20] Z. Huang et al., "Enhanced permeability and antifouling performance of polyether sulfone (PES) membrane via elevating magnetic Ni@MXene NPs to upper layer in phase inversion process," *J. Memb. Sci.*, vol. 623, no. January, p. 119080, 2021, doi: 10.1016/j.memsci.2021.119080.

[21] C. Xing et al., "Anti-swelling polyethyleneimine-modified MXene nanofiltration membranes for efficient and selective molecular separation," *EcoMat*, vol. 5, no. 3, pp. 1–11, 2023, doi: 10.1002/eom2.12300.

[22] A. Dashtbozorg, E. Saljoughi, S. M. Mousavi, and S. Kiani, "High-performance and robust polysulfone nanocomposite membrane containing 2D functionalized MXene nanosheets for the nanofiltration of salt and dye solutions," *Desalination*, vol. 527, no. January, p. 115600, 2022, doi: 10.1016/j.desal.2022.115600.

[23] Y. Li et al., "Highly permeable and stable hyperbranched polyethyleneimine crosslinked AgNP@$Ti_3C_2Tx$ MXene membranes for nanofiltration," *J. Memb. Sci.*, vol. 670, no. November 2022, p. 121376, 2023, doi: 10.1016/j.memsci.2023.121376.

[24] R. Han, Y. Xie, and X. Ma, "Crosslinked P84 copolyimide/MXene mixed matrix membrane with excellent solvent resistance and permselectivity," *Chinese J. Chem. Eng.*, vol. 27, no. 4, pp. 877–883, 2019, doi: 10.1016/j.cjche.2018.10.005.

[25] X. Zhu et al., "Crumple-textured polyamide membranes via MXene nanosheet-regulated interfacial polymerization for enhanced nanofiltration performance," *J. Memb. Sci.*, vol. 635, no. April, p. 119536, 2021, doi: 10.1016/j.memsci.2021.119536.

[26] R. Han, X. Ma, Y. Xie, D. Teng, and S. Zhang, "Preparation of a new 2D MXene/PES composite membrane with excellent hydrophilicity and high flux," *RSC Adv.*, vol. 7, no. 89, pp. 56204–56210, 2017, doi: 10.1039/c7ra10318b.

[27] S. Zhang et al., "Direct deposition of two-dimensional MXene nanosheets on commercially available filter for fast and efficient dye removal," *J. Hazard. Mater.*, vol. 384, no. July, p. 121367, 2020, doi: 10.1016/j.jhazmat.2019.121367.

[28] Z. Huang et al., "Facile synthesis of 2D $TiO_2$@MXene composite membrane with enhanced separation and antifouling performance," *J. Memb. Sci.*, vol. 640, no. September, p. 119854, 2021, doi: 10.1016/j.memsci.2021.119854.

[29] Y. Li, R. Dai, H. Zhou, X. Li, and Z. Wang, "Aramid Nanofiber Membranes Reinforced by MXene Nanosheets for Recovery of Dyes from Textile Wastewater," *ACS Appl. Nano Mater.*, vol. 4, no. 6, pp. 6328–6336, 2021, doi: 10.1021/acsanm.1c01217.

[30] G. K. Inwati et al., "applied sciences 2D Personality of Multifunctional Carbon Nitrides towards Enhanced Catalytic Performance in Energy Storage and Remediation," 2022.

[31] G. K. Inwati et al., "Enhanced Plasmon Based Ag and Au Nanosystems and Their Improved Biomedical Impacts," 2022.

[32] G. K. Inwati et al., "Study of photoluminescence and nonlinear optical behaviour of AgCu NPs for nanophotonics," *Nano-Structures and Nano-Objects*, vol. 28, p. 100807, 2021, doi: 10.1016/j.nanoso.2021.100807.

[33] T. Kashyap, S. Biswas, S. Ahmed, D. Kalita, P. Nath, and B. Choudhury, "Plasmon activation versus plasmon quenching on the overall photocatalytic performance of Ag/Au bimetal decorated g-C3N4 nanosheets under selective photoexcitation: A mechanistic understanding with experiment and theory," *Appl. Catal. B Environ.*, vol. 298, no. August, p. 120614, 2021, doi: 10.1016/j.apcatb.2021.120614.

[34] Y. Zhao, X. Zhang, T. Wang, T. Song, and P. Yang, "Fabrication of rGO/CdS@2H, 1T, amorphous MoS2 heterostructure for enhanced photocatalytic and electrocatalytic activity," *Int. J. Hydrogen Energy*, vol. 45, no. 41, pp. 21409–21421, 2020, doi: 10.1016/j.ijhydene.2020.05.217.

[35] P. Malik, G. K. Inwati, T. K. Mukherjee, S. Singh, and M. Singh, "Green silver NP and Tween-20 modulated pro-oxidant to antioxidant curcumin transformation in aqueous CTAB stabilized peanut oil emulsions," *J. Mol. Liq.*, vol. 291, 2019, doi: 10.1016/j.molliq.2019.111252.

[36] S. Modi et al., "Recent Trends in Fascinating Applications of Nanotechnology in Allied Health Sciences," 2022.

[37] V. K. Yadav et al., "A Noble and Economical Method for the Synthesis of Low Cost Zeolites From Coal Fly Ash Waste," *Adv. Mater. Process. Technol.*, vol. 00, no. 00, pp. 1–19, 2021, doi: 10.1080/2374068X.2021.1927640.

[38] Y. Rao, G. Inwati, A. Kumar, J. Meena, and M. Singh, "Metal Oxide based NPs use for Pressure Sensor," *Int. J. Curr. Eng. Technol.*, vol. 4, no. 4, pp. 2492–2498, 2014.

[39] P. Malik, V. Katyal, V. Malik, A. Asatkar, G. Inwati, and T. K. Mukherjee, "Nanobiosensors: Concepts and Variations," *ISRN Nanomater.*, vol. 2013, pp. 1–9, 2013, doi: 10.1155/2013/327435.

[40] N. Palaniappan, G. K. Inwati, and M. Singh, "Biomaterial Co-Cr-Mo alloys nano coating calcium phosphate orthopedic treatment," *IOP Conf. Ser. Mater. Sci. Eng.*, vol. 64, no. 1, 2014, doi: 10.1088/1757-899X/64/1/012026.

[41] X. Huang et al., "Facile Preparation of Hierarchical AgNP-Loaded MXene/Fe3O4 Polymer Nanocomposites by Electrospinning with Enhanced Catalytic Performance for Wastewater Treatment," *ACS Omega*, vol. 4, no. 1, pp. 1897–1906, 2019, doi: 10.1021/acsomega.8b03615.

[42] P. Kumar et al., "Optical and surface properties of Zn doped CdO nanorods and antimicrobial applications," *Colloids Surfaces A*, vol. 605, no. July, p. 125369, 2020, doi: 10.1016/j.colsurfa.2020.125369.

[43] Y. Zhuang, Y. Liu, and X. Meng, "Fabrication of TiO2 nanofibers/MXene Ti3C2 nanocomposites for photocatalytic H2 evolution by electrostatic self-assembly," *Appl. Surf. Sci.*, vol. 496, no. August, 2019, doi: 10.1016/j.apsusc.2019.143647.

[44] T. Zhu, Y. Liang, Y. Wang, J. Wang, W. Wang, and J. Fu, "Applied Catalysis B : Environmental P-type β $Ni(OH)_2$ NPs sensitize CdS nanorod array photoanode to prolong charge carrier lifetime and highly improve bias-free visible-light- driven $H_2$ evaluation," *Appl. Catal. B Environ.*, vol. 271, no. December 2019, p. 118945, 2020, doi: 10.1016/j.apcatb.2020.118945.

[45] B. Deepa and V. Rajendran, "Nano-Structures & Nano-Objects Pure and Cu metal doped $WO_3$ prepared via co-precipitation method and studies on their structural , morphological , electrochemical and optical properties," *Nano-Structures & Nano-Objects*, vol. 16, pp. 185–192, 2018, doi: 10.1016/j.nanoso.2018.06.005.

[46] S. Zong et al., "Mxene-$TiO_2$ composite with exposed {101} facets for the improved photocatalytic hydrogen evolution activity," *J. Alloys Compd.*, vol. 896, p. 163039, 2022, doi: 10.1016/j.jallcom.2021.163039.

[47] S. Sukidpaneenid, C. Chawengkijwanich, C. Pokhum, T. Isobe, P. Opaprakasit, and P. Sreearunothai, "Multi-function adsorbent-photocatalyst MXene-$TiO_2$ composites for removal of enrofloxacin antibiotic from water," *J. Environ. Sci. (China)*, vol. 124, pp. 414–428, 2023, doi: 10.1016/j.jes.2021.09.042.

[48] A. T. Kuvarega, R. W. M. Krause, and B. B. Mamba, "Nitrogen / Palladium-Codoped $TiO_2$ for Efficient Visible Light Photocatalytic Dye Degradation," pp. 22110–22120, 2011, doi: 10.1021/jp203754j.

[49] S. Marimuthu et al., "Silver NPs in dye effluent treatment: A review on synthesis, treatment methods, mechanisms, photocatalytic degradation, toxic effects and mitigation of toxicity," *J. Photochem. Photobiol. B Biol.*, vol. 205, no. August 2019, p. 111823, 2020, doi: 10.1016/j.jphotobiol.2020.111823.

# Chapter 12

# Microalgae in Wastewater Treatment: Benefits and Challenges

**Saivenkatesh Korlam[1,*]
and Sankara Rao Miditana[2]**

[1]Department of Botany, S.V.A. Government College, Srikalahasti, Tirupati Dt., A.P., India
[2]Department of Chemistry, Government Degree College, Puttur, Tirupati Dt., A.P., India

## Abstract

Microscopic photosynthetic organisms known as microalgae have gained attention as effective agents for treating wastewater. Their capacity to absorb nutrients, degrade organic pollutants, and eliminate contaminants makes them valuable for both environmental cleanup and biomass production.This chapter presents an in-depth assessment of the effectiveness of various microalgal strains such as *Chlorella, Scenedesmus,* and *Spirulina* in extracting critical eutrophication agents like nitrogen and phosphorus, along with organic substances and toxic metals from wastewater systems. The mechanisms by which microalgae facilitate the removal of these pollutants, including nutrient uptake, adsorption, and bioaccumulation, are critically examined. Additionally, the role of microalgae in sequestering carbon dioxide and reducing pathogen loads is explored. Various cultivation systems, such as open ponds and photobioreactors, are analyzed for their efficiency and scalability. The chapter also discusses the environmental and economic benefits of integrating microalgae with conventional wastewater treatment processes, emphasizing resource recovery and energy generation. Beyond pollution removal, the produced biomass offers opportunities for valorization into biofuels, animal feed, and high-value compounds, thus transforming wastewater treatment into a resource recovery process. However, challenges such as low biomass productivity, susceptibility to contamination, and the need for efficient harvesting and downstream processing remain obstacles to large-scale implementation. This chapter aims to provide a comprehensive overview of microalgal technology, identify knowledge gaps, and propose future research directions to optimize microalgal wastewater treatment for sustainable and resource-efficient applications.

**Keywords:** bioaccumulation, biofuels, biomass production, contaminants, eutrophication, greenhouse gas, microalgae

---

* Corresponding Author's Email: k.saivenkatesh@gmail.com

In: Advancements in Wastewater Treatment
Editors: Neha Agarwal, Maulin P. Shahl, Vijendra Singh Solankil et al.
ISBN: 979-8-90134-007-3

## 1. Introduction

Microalgae are unicellular, photosynthetic organisms that exhibit remarkable diversity, with over 100,000 species identified, including both eukaryotic and prokaryotic forms such as cyanobacteria [1]. They possess high photosynthetic efficiency, enabling them to convert light and carbon dioxide into biomass, and some species can also grow heterotrophically in the absence of light [1, 2]. Microalgae exhibit remarkable environmental adaptability, thriving in diverse aquatic ecosystems, including extreme conditions found in isolated and confined environments (IACEs). Their ability to withstand stress and rapid growth rate makes them a promising food resource. This chapter highlights their favorable physiological and nutritional characteristics, which are essential for sustainable food supply and health enhancement in IACEs. This adaptability and growth potential are critical for addressing the challenges of long-term food security in such environments [3]. They are rich in nutrients and bioactive compounds, including long-chain fatty acids and pigments, which offer significant health benefits [3]. Their applications span multiple sectors, including biofuels, nutraceuticals, wastewater treatment, and pharmaceuticals, highlighting their potential as a sustainable resource for the 21$^{st}$ century [1, 2].

Microalgae have a rich historical background, evolving from traditional uses in food and feed to emerging applications across various fields, including biofuels, nutraceuticals, and cosmetics. Microalgae species like *Spirulina* and *Chlorella* are recognized for their diverse bioactive compounds, which exhibit potent antioxidant, antimicrobial, and anti-inflammatory properties. These compounds contribute to their potential health benefits, including protection against oxidative damage and modulation of immune responses [4]. The recent surge in interest is driven by advancements in biotechnology, enabling the large-scale production of microalgal biomass and its derivatives, which are now integral to the biorefinery concept aimed at sustainable energy solutions [5, 6]. Furthermore, microalgae are being explored as eco-friendly alternatives in the cosmetics industry and as natural preservatives, reflecting their versatility and potential to meet growing market demands [7].

Wastewater treatment is crucial in addressing global water pollution, particularly as urbanization and industrialization escalate the presence of contaminants. Innovative technologies play a pivotal role in enhancing treatment processes, with advancements such as membrane bioreactors, anaerobic reactors, and phytoremediation techniques demonstrating significant potential for improving water quality and reducing environmental impacts [8]. The European Parliament's recent directive emphasizes the need for reducing pollutants like BOD5, nitrogen, and phosphorus in treated wastewater, highlighting the sector's contribution to greenhouse gas emissions and the potential for energy efficiency improvements [8]. Furthermore, integrating these technologies into a circular economy framework can optimize resource recovery, such as nitrogen and phosphorus reuse, while addressing emerging pollutants through bioremediation strategies that utilize nanomaterials [9]. Thus, the adoption of innovative wastewater treatment technologies is essential for sustainable environmental management and the protection of water resources.

The emerging interest in microalgae for wastewater treatment is driven by their ability to efficiently remove pollutants while simultaneously producing valuable biomass. Microalgae, such as *Chlorella vulgaris*, have demonstrated significant capabilities in nutrient removal, particularly nitrogen and phosphorus, which are prevalent in agricultural and industrial wastewater [10]. Studies indicate that microalgae can achieve nutrient removal rates exceeding 99% in municipal wastewater, highlighting their effectiveness in enhancing water quality [11]. Furthermore, the integration of microalgae into wastewater treatment systems not only addresses the challenges of conventional methods, which often fall short in organic matter removal and incur high operational costs, but also offers a sustainable solution by utilizing wastewater as a nutrient source for algal growth [12, 13]. The potential for downstream applications of harvested microalgal biomass, including biofuel production, further underscores the economic viability of this approach [13, 14].

Microalgae have emerged as a dual-benefit solution for wastewater treatment, promoting environmental sustainability while enabling resource recovery. Their ability to absorb nutrients such as nitrogen and phosphorus from wastewater significantly enhances water quality, achieving removal rates exceeding 90% in some studies [15]. This bioremediation process not only purifies water but also generates valuable biomass that can be utilized for various applications, including biofuels and animal feed [13, 14]. Furthermore, microalgae cultures can effectively sequester heavy metals and reduce biochemical oxygen demand (BOD) and chemical oxygen demand (COD), thus preventing secondary pollution [12, 14]. Despite challenges in commercialization, such as high operational costs, innovative systems like suspended-solid reactors have shown promise in optimizing microalgal growth and reducing treatment expenses [11]. Overall, integrating microalgae into wastewater management presents a sustainable approach that addresses both ecological and economic concerns [13, 15].

## 2. Types of Microalgae Commonly Used in Microalgae-Based Wastewater Treatment

Microalgae are increasingly recognized for their sustainable and cost-effective role in wastewater treatment, as they can effectively absorb nutrients, lower pollutant levels, and generate biomass of commercial value. Certain microalgal species are prominent in treatment applications due to their robustness, high efficiency in nutrient uptake, and suitability for large-scale cultivation.

### 2.1. Commonly Used Microalgal Species

#### *2.1.1. Chlorella Species*

*Chlorella species* are extensively utilized in wastewater management because of their resilience and exceptional ability to eliminate nutrients. They are highly proficient in extracting nitrogen, phosphorus, and heavy metals from both municipal and industrial effluents. *Chlorella vulgaris*, in particular, is well-suited for secondary and tertiary treatment stages due to its adaptability to nutrient-dense conditions [16, 17].

#### *2.1.2. Spirulina Species*

*Spirulina*, specifically *Arthrospira platensis*, is known for its high tolerance to extreme pH and salinity conditions, which makes it ideal for treating industrial wastewater such as that from textile and tannery industries. Additionally, its high biomass productivity and protein-rich composition make it attractive for resource recovery [18].

#### *2.1.3. Scenedesmus Species*

*Scenedesmus* species are preferred for their rapid growth rates and ability to form colonies, which facilitate easy harvesting. They are particularly efficient at removing ammonia, nitrates, and phosphates, which are common in agricultural and municipal wastewater. The species *Scenedesmus obliquus* has been extensively studied for its potential to treat wastewater while producing high-value biofuels [19, 20].

#### *2.1.4. Other Species*

Microalgae like *Nannochloropsis, Dunaliella,* and *Haematococcus* can effectively remove emerging contaminants from wastewater. Their growth rates and contaminant removal efficiencies are influenced by factors such as light intensity, $CO_2$ addition, and nutrient presence, enhancing wastewater treatment processes [18, 20].

### 2.2. Selection Criteria for Specific Wastewater Types

The choice of microalgal species for effective wastewater treatment is influenced by multiple considerations, such as the specific characteristics of the wastewater, prevailing environmental parameters, and the intended outcome or by-products of the treatment process.

#### *2.2.1. Nutrient Profile of Wastewater*

The nutrient composition, particularly nitrogen and phosphorus concentrations, is a critical determinant. For instance, *Chlorella* species are often chosen for municipal wastewater with high organic and nutrient loads, while *Scenedesmus* species are better suited for agricultural runoff due to their ammonia tolerance [21].

#### *2.2.2. Environmental Conditions*

Factors such as light availability, temperature, and pH also influence species selection. *Spirulina*, for instance, thrives in alkaline and saline conditions, making it ideal for wastewater from soda ash or tannery industries [22].

#### *2.2.3. Specific Contaminants*

Some microalgal strains possess specialized capabilities for eliminating heavy metals and organic contaminants. *Chlorella,* for instance, is frequently utilized in the treatment of wastewater polluted with toxic metals such as cadmium and lead [23].

#### *2.2.4. End-Use of Biomass*

The potential applications of algal biomass, such as biofuel production, animal feed, or biofertilizer, may also guide species selection. For example, protein-rich species like *Spirulina* are valuable for animal feed, while lipid-rich species like *Scenedesmus* are suited for biofuel production [24].

### 2.3. Operational Systems Involved in Microalgae-based Wastewater Treatment

The operational systems utilized in microalgae-based wastewater treatment are designed to optimize the growth of microalgae and enhance their pollutant removal efficiency. These systems vary in design, cost, and efficiency, with each having unique advantages and limitations depending on the wastewater type and treatment objectives. Their applications, advantages and limitations are described below.

#### *2.3.1. Open Pond Systems*

Open pond configurations often referred to as raceway ponds or lagoons are widely adopted in wastewater treatment involving microalgae. These systems feature shallow basins equipped with circulation mechanisms to promote even algal proliferation and balanced nutrient dispersion.

- *Advantages:* Open ponds are cost-effective, easy to construct, and require minimal energy inputs. They are particularly suitable for large-scale operations where land availability is not a constraint [25].
- *Limitations:* These systems are susceptible to contamination by other microorganisms, variable environmental conditions, and water loss due to evaporation. Furthermore, their efficiency can be reduced by insufficient light penetration in dense cultures [26].
- *Applications:* Open pond systems are typically used for treating municipal and agricultural wastewater, where large volumes need to be processed cost-effectively [27].

#### *2.3.2. Photobioreactors (PBRs)*

Photobioreactors are enclosed systems that provide a controlled environment for microalgae cultivation. They are available in various designs, such as tubular, flat-panel, and column reactors.

- *Advantages:* PBRs offer higher biomass productivity, enhanced nutrient removal efficiency, and better control over environmental parameters like light intensity, $CO_2$ concentration, and temperature. They minimize contamination risks and water loss, making them suitable for high-value applications [28].
- *Limitations:* The primary drawbacks of PBRs include high capital and operational costs, as well as the need for regular cleaning to prevent fouling and maintain light transparency [24].
- *Applications:* PBRs are widely used for industrial wastewater containing high concentrations of specific pollutants, such as heavy metals or organic compounds. They are also suitable for the production of high-value algal products alongside wastewater treatment [29].

#### *2.3.3. Hybrid Systems and Their Applications*

To address the shortcomings of standalone systems, hybrid configurations integrate the advantages of both open ponds and photobioreactors, offering a more efficient and balanced approach to microalgal wastewater treatment. For instance, they may involve using open ponds for initial biomass cultivation and PBRs for enhancing biomass productivity or nutrient removal efficiency.

- *Advantages:* Hybrid systems allow scalability while maintaining a balance between cost-effectiveness and operational efficiency. They also enable the integration of different treatment stages, such as pre-treatment in open ponds followed by advanced polishing in PBRs [30].
- *Limitations:* These systems involve greater complexity in design and operation compared to standalone systems.
- *Applications:* These systems are particularly useful for treating complex wastewater streams, such as municipal wastewater mixed with industrial effluents. They are also employed in scenarios where the recovery of algal biomass for biofuel or other high-value products is a priority [31].

## 3. Mechanisms of Action Involved in Microalgae-Based Wastewater Treatment

Microalgae have proven to be a powerful tool in wastewater treatment, utilizing their innate ability to capture and convert a wide range of pollutants both organic and inorganic as well as heavy metals and harmful microorganisms. Various studies highlight their efficiency in nutrient removal, particularly nitrogen and phosphorus, which are critical in preventing eutrophication in water bodies [13, 14]. For instance, *Chlorella vulgaris* has demonstrated significant reductions in ammonium and phosphate levels in agricultural wastewater, showcasing its potential for both purification and biodiesel production [10].

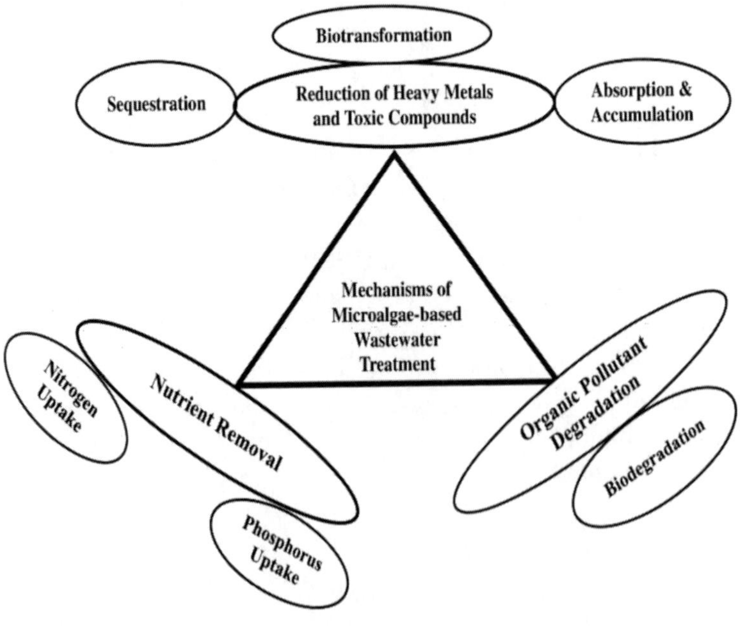


**Figure 1.** Mechanisms of Action Involved in Microalgae-based Wastewater Treatment.

Additionally, microalgae can be cultivated in photobioreactors, optimizing growth conditions and enhancing pollutant degradation [32]. The biomass generated during treatment not only contributes to bioremediation but also holds promise for diverse applications in agriculture, pharmaceuticals, and biofuels, thus addressing both environmental and economic challenges associated with wastewater management [33].

Owing to their diverse metabolic capabilities, microalgae have become a promising and eco-friendly approach for treating wastewater. These organisms play multiple roles in purifying effluents by engaging in nutrient uptake, breaking down organic contaminants, and minimizing the presence of heavy metals and other hazardous substances.

### 3.1. Nutrient Removal (Nitrogen and Phosphorus Uptake)

A major advantage of using microalgae in wastewater treatment lies in their capacity to absorb and utilize surplus nitrogen (N) and phosphorus (P) present in effluents. Nitrogen commonly appears as ammonium ($NH_4^+$), nitrate ($NO_3^-$), or organic nitrogen compounds, while phosphorus is predominantly found as phosphate ($PO_4^{3-}$). These elements are vital for microalgal metabolism, serving as building blocks for amino acids, nucleic acids, and phospholipids during growth.

- *Nitrogen Uptake:* Microalgae utilize nitrogen through two primary pathways: the assimilation of ammonium via glutamine synthetase (GS) and the reduction of nitrates through nitrate reductase (NR). Both processes are energy-intensive, but they allow the algae to convert nitrogen into forms that contribute to their cellular structures [11]. Studies have demonstrated that species such as *Chlorella vulgaris* and *Scenedesmus obliquus* are particularly effective in nutrient uptake due to their rapid growth rates and high nutrient assimilation capacities [34].
- *Phosphorus Uptake:* Phosphorus is absorbed as phosphate ions and incorporated into vital cellular components, such as DNA, RNA, and ATP. In nutrient-rich wastewater, microalgae can significantly reduce phosphate concentrations by assimilating these compounds into their biomass [1].

These nutrient removal processes are crucial not only for improving water quality but also for mitigating the risks of eutrophication, which is caused by excess nitrogen and phosphorus in aquatic ecosystems [11].

### 3.2. Organic Pollutant Degradation

Microalgae are recognized for their capacity to break down various organic contaminants, including oils, fats, and industrial chemicals, via their metabolic activities. These pollutants can be removed through both direct assimilation into the algal biomass and enzymatic breakdown.

- *Biodegradation of Organic Pollutants:* Certain species of microalgae possess enzymes that can break down complex organic pollutants into simpler, less toxic substances.

For instance, the enzymatic activities of oxidative enzymes like laccases and peroxidases enable algae to degrade phenolic compounds, pesticides, and other synthetic organic chemicals [12].

- *Metabolic Pathways:* In some cases, microalgae can metabolize organic pollutants as a carbon source, particularly in the absence of more readily available organic matter. Through this process, algae convert the pollutants into biomass or simpler organic molecules, which are less harmful to the environment [8].

This biodegradation ability makes microalgae an attractive option for removing organic pollutants from industrial effluents, such as those from food processing, textiles, and pharmaceuticals [2].

### 3.3. Reduction of Heavy Metals and Toxic Compounds

Microalgae are also capable of reducing heavy metals and other toxic compounds from wastewater through both physical adsorption and biochemical transformations. These processes not only help detoxify water but also have the potential to recover valuable metals for industrial use.

- *Absorption and Accumulation:* Microalgae can absorb heavy metals such as lead (Pb), cadmium (Cd), mercury (Hg), and chromium (Cr) from wastewater. This is typically achieved through passive absorption via cell walls, which can bind these metals through various functional groups like carboxylates, hydroxyls, and sulfhydryls [35].
- *Biotransformation:* Some microalgae can actively transform toxic metals into less harmful forms through enzymatic processes, such as reducing metal ions to a lower oxidation state. For instance, certain algae species can reduce hexavalent chromium (Cr(VI)) to trivalent chromium (Cr(III)), which is less toxic and more stable [9].
- *Sequestration:* Algae can also sequester heavy metals by incorporating them into their cellular structure, making them unavailable for further contamination. This makes microalgae a potential tool not only for wastewater treatment but also for remediating contaminated sites [11].

These processes underscore the potential of microalgae as an effective solution for treating wastewater contaminated with organic and inorganic pollutants, providing an environmentally friendly alternative to conventional water purification techniques. Microalgae possess high surface-area-to-volume ratios and contain functional groups, like carboxyl, hydroxyl, and amino groups, which can bind metal ions. Species such as *Spirulina platensis* and *Chlorella pyrenoidosa* have demonstrated remarkable ability to accumulate and remove heavy metals from wastewater [36].

## 4. Benefits of Using Microalgae in Wastewater Treatment

Microalgae-based wastewater treatment systems provide multiple environmental, economic, and operational advantages. These benefits make microalgae an attractive alternative to traditional wastewater treatment methods, offering solutions to pollution and resource recovery challenges.

### 4.1. Environmental Benefits

#### *4.1.1. Carbon Dioxide Sequestration*

A major environmental advantage of microalgae in wastewater treatment is their ability to absorb and store carbon dioxide ($CO_2$). Microalgae utilize $CO_2$ during photosynthesis to grow and reproduce, helping to mitigate greenhouse gas emissions. This process is particularly beneficial in wastewater treatment facilities where $CO_2$ emissions are a byproduct of energy consumption. Studies have shown that microalgal systems can absorb significant amounts of $CO_2$, reducing the carbon footprint of wastewater treatment [37]. Furthermore, microalgae can sequester $CO_2$ while simultaneously removing nutrients like nitrogen and phosphorus from wastewater, contributing to nutrient recycling and mitigating eutrophication in aquatic ecosystems [38].

#### *4.1.2. Minimal Chemical Usage Compared to Conventional Methods*

In contrast to traditional chemical treatments that use harsh chemicals such as chlorine, alum, and coagulants in wastewater treatment, microalgae-based systems rely on minimal chemical inputs. Microalgae naturally absorb excess nutrients, such as nitrogen and phosphorus, along with pollutants from the wastewater, thereby reducing the need for chemical additives. This approach lowers the environmental impact associated with chemical handling and disposal [39]. Additionally, the lack of toxic chemicals in the treatment process makes microalgal systems a more sustainable and environmentally friendly option [40].

### 4.2. Economic and Resource Recovery Advantages

#### *4.2.1. Generation of Biomass for Biofuels, Fertilizers, and Animal Feed*

A notable benefit of using microalgae in wastewater treatment is the potential for biomass production. The microalgae cultivated during the treatment process can be harvested to produce a range of valuable by-products, including biofuels, fertilizers, and animal feed. Algal biomass is rich in lipids, proteins, and carbohydrates, making it a prime source for biofuel production, especially biodiesel [41]. Furthermore, algal residues can be repurposed as organic fertilizers or animal feed, creating an integrated approach to waste management and resource recovery [31]. This dual function of wastewater treatment and biomass production positions microalgae as a crucial element in the circular economy.

#### *4.2.2. Potential for Energy Recovery via Anaerobic Digestion*

The biomass produced from microalgae during wastewater treatment can also be processed via anaerobic digestion to produce biogas, primarily methane. This process not only generates renewable energy but also decreases the volume of organic waste, improving the overall efficiency of wastewater treatment. Combining anaerobic digestion with microalgae-based systems holds significant potential for energy recovery, making the system partially self-sustaining [42]. This synergy between energy production and wastewater treatment increases the economic viability of microalgae-based systems by lowering operational costs and contributing to renewable energy generation.

### 4.3. Scalability and Flexibility

#### *4.3.1. Adaptability to Different Wastewater Types*

Microalgae systems exhibit remarkable adaptability to different types of wastewater, including municipal, industrial, and agricultural effluents. They can effectively treat wastewater with varying nutrient and pollutant concentrations. For instance, *Chlorella species* are often used to treat nutrient-rich municipal wastewater, while *Spirulina* and *Scenedesmus* are more suited for saline or industrial effluents [16]. This adaptability allows microalgae to be applied across a wide array of wastewater treatment scenarios [27]. Furthermore, their ability to thrive under varying environmental conditions such as pH, temperature, and light further enhances their versatility in treatment systems.

#### *4.3.2. Feasibility in Both Urban and Rural Settings*

Microalgae-based wastewater treatment systems can be implemented in both urban and rural settings, offering flexibility in their deployment. In urban areas, where space and resources are limited, compact photobioreactors can be used for efficient wastewater treatment and biomass production. On the other hand, open pond systems, which require larger areas, are well-suited for rural or agricultural settings where land availability is less constrained [25]. This scalability and adaptability make microalgae-based systems suitable for a variety of geographic regions and population densities, addressing the wastewater treatment needs of both dense urban populations and sparsely populated rural areas.

## 5. Microalgae in Wastewater Treatment: Challenges and Limitations

Despite the many advantages of utilizing microalgae in wastewater treatment, several challenges and limitations hinder the widespread adoption of this technology. These challenges can be broadly classified into technical, economic, and regulatory concerns, all of which influence the scalability and successful implementation of microalgae-based treatment systems.

## 5.1. Technical Challenges

### *5.1.1. Maintaining Optimal Growth Conditions (e.g., Light, Temperature, $p^H$)*

A major technical challenge in microalgae-based wastewater treatment is maintaining optimal growth conditions for the algae. Microalgae require specific environmental factors, such as appropriate light intensity, temperature, and $p^H$, to effectively remove nutrients and produce biomass. Different species have distinct light and temperature requirements; for example, excessive light can lead to photoinhibition, while insufficient light can limit photosynthesis [28]. Additionally, maintaining a stable $p^H$ is critical for nutrient uptake and overall algal growth. Variations in the composition of wastewater can cause $p^H$ fluctuations, which may stress the algae and reduce their efficiency in wastewater treatment [38]. The need for precise control of these environmental factors adds complexity and increases operational costs.

### *5.1.2. Contamination Risks in Open Systems*

Open pond systems, which are among the most cost-effective options for microalgae-based wastewater treatment, face significant contamination risks. These systems are susceptible to the introduction of unwanted microorganisms, such as bacteria, protozoa, and other algae species, which can outcompete the desired microalgae and reduce overall treatment efficiency [25]. Furthermore, contamination can lead to the accumulation of harmful pathogens, posing a risk to human and environmental health. Although closed systems such as photobioreactors can mitigate these risks, they are more expensive and complex to maintain [29].

## 5.2. Economic Constraints

### *5.2.1. High Initial Capital Costs for Advanced Systems*

The initial capital investment for advanced microalgae-based wastewater treatment systems, especially photobioreactors, can be prohibitively high. Photobioreactors, while offering higher efficiency and better control over environmental conditions, require significant upfront costs for infrastructure, materials, and installation [24]. Additionally, the maintenance of these systems, including cleaning, monitoring, and replacement of parts, contributes to the ongoing operational expenses. This high capital cost presents a major barrier to the widespread adoption of microalgae-based wastewater treatment, particularly in low-income or developing regions where funding for such technologies may be limited.

### *5.2.2. Scaling Difficulties in Industrial Applications*

Scaling up microalgae-based wastewater treatment systems from laboratory or pilot-scale studies to full-scale industrial applications is another significant challenge. Many of the technical advantages seen in smaller systems, such as high nutrient removal efficiency and low energy consumption, are often not replicated at larger scales. Factors such as light distribution, temperature control, and nutrient availability become more difficult to manage in larger systems [40]. Moreover, large-scale systems may require substantial land area, especially when using open pond systems, which could limit their applicability in urban areas with limited space [27]. Scaling difficulties can lead to inefficiencies and increased costs, reducing the overall feasibility of large-scale microalgae-based treatment facilities.

### 5.3. Regulatory and Social Acceptance

#### *5.3.1. Need for Standardized Protocols*

The lack of standardized protocols for microalgae-based wastewater treatment is another significant limitation. There is no universally accepted set of guidelines for the design, operation, and performance assessment of microalgae systems. The variability in wastewater types, microalgae species, and environmental conditions means that treatment strategies must be tailored to specific applications, but without standardized protocols, comparing performance across different systems or ensuring consistent results is challenging [43]. Furthermore, the absence of regulatory frameworks for the use of microalgae in wastewater treatment can hinder investment and development in this area, as stakeholders may be hesitant to adopt the technology without clear regulations and performance benchmarks.

#### *5.3.2. Public Perception and Awareness*

Public perception and social acceptance of microalgae-based wastewater treatment systems can present significant challenges. Skepticism or resistance may arise, particularly concerning the safety and quality of the treated water, despite the proven effectiveness and safety of these systems. Concerns about potential health risks or contamination could hinder the widespread adoption of microalgae-based technologies, especially in urban or densely populated areas [44]. Additionally, the general lack of awareness and understanding of the benefits of these systems may slow their adoption. Both decision-makers and the public may be unaware of the environmental and economic advantages that microalgae systems offer for wastewater treatment.

## 6. Recent Advances and Future Directions

Microalgae-based wastewater treatment has made significant strides in recent years. Advances in microalgae technology, its integration with other treatment systems, and alignment with circular economy principles have greatly enhanced the potential of microalgae as a sustainable solution for wastewater treatment. These developments offer the promise of more efficient, cost-effective, and environmentally friendly wastewater treatment systems.

### 6.1. Innovations in Microalgae Technology

#### *6.1.1. Genetic Engineering for Enhanced Pollutant Uptake*

One of the key recent advances in microalgae technology is the use of genetic engineering to enhance the pollutant uptake capabilities of algae. Through genetic modification, researchers have developed microalgae strains with improved efficiencies in removing nutrients like nitrogen, phosphorus, and heavy metals from wastewater. For example, transgenic microalgae have been designed to express specific genes that enhance their ability to absorb and metabolize pollutants, thus improving their overall treatment performance [45]. Genetic engineering also enables the development of algae that can tolerate harsher environmental conditions or more concentrated pollutants, expanding the range of wastewater types that microalgae can treat [46].

These advances in genetic engineering have the potential to make microalgae-based systems more versatile, scalable, and effective in wastewater treatment applications.

#### *6.1.2. Improved Bioreactor Designs*

Another significant development in microalgae-based wastewater treatment is the improvement of bioreactor designs. Photobioreactors (PBRs) have undergone various innovations to optimize light penetration, carbon dioxide supply, and nutrient availability. Recent advancements in PBR designs, such as tubular and flat-panel systems, have improved mass transfer, reduced energy consumption, and enhanced microalgal productivity [47]. Hybrid systems that combine open ponds with photobioreactors have also been explored to balance cost-effectiveness with efficiency. Additionally, automated control systems for monitoring and adjusting parameters such as light intensity, pH, and temperature are making these systems more efficient and easier to manage [48]. These design improvements contribute to the overall efficiency of microalgae-based wastewater treatment, enabling better resource utilization and wastewater management.

### 6.2. Integrating Microalgae with Other Treatment Systems

#### *6.2.1. Coupling with Anaerobic Digestion or Constructed Wetlands*

A promising avenue for microalgae-based wastewater treatment is the integration of microalgae systems with other treatment technologies, such as anaerobic digestion and constructed wetlands. Combining microalgae with anaerobic digestion can improve both nutrient removal and energy recovery. The algal biomass can be processed through anaerobic digestion to generate biogas, primarily methane, which can then be utilized as a renewable energy source [49]. Additionally, integrating microalgae with constructed wetlands can improve nutrient uptake and pollutant removal, as wetlands can provide a stable environment for algae growth while reducing the risk of contamination in open systems [50]. This synergistic approach can increase treatment efficiency, reduce costs, and improve overall system sustainability.

#### *6.2.2. Synergistic Approaches to Maximize Efficiency*

Other synergistic approaches include integrating microalgae-based systems with traditional activated sludge processes or membrane bioreactors. Such combinations can enhance nutrient removal and improve the overall treatment capacity of the system. In these hybrid systems, microalgae absorb excess nutrients like nitrogen and phosphorus, while bacteria in the activated sludge or membrane bioreactor degrade organic pollutants, providing a comprehensive solution to wastewater treatment [25]. These synergistic approaches maximize efficiency by combining the strengths of both biological and physical processes, allowing for higher treatment performance with reduced energy requirements.

### 6.3. Potential for Circular Economy

#### *6.3.1. Emphasizing Resource Recovery and Reuse*

Microalgae-based wastewater treatment systems have significant potential for contributing to the circular economy, particularly through resource recovery and reuse. The biomass produced from algae during wastewater treatment can be used for various purposes, such as biofuel production, fertilizers, and animal feed, contributing to the recycling of nutrients and organic matter [31]. Furthermore, the treated water can be reused for irrigation or other non-potable domestic applications, thereby reducing freshwater consumption and promoting resource sustainability. The ability to recover valuable by-products such as algal biomass, biofuels, and fertilizers makes microalgae-based systems an essential component of sustainable waste management and resource recovery strategies.

#### *6.3.2. Aligning with Global Sustainability Goals*

Microalgae-based wastewater treatment aligns closely with global sustainability goals, including those outlined in the United Nations Sustainable Development Goals (SDGs). Specifically, it supports SDG 6 (Clean Water and Sanitation), SDG 7 (Affordable and Clean Energy), and SDG 13 (Climate Action) by integrating microalgae technology into wastewater treatment systems. This approach not only addresses water purification needs but also contributes to renewable energy production and mitigates climate change impacts. By enabling nutrient removal, reducing energy consumption, and producing renewable biofuels, microalgae-based systems contribute to both environmental protection and resource sustainability [51]. These systems support a circular approach to wastewater management, where waste is treated, valuable resources are recovered, and environmental impacts are minimized, contributing to a more sustainable and resilient global economy.

## Conclusion

Microalgae have gained recognition as an innovative and sustainable alternative to traditional wastewater treatment methods. These microorganisms exhibit exceptional efficiency in removing nutrients particularly nitrogen and phosphorus, which are key contributors to water pollution. Through the process of photosynthesis, microalgae not only purify wastewater but also aid in carbon sequestration, thereby lowering greenhouse gas emissions and contributing to climate change mitigation. This dual role in both wastewater treatment and environmental protection highlights the potential of microalgae as a comprehensive and eco-friendly strategy for integrated water and resource management.

A major advantage of microalgae-based treatment systems is their capacity to produce valuable by-products, such as biomass, which can be used for biofuel production, organic fertilizers, and animal feed. This potential for resource recovery supports the principles of the circular economy, where waste is viewed not as a burden but as a resource to be repurposed and reintegrated into the production cycle. Additionally, the adaptability of microalgae to various wastewater types, ranging from municipal to industrial effluents, enhances their scalability and applicability in diverse settings, including urban and rural environments. Innovations in bioreactor designs and hybrid systems are further improving the efficiency and feasibility of these technologies, making them a promising option for sustainable wastewater treatment. Despite their potential, the adoption of microalgae-based systems faces challenges, including the need for optimized growth conditions, high initial capital costs, and the risk of contamination in open systems. However, recent advancements in genetic engineering, bioreactor technology, and integrated treatment approaches are addressing these limitations. For instance, coupling microalgae systems with anaerobic digestion and constructed wetlands has proven effective in enhancing nutrient removal and energy recovery. These synergistic solutions not only improve efficiency but also expand the applicability of microalgae-based systems in industrial and decentralized wastewater treatment settings.

In summary, microalgae offer a transformative pathway for wastewater treatment by combining environmental benefits with resource recovery. Their ability to address pressing issues like nutrient pollution, carbon sequestration, and sustainable energy production makes them a key player in the future of wastewater management. As innovations continue to refine these systems and address existing challenges, microalgae-based wastewater treatment is poised to become an integral part of global efforts to achieve sustainability and environmental resilience.

## References

[1] Papapolymerou, G. (2019). Current and potential applications of microalgae: A mini review. *Open Access Journal*, 11, 555811. https://doi.org/10.19080/OFOAJ.2019.11.555811.

[2] Kukwa, D. T., & Chetty, M. (2021). Microalgae: The Multifaceted Biomass of the 21st Century. In *IntechOpen eBooks*. https://doi.org/10.5772/intechopen.94090.

[3] Zhang, J., Yang, H., Sun, Y., Yan, B., Chen, W., & Fan, D. (2024). The potential use of microalgae for nutrient supply and health enhancement in isolated and confined environments. *Comprehensive Reviews in Food Science and Food Safety,* 23(4). https://doi.org/10.1111/1541-4337.13418.

[4] Pereira, L., Cotas, J., & Valado, A. (2024). Antioxidants from microalgae and their potential impact on human well-being. *Exploratory Drug Science, 2*, 292–321. https://doi.org/10.37349/eds.2024.00048.

[5] Cadoret, J. P., Garnier, M., & Saint-Jean, B. (2012). Microalgae, functional genomics, and biotechnology. *In Advances in Botanical Research* (Vol. 64, pp. 189-213). https://doi.org/10.1016/B978-0-12-391499-6.00008-6.

[6] Tang, Y., Rosenberg, J. N., Bohutskyi, P., Yu, G., Betenbaugh, M. J., & Wang, F. (2015). Microalgae as a feedstock for biofuel precursors and value-added products: Green fuels and golden opportunities. *Bioresources*, 11(1), 81. https://doi.org/10.15376/BIORES.11.1.TANG.

[7] Sastre, R. R., & Posten, C. (2010). The variety of microalgae applications as a renewable resource. *Chemie Ingenieur Technik, 82*(11), 1925–1939. https://doi.org/10.1002/cite.201000124.

[8] Derco, J., Guľašová, P., Legan, M., Zakhar, R., & Žgajnar Gotvajn, A. (2024). Sustainability strategies in municipal wastewater treatment. *Sustainability*, 16(20), 9038. https://doi.org/10.3390/su16209038.

[9] Ahmad, H. W., Bibi, M., Chandrasekaran, M., Ahmad, S., & Kyriakopoulos, G. L. (2024). Sustainable wastewater treatment strategies in effective abatement of emerging pollutants. *Water*, 16(20), 2893. https://doi.org/10.3390/w16202893 .

[10] Meshcheryakova, Y., & Nagornov, S. (2024b). Chlorella Microalgae for Wastewater Treatment in Agriculture. *Science in the Central Russia*, 4, 158–166. https://doi.org/10.35887/2305-2538-2024-4-158-166.

[11] Serrano-Blanco, S., Zan, R., Harvey, A., & Velásquez-Orta, S. B. (2024). Intensified microalgae production and development of microbial communities on suspended carriers and municipal wastewater. *Journal of Environmental Management.* https://doi.org/10.1016/j.jenvman.2024.122717.

[12] Bhardwaj, A. K., Vij, D., Prasannakumaram, K. M., & Vasistha, S. (2024). Wastewater treatment coupled with multifaceted microalgae: Benefits and challenges. *In Microalgae for Wastewater Treatment* (pp. 17-30). https://doi.org/10.1201/9781003515029-17.

[13] Liberti, D., Pinheiro, F., Simões, B. D., Varela, J., & Barreira, L. (2024). Beyond bioremediation: The untapped potential of microalgae in wastewater treatment. *Water*, 16(19), 2710. https://doi.org/10.3390/w16192710.

[14] Kaparapu, J., & Prasad, M. K. (2024). Microalgae for Wastewater Treatment. In *CRC Press eBooks* (pp. 107–124). https://doi.org/10.1201/9781003219194-8.

[15] Xu, K., Du, M., Rao, Y., Luo, J., Chen, Z., Li, C., Lei, A., & Wang, J. (2024). Microalgae-mediated heavy metal removal in wastewater treatment: Mechanisms, influencing factors, and novel techniques. *Algal Research-Biomass Biofuels and Bioproducts*. https://doi.org/10.1016/j.algal.2024.103645.

[16] Razzak, S. A., Hossain, M. M., Lucky, R. A., Bassi, A. S., & de Lasa, H. (2013). Integrated CO2 capture, wastewater treatment, and biofuel production by microalgae culturing-A review. *Renewable and Sustainable Energy Reviews, 27*, 622–653. https://doi.org/10.1016/j.rser.2013.05.063.

[17] Wu, H., Liu, Z., Liu, S., Zhao, X., & Fang, Z. (2020). Microalgae-based systems for wastewater treatment: A critical review. *Science of the Total Environment, 716*, 137080. https://doi.org/10.1016/j.scitotenv.2020.137080.

[18] Abinandan, S., & Shanthakumar, S. (2015). Challenges and opportunities in application of microalgae (*Chlorella vulgaris*) for wastewater treatment: A review. *Renewable and Sustainable Energy Reviews, 52*, 123–132. https://doi.org/10.1016/j.rser.2015.07.116.

[19] Martínez, M. E., Sánchez, S., Jiménez, J. M., Yousfi, F. E., & Muñoz, L. (2000). Nitrogen and phosphorus removal from urban wastewater by the microalga *Scenedesmus obliquus*. *Bioresource Technology, 73*(3), 263–272. https://doi.org/10.1016/S0960-8524(99)00121-2.

[20] Sydney, E. B., da Silva, T. E., Tokarski, A., Novak, A. C., de Carvalho, J. C., Woiciechowski, A. L., & Soccol, C. R. (2011). Screening of microalgae with potential for biodiesel production and nutrient removal from treated domestic sewage. *Applied Energy, 88*(10), 3291–3294. https://doi.org/10.1016/j.apenergy.2011.02.023.

[21] Gonçalves, A. L., Pires, J. C. M., & Simões, M. (2017). A review on the use of microalgal consortia for wastewater treatment. *Algal Research, 24*, 403–415. https://doi.org/10.1016/j.algal.2016.11.008.

[22] Singh, R. N., & Dhar, D. W. (2019). Microalgae as second-generation biofuel: A review. *Agricultural Engineering International: CIGR Journal, 21*(1), 1–17.

[23] Mulbry, W., Kondrad, S., Pizarro, C., & Kebede-Westhead, E. (2008). Treatment of dairy manure effluent using freshwater algae: Algal productivity and recovery of manure nutrients using pilot-scale algal turf scrubbers. *Bioresource Technology, 99*(17), 8137–8142. https://doi.org/10.1016/j.biortech.2008.03.073.

[24] Rawat, I., Ranjith Kumar, R., Mutanda, T., & Bux, F. (2011). Dual role of microalgae: Phycoremediation of domestic wastewater and biomass production for sustainable biofuels production. *Applied Energy, 88*(10), 3411–3424. https://doi.org/10.1016/j.apenergy.2010.11.025.

[25] Park, J. B. K., Craggs, R. J., & Shilton, A. N. (2011). Wastewater treatment high rate algal ponds for biofuel production. *Bioresource Technology, 102*(1), 35–42. https://doi.org/10.1016/j.biortech.2010.06.158.

[26] Borowitzka, M. A. (1999). Commercial production of microalgae: Ponds, tanks, tubes, and fermenters. *Journal of Biotechnology, 70*(1-3), 313–321. https://doi.org/10.1016/S0168-1656(99)00083-8.

[27] Christenson, L., & Sims, R. (2011). Production and harvesting of microalgae for wastewater treatment, biofuels, and bioproducts. *Biotechnology Advances, 29*(6), 686–702. https://doi.org/10.1016/j.biotechadv.2011.05.015.

[28] Ugwu, C. U., Aoyagi, H., & Uchiyama, H. (2008). Photobioreactors for mass cultivation of algae. *Bioresource Technology, 99*(10), 4021–4028. https://doi.org/10.1016/j.biortech.2007.01.046.

[29] Brennan, L., & Owende, P. (2010). Biofuels from microalgae—A review of technologies for production, processing, and extractions of biofuels and co-products. *Renewable and Sustainable Energy Reviews, 14*(2), 557–577. https://doi.org/10.1016/j.rser.2009.10.009.

[30] Morales-Sánchez, D., Martinez-Rodriguez, O. A., & Martinez, A. (2017). Heterotrophic microalgal cultivation in the production of biodiesel. *Biofuels, Bioproducts and Biorefining, 11*(6), 923–935. https://doi.org/10.1002/bbb.1815.

[31] Pittman, J. K., Dean, A. P., & Osundeko, O. (2011). The potential of sustainable algal biofuel production using wastewater resources. *Bioresource Technology, 102*(1), 17–25. https://doi.org/10.1016/j.biortech.2010.06.035.

[32] Parra, Karen & Hernández, María & García, Víctor & Melchor, Dulce & Martínez-Roldán, Alfredo. (2024). Microalgae and Cyanobacteria in the Biological Treatment of Effluents. *Revista de Gestão Social e Ambiental.* 18. e08682. 10.24857/rgsa.v18n10-010.

[33] Bagul, S. Y., Pandiyan, K., Khadke, G., & Dhar, D. W. (2024). Role of Microalgae in Value Addition and Bioremediation of Wastewater. In *CRC Press eBooks* (pp. 144–154). https://doi.org/10.1201/9781003219156-13.

[34] Wang, H., Ji, G., Bai, X., & Li, W. (2018). Nutrient removal from wastewater by different microalgae in photobioreactors. *Environmental Technology*, 39(24), 3174-3184. https://doi.org/10.1080/09593330.2017.1397762.

[35] Waqas, A., Bibi, H., Chandrasekaran, M., Ahmad, S., & Kyriakopoulos, G. L. (2024). Sustainable wastewater treatment strategies in effective abatement of emerging pollutants. *Water*, 16(20), 2893. https://doi.org/10.3390/w16202893.

[36] Gupta, S. K., Ansari, F. A., Shriwastav, A., & Bux, F. (2019). Microalgae for bioremediation of wastewater pollutants. *Microbial Wastewater Treatment*, 59-78. https://doi.org/10.1007/978-3-319-70069-7_4.

[37] Zhu, X., Liang, J., & Zhang, Y. (2021). Advances in carbon dioxide sequestration using microalgae: A review. *Biological CO2 Sequestration and Utilization, 39*, 289–314. https://doi.org/10.1007/s11356-021-13124-w.

[38] Liu, Y., Yang, Z., & Zhang, L. (2022). Microalgae in wastewater treatment and nutrient removal: A review. *Environmental Science and Pollution Research, 29*, 1–15. https://doi.org/10.1007/s11356-022-19811-3.

[39] Rath, A., Venkatramanan, M., & Singh, M. (2017). Microalgae for wastewater treatment: A review. *Environmental Science and Pollution Research, 24*(10), 9141–9154. https://doi.org/10.1007/s11356-017-8697-5.

[40] Lundquist, T. J., Woertz, I. C., Quinn, J. C., & Benemann, J. R. (2010). A realistic technology and engineering assessment of algae biofuels. *Environmental Energy Technologies Division, Lawrence Berkeley National Laboratory.*

[41] Lardon, L., Helias, A., Sialve, B., Steyer, J. P., & Bernard, O. (2009). Life-cycle assessment of biodiesel production from microalgae. *Environmental Science & Technology, 43*(17), 6475–6481. https://doi.org/10.1021/es900705j.

[42] Hannon, M., Gimpel, J., Tran, M., Rasala, B., & Mayfield, S. (2010). Biofuels from algae: Challenges and potential. *Bioresource Technology, 102*(1), 1–15. https://doi.org/10.1016/j.biortech.2010.03.051.

[43] Sialve, B., Bernard, O., & Steyer, J. P. (2009). Anaerobic digestion of microalgae as a necessary step to make microalgal biodiesel sustainable. *Biotechnology Advances, 27*(4), 409–416. https://doi.org/10.1016/j.biotechadv.2009.01.001.

[44] Hernández, L. F., Téllez, M., & Noyola, A. (2014). Microalgae-based treatment of wastewater: A review of challenges and limitations. *Applied Microbiology and Biotechnology, 98*(9), 3723–3734. https://doi.org/10.1007/s00253-014-5745-4.

[45] Luo, L., Zhao, L., & Yu, X. (2020). Genetic modification of microalgae for enhanced pollutant removal from wastewater: Advances and prospects. *Frontiers in Environmental Science, 8*, 43. https://doi.org/10.3389/fenvs.2020.00043.

[46] Nagarajan, D., Nadar, S. M., & Rajendran, R. (2021). Advances in the genetic engineering of microalgae for wastewater treatment. *Bioresource Technology, 318*, 123991. https://doi.org/10.1016/j.biortech.2020.123991.

[47] Hu, Q., Sharma, P., & Liu, J. (2021). Recent advances in microalgae-based wastewater treatment technologies: Bioengineering, harvesting, and applications. *Bioresource Technology, 321*, 124421. https://doi.org/10.1016/j.biortech.2020.124421.

[48] Mata, T. M., Martins, A. A., & Caetano, N. S. (2010). Microalgae for biodiesel production and other applications: A review. *Renewable and Sustainable Energy Reviews, 14*(1), 217–232. https://doi.org/10.1016/j.rser.2009.07.020.

[49] Zhu, X., Liang, J., & Zhang, Y. (2020). Integrating microalgae-based wastewater treatment with anaerobic digestion for sustainable biogas production: A review. *Environmental Science and Pollution Research, 27*, 29656–29673. https://doi.org/10.1007/s11356-020-08262-5.

[50] Vymazal, J. (2011). Constructed wetlands for wastewater treatment. *Water, 3*(2), 125–147. https://doi.org/10.3390/w3020125.

[51] Mussgnug, J. H., Klassen, V., Schläwinkel, J., Knoop, H., & Kruse, O. (2010). Genetic engineering of algae for enhanced biofuel production. *Current Opinion in Biotechnology, 21*(3), 9–16. https://doi.org/10.1016/j.copbio.2009.11.010.

Chapter 13

# Photocatalysis: A Green Approach for the Removal of Contaminants from Wastewater

**Winnie Teja Dokka**[1,*]
**and Nalini Simma**[2]
[1]Department of Chemistry, Government Degree College, Chintalapudi, A.P., India

[2]Department of Chemistry, Government Degree College(M), Srikakulam, A.P., India

## Abstract

Photocatalysis is a promising and eco-friendly technique that provides an effective solution for treating wastewater, sustainably. Utilizing light energy, photocatalysts produce reactive oxygen species that can break down various pollutants such as organic compounds, toxic metals, and harmful microbes with high efficiency. Owing to the characteristics like large surface area, nanoscale dimensions, and unique morphology, photocatalysts are highly effective in the removal of contaminants from wastewater. Additionally, the potential of photocatalysis is enhanced through innovations like tree-like nanofiber membranes that improve mass transfer and adsorption, and co-doped $TiO_2$ catalysts achieving up to 98% degradation of specific contaminants. Photocatalysis can also be synergistically combined with other advanced oxidation processes, such as sonophotolysis, to further enhance degradation rates. This chapter focuses on the key processes responsible for eliminating pollutants through various types of photocatalysts, with a particular emphasis on semiconductor like titanium dioxide and zinc oxide. Recent advancements and modification strategies aimed at enhancing photocatalytic efficiency are also discussed. It also provides valuable insights into photocatalysis as a promising, economically viable, and versatile green strategy for tackling the growing problem of water contamination intensified by rapid industrial growth and expanding urban areas.

**Keywords:** advanced oxidation process, co-doping, photocatalysis, dye, semiconductor, nanofibre

* Corresponding Author's Email: winnie.teja35@gmail.com

In: Advancements in Wastewater Treatment
Editors: Neha Agarwal, Maulin P. Shahl, Vijendra Singh Solankil et al.
ISBN: 979-8-90134-007-3

# 1. Introduction

Wastewater contamination is the deterioration of water resources resulting from wastewater that is either not treated or insufficiently processed. This category includes several sources, such as municipal sewage, industrial effluents, agricultural run-off, and stormwater [1]. Wastewater is a complex mixture of several contaminants, including organic waste, heavy metals, toxic compounds, pathogens, and nutrients such as nitrogen and phosphorus. These pollutants negatively affect ecosystems, diminish water quality, and pose significant risks to human health and environmental sustainability. Mitigating wastewater contamination is essential for preserving ecological balance and safeguarding human health.

## 1.1. Environmental and Health Implications of Wastewater Pollution

Wastewater pollution has significant environmental consequences that threaten ecosystems, biodiversity, and human health.

- *Eutrophication:* Eutrophication, characterized by excessive algal proliferation in aquatic environments, is stimulated by nutrient-laden wastewater with elevated levels of nitrogen and phosphorus. This disrupts aquatic ecosystems by blocking sunlight and reducing dissolved oxygen levels during algal bloom decomposition. This leads to hypoxia, fish mortality, and a reduction in biodiversity. Some algae produce toxins harmful to aquatic life and human health [2].
- *Habitat Degradation:* Toxic substances in wastewater, including heavy metals such as mercury, lead, and arsenic, contaminate aquatic environments. These contaminants accumulate in organisms through biomagnification, impacting predators and humans. Habitat degradation occurs when water quality declines, resulting in stress and population reduction in vulnerable species. Aquatic vegetation, crucial for environmental balance, is also affected, diminishing biodiversity [3].
- *Soil and Groundwater Contamination:* Untreated wastewater permeating the soil introduces contaminants such as nitrates, phosphates, and toxic substances. This process reduces soil quality and pollutes groundwater, adversely affecting plant growth and potable water sources. Elevated nitrate concentrations in groundwater are associated with health complications, including methemoglobinemia, popularly referred to as blue baby syndrome, in babies [4]. Furthermore, terrestrial ecosystems dependent on pristine water and fertile soil are profoundly affected by this pollution.
- *Microplastic Pollution:* Wastewater frequently contains microplastics, which derive from sources like synthetic textile fibres and personal care items [5]. These little plastic particles are consumed by marine species, resulting in physical damage and exposure to hazardous compounds. Microplastics ascend the food chain through trophic transfer, ultimately entering the human diet through seafood intake, hence presenting possible health hazards.
- *Dissemination of Waterborne Diseases:* Raw sewage often contains harmful microorganisms like bacteria, viruses, and parasites that can cause severe illnesses, including typhoid, dysentery, and cholera. In regions with inadequate effluent

treatment infrastructure and inadequate sanitation, these diseases are particularly prevalent [6].

- *Toxic Exposure:* Contaminated wastewater containing toxic substances like lead, mercury and other hazardous chemicals can enter the food chain and pose significant health risks. Prolonged exposure to these toxins can lead to neurological, reproductive, and developmental health problems in humans, posing major public health challenges [7].
- *Antibiotic Resistance:* Residual antibiotics often seen in wastewater can promote the emergence of antibiotic-resistant microorganisms. These microorganisms represent an escalating hazard to world health by diminishing the efficacy of modern treatments for bacterial infections [8]. The consequences of wastewater pollution, along with their causes, environmental impacts, and associated health issues are outlined in Table 1.

**Table 1.** Causes, environmental impacts, and health issues of wastewater pollution

| Implication | Causes | Environmental Impact | Health Issues |
|---|---|---|---|
| Eutrophication | Nutrient-rich wastewater (nitrogen, phosphorus) | Excessive algal growth, reduced oxygen levels, mass fish mortality, and a decline in species diversity | Contamination of drinking water with algal toxins, gastrointestinal and liver-related illnesses |
| Habitat Degradation | Toxic chemicals, heavy metals in effluents | Poisoning of aquatic life, reduction in biodiversity, habitat loss | Bioaccumulation of toxins leading to neurological, developmental, and reproductive disorders |
| Soil and Groundwater Contamination | Leaching of pollutants into soil and groundwater | Degraded soil quality, contamination of aquifiers, harm to terrestrial and underground ecosystems | Nitrate poisoning (methemoglobinemia), exposure to hazardous chemicals through drinking water |
| Microplastic Pollution | Microplastics from textiles, personal care products | Disruption of marine ecosystems, ingestion by aquatic life, transfer through food chains | Potential health risks through seafood consumption, exposure to toxins, unknown long-term impacts |
| Spread of Waterborne Diseases | Pathogens in untreated sewage | Contamination of water bodies, disruption of ecosystems | Outbreaks of cholera, typhoid, dysentery, and other infectious diseases |
| Antibiotic Resistance | Antibiotics in wastewater | Proliferation of resistant bacteria in aquatic and terrestrial ecosystems | Reduced effectiveness of medical treatments, spread of antibiotic-resistant bacterial infections |
| Toxic Exposure | Heavy metals, hazardous chemicals | Contamination of water, bioaccumulation in organisms | Chronic diseases from metal poisoning, such as mercury or lead toxicity |

## 1.2. Photocatalysis

Photocatalysis involves the activation of a substance called a photocatalyst using light energy, usually from UV or visible light sources. This activation speeds up chemical reactions without the photocatalyst being consumed or undergoing permanent changes. Upon absorbing light with adequate energy, the photocatalyst's electrons are promoted from the valence band to the conduction band, resulting in the creation of electron-hole pairs. These reactive species initiate redox reactions that facilitate the breakdown or transformation of various chemical substances. This technique is widely acknowledged for its diverse environmental and industrial applications. In environmental remediation, photocatalysis is highly effective in breaking down organic and inorganic pollutants, making it a valuable tool for treating wastewater and purifying air. Its role in renewable energy is equally significant, aiding processes including generating hydrogen via water splitting and transforming carbon dioxide into valuable fuel sources, which are critical for sustainable energy solutions.

In synthetic chemistry, photocatalysis facilitates targeted chemical reactions under mild conditions, reducing the reliance on harsh reagents and energy-intensive methods. By harnessing light energy, often derived from sunlight, photocatalysis offers an eco-friendly, versatile, and efficient approach to tackling global challenges, including pollution control, clean energy generation, and sustainable chemical synthesis.

### *1.2.1. Fundamental Principles of Photocatalysis*

Photocatalysis is a technique that utilizes light-often in the UV or visible spectrum-to stimulate a photocatalyst, initiating chemical reactions [9]. The fundamental principles of photocatalysis include:

- *Light Absorption:* The photocatalytic process begins when the photocatalyst absorbs light energy at a specific wavelength. The energy absorbed must be equal to or greater than the bandgap energy of the material, which is the energy gap between its valence and conduction bands.
- *Electron Excitation and Charge Separation:* When light is absorbed by the photocatalyst, electrons ($e^-$) are energized and move from the valence band to the conduction band, creating vacant positive sites (holes ($h^+$)) in the valence band. This results in the formation of electron-hole pairs, which play a crucial role in driving photocatalytic reactions.
- *Charge Migration:* The negatively charged electrons and positively charged holes ($e^-$ and $h^+$) generated move toward the photocatalyst's surface. To achieve efficient photocatalysis, it is essential to maintain proper separation and movement of charges, as recombination of paired excited electrons and corresponding positive holes can release energy in the form of heat, thereby lowering the overall efficiency of the process.
- *Redox Reactions:* Charge carriers initiate redox reactions on the surface of the photocatalyst. Electrons in the conduction band are involved in reduction, whereas holes in the valence band facilitate oxidation. For instance:
  - Oxidation: Breaking down organic pollutants into harmless by-products.
  - Reduction: Splitting water molecules to produce hydrogen or converting carbon dioxide into valuable fuels.
- *Reaction Products:* The final stage entails the formation of reaction products. In environmental applications, harmful substances are transformed into benign or less harmful substances, including carbon dioxide and water.

The effectiveness of photocatalysis is influenced by various factors, including the characteristics of the photocatalyst, the light's intensity and wavelength, and the concentration of reactants. By leveraging these principles, photocatalysis serves as an eco-friendly, sustainable, and versatile solution for environmental cleanup, renewable energy production, and industrial chemical synthesis.

### 1.3. Significance of Photocatalysis as an Eco-Friendly Technology for Treating Wastewater

Photocatalysis has gained recognition as an effective eco-friendly approach for tackling the globe issue of wastewater treatment. This innovative approach utilizes light energy, typically from sunlight or ultraviolet light, to activate photocatalysts, driving chemical reactions that degrade contaminants into harmless byproducts. Its importance stems from offering a sustainable, eco-friendly, and highly effective approach to addressing water pollution.

- *Environmental Sustainability:* Photocatalysis relies on abundant natural resources-light energy and photocatalysts-minimizing the need for harmful chemicals or energy-intensive processes typically associated with conventional wastewater treatment methods. It operates at ambient temperature and pressure, reducing energy consumption and the associated carbon footprint [10, 11].
- *Degradation of Refractory Pollutants:* One of the major advantages of photocatalysis is its ability to break down complex and refractory organic and inorganic pollutants, including pesticides, dyes, pharmaceuticals, and industrial effluents, those are resistant to conventional biological or chemical treatments. This method converts harmful pollutants into harmless substances like carbon dioxide, water, and basic inorganic ions [10].
- *Disinfection Capability:* Photocatalysis effectively inactivates pathogens, such as bacteria, viruses, and protozoans present in wastewater. The generation of Reactive Oxygen Species (ROS), including hydroxyl radicals, during the photocatalytic process plays a key role in microbial disinfection, ensuring safe water for reuse or discharge into the environment [12].
- *Minimal Secondary Pollution:* Unlike conventional treatment methods that may produce sludge or chemical residues requiring further disposal, photocatalysis generates minimal secondary waste [13]. This characteristic enhances its appeal as a green technology for sustainable wastewater management.
- *Adaptability and Scalability:* Photocatalytic systems are adaptable to diverse wastewater types and can be integrated into decentralized or centralized treatment facilities. Advances in photocatalyst materials and reactor designs are making the technology increasingly cost-effective and scalable for industrial and municipal applications [14].
- *Utilization of Renewable Energy*: By leveraging solar energy as a primary energy source, photocatalysis aligns with global efforts to transition toward renewable energy technologies. This feature helps reduce reliance on non-renewable energy sources, thereby improving the sustainability of the entire process [15].
- *Potential for Circular Economy*: Photocatalysis aligns with circular economy concepts by facilitating water recycling and the extraction of useful resources [16]. For instance,

it can assist in treating industrial effluents to remove harmful substances, allowing water to be recycled within production processes.

This chapter highlights the integration of photocatalysis as a green technology into wastewater management systems, emphasizing its potential to achieve cleaner water resources while contributing to global sustainability goals. With ongoing research and advancements, photocatalysis is set to have a significant impact on addressing the challenges of modern wastewater treatment.

## 2. Types of Photocatalysts

Photocatalysts are substances that enable chemical reactions when exposed to light, yet they remain unchanged during the process. They are essential in photocatalysis, especially for applications such as treating wastewater, purifying air, and generating renewable energy. The effectiveness of photocatalysts depends on their optical, electronic, and surface properties. The major types of photocatalysts can be broadly categorized (Table 2) as follows:

**Table 2.** Types of photocatalysts

| Type | Key properties | Example materials | Common Applications |
|---|---|---|---|
| Semiconductors | High Stability, UV absorption | $TiO_2$, ZnO, CdS | Water/Air Purification, $H_2$ Production |
| Metal-Organic Frameworks | Porous, tunable, high surface area | MIL-125, ZIF-8 | Adsorption, Organic pollutant degradation |
| Carbon-based Materials | Conductive, High surface area, Stability | Graphene, CNTs | Enhanced photocatalytic degradation |
| Hybrid Photocatalysts | Combined properties, extended light range | $TiO_2$/CdS, $TiO_2$/rGO | Visible light driven pollutant breakdown |

### 2.1. Semiconductors

Semiconductors are the most commonly used photocatalysts due to their effective charge separation properties, high stability, and relatively low cost [17]. Their effectiveness is largely attributed to their ability to harness light and initiate electron-hole pair creation for driving redox reactions in various photocatalytic processes. When illuminated with light possessing energy equal to or exceeding their band gap, semiconductors enable the excitation of electrons from the valence band to the conduction band, resulting in the formation of positively charged holes. These charge carriers drive redox reactions, which support processes such as pollutant breakdown, hydrogen production, and carbon dioxide conversion [18].

Among numerous semiconductors, a few have emerged as the most widely studied and applied due to their distinctive attributes and diverse functionality. Below is a detailed description of these materials:

### 2.1.1. Titanium Dioxide ($TiO_2$)

Titanium Dioxide ($TiO_2$) is considered a prominent and important semiconductor photocatalyst, recognized for its chemical stability, non-toxicity, and impressive photocatalytic activity. Its wide bandgap (~3.2 eV for anatase and ~3.0 eV for rutile) enables effective utilization under ultraviolet (UV) light, making it suitable for various environmental applications, particularly in pollutant degradation. It exhibits remarkable resistance to chemical degradation, ensuring longevity in photocatalytic applications. As an abundant material on Earth, $TiO_2$ poses minimal environmental and health risks, making it a preferred choice for sustainable technologies [19].

#### *2.1.1.1. Applications of $TiO_2$ in Photocatalysis*

- *Wastewater Treatment*: $TiO_2$ is widely used to degrade organic pollutants, including dyes, pesticides, and pharmaceuticals, into harmless byproducts, ensuring cleaner water resources.
- *Hydrogen Production*: It plays a significant role in water splitting processes to generate hydrogen, a clean and renewable energy source.
- *Air Purification*: $TiO_2$ degrades toxic gases and airborne volatile organic compounds (VOCs), contributing to improved air quality and a healthier environment.
- *Antimicrobial Applications*: It is utilized in sterilization processes to inactivate bacteria, viruses, and other pathogens, enhancing hygiene and safety.
- *Self-Cleaning Surfaces*: $TiO_2$ coatings are applied to glass, tiles, and building materials, enabling self-cleaning properties by breaking down dirt and organic matter when exposed to light.
- *Photocatalytic Coatings:* It is used in paints and protective coatings to degrade pollutants and resist microbial growth, extending surface durability.
- *Carbon Dioxide Reduction*: $TiO_2$ facilitates repurposing $CO_2$ for the production of useful fuels and chemical substances contributing to carbon capture and utilization technologies.
- *Energy Applications*: It is integrated into dye-sensitized solar cell technologies and other energy-related devices to enhance efficiency and performance.

#### *2.1.1.2. Challenges and Modifications of $TiO_2$ in Photocatalysis*

- *Limited Visible Light Activity*: The light absorption of $TiO_2$ is primarily limited to the ultraviolet (UV) range, a small portion of the solar spectrum, which hinders its effectiveness in visible sunlight.
- *Doping with Transition Metals*: Incorporating metals such as Fe, Cu, Ni, or Ag into the $TiO_2$ lattice improves the efficiency of its photocatalytic behaviour under visible light irradiation by narrowing the band gap and enhancing light absorption.
- *Doping with Alkali Metals*: Adding alkali metals like Li, Na, or K can improve the surface charge characteristics of $TiO_2$, facilitating better adsorption of pollutants and enhancing photocatalytic efficiency.

- *Doping with Alkaline Earth Metals*: Elements such as Mg, Ca, or Sr are used in order to tune the electronic properties and stability of $TiO_2$, improving its performance and resistance to photo-corrosion.
- *Doping with Non-Metals:* Elements such as nitrogen (N) and sulfur (S) are introduced into $TiO_2$ aimed at increasing visible light uptake and photocatalytic activity.
- *Hybridization with Other Compounds*: Incorporating materials like graphene or carbon nanotubes with $TiO_2$ improves the separation of charge carriers, reduces recombination losses, and improves overall efficiency.
- *Structural Modifications*: Engineering nanostructures, such as nanoparticles, nanotubes, or thin films, enhances light absorption and increases the functional surface area involved in photocatalytic activity.
- *Surface Sensitization:* Coating $TiO_2$ with photosensitive dyes or semiconductors with narrower band gaps facilitates visible light absorption and extends its application range.

### *2.1.2. Zinc Oxide (ZnO)*

Known for its strong stability, Zinc Oxide (ZnO) possesses a bandgap of about 3.3 eV, making it highly active as a photocatalyst when exposed to UV light. Its superior electron mobility compared to $TiO_2$ enhances charge transport efficiency, making it a valuable material across multiple fields, especially for environmental remediation. Oxygen vacancies and surface defects play a crucial role in boosting photocatalytic efficiency by facilitating better separation of charge carriers. ZnO is effectively used in water treatment processes, demonstrating robust photocatalytic activity for the decomposition of harmful organic pollutants [20].

#### *2.1.2.1. Practical-Uses of Zinc Oxide (ZnO) in Photocatalysis*

- *Water Treatment*: ZnO is highly effective in degrading toxic organic materials like dyes, pesticides, and pharmaceuticals, as well as microbial contaminants in water, ensuring cleaner water resources.
- *Self-Cleaning Surfaces:* ZnO is applied to surfaces like glass and tiles to create self-cleaning materials that break down dirt and organic matter under light exposure.
- *Antibacterial Coatings*: It is used in coatings to prevent the proliferation of bacteria, fungi, and other pathogens, enhancing hygiene in various applications.
- *Air Purification*: ZnO aids in removing volatile organic compounds (VOCs) and harmful gases from the air, contributing to improved air quality.
- *Energy Applications*: It is employed in photocatalytic systems for hydrogen production through water splitting and in energy devices like dye-sensitized solar cells.
- *Degradation of Persistent Pollutants*: ZnO is effective in breaking down resistant contaminants like pharmaceutical residues and hormone-disrupting compounds found in wastewater.
- *Food Packaging*: ZnO-based photocatalytic coatings are explored in food packaging to inhibit microbial contamination and extend shelf life.
- *Carbon dioxide Conversion*: ZnO is also utilized in photocatalytic reduction-driven synthesis of beneficial fuels and chemicals from $CO_2$.

### 2.1.2.2. *Challenges and Modifications of ZnO in Photocatalysis*

- *Photo-Corrosion in Aqueous Environments*: ZnO tends to degrade under prolonged exposure to light and water, limiting its long-term usability in photocatalytic applications.
- *Doping with Metals:* Incorporating elements like aluminum (Al), magnesium (Mg), or transition metals improves ZnO's structural stability, charge separation, and overall photocatalytic efficiency.
- *Surface Modification*: Applying coatings or functional layers on ZnO particles helps to reduce photo-corrosion and enhance its durability in water-based systems.
- *Bandgap Engineering*: Narrowing the bandgap through doping with non-metalllic elements like nitrogen (N) or sulfur (S), extends ZnO's absorptivity of light in the visible wavelength range, increasing its activity under sunlight.
- *Integrating with Other Semiconductors*: Combining ZnO with materials like $TiO_2$ or graphene oxide improves charge carrier separation and enhances photocatalytic performance.
- *Nano structuring*: Designing ZnO nanostructures such as nanorods, nanotubes, or nanoparticles, increases surface area and light absorption, thereby boosting efficiency.
- *Incorporation of Co-Catalysts*: Employing co-catalysts like noble metals (Pt, Au) or carbon materials increases photocatalytic efficiency by enhancing charge mobility and reducing charge carrier recombination.
- *Stability in Harsh Conditions*: Modifications are implemented to improve ZnO's resistance to extreme pH or temperature variations, making it suitable for diverse environmental applications.

### ***2.1.3. Perovskite-Based Photocatalysts***

Perovskite-based photocatalysts, such as $SrTiO_3$, $BaTiO_3$, and $CsPbBr_3$, are distinguished by their unique $ABX_3$ crystal structure, where 'A' is a large cation, 'B' is a smaller metal cation, and 'X' represents an anion, typically oxygen or halides. These materials possess exceptional light-harvesting capabilities due to their tunable and narrow bandgaps, which enable efficient absorption of visible light. This property makes perovskites highly suitable for photocatalytic applications under solar irradiation. Additionally, their high carrier mobility, excellent charge separation efficiency, and adjustable electronic properties contribute to their effectiveness in promoting redox reactions. These characteristics, combined with their chemical flexibility, allow perovskites to be tailored for broad applications in environmental remediation, hydrogen evolution, and $CO_2$ reduction, establishing them as a potential class of materials within the domain of photocatalysis [21].

#### 2.1.3.1. *Applications*

- *Photocatalytic $CO_2$ Reduction*: Perovskite photocatalysts are employed to convert carbon dioxide into valuable fuels or compounds, such as methane or methanol, through light-driven processes, contributing to carbon capture and utilization technologies.

- *Hydrogen Production via Water Splitting*: These materials facilitate the photocatalytic water splitting to generate hydrogen and oxygen, offering a sustainable and efficient method for clean energy production.
- *Degradation of Pollutants*: Perovskites are used to break down organic and inorganic pollutants in water and air, aiding in environmental cleanup efforts.
- *Solar Fuel Generation*: Their ability to harvest visible light efficiently makes them suitable for generating solar fuels in renewable energy applications.
- *Energy Storage and Conversion*: Perovskites are integrated into photocatalytic systems for storing and converting solar energy into chemical energy.
- *Selective Chemical Synthesis:* They enable light-driven synthesis of valuable chemical compounds under mild conditions, reducing the need for harsh reagents and high energy inputs.

#### *2.1.3.2. Challenges and Modifications*

- *Instability in Water and Air*: Perovskites degrade easily when exposed to moisture, oxygen, or ambient environmental conditions, limiting their practical applications.
- *Structural Engineering*: Modifying the perovskite composition by incorporating robust cations, anions, or alternative halides enhances their chemical and structural stability.
- *Protective Coatings*: Applying protective layers such as polymers, metal oxides, or graphene shields the perovskite material from environmental degradation, improving its longevity.
- *Doping with Elements*: Introducing substituents like transition metals used for doping or rare earth elements, into the perovskite structure enhances its electronic properties, stability, and photocatalytic efficiency.
- *Coupling with Stable Materials*: Combining perovskites with stable photocatalysts, like $TiO_2$ or ZnO, helps improve charge separation and overall durability while leveraging their complementary properties.
- *Surface Passivation*: Treating the surface of perovskites with passivating agents reduces defect sites and minimizes degradation under operational conditions.
- *Nanostructure Design*: Creating nanostructured perovskites, such as nanoparticles or nanocomposites, improves light absorption and increases active surface areas while addressing stability concerns.
- *Environment-Specific Modifications*: Tailoring perovskite photocatalysts for specific operational environments, such as controlled humidity or inert atmospheres, enhances performance and usability.

### **2.2. Metal Organic Frameworks (MOFs)**

Metal-Organic Frameworks (MOFs) represent a group of porous compounds formed through the coordination of metal ions or clusters with organic ligands. Due to their tunable architectures, large surface areas, and excellent light absorption properties, they are a promising platform for photocatalysis [22]. MOFs offer an innovative approach to photocatalysis, offering a sustainable and efficient pathway for addressing global challenges related to energy and environmental remediation. Their versatility and tunability continue to make them a focal point of research in green chemistry and renewable energy [23]. Notable examples of MOFs used in photocatalysis include MIL-125 (Ti), ZIF-8, UiO-66 (Zr), MOF-5 (Zn), PCN-222 (Zr), $NH_2$-MIL-101 (Fe), MIL-53 (Al), CPO-27 (Ni/Mg), and HKUST-1 (Cu).

### 2.2.1. Properties of MOFs in Photocatalysis

- *High Surface Area and Porosity*: The large surface area and tunable pore sizes of MOFs allow them to adsorb and concentrate reactants near active sites, enhancing photocatalytic efficiency.
- *Tailored Bandgap*: The electronic properties of MOFs can be tuned by selecting different metal nodes and organic linkers, enabling precise control over light absorption and charge transfer.
- *Light Harvesting Capability*: Many MOFs exhibit strong light absorption in the ultraviolet and visible spectra, making them efficient for solar-driven photocatalytic reactions.
- *Charge Separation and Transfer*: MOFs can enhance charge carrier separation and suppress the recombination of electrons and holes by incorporating conductive linkers or doping with other materials.

### 2.2.2. Applications

- *Water Splitting for Hydrogen Production*: MOFs with photoactive centers, such as Ti, Zn, or Zr-based frameworks, are used to generate hydrogen and oxygen through water splitting under light irradiation, contributing to clean energy production.
- *Carbon Dioxide Reduction*: MOFs act as photocatalysts to convert $CO_2$ into valuable fuels or chemicals like methanol, reducing greenhouse gas emissions and supporting sustainable energy goals.
- *Degradation of Pollutants:* Metal-Organic Frameworks (MOFs) are capable of breaking down organic contaminants in water-such as dyes, pesticides, and pharmaceutical residues-and can also remove airborne pollutants like volatile organic compounds (VOCs).
- *Selective Organic Synthesis*: MOFs enable light-driven, selective chemical reactions in a mild environment, without relying on toxic or corrosive substances or extreme reaction environments.

#### *2.2.2.1. Challenges and Modifications*

Despite their potential, MOFs face challenges, such as low stability under harsh environmental conditions and limited efficiency in charge transfer. Modifications like integrating MOFs with semiconductors (e.g., $TiO_2$ or ZnO), doping with metallic or non-metallic elements, and generating hybrid composites improve their stability and performance. Additionally, designing MOFs with photoactive organic linkers or incorporating co-catalysts can further enhance their photocatalytic capabilities.

## 2.3. Carbon-Based Photocatalysts

Carbon-based photocatalysts represent a wide array of materials, such as carbon nanotubes (CNTs), graphene and its oxidized form (graphene oxide), carbon quantum dots, fullerene compounds like C-60 and C-70, mesoporous carbon structures (e.g., CMK-3), activated carbon, and graphite carbon nitride (g-$C_3N_4$). These materials have attracted considerable attention owing to their exceptional characteristics, including excellent electrical conductivity, expansive surface area, and adjustable optical features. They serve as standalone photocatalysts or in combination with semiconductors to enhance photocatalytic efficiency. Their versatility and sustainability make them ideal for addressing environmental and energy-related challenges [24].

### *2.3.1. Properties*

- *High Electrical Conductivity*: Carbon-based materials enhance photocatalytic efficiency by promoting effective charge separation and transport, which helps reduce the recombination of electron-hole pairs during the reaction process.
- *Large Surface Area:* Their large surface area offers numerous active sites, enhancing adsorption capacity and catalytic activity during reactions.
- *Chemical Stability:* These materials exhibit excellent chemical and thermal stability, ensuring durability under various reaction conditions.
- *Light Absorption:* Carbon materials can absorb light across the UV and visible spectra. Their optical properties can be further tailored for enhanced performance.
- *Sustainability:* Carbon-based photocatalysts are derived from abundant and renewable resources, supporting green and sustainable applications.

### *2.3.2. Applications of Carbon-Based Photocatalysts*

- *Water Purification*: They enable effective breakdown of organic and microbial contaminants in wastewater, as well as the elimination of heavy metals through combined adsorption and photodegradation processes.
- *Air Purification*: Photocatalysis is effective in eliminating volatile organic compounds (VOCs) and toxic gases like nitrogen oxides ($NO_X$) and sulfur dioxide ($SO_2$) from the environment.
- *Hydrogen Production*: Facilitation of water splitting for hydrogen generation, a clean energy source.
- *Carbon Dioxide Reduction*: Photocatalytic techniques enable the transformation of carbon dioxide ($CO_2$) into valuable fuels like methane and methanol.
- *Energy Storage and Conversion*: Carbon materials improve the performance of systems designed for harnessing solar energy conversion and storage.
- *Self-Cleaning Surfaces*: Incorporation into coatings for self-cleaning and antibacterial applications.

### *2.3.3. Challenges and Modifications*

- Limited Visible Light Absorption
  *Challenge:* Some carbon-based materials exhibit low efficiency under visible light.
  *Modification:* Incorporating elements such as nitrogen, sulphur, or boron through doping enhances their light absorption and narrows the bandgap.
- Charge Recombination Issues
  *Challenge:* Recombination of electron-hole pairs reduces photocatalytic efficiency.
  *Modification:* Combining carbon materials with semiconductors like $TiO_2$ or ZnO facilitates charge separation and improves overall performance.
- Tendency to Aggregate
  *Challenge:* Aggregation of carbon materials in aqueous systems limits their accessible surface area.
  *Modification:* Surface functionalization with hydrophilic groups or the use of dispersants improves dispersion and reactivity.
- Complex Fabrication and High Costs
  *Challenge:* Producing high-purity carbon-based photocatalysts is expensive and resource-intensive.
  *Modification:* Developing cost-effective synthesis methods, such as green approaches using biomass-derived precursors, reduces production costs.
- Environmental and Stability Concerns
  *Challenge:* Carbon materials may degrade or pose environmental risks when improperly disposed of.
  *Modification:* Employing protective coatings and recycling methods enhances material stability and reduces environmental impact.

## 2.4. Hybrid Photocatalysts

Hybrid photocatalysts are composite materials that combine the strengths of multiple types of photocatalysts to improve efficiency, stability, and light absorption.

- *$TiO_2$/CdS Hybrids*: CdS, known for its ability to absorb visible light, is combined with $TiO_2$ to broaden the photocatalyst's light responsiveness into the visible spectrum. This pairing boosts photocatalytic performance under sunlight while preserving the inherent stability of $TiO_2$ [25].
- *MOF-Semiconductor Hybrids*: Combining MOFs with semiconductors (e.g., MOF-$TiO_2$) enhances pollutant adsorption from the MOF's porous structure while benefiting from the semiconductor's photocatalytic power. This combination improves charge separation and light absorption, making it ideal for pollutant degradation [26].
- *Carbon-Based Hybrids:* Incorporating graphene or CNTs with semiconductors (e.g., graphene/ZnO, $TiO_2$/CNT) enhances electron mobility, and minimizing electron-hole recombination increasing photocatalytic efficiency [27].

## 3. Mechanisms of Photocatalytic Degradation

Photocatalytic degradation is a process that utilizes light energy, typically in the ultraviolet (UV) or visible spectrum, to activate a photocatalyst, leading to the decomposition of organic and inorganic pollutants in aqueous or gaseous phases. Advanced Oxidative Processes (AOPs) in photocatalysis are sophisticated methods designed to produce highly reactive oxidative species, particularly hydroxyl radicals (OH·), for breaking down organic and inorganic pollutants [28]. These processes are characterized by their efficiency in mineralizing pollutants into non-toxic byproducts, such as carbon dioxide ($CO_2$) and water ($H_2O$), under normal environmental conditions [29]. AOPs often enhance the efficiency of photocatalytic systems by integrating additional chemical or physical agents.

### 3.1. Principle of AOPs in Photocatalysis

AOPs leverage the formation of oxidative radicals, especially hydroxyl radicals (OH·), which exhibit strong oxidation potential (2.8 V) and can non-selectively oxidize broad spectrum of pollutants. The key mechanism involves the generation of radicals through light-activated processes, often mediated by semiconductors such as titanium dioxide ($TiO_2$) or modified photocatalysts.

### 3.2. Photocatalytic Activation and Radical Generation

Advanced oxidation processes (AOPs) begin when the photocatalyst absorbs photons with energy equal to or exceeding its bandgap, resulting in the generation of electron-hole pairs ($e^-/h^+$)

$$TiO_2 + h\upsilon \longrightarrow TiO_2\,(e^- + h^+)$$

The holes ($h^+$) present in the valence band function as powerful oxidants, whereas the electrons ($e^-$) in the conduction band are involved in reduction processes. These charge carriers contribute to the creation of reactive oxygen species (ROS), which play a vital role in the breakdown of pollutants.

### 3.3. Key Oxidative Processes in AOPs

AOPs in photocatalysis are enhanced by introducing external agents or combining with other processes to generate radicals more efficiently. The primary methods include:

*(a) Hydroxyl Radical Formation*

Hydroxyl radicals serve as the fundamental agents in AOPs because of their exceptional reactivity. They are formed via several pathways:

- Water Oxidation by Holes:

$$h^+ + H_2O \longrightarrow OH\cdot + H^+$$
$$h^+ + OH^- \longrightarrow OH\cdot$$

- Hydrogen Peroxide Photodecomposition:

$$H_2O_2 + h\upsilon \longrightarrow 2OH\cdot$$

*(b) Superoxide Radical Formation*

Electrons reduce oxygen molecules, producing superoxide radicals ($O_2^{\bullet -}$)

$$e^- + O_2 \longrightarrow O_2^{\bullet -}$$

Subsequent reactions involving superoxide radicals contribute to the formation of additional oxidative species:

$$O_2^{\bullet -} + H^+ \longrightarrow HO_2^{\bullet}$$

*(c) Fenton-Like Reactions*

Fenton-like reactions involve transition metals such as Fe(II) to promote hydroxyl radical production.

$$H_2O_2 + Fe^{2+} \longrightarrow OH\cdot + OH^- + Fe^{3+}$$
$$Fe^{3+} + e^- \longrightarrow Fe^{2+}$$

*(d) Ozone Integration*

Ozone ($O_3$) can be introduced to generate additional oxidative species. Under UV irradiation, ozone decomposes to produce hydroxyl radicals:

$$O_3 + H_2O + h\upsilon \longrightarrow 2OH\cdot + O_2$$

## 3.4. Pollutant Degradation Mechanisms

The oxidative species generated during AOPs react with pollutants in the following ways:

- Direct Oxidation by Holes:

$$R + h^+ \longrightarrow R\cdot$$

- Radical-Mediated Oxidation: Hydroxyl radicals oxidize organic pollutants (R-H) by removing hydrogen atoms or reacting with unsaturated bonds:

$$R\text{-}H + OH\cdot \longrightarrow R\cdot + H_2O$$

This process continues through intermediate reactions, ultimately breaking down the pollutant into $CO_2$ and $H_2O$.

For example, the degradation of a chlorinated compound, such as trichloroethylene ($CHCl{=}CCl_2$), proceeds as:

$$CHCl{=}CCl_2 + OH\cdot \longrightarrow \text{Intermediates} \longrightarrow CO_2 + HCl$$

### 3.5. Enhancements in AOPs

Advanced oxidative processes can be enhanced through various modifications:

- *Doping of Photocatalysts*: Incorporating elements like nitrogen (N), sulfur (S), or metals (e.g., copper, magnesium, silver, platinum) into semiconductors reduces the bandgap, enabling visible light activation.

$$TiO_2\,(\text{doped}) + h\upsilon \longrightarrow TiO_2\,(e^- + h^+)$$

- *Combination with External Agents*: Adding hydrogen peroxide ($H_2O_2$), ozone ($O_3$), or ultrasound can synergistically increase radical generation.
- *Heterojunction Formation*: Combining two semiconductors with complementary band structures facilitates charge separation and enhances photocatalytic activity.

## 4. Photocatalytic Applications in Wastewater Purification

Photocatalysis has gained attention as an effective method for wastewater treatment, owing to its capability to break down various organic contaminants [30]. The rapid development of novel photocatalytic materials has positioned this technology as a leading approach for sustainable wastewater treatment [31]. Photocatalytic degradation can effectively remove toxic organic substances such as dyes, pharmaceutical residues, pesticides, and other micropollutants in wastewater [30, 32].

A major benefit of photocatalysis is its capacity to break down organic pollutants, transforming them into non-toxic byproducts such as carbon dioxide and water. Furthermore, photocatalytic systems can be designed to operate under ambient conditions without the need for harsh chemicals or high energy inputs, making them an eco-friendly alternative to conventional wastewater treatment techniques. Photocatalytic reactors can be used to treat various types of industrial and municipal wastewater [31]. For example, the photocatalytic degradation of dyes from textile industry effluents has been extensively studied, demonstrating the potential of this technology to address a major source of water pollution. Photocatalysis has also shown promise in removing antibiotics, personal care products, and other emerging contaminants from wastewater.

## 4.1. Organic Pollutant Degradation Using Photocatalysis

The core concept behind photocatalytic wastewater treatment involves the production of highly reactive oxygen species such as hydroxyl radicals (•OH), superoxide radicals ($O_2^{\bullet-}$), and hydrogen peroxide ($H_2O_2$), when a semiconductor photocatalyst is activated by light energy [33]. These reactive species can then oxidize and mineralize a wide range of hazardous organic contaminants, transforming them into less harmful substances.

Photocatalysts, including titanium dioxide ($TiO_2$) and other semiconductor compounds, have been thoroughly investigated for their capability to decompose numerous organic contaminants in wastewater, such as dyes, pesticides, pharmaceutical residues, and personal care products [34]. Photocatalytic degradation begins with the adsorption of pollutants onto the surface of the photocatalyst, after which reactive oxygen species are produced to target and decompose the organic compounds.

However, several challenges still need to be addressed to enable the widespread adoption of photocatalytic wastewater treatment. These include improving the photocatalytic efficiency, enhancing the separation and reuse of the catalysts, and optimizing the operating conditions to achieve thorough mineralization of the organic impurities into stable non-toxic byproducts. The performance of various photocatalysts in degrading different dyes is summarized in the Table 3:

**Table 3.** Performance of different catalysts on the dyes

| S.No. | Name of the Catalyst | Name of the Dye | Time for Degradation | Reference |
|---|---|---|---|---|
| 1 | $CoFe_2O_4$ | Methylene Blue | 120 min | [35] |
| 2 | $NiFe_2O_4$ | Rhodamine B | 90 min | [36] |
| 3 | $CuFe_2O_4$ | Methyl Orange | 150 min | [35] |
| 4 | $TiO_2$ | Methylene Blue | 30 min | [37] |
| 5 | ZnO | Rhodamine B | 60 min | [38] |
| 6 | $TiO_2$ | Methyl Orange | 45 min | [39] |
| 7 | $TiO_2$-rGO | Direct Blue 6 | 60 min | [40] |
| 8 | ZnO-CuO | Brilliant Green | 80 min | [41] |

## 4.2. Heavy Metal Removal via Photocatalytic Processes

In addition to organic pollutant degradation, photocatalysis has been studied for removing heavy metal ions from wastewater. The photocatalytic process can facilitate the oxidation, reduction, or precipitation of heavy metal ions, leading to their removal from the aqueous environment. Recent studies have demonstrated the potential of photocatalysts, such as metal-doped titanium dioxide and other composite materials, to effectively remove heavy metals like cadmium (Cd), lead (Pb), chromium (Cr), and arsenic (As) from wastewater [42, 43]. The photocatalytic conversion of metal ions to less soluble forms, or the adsorption of metal ions onto the photocatalyst surface, enables the subsequent separation and removal of these contaminants.

### 4.3. Disinfection and Pathogen Inactivation using Photocatalysis

Another important application of photocatalysis in wastewater treatment is the inactivation of pathogenic microorganisms, such as bacteria, viruses, and protozoa. Photocatalytic processes can generate reactive oxygen species that damage the cell walls, membranes, and genetic material of microorganisms, leading to their inactivation and sterilization. The use of photocatalysts, like titanium dioxide, for water disinfection has been extensively researched, demonstrating its effectiveness in eliminating a wide range of waterborne pathogens, including *Escherichia coli, Salmonella, Legionella,* and *Cryptosporidium*.

Significant progress has been made in recent years to advance photocatalytic systems for wastewater treatment, improving their efficiency, stability, and practical applicability in eliminating organic contaminants, toxic metals, and pathogens [44, 45].

### 4.4. Integrated Photocatalytic Systems for Wastewater Treatment

To further enhance the efficiency and applicability of photocatalytic wastewater treatment, researchers have explored the integration of photocatalysis with techniques like membrane filtration, adsorption, and biological processes. Photocatalytic membrane reactors, for instance, combine the pollutant degradation capabilities of photocatalysis with the separation and concentration abilities of membrane filtration, enabling a more comprehensive treatment approach.

## 5. Factors Influencing Photocatalytic Efficiency

### 5.1. Photocatalyst Properties (Morphology, Surface Area, Bandgap)

The properties of the photocatalyst significantly influence photocatalytic efficiency. Morphology, such as nanoparticle size and shape, determines the surface area and active sites available for reactions. Higher surface areas enhance pollutant adsorption and catalytic activity. The bandgap energy of the photocatalyst dictates its activation under specific wavelengths of light. Narrower bandgaps allow visible light utilization, while wide-bandgap materials like $TiO_2$ predominantly operate under UV light. Modifications, such as doping with metals (e.g., silver) or non-metals (e.g., nitrogen), improve light absorption and charge separation efficiency, reducing recombination of electrons and holes. Nanostructured catalysts like nanotubes and nanowires also exhibit enhanced performance due to their unique morphologies [46, 47].

### 5.2. Light Source (Intensity, Wavelength)

The intensity and wavelength of the light source directly affect photocatalytic reactions. Higher light intensity generates more electron-hole pairs, increasing ROS production and pollutant degradation rates. The wavelength must correspond to the bandgap energy of the photocatalyst in order to trigger the reaction. UV light is typically required for materials like $TiO_2$, but visible-light-responsive photocatalysts, such as those doped with noble metals, can utilize solar energy. This broadens their application, especially in areas with high sunlight availability [48].

### 5.3. Wastewater Characteristics ($p^H$, Temperature, Pollutant Concentration)

The $p^H$ of wastewater affects the surface charge characteristics of the photocatalyst and pollutant speciation. Optimal $p^H$ ensures maximum interaction between the catalyst and pollutants. Temperature influences reaction kinetics, with moderate increases enhancing efficiency without destabilizing the photocatalyst. Pollutant concentration affects reaction rates; low concentrations allow faster degradation, while high loads may saturate active sites, slowing the process [49].

### 5.4. Operating Conditions (Reactor Design, Flow Rate)

Reactor design plays a critical role in optimizing light utilization and photocatalyst exposure. Fixed-bed, slurry, and fluidized-bed reactors each offer unique advantages for specific applications. The flow rate of wastewater determines duration of contact between pollutants and the photocatalyst. Optimal flow rates ensure efficient degradation without overloading the system. Innovations such as photoreactors with enhanced light distribution and immobilized catalysts improve scalability and efficiency [50].

## Conclusion

Photocatalysis represents a transformative green technology for wastewater treatment, leveraging light energy to effectively degrade a wide range of contaminants. Its capacity to mineralize organic pollutants, remove heavy metals, and inactivate pathogens underscores its versatility and environmental significance. Recent advancements in photocatalyst materials, including semiconductors and hybrid systems, have enhanced their efficiency, scalability, and applicability under natural conditions. Despite challenges such as limited light absorption and recombination of electron-hole pairs, innovative approaches like material doping and nano structuring have addressed these limitations. As a sustainable and adaptable solution, photocatalysis holds great promise for mitigating water pollution and supporting global efforts toward ecological preservation and resource sustainability.

## References

[1] Fn, C., & Mf, M. (2017). Factors affecting water pollution: a review. *Journal of Ecosystem & Ecography*, *07*(01). https://doi.org/10.4172/2157-7625.1000225.

[2] Ansari, A. A., Gill, S. S., & Khan, F. A. (2010). Eutrophication: threat to aquatic ecosystems. In *Springer eBooks* (pp. 143–170). https://doi.org/10.1007/978-90-481-9625-8_7.

[3] Ogidi, O. I., & Akpan, U. M. (2022). Aquatic biodiversity loss: Impacts of pollution and anthropogenic activities and strategies for conservation. In *Sustainable development and biodiversity* (pp. 421–448). https://doi.org/10.1007/978-981-19-3326-4_16.

[4] Majumdar, D. (2003). The Blue Baby Syndrome. *Resonance*, *8*(10), 20–30. https://doi.org/10.1007/bf02840703.

[5] Galvão, A., Aleixo, M., De Pablo, H., Lopes, C., & Raimundo, J. (2020). Microplastics in wastewater: microfiber emissions from common household laundry. *Environmental Science and Pollution Research*, *27*(21), 26643–26649. https://doi.org/10.1007/s11356-020-08765-6.

[6] Abraham, W. (2010). Megacities as sources for pathogenic bacteria in rivers and their fate downstream. *International Journal of Microbiology*, *2011*, 1–13. https://doi.org/10.1155/2011/798292.

[7] Vardhan, K. H., Kumar, P. S., & Panda, R. C. (2019). A review on heavy metal pollution, toxicity and remedial measures: Current trends and future perspectives. *Journal of Molecular Liquids*, *290*, 111197. https://doi.org/10.1016/j.molliq.2019.111197.

[8] Baruah, J., Singh, L. S., Salvia, T., & Sarma, J. (2024). Antimicrobial resistance a continued global threat to public health – A perspective and mitigation strategies. *Journal of Laboratory Physicians*, *0*, 1–12. https://doi.org/10.25259/jlp_24_2024.

[9] Hassaan, M. A., El-Nemr, M. A., Elkatory, M. R., Ragab, S., Niculescu, V., & Nemr, A. E. (2023). Principles of Photocatalysts and their Different Applications: a review. *Topics in Current Chemistry*, *381*(6). https://doi.org/10.1007/s41061-023-00444-7.

[10] Elumalai, P., Parthipan, P., Sivakumar, L., Thanigaivel, S., Gnanasekaran, L., Gao, X., Cui, J., Moon, C. J., Theerthagiri, J., & Choi, M. Y. (2024). Photocatalysis: a sustainable approach for removing hazardous polyaromatic hydrocarbons. *ChemistrySelect*, *9*(34). https://doi.org/10.1002/slct.202402459.

[11] Miad, A. A., Saikat, S. P., Alam, M. K., Hossain, M. S., Bahadur, N. M., & Ahmed, S. (2024). Metal oxides-based photocatalyst for the efficient degradation of organic pollutants for a sustainable environment: a review. *Nanoscale Advances*, *6*(19), 4781–4803. https://doi.org/10.1039/d4na00517a.

[12] Kozak, M., Mazierski, P., Żebrowska, J., Klimczuk, T., Lisowski, W., Żak, A. M., Skowron, P. M., & Zaleska-Medynska, A. (2024). Detailed Insight into Photocatalytic Inactivation of Pathogenic Bacteria in the Presence of Visible-Light-Active Multicomponent Photocatalysts. *Nanomaterials*, *14*(5), 409. https://doi.org/10.3390/nano14050409.

[13] Kumar, N., Mittal, H., Alhassan, S. M., & Ray, S. S. (2018). Bionanocomposite hydrogel for the adsorption of dye and reusability of generated waste for the photodegradation of ciprofloxacin: A demonstration of the circularity concept for water purification. *ACS Sustainable Chemistry & Engineering*, *6*(12), 17011–17025. https://doi.org/10.1021/acssuschemeng.8b04347.

[14] Wang, D., Mueses, M. A., Márquez, J. a. C., Machuca-Martínez, F., Grčić, I., Moreira, R. P. M., & Puma, G. L. (2021). Engineering and modeling perspectives on photocatalytic reactors for water treatment. *Water Research*, *202*, 117421. https://doi.org/10.1016/j.watres.2021.117421.

[15] Zhang, J., Wu, H., Shi, L., Wu, Z., Zhang, S., Wang, S., & Sun, H. (2023). Photocatalysis coupling with membrane technology for sustainable and continuous purification of wastewater. *Separation and Purification Technology*, *329*, 125225. https://doi.org/10.1016/j.seppur.2023.125225.

[16] Omran, B. A., & Baek, K. (2022). Valorization of agro-industrial biowaste to green nanomaterials for wastewater treatment: Approaching green chemistry and circular economy principles. *Journal of Environmental Management*, *311*, 114806. https://doi.org/10.1016/j.jenvman.2022.114806.

[17] Serpone, N., & Emeline, A. V. (2012). Semiconductor Photocatalysis — Past, present, and future outlook. *The Journal of Physical Chemistry Letters*, *3*(5), 673–677. https://doi.org/10.1021/jz300071j.

[18] Ahmad, I., Zou, Y., Yan, J., Liu, Y., Shukrullah, S., Naz, M. Y., Hussain, H., Khan, W. Q., & Khalid, N. (2022). Semiconductor photocatalysts: A critical review highlighting the various strategies to boost the photocatalytic performances for diverse applications. *Advances in Colloid and Interface Science*, *311*, 102830. https://doi.org/10.1016/j.cis.2022.102830.

[19] Marcelis, E. J., Elshof, J. E. T., & Morales-Masis, M. (2024). Titanium dioxide: a versatile Earth-Abundant optical material for photovoltaics. *Advanced Optical Materials*. https://doi.org/10.1002/adom.202401423.

[20] Alshammari, A. S., Mohamed, M., Khan, Z. R., Bouzidi, M., & Gandouzi, M. (2024). Effect of precursors concentrations on the photocatalysis performance stability of electrodeposited ZNO nanorods

and their robustness in aqueous environments. *Crystals*, *14*(5), 393. https://doi.org/10.3390/cryst14050393.

[21] Kong, J., Yang, T., Rui, Z., & Ji, H. (2018). Perovskite-based photocatalysts for organic contaminants removal: Current status and future perspectives. *Catalysis Today*, *327*, 47–63. https://doi.org/10.1016/j.cattod.2018.06.045.

[22] Wang, S., & Wang, X. (2015). Multifunctional Metal-Organic frameworks for photocatalysis. *Small*, *11*(26), 3097–3112. https://doi.org/10.1002/smll.201500084.

[23] Zhang, T., & Lin, W. (2014). Metal–organic frameworks for artificial photosynthesis and photocatalysis. *Chemical Society Reviews*, *43*(16), 5982–5993. https://doi.org/10.1039/c4cs00103f.

[24] Tripathi, M., Agarwa, A., Singh, M., & Mubarak, N. M. (2024). Carbon-based materials as a promising route for water treatment. In *Elsevier eBooks* (pp. 63–96). https://doi.org/10.1016/b978-0-443-18524-3.00011-8.

[25] Zhou, P., Yu, J., & Jaroniec, M. (2014). All-Solid-State Z-Scheme photocatalytic Systems. *Advanced Materials*, *26*(29), 4920–4935. https://doi.org/10.1002/adma.201400288.

[26] Dong, P., Gao, K., Zhang, L., Huan, H., Xie, M., Yang, X., & Zhang, J. (2024). Hydrogen bond-assisted construction of MOF/semiconductor heterojunction photocatalysts for highly efficient electron transfer. *Applied Catalysis B Environment and Energy*, *357*, 124297. https://doi.org/10.1016/j.apcatb.2024.124297.

[27] Hu, M., Yao, Z., & Wang, X. (2017). Graphene-Based nanomaterials for catalysis. *Industrial & Engineering Chemistry Research*, *56*(13), 3477–3502. https://doi.org/10.1021/acs.iecr.6b05048.

[28] Bethi, B., Sonawane, S. H., Bhanvase, B. A., & Gumfekar, S. P. (2016). Nanomaterials-based advanced oxidation processes for wastewater treatment: A review. *Chemical Engineering and Processing - Process Intensification*, *109*, 178–189. https://doi.org/10.1016/j.cep.2016.08.016.

[29] Saravanan, A., Deivayanai, V., Kumar, P. S., Rangasamy, G., Hemavathy, R., Harshana, T., Gayathri, N., & Alagumalai, K. (2022). A detailed review on advanced oxidation process in treatment of wastewater: Mechanism, challenges and future outlook. *Chemosphere*, *308*, 136524. https://doi.org/10.1016/j.chemosphere.2022.136524.

**[30]** Mandade, P. (2021). Introduction, basic principles, mechanism, and challenges of photocatalysis. In Elsevier eBooks (p. 137). Elsevier BV. https://doi.org/10.1016/b978-0-12-821496-1.00016-7.

[31] Ren, G., Han, H., Wang, Y., Liu, S., Zhao, J., Meng, X., & Li, Z. (2021). Recent Advances of Photocatalytic Application in Water Treatment: A Review [Review of Recent Advances of Photocatalytic Application in Water Treatment: A Review]. *Nanomaterials,* 11(7), 1804. Multidisciplinary Digital Publishing Institute. https://doi.org/10.3390/nano11071804.

[32] Nasir, A. M., Awang, N., Jaafar, J., Ismail, A. F., Othman, M. H. D., Rahman, M. A., Aziz, F., & Yajid, M. A. M. (2020). Recent progress on fabrication and application of electrospun nanofibrous photocatalytic membranes for wastewater treatment: A review. *Journal of Water Process Engineering*, 40, 101878. Elsevier BV. https://doi.org/10.1016/j.jwpe.2020.101878.

[33] Miditana, S. R., Tirukkovalluri, S. R., Raju, I. M. et al. Gemini surfactant assisted synthesis of mesoporous Mn/Mg bimetal doped $TiO_2$ nanomaterial: characterization and photocatalytic activity studies under visible light irradiation. *Sustain Environ Res* 31, 6 (2021). https://doi.org/10.1186/s42834-021-00078-8.

[34] Anucha C. B., Altin I, Bacaksiz E, Stathopoulos V. N. Titanium dioxide ($TiO_2$)-based photocatalyst materials activity enhancement for contaminants of emerging concern (CECs) degradation: In the light of modification strategies. *Chemical Engineering Journal Advances*, 10 (2022) 100262. https://doi.org/10.1016/j.ceja.2022.100262.

[35] Gupta, N. K., Ghaffari, Y., Kim, S., Bae, J., Kim, K. S., & Saifuddin, M. Photocatalytic Degradation of Organic Pollutants over $MFe_2O_4$ (M = Co, Ni, Cu, Zn) Nanoparticles at Neutral $p^H$. In *Scientific Reports. Nature Portfolio.* 10, 1(2020). https://doi.org/10.1038/s41598-020-61930-2.

[36] Alzahrani F. H, Irshad A,Zulfiqar S, Alrowaili Z.A, Ain N.U, Anwar A, Warsi M. F, M. S. Al-Buriahi. Nd-doped $NiFe_2O_4$ and its composite with CNTs to tune photocatalytic activity. *Journal of Rare Earths*, 41,12 (2023) 1919-1928. https://doi.org/10.1016/j.jre.2023.07.020.

[37] Kaur, S., & Singh, V. $TiO_2$ mediated photocatalytic degradation studies of Reactive Red 198 by UV irradiation. In *Journal of Hazardous Materials*, 141, 1 (2006) 230. https://doi.org/10.1016/j.jhazmat.2006.06.123.

[38] Saidani A, Boudraa R, Fendi K, Benouadah L, Benabbas A, Djermoune A, Salvestrini S, Bollinger J-C, Alayyaf AA, Mouni L. Effect of Calcination Temperature on the Photocatalytic Activity of Precipitated ZnO Nanoparticles for the Degradation of Rhodamine B Under Different Light Sources. *Water*. 2025; 17(1):32. https://doi.org/10.3390/w17010032.

[39] Heng, J. Z. X., Tang, K. Y., Regulacio, M. D., Lin, M., Loh, X. J., Li, Z., & Ye, E. (2021). Solar-Powered Photodegradation of Pollutant Dyes Using Silver-Embedded Porous TiO2 Nanofibers. In *Nanomaterials* (Vol. 11, Issue 4, p. 856). Multidisciplinary Digital Publishing Institute. https://doi.org/10.3390/nano11040856.

[40] Quan, Y., Su, R., Hu, M., Lang, J., Fan, H., Shen, H., Gao, M., Li, B., Liu, Y., & Yang, J. (2020). Construction of an MZO heterojunction system with improved photocatalytic activity for degradation of organic dyes. In *CrystEngComm* (Vol. 22, Issue 42, p. 7059). Royal Society of Chemistry. https://doi.org/10.1039/d0ce00581a.

[41] Chauhan, M., Kaur, N., Bansal, P., Kumar, R., Srinivasan, S. S., & Chaudhary, G. R. (2020). Proficient Photocatalytic and Sonocatalytic Degradation of Organic Pollutants Using CuO Nanoparticles. In *Journal of Nanomaterials* (Vol. 2020, p. 1). Hindawi Publishing Corporation. https://doi.org/10.1155/2020/6123178.

[42] Narewadikar, N. A., Rajpure, K. Y. Recent Advancements in Doped Titanium Dioxide (TiO2) Nanostructures for Photocatalytic Dye Degradation. *Nanotechnol Russia* 17, 39–58 (2022). https://doi.org/10.1134/S2635167622010104.

[43] Akhter P, Arshad A, Saleem A, Hussain M. Recent Development in Non-Metal-Doped Titanium Dioxide Photocatalysts for Different Dyes Degradation and the Study of Their Strategic Factors: A Review. *Catalysts*. 2022; 12(11):1331. https://doi.org/10.3390/catal12111331.

[44] Mashuri, S. I. S., Ibrahim, M. L., Kasim, M. F., Mastuli, M. S., Rashid, U., Abdullah, A. H., Islam, M. S., Asikin-Mijan, N., Tan, Y. H., Mansir, N., Kaus, N. H. M., & Hin, T.-Y. Y. (2020). Photocatalysis for Organic Wastewater Treatment: From the Basis to Current Challenges for Society. In *Catalysts* (Vol. 10, Issue 11, p. 1260). Multidisciplinary Digital Publishing Institute. https://doi.org/10.3390/catal10111260.

[45] Han, X., Sun, M., Chai, X., Li, J., Wu, Y., & Sun, W. (2022). Progress in Synthesis and Photocatalytic Activity of $MAl_2O_4$(M=Mg, Sr, Ba) Based Photocatalysts. In *Frontiers in Materials* (Vol. 9). Frontiers Media. https://doi.org/10.3389/fmats.2022.845664.

[46] Chen X, Mao SS. Titanium dioxide nanomaterials: Synthesis and applications. *Chem Rev*. 2007;107(7):2891-2959. https://doi.org/10.1021/cr0500535.

[47] Kumar, A., Kumar, L. (2024). Nanostructured catalysts for $CO_2$ reduction: systematic insights and emerging strategies. *Res Chem Intermed* 50, 195–217. https://doi.org/10.1007/s11164-023-05190-5.

[48] Pelaez M, Nolan N. T, Pillai S. C, Seery M. K, Falaras P, Kontos A. G, Dunlop P. S. M, Hamilton J. A review on the visible light active titanium dioxide photocatalysts for environmental applications. *Appl Catal B*. 2012; 125:331-349. https://doi.org/10.1016/j.apcatb.2012.05.036.

[49] Hoffmann, M. R., Martin, S. T., Choi, W., & Bahnemann, D. W. (1995). Environmental Applications of Semiconductor Photocatalysis. *Chemical Reviews*, 95, 69-96. https://doi.org/10.1021/CR00033A004.

[50] Chong, M. N., Jin, B., Chow, C. W., & Saint, C. (2010). Recent developments in photocatalytic water treatment technology: a review. *Water research,* 44(10), 2997–3027. https://doi.org/10.1016/j.watres.2010.02.039.

## Chapter 14

# Turning Waste into Wealth: Wastewater Resource Recovery Strategies (Energy, Nutrients, and Water)

**R. Dhinesh***
**and S. S. Kirthiga**
Department of Aquatic Environment Management, Faculty of Fisheries,
Kerala University of Fisheries and Ocan Studies, Kochi, Kerala, India

## Abstract

Wastewater, being an inevitable byproduct originating from diverse sources of human activities can be sustainably used as a renewable resource that incorporates the concept of resource recovery transforming waste into valuable resources. Nutrient recovery techniques, including phosphorus precipitation, nitrogen stripping, and advanced integrated systems, recycle essential nutrients like nitrogen and phosphorus, helping to mitigate the excessive use of synthetic fertilizers and address nutrient pollution. Water recovery strategies that focus on advanced filtration technologies such as reverse osmosis, ultrafiltration, and membrane distillation, ultimately produce high-quality water with diverse uses.

Given the importance of these strategies, this chapter provides a detailed overview of efficient methods for recovering energy, nutrients, and water from wastewater, highlighting their potential to enhance sustainability and support the circular economy. It explores energy recovery techniques such as anaerobic digestion for biogas production, microbial fuel cells (MFCs), and thermochemical processes, underscoring their role in generating renewable energy from organic matter. Furthermore, the emerging technologies like algae-based systems, electrochemical methods, and resource recovery from industrial wastewater and the integration of these technologies with existing wastewater treatment infrastructure, are thoroughly evaluated for their environmental and economic impacts. Moreover, the current policy frameworks, financial criteria, and the challenges faced during the practical application of these technologies are also examined.

**Keywords:** resource recovery, wastewater, energy, nutrients, eco-friendly, Wastewater treatment, Microbial fuel cells

---

* Corresponding Author's Email: dhinesh7143@gmail.com

In: Advancements in Wastewater Treatment
Editors: Neha Agarwal, Maulin P. Shahl, Vijendra Singh Solankil et al.
ISBN: 979-8-90134-007-3

## 1. Introduction

Wastewater, traditionally considered a disposal challenge, is now recognized as a valuable resource with immense potential for recovery and reuse, reflecting the global shift towards sustainability and resource optimization. With growing water demand and the depletion of finite natural resources, sustainable wastewater management strategies have become imperative for addressing environmental and economic challenges. Wastewater is not merely a byproduct; it is a reservoir of recoverable resources, including water, nutrients, and energy, which can be harnessed through innovative technologies and integrated frameworks [1]. Unlocking the potential of wastewater requires a paradigm shift that prioritizes recovery and reuse over disposal, supported by advancements in science, engineering, and policy-making. Energy recovery from wastewater has emerged as a transformative area, offering the dual benefits of reducing the environmental footprint of treatment facilities and contributing to global energy demands [2]. Organic matter within wastewater serves as a concentrated source of chemical energy that can be effectively harnessed through processes like anaerobic digestion, MFCs, and thermochemical conversion [3]. Anaerobic digestion is particularly effective, producing biogas, a renewable energy source composed of methane and carbon dioxide, used for power generation, heating, or transportation fuel [4]. Meanwhile, MFCs represent a cutting-edge technology, leveraging microbial activity to directly convert organic pollutants into electricity. Gasification and pyrolysis, thermochemical processes that break down organic matter under controlled conditions, further expand the spectrum of bioenergy products, including syngas and biochar [5]. These advancements foster energy self-sufficiency in wastewater treatment plants and contribute to global decarbonization efforts and sustainable development goals.

The nutrient content of wastewater, particularly nitrogen and phosphorus, represents another critical resource with immense potential for agricultural and industrial applications [6]. Left untreated, these nutrients contribute to water pollution and eutrophication, disrupting aquatic ecosystems [7]. However, innovative technologies, such as struvite crystallization, ammonia stripping, and advanced filtration, are revolutionizing nutrient recovery [8]. Struvite, a phosphate mineral recovered from wastewater, is gaining prominence as a slow-release, eco-friendly fertilizer, addressing the dual challenge of nutrient pollution and the depletion of phosphate rock reserves [9]. These technologies not only enable the recovery of essential nutrients but also align with circular economy principles by reintegrating recovered materials into agricultural and industrial cycles. Nutrient recovery also reduces reliance on energy-intensive synthetic fertilizers and mitigates environmental risks associated with conventional wastewater discharge [10]. Water reuse is arguably the most direct and impactful aspect of wastewater resource recovery, with the potential to alleviate water scarcity and improve resilience to climate change. Advanced treatment technologies, such as membrane bioreactors, reverse osmosis, and ultraviolet disinfection, enable wastewater to be purified to stringent quality standards [11]. This reclaimed water is increasingly being used across diverse applications, including agricultural irrigation, industrial cooling, groundwater recharge, and even direct potable reuse [12]. In arid and semi-arid regions, direct potable reuse has emerged as a critical strategy, where treated wastewater is introduced into the drinking water supply after advanced treatment. Indirect potable reuse, which involves blending treated wastewater with natural water sources, further enhances water security in regions facing chronic water shortages. The adoption of water reuse strategies not only reduces dependence on traditional freshwater sources but also supports integrated water resource management frameworks [13]. Emerging innovations are pushing the boundaries of wastewater resource recovery, driving the development of hybrid and multifunctional systems. For instance, integrating anaerobic digestion with struvite precipitation simultaneously recovers biogas and phosphorus, maximizing resource extraction [14]. Algal-based treatment systems utilize wastewater nutrients to cultivate algae, which can be harvested for biofuel production, bioplastics, and animal feed, establishing a nexus between wastewater management and the bioeconomy [15]. Additionally, bioelectrochemical systems and advanced oxidation processes are enabling decentralized and highly efficient recovery of valuable compounds, including rare earth elements and heavy metals [16]. These cutting-edge technologies demonstrate the potential of wastewater treatment plants to evolve into "resource recovery hubs," contributing to the sustainable management of energy, water, and nutrients while minimizing waste generation.

This chapter provides a comprehensive exploration of wastewater resource recovery strategies, emphasizing the recovery of energy, nutrients, and water. By delving into the technological innovations, economic opportunities, and environmental benefits associated with resource recovery, the chapter underscores the transformative potential of wastewater in driving sustainability. As global water, energy, and nutrient demands continue to rise, adopting a resource-centric approach to wastewater management offers a viable pathway to address the interconnected challenges of resource scarcity, environmental degradation, and climate change. Through scientific advancements and interdisciplinary collaboration, wastewater can be transformed from an environmental liability into a cornerstone of circular economies and sustainable development.

## 2. Wastewater and Resource Recovery

Wastewater, often seen as a pollutant or environmental burden, is increasingly recognized as a valuable resource capable of contributing to environmental sustainability, economic growth, and resource conservation. It contains a variety of components that can be recovered and repurposed for various beneficial uses, making it a significant player in resource recovery initiatives [17]. Water, energy, and nutrients, which are typically extracted from natural resources, are abundant in wastewater and can be harnessed through innovative treatment processes [18]. By viewing wastewater as a potential resource rather than waste, society can transform the way it manages water systems, promoting circular economies and reducing dependence on raw materials (Figure 1). This shift in perspective is crucial, particularly as global populations rise and freshwater supplies become more strained due to overuse, pollution, and climate change.

Water recovery from wastewater is one of the most direct and impactful aspects of resource recovery. Advanced treatment technologies, including membrane filtration, reverse osmosis, and ultraviolet disinfection, enable the purify wastewater to meet the standards required for a wide range of applications. These systems can produce treated wastewater suitable for agricultural irrigation, industrial processes, and, in some cases, potable water reuse [19]. This process not only alleviates pressure on freshwater resources but also enhances water security, particularly in water-scarce regions. Furthermore, water reuse minimizes the need for constructing new water treatment infrastructure and reduces the environmental impact of wastewater discharge into natural water bodies, thereby addressing both water scarcity and pollution concerns [20]. In addition to water, wastewater is a rich source of nutrients, including nitrogen, phosphorus, and trace elements, which are vital for agricultural productivity. These nutrients are often lost when wastewater is disposed of untreated or treated using conventional methods that do not prioritize recovery [21]. However, with the right technologies, wastewater can be a sustainable source of fertilizers and soil conditioners. Struvite crystallization, ammonia stripping, and membrane filtration systems allow for the extraction of these nutrients, which can then be converted into slow-release fertilizers that reduce the environmental impact of chemical fertilizers [22]. Nutrient recovery not only improves soil health but also addresses the global challenge of nutrient imbalances and the depletion of natural phosphorus reserves, while preventing the eutrophication of aquatic ecosystems caused by nutrient runoff from agriculture [23]. Energy recovery from wastewater represents another significant opportunity for resource recovery. Wastewater contains organic matter that can be converted into biogas, which is primarily composed of methane, through processes like anaerobic digestion. Biogas can be used to generate electricity, heat, or be upgraded to biomethane, a renewable energy source that can substitute for natural gas [24]. Technologies such as microbial fuel cells (MFCs) and microbial electrolysis cells (MECs) offer more innovative approaches, where the organic matter in wastewater is directly converted into electricity through microbial activity [25]. These technologies provide a dual benefit by generating energy while reducing the environmental footprint of wastewater treatment. Additionally, the recovery of energy from wastewater decreases the operational costs of treatment plants and can potentially make them energy-positive, contributing to sustainable energy solutions.

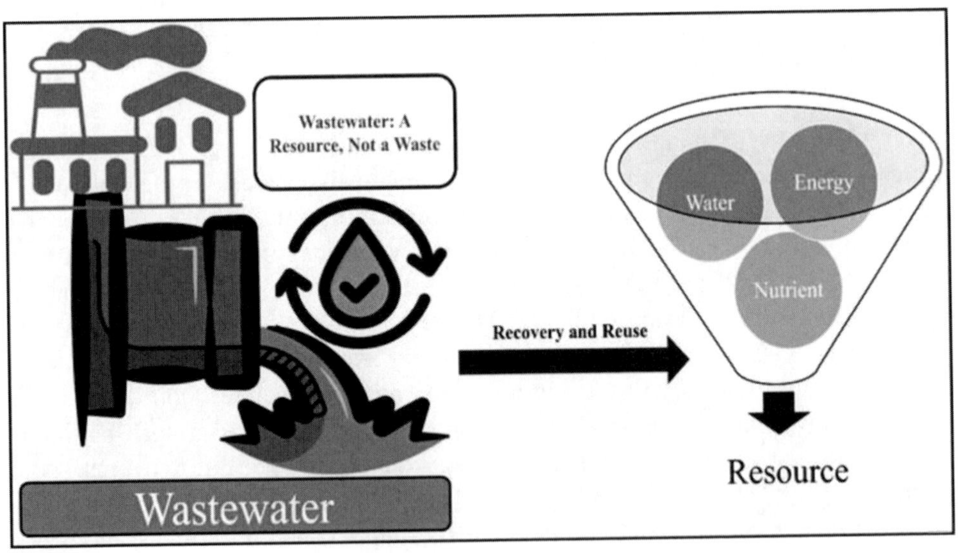


**Figure 1.** Wastewater as a resource and the potential of wastewater recovery.

The future potential of wastewater as a resource extends beyond traditional recovery strategies, with emerging innovations poised to expand its role in resource conservation. Hybrid systems that integrate energy, water, and nutrient recovery are gaining traction, offering integrated solutions that maximize the recovery of multiple resources simultaneously. Algal-based systems, for example, can utilize wastewater nutrients to cultivate biomass that can be converted into biofuels, while also improving water quality by removing excess nitrogen and phosphorus [26]. Furthermore, research into bioelectrochemical systems, advanced oxidation processes, and other cutting-edge technologies is paving the way for more efficient, decentralized, and cost-effective wastewater treatment solutions [27]. These developments highlight the vast potential of wastewater not just as a waste product, but as a renewable resource that can play a critical role in addressing global challenges related to water, energy, and food security.

## 3. Strategies for Energy Recovery from Wastewater

The transformation of wastewater into a renewable energy source represents a critical component of sustainable wastewater management, providing solutions for reducing energy dependency while mitigating environmental pollution. Wastewater, rich in organic matter, serves as a potential feedstock for various energy recovery technologies [28]. These processes harness the biochemical and thermochemical properties of wastewater to generate bioenergy, including biogas, electricity, and other valuable energy carriers.

Beyond offsetting the energy demands of treatment facilities, these technologies contribute to greenhouse gas reduction and resource circularity, making energy recovery a vital aspect of achieving sustainability in water treatment systems [29]. Strategies for energy recovery from wastewater are illustrated in Figure 2.

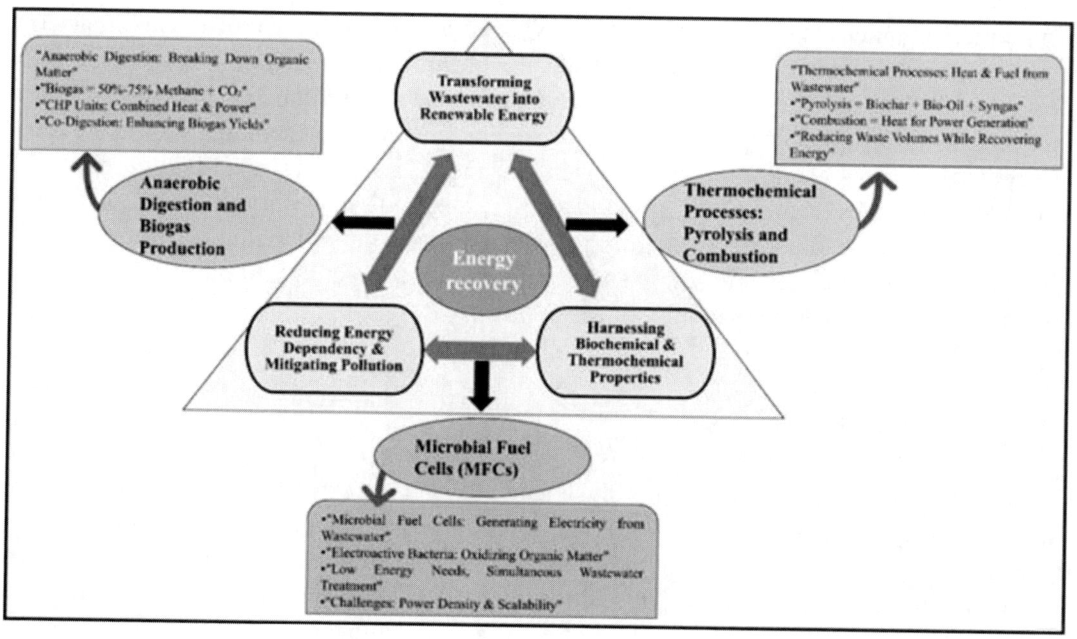


**Figure 2.** Strategies for Energy Recovery from Wastewater.

### 3.1. Anaerobic Digestion and Biogas Production

Anaerobic digestion is a widely implemented biological process that utilizes anaerobic microorganisms to degrade organic matter in wastewater under oxygen-free conditions, resulting in the production of biogas and digestate [30]. Biogas primarily consists of methane ($CH_4$) and carbon dioxide ($CO_2$), with trace amounts of hydrogen sulfide ($H_2S$) and other gases. The methane content, typically ranging from 50% to 75%, makes biogas a highly efficient renewable energy source [31]. This process is extensively employed in wastewater treatment plants (WWTPs) to convert sludge into energy, significantly reducing treatment costs and carbon footprint. Advanced configurations, such as co-digestion, enhance biogas yields by blending wastewater sludge with other organic wastes, such as food waste or agricultural residues, optimizing the substrate composition for microbial activity [32]. The integration of anaerobic digestion with energy recovery systems, such as combined heat and power (CHP) units, further maximizes energy efficiency by simultaneously generating electricity and thermal energy.

### 3.2. Microbial Fuel Cells (MFCs)

Microbial fuel cells (MFCs) are an innovative technology for energy recovery from wastewater, utilizing electroactive microorganisms to directly convert the chemical energy in organic matter into electricity [25]. MFCs consist of an anode and a cathode separated by a proton exchange membrane. Organic matter in wastewater is oxidized at the anode by electroactive bacteria, releasing electrons and protons. The electrons travel through an external circuit to generate electricity, while the protons diffuse through the membrane to the cathode, where they combine with oxygen to form water [33]. MFCs offer several advantages, including low operational energy requirements, concurrent wastewater treatment, and direct electricity production. However, challenges such as low power densities, electrode material costs, and scalability hinder their widespread adoption. Ongoing research focuses on optimizing electrode designs, improving electron transfer efficiency, and developing cost-effective materials to enhance the performance and commercial viability of MFCs [34].

### 3.3. Thermochemical Processes

Thermochemical processes such as pyrolysis and combustion offer alternative approaches to energy recovery from wastewater by converting organic matter into heat and fuel products. Pyrolysis involves the thermal decomposition of organic materials in the absence of oxygen, producing biochar, bio-oil, and syngas and these products have versatile applications like the biochar can be used as a soil amendment, bio-oil as a renewable fuel, and syngas as a precursor for energy generation. These products have versatile applications: biochar as a soil amendment, bio-oil as a renewable fuel, and syngas as a precursor for energy generation [35]. Combustion, in contrast, is a complete oxidation process where organic matter is burned in the presence of oxygen, releasing heat energy that can be harnessed for power generation [36]. Both processes are particularly effective for treating the solid fraction of wastewater sludge, reducing waste volumes and recovering energy simultaneously. Advanced technologies, such as fluidized bed reactors and gasification systems, enhance the efficiency and environmental performance of thermochemical processes, minimizing emissions and maximizing energy yields [37].

### 3.4. Emerging Innovations in Wastewater Energy Recovery

Emerging innovations in wastewater energy recovery are revolutionizing the efficiency and sustainability of wastewater treatment plants by integrating hybrid and advanced technologies. One such approach is algal biofuel production, where wastewater rich in nitrogen and phosphorus serves as a nutrient source for algae cultivation. The harvested algae can be processed into biofuels, bioplastics, and other high-value bioproducts, contributing to renewable energy generation while simultaneously aiding in wastewater treatment [26]. This dual-purpose system not only reduces nutrient loads in effluents but also provides a sustainable feedstock for biofuel production, making it a promising avenue for energy recovery. Another significant advancement is the development of bioelectrochemical systems (BES), which integrate microbial fuel cells (MFCs) and microbial electrolysis cells (MECs) to facilitate simultaneous energy generation and pollutant degradation. In these systems, electroactive bacteria break down organic matter in wastewater, releasing electrons that can be harnessed for electricity production in MFCs or hydrogen generation in MECs [27]. Although scalability has been a challenge for BES technologies, recent improvements in electrode materials, biofilm formation, and system design are enhancing their efficiency and viability for practical applications in wastewater treatment plants. Thermoelectric waste heat recovery presents another emerging strategy, focusing on harnessing the thermal energy present in treated effluent. Wastewater treatment processes generate substantial heat, which is often lost to the environment, if not recovered [13]. Advanced heat exchangers are now being employed in wastewater treatment plants to capture and repurpose this waste heat for district heating, industrial applications, or even electricity generation through thermoelectric conversion [2]. This approach not only enhances the overall energy efficiency of treatment plants but also contributes to reducing their reliance on external energy sources.

The integration of these emerging technologies into modern wastewater treatment plants is shifting the paradigm from traditional energy-intensive operations toward energy-neutral and even energy-positive facilities. By leveraging algal biofuels, bioelectrochemical systems, and thermoelectric waste heat recovery, wastewater treatment plants can become key players in the circular economy, reducing waste while maximizing resource recovery [3,15]. These innovations align with sustainable development goals (SDGs), promoting cleaner energy alternatives and mitigating environmental impacts associated with wastewater management. As research and technological advancements continue, the future of wastewater treatment is increasingly focused on energy sustainability and resource efficiency, transforming wastewater from a burden into a valuable energy asset.

## 4. Nutrient Recovery Strategies

Nutrient recovery strategies are critical for addressing nutrient pollution and reducing dependence on synthetic fertilizers. Techniques such as phosphorus recovery via struvite formation and nitrogen recovery through stripping and absorption have shown promise in recovering valuable nutrients from wastewater [22] (Figure 3). Integrated systems combining biological, chemical, and physical processes are increasingly used to achieve efficient nutrient recycling [38]. These strategies help mitigate nutrient loading into water bodies, reduce eutrophication, and promote circular economy principles by recycling vital elements like nitrogen and phosphorus.

### 4.1. Phosphorus Recovery

Phosphorus recovery from wastewater has gained significant attention due to its critical role as a finite nutrient essential for agriculture. Struvite (magnesium ammonium phosphate hexahydrate) precipitation is one of the most effective methods for phosphorus recovery [22]. In this process, phosphorus is precipitated from wastewater by adding magnesium salts under controlled pH conditions, forming crystalline struvite. The recovered struvite is a slow-release fertilizer with high agronomic value, making it a sustainable alternative to conventional phosphate fertilizers [39]. This technique has been successfully applied in municipal and industrial wastewater treatment plants, where sludge dewatering liquor and anaerobic digestate are treated to extract phosphorus. Studies indicate that struvite formation not only recovers valuable nutrients but also mitigates issues such as pipe scaling and equipment corrosion caused by excessive phosphorus in wastewater systems [40]. Advancements in reactor designs, such as fluidized bed reactors, have enhanced process efficiency and product quality, increasing their feasibility for large-scale applications.

### 4.2. Nitrogen Recovery

Nitrogen recovery is vital for addressing the environmental and economic challenges posed by excess nitrogen in wastewater. Ammonia stripping and absorption are among the most widely utilized techniques for nitrogen recovery [41]. In this process, wastewater is heated and aerated under alkaline conditions, causing ammonia to volatilize into the gas phase. The volatilized ammonia is then captured using an acid solution, typically sulfuric acid, forming ammonium sulfate, which is a valuable fertilizer. The efficiency of ammonia stripping depends on parameters such as pH, temperature, and air-to-liquid ratio, with higher pH and temperature favoring the removal process [42]. Recent advancements include integrating membrane-based ammonia stripping systems, which improve energy efficiency and selectivity while reducing operational costs. Additionally, ammonia recovery reduces the need for energy-intensive nitrification-denitrification processes in wastewater treatment, thus lowering greenhouse gas emissions and energy consumption [43].

### 4.3. Integrated Systems for Nutrient Recycling

Integrated systems combine physical, chemical, and biological processes to achieve comprehensive nutrient recovery and recycling from wastewater. These systems leverage the synergistic effects of multiple technologies, such as combining anaerobic digestion with struvite precipitation and ammonia stripping, to optimize resource recovery [44]. For instance, anaerobic digestion not only generates biogas but also produces nutrient-rich digestate, which can be treated further for phosphorus and nitrogen recovery. Algae-based systems represent another promising integrated approach, where microalgae assimilate nutrients from wastewater and are subsequently harvested to produce biofertilizers or bioenergy [45]. Integrated membrane bioreactors, incorporating ultrafiltration and advanced oxidation processes, offer high efficiency in nutrient recovery while simultaneously improving effluent quality [46]. These integrated systems align with the principles of the circular economy by maximizing resource recovery, reducing waste generation, and enhancing the economic viability of wastewater treatment operations.

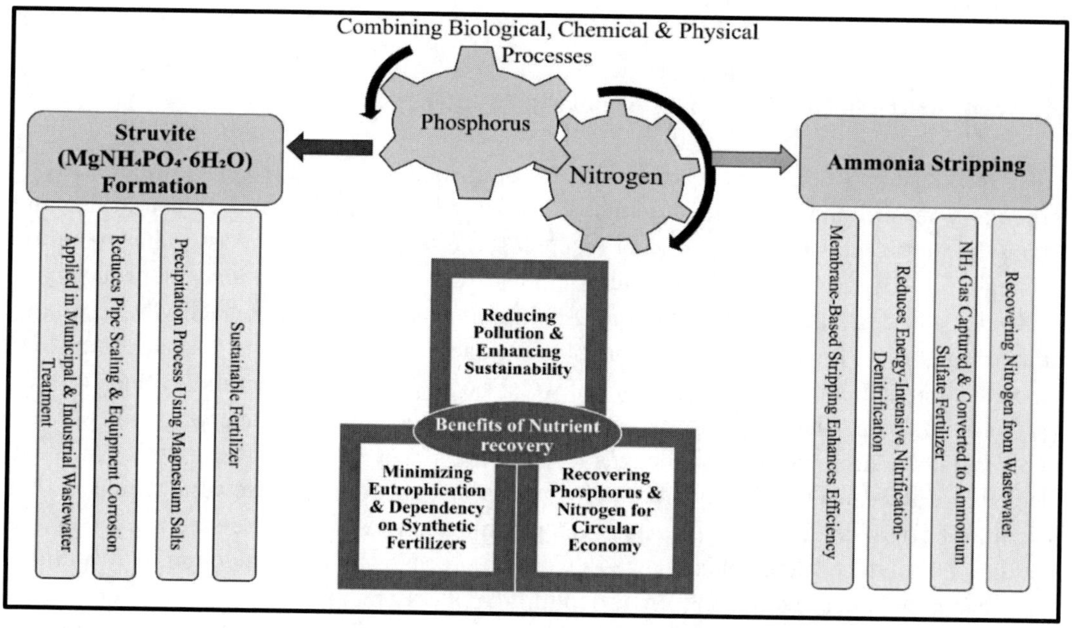


**Figure 3.** Strategies involved in the recovery of nutrients and the benefits involved.

### 4.4. Role in Reducing Synthetic Fertilizer Use and Nutrient Pollution

Nutrient recovery from wastewater plays a crucial role in mitigating the environmental and economic impacts of synthetic fertilizer use and nutrient pollution. Recovered nutrients such as struvite and ammonium sulfate can replace conventional fertilizers, thereby reducing dependency on energy-intensive fertilizer production processes and the depletion of finite phosphate rock reserves [22]. Additionally, nutrient recovery prevents the discharge of excessive nitrogen and phosphorus into aquatic ecosystems, which can lead to eutrophication a process characterized by algal blooms, oxygen depletion, and loss of biodiversity [47]. By addressing nutrient imbalances in wastewater, nutrient recovery technologies contribute to the restoration of ecological balance and the sustainability of agricultural systems. Policies promoting the use of recovered fertilizers and the development of nutrient recycling infrastructure are essential for realizing the full potential of these technologies [48].

## 5. Water Recovery and Reuse

Water recovery and reuse are essential strategies for addressing water scarcity, particularly in water-stressed regions. Advanced filtration technologies such as reverse osmosis, ultrafiltration, and membrane distillation are commonly used to recover high-quality water from wastewater [49]. These technologies enable the treatment and recycling of water for diverse applications, including irrigation, industrial processes, and even potable reuse. By promoting water recycling, these strategies alleviate pressure on freshwater resources and contribute to sustainable water management, while simultaneously reducing environmental pollution and supporting the circular economy [50].

### 5.1. Need for High-Quality Water Recovered Water

Water recovery and reuse are pivotal in addressing the global challenge of water scarcity, particularly in arid and semi-arid regions. With growing population and industrial demands placing immense pressure on freshwater resources, the reuse of treated wastewater presents a sustainable alternative to augment water supplies. Recycling water reduces dependency on freshwater sources, alleviating stress on overexploited aquifers and surface water systems. Furthermore, it mitigates the environmental impacts of wastewater discharge, such as eutrophication and contamination of natural water bodies [47]. By integrating water recovery systems with urban planning and industrial strategies, societies can ensure a resilient water supply while addressing the challenges of climate change and growing water demand. Recovered water must meet stringent quality standards to support diverse applications such as agricultural irrigation, industrial use, and even potable reuse. Advanced treatment technologies, including multi-stage filtration, advanced oxidation processes, and disinfection, play a critical role in ensuring water quality [49]. In agricultural applications, recovered water can be treated to remove pathogens and heavy metals while retaining essential nutrients, providing a sustainable water source for crop irrigation [10]. For industrial purposes, water recovery systems are tailored to remove specific contaminants, ensuring compatibility with industrial processes like cooling and boiler feed [16]. Advanced sensor technologies and real-time monitoring systems are increasingly employed to maintain water quality throughout the recovery process, addressing concerns related to health and environmental safety.

### 5.2. Advanced Filtration Technologies

Advanced filtration technologies, such as reverse osmosis (RO), ultrafiltration (UF), and membrane distillation (MD), play a crucial role in water recovery and reuse. RO, with its semipermeable membranes, effectively removes salts and contaminants from wastewater. UF, operating under lower pressure, is effective for removing suspended solids and larger molecules [51]. MD, a thermal-driven process, is particularly suitable for high-salinity wastewater treatment [52]. These technologies, through their ability to produce high-quality water, are essential for meeting the increasing demand for water in industrial, agricultural, and municipal applications.

### *5.2.1. Reverse Osmosis*

Reverse osmosis (RO) is a widely adopted technology for water recovery, particularly in wastewater treatment and desalination. This process operates by applying high pressure to force water through a semi-permeable membrane, which blocks contaminants such as salts, heavy metals, and organic compounds [49]. RO membranes have pore sizes as small as 0.0001 microns, enabling the removal of up to 99% of dissolved solutes and microorganisms [53]. One of its significant applications includes recovering potable water from industrial and municipal wastewater. However, challenges such as membrane fouling and high energy consumption limit its scalability, necessitating the integration of pre-treatment steps and energy-efficient designs [54]. Advances in materials science, such as the development of thin-film composite membranes, have improved the efficiency and durability of RO systems, making them more viable for large-scale applications.

### *5.2.2. Ultrafiltration*

Ultrafiltration (UF) is another effective filtration method used in wastewater recovery, relying on membranes with pore sizes ranging from 0.01 to 0.1 microns. It is primarily employed for the removal of suspended solids, bacteria, and larger organic molecules from water [51]. UF operates under lower pressure than RO, making it more energy-efficient and cost-effective for pre-treatment in water recovery systems. Its applications include treating industrial effluents, tertiary wastewater polishing, and pretreatment for RO systems. Despite its advantages, UF membranes are prone to fouling, necessitating regular cleaning and maintenance [55]. Recent innovations, such as hydrophilic membrane coatings and self-cleaning systems, have enhanced the operational efficiency of UF, ensuring consistent performance in diverse applications [56].

### *5.2.3. Membrane Distillation*

Membrane distillation (MD) is an emerging thermal-driven process for water recovery, leveraging hydrophobic membranes that allow water vapor to pass while rejecting liquid-phase contaminants [52]. This technology is particularly suitable for treating high-salinity wastewater and recovering clean water from industrial processes. MD operates at lower temperatures compared to conventional distillation, making it energy-efficient when integrated with renewable heat sources such as solar energy or waste heat [57]. Its ability to achieve high rejection rates for salts, heavy metals, and organic pollutants has positioned MD as a promising alternative for water recovery in challenging scenarios. Ongoing research focuses on developing advanced membrane materials and hybrid systems to enhance MD's efficiency and scalability for industrial applications.

## 6. Emerging Technologies in Wastewater Resource Recovery

Algae-based systems have emerged as a promising approach for simultaneous energy and nutrient recovery from wastewater. Algae, particularly microalgae, can assimilate nitrogen and phosphorus from wastewater, making them an effective tool for nutrient recovery [45]. These systems not only mitigate nutrient pollution but also produce valuable biomass that can be converted into biofuels through processes like anaerobic digestion or transesterification. Algae-based systems, often coupled with photobioreactors, capitalize on photosynthesis to enhance pollutant removal and biomass production [15]. The integration of algal systems into wastewater treatment plants for resource recovery presents an opportunity to create a more sustainable and circular approach to water and nutrient management. Electrochemical methods, including electrocoagulation, electrooxidation, and electrochemical cells, are gaining traction in wastewater treatment for resource recovery. These methods exploit electrical energy to drive chemical reactions that aid in pollutant removal and the recovery of valuable resources [58]. For instance, electrocoagulation can remove heavy metals and organic contaminants, while electrooxidation can degrade refractory organic pollutants into simpler, less harmful compounds [59]. Electrochemical cells, such as microbial fuel cells (MFCs), not only degrade organic matter but also generate electricity, thus contributing to energy recovery [25]. These technologies provide promising solutions for integrating resource recovery with wastewater treatment, offering both environmental and economic benefits.

Industrial wastewater treatment and recovery have seen significant technological advancements, driven by the need to comply with stringent environmental regulations and reduce operational costs. Advanced treatment processes, such as membrane filtration, advanced oxidation, and zero-liquid discharge technologies, are now employed to treat and recover resources from industrial effluents [60]. Membrane bioreactors (MBRs), for example, combine biological treatment with membrane filtration to produce high-quality effluent while recovering water for reuse [61]. Innovations in resource recovery from industrial wastewater not only improve sustainability by reusing water but also recover valuable byproducts, such as metals, chemicals, and nutrients, thus supporting the circular economy. These developments highlight the growing importance of integrating advanced technologies to minimize waste and maximize resource efficiency in industrial settings. The integration of emerging technologies into existing wastewater treatment infrastructure is a critical challenge for maximizing the potential of resource recovery. Many emerging technologies, such as algae-based systems and electrochemical methods, require significant infrastructure adaptations and operational adjustments. However, when integrated properly, these technologies can enhance the resource recovery capabilities of traditional wastewater treatment plants, for example, by improving energy efficiency or recovering nutrients more effectively [62]. The integration process involves designing hybrid systems that leverage both conventional and emerging technologies while ensuring economic feasibility and long-term sustainability.

Nanotechnology has the potential to revolutionize wastewater resource recovery by offering enhanced capabilities for contaminant removal, resource recovery, and process efficiency. Nanomaterials, such as carbon-based nanoparticles, metal oxides, and magnetic nanoparticles, exhibit unique properties like high surface area and reactivity, which enable them to efficiently remove contaminants including heavy metals, organic pollutants, and pathogens [63]. In nutrient recovery, nanoparticles can be used to selectively adsorb nitrogen and phosphorus from wastewater, enabling their recycling. Moreover, nanotechnology plays a significant role in improving membrane performance in filtration systems, enhancing the flux and fouling resistance of membranes used in water treatment [64]. The development of nanomaterials for resource recovery represents a promising frontier in wastewater treatment, potentially enhancing both efficiency and sustainability. Artificial intelligence (AI) and automation are increasingly being utilized to optimize wastewater resource recovery processes, enhancing operational efficiency and decision-making. AI algorithms can analyze large datasets from monitoring systems to predict treatment performance, optimize energy consumption, and identify opportunities for resource recovery [65]. For example, AI-based models can forecast the performance of resource recovery technologies like microbial fuel cells or membrane bioreactors under varying operational conditions, leading to better resource management and energy utilization [66]. Additionally, automation improves process control, reducing the need for manual intervention and enabling real-time adjustments to maintain optimal performance. By integrating AI and automation into wastewater treatment plants, operators can achieve higher recovery rates, reduce operational costs, and contribute to the circular economy in a more efficient and sustainable manner [65].

## 7. Environmental and Economic Impacts of Resource Recovery

The primary environmental and economic impacts of wastewater resource recovery are illustrated in Figure 4. The environmental benefits of wastewater resource recovery are vast and multifaceted. By recovering energy, nutrients, and water from wastewater, treatment plants can significantly reduce their reliance on external energy sources, lower greenhouse gas emissions, and mitigate pollution [67]. Energy recovery, for example, through anaerobic digestion and microbial fuel cells, not only provides renewable energy but also reduces the carbon footprint of wastewater treatment processes. This shift from conventional energy-intensive methods to energy-positive systems reduces the overall environmental impact of wastewater treatment. Additionally, nutrient recovery prevents the over-application of chemical fertilizers in agriculture, thus curbing nutrient pollution in water bodies, which is a major cause of eutrophication and aquatic ecosystem degradation [68]. Furthermore, the reuse of treated wastewater decreases pressure on freshwater ecosystems, conserving vital water resources, particularly in water-scarce regions, and promoting the sustainable use of water.

On the economic front, wastewater resource recovery presents significant opportunities for cost savings, revenue generation, and economic development. By recovering energy from wastewater, treatment facilities can lower their operational costs, offsetting the need for external energy inputs. The sale of biogas or electricity generated from wastewater treatment can further contribute to the economic viability of treatment plants, transforming them from cost centers into revenue-generating assets. Nutrient recovery, particularly the production of struvite as a slow-release fertilizer, offers farmers a cost-effective alternative to synthetic fertilizers, which are often expensive and energy-intensive to produce [69]. The recovery of valuable minerals and compounds, including rare earth elements, provides economic incentives for industries reliant on these materials. Additionally, implementing advanced water reuse systems can significantly reduce freshwater procurement costs in water-scarce regions, lowering expenses for urban and industrial water supplies. The broader economic impact of wastewater resource recovery extends beyond the treatment facility itself, contributing to the development of green industries, job creation, and the growth of the circular economy. The development and deployment of resource recovery technologies create new markets for innovation, offering opportunities for entrepreneurship and investment in the environmental sector. Industries involved in the manufacture of recovery technologies, from biogas production systems to advanced filtration membranes, are poised for growth as global demand for sustainable solutions increases [70]. Additionally, resource recovery initiatives align with circular economy principles by closing the loop between waste generation and resource utilization, promoting long-term sustainability. This creates a more resilient and sustainable economic model that reduces dependency on finite resources and fosters a transition to a low-carbon, resource-efficient society. Thus, wastewater resource recovery offers a comprehensive pathway to both environmental sustainability and economic growth, benefiting a wide range of stakeholders from municipalities to industries and farmers [71].

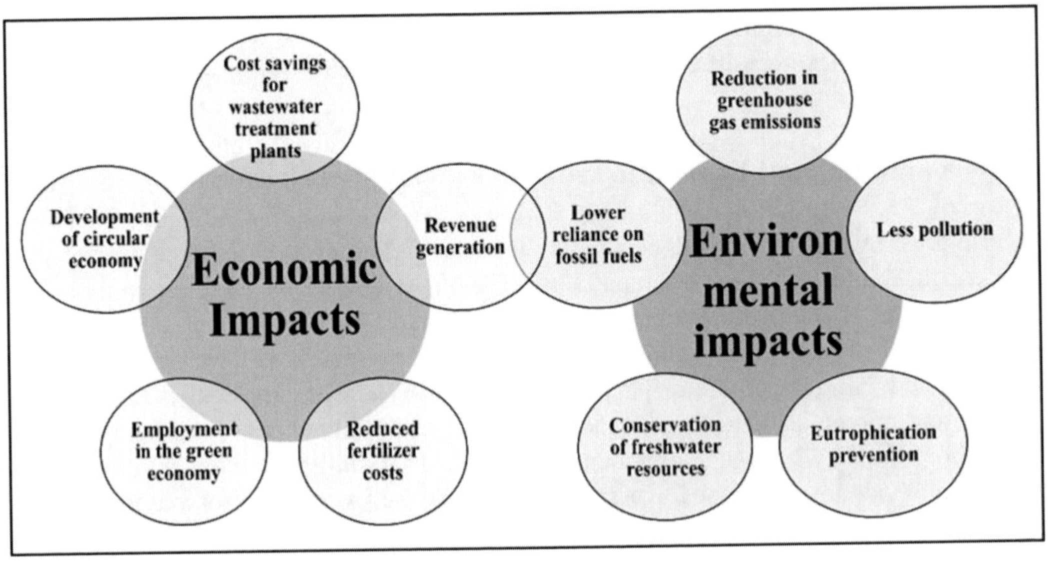


**Figure 4.** Major Environmental and Economic Impacts of Resource Recovery.

## 8. Future Trends in Wastewater Resource Recovery

The future of wastewater resource recovery is marked by continued innovation and the integration of advanced technologies that promise to enhance the efficiency and scope of resource extraction. One of the most promising developments is the ongoing progress in bio electrochemical systems, such as microbial fuel cells (MFCs) and microbial electrolysis cells (MECs), which allow for the direct conversion of organic pollutants into electricity or hydrogen [72]. These technologies have the potential to transform wastewater treatment facilities into self-sustaining, energy-positive systems that not only recover energy but also contribute to the generation of renewable energy. As these systems evolve, they are expected to become more efficient, affordable, and scalable, enabling widespread adoption across both developed and developing regions. Hybrid systems that combine multiple resource recovery processes, such as energy generation, nutrient recovery, and water purification, will become more prevalent, offering integrated solutions that maximize resource extraction while minimizing waste generation [38].

Another critical area of future innovation lies in the recovery of nutrients from wastewater, particularly nitrogen, phosphorus, and trace elements. Struvite crystallization, ammonia stripping, and advanced filtration technologies are rapidly advancing, making nutrient recovery more economically viable and technically feasible [73]. These recovered nutrients can be used as eco-friendly fertilizers, offering an alternative to synthetic fertilizers, which are energy-intensive and contribute to environmental degradation. As global demand for fertilizers grows and phosphate rock reserves continue to dwindle, nutrient recovery will become increasingly essential to ensure food security and reduce the environmental impact of conventional farming practices. Additionally, the recovery of rare earth elements and other high-value minerals from wastewater streams is emerging as a viable solution to meet the growing demand for critical materials in electronics, renewable energy technologies, and other industries [74]. Research into selective recovery methods and the use of advanced filtration and extraction technologies will play a key role in addressing this need. Water reuse will continue to be a central focus in the future of wastewater resource recovery, especially in regions facing water scarcity due to population growth, climate change, and over-extraction of freshwater resources. As water scarcity intensifies, advanced treatment technologies such as reverse osmosis, membrane bioreactors, and ultraviolet disinfection will evolve to make wastewater reuse more cost-effective and reliable [75]. In particular, the development of decentralized treatment systems will offer more flexibility and resilience in water reuse strategies, enabling communities to independently manage and purify wastewater on-site. This shift toward decentralized systems will be accompanied by a greater emphasis on improving public acceptance of water reuse, including direct potable reuse, through the use of advanced monitoring and quality control systems. Innovations in sensor technologies, real-time data analytics, and smart water management systems will ensure the safety and quality of reused water, thereby increasing its acceptance for a wide range of applications, from agricultural irrigation to potable water. Looking further ahead, the integration of wastewater resource recovery into circular economy frameworks will shape the future of wastewater management. Wastewater treatment plants will evolve from mere waste disposal facilities into fully integrated resource recovery hubs, where waste is converted into valuable products such as biogas, fertilizer, and clean water. This transition will be supported by the growing trend of circular economy principles, where resources are reused, recycled, and reintegrated into the economy rather than discarded.

## Conclusion

The future of wastewater management lies in the paradigm shift from viewing wastewater as a burden to recognizing it as a valuable resource. As the global demand for water, energy, and nutrients continues to rise, wastewater resource recovery presents a promising solution to meet these demands while addressing environmental concerns. Through the development and integration of advanced technologies, such as microbial fuel cells, nutrient recovery systems, and water reuse strategies, wastewater treatment facilities are evolving into resource recovery hubs that contribute significantly to sustainability. This transformation is key to supporting the transition to a circular economy, where waste is minimized, and resources are continuously cycled back into productive use. The ongoing innovation in wastewater treatment processes, along with the growing recognition of the importance of resource recovery, will shape a more sustainable future for both environmental management and economic growth. The successful implementation of wastewater resource recovery strategies requires a collaborative effort from policymakers, industries, researchers, and communities. As the technologies continue to mature and become more cost-effective, their widespread adoption will be crucial in achieving sustainability goals, particularly in water-scarce and resource-depleted regions. By unlocking the potential of wastewater as a resource for energy, nutrients, and water, society can create a more resilient infrastructure that reduces dependency on finite resources, enhances environmental protection, and supports economic development. Ultimately, wastewater resource recovery is not only a vital component of modern wastewater management but also a strategic approach to securing a sustainable and resource-efficient future for generations to come.

## References

[1] Sharma, P., Singh, S., Ramamurthy, P. C., Singh, J. and Biswas, J. K., 2024. Regenerative resource recovery from wastewater: State-of-the-art bio-based soft technology. *Current Opinion in Environmental Science & Health*, p.100587.

[2] Achi, C. G., Snyman, J., Ndambuki, J. M. and Kupolati, W. K., 2024. Advanced Waste-to-Energy Technologies: A Review on Pathway to Sustainable Energy Recovery in a Circular Economy. *Nature Environment & Pollution Technology*, *23*(3).

[3] Bhowmik, D., Chetri, S., Enerijiofi, K. E., Naha, A., Kanungo, T. D., Shah, M. P. and Nath, S., 2023. Multitudinous approaches, challenges and opportunities of bioelectrochemical systems in conversion of waste to energy from wastewater treatment plants. *Cleaner and Circular Bioeconomy*, *4*, p. 100040.

[4] Obileke, K., Nwokolo, N., Makaka, G., Mukumba, P. and Onyeaka, H., 2021. Anaerobic digestion: Technology for biogas production as a source of renewable energy—A review. *Energy & Environment*, *32*(2), pp. 191-225.

[5] Shen, Y., Wang, J., Ge, X. and Chen, M., 2016. By-products recycling for syngas cleanup in biomass pyrolysis–An overview. *Renewable and Sustainable Energy Reviews*, *59*, pp.1246-1268.

[6] Rosemarin, A., Macura, B., Carolus, J., Barquet, K., Ek, F., Järnberg, L., Lorick, D., Johannesdottir, S., Pedersen, S. M., Koskiaho, J. and Haddaway, N. R., 2020. Circular nutrient solutions for agriculture and wastewater–a review of technologies and practices. *Current Opinion in Environmental Sustainability*, *45*, pp. 78-91.

[7] Ngatia, L., Grace III, J. M., Moriasi, D. and Taylor, R., 2019. Nitrogen and phosphorus eutrophication in marine ecosystems. *Monitoring of marine pollution*, *1*, pp. 1-17.

[8] Sheik, A. G., Krishna, S. B. N., Patnaik, R., Ambati, S. R., Bux, F. and Kumari, S., 2024. Digitalization of Phosphorous Removal Process in Biological Wastewater Treatment Systems: Challenges, and Way Forward. *Environmental Research*, p. 119133.

[9] Chandrasekaran, S., Zaffar, A. and Balasubramanian, P., 2024. Struvite in circular economy: Production techniques, emerging applications and market opportunities. *Wiley Interdisciplinary Reviews: Energy and Environment*, *13*(4), p. e529.

[10] Ahmed, M., Ahmad, S., Qadir, G., Hayat, R., Shaheen, F. A. and Raza, M. A., 2019. Innovative processes and technologies for nutrient recovery from wastes: a comprehensive review. *Sustainability*, *11*(18), p. 4938.

[11] Rahman, T. U., Roy, H., Islam, M. R., Tahmid, M., Fariha, A., Mazumder, A., Tasnim, N., Pervez, M. N., Cai, Y., Naddeo, V. and Islam, M. S., 2023. The advancement in membrane bioreactor (MBR) technology toward sustainable industrial wastewater management. *Membranes*, *13*(2), p. 181.

[12] Fito, J. and Van Hulle, S. W., 2021. Wastewater reclamation and reuse potentials in agriculture: towards environmental sustainability. *Environment, Development and Sustainability*, *23*(3), pp. 2949-2972.

[13] Siddique, I., 2021. Sustainable Water Management in Urban Areas: Integrating Innovative Technologies and Practices to Address Water Scarcity and Pollution. *The Pharmaceutical and Chemical Journal*, *8*(1), pp. 172-178.

[14] Lin, H., Gan, J., Rajendran, A., Reis, C. E. R. and Hu, B., 2015. Phosphorus removal and recovery from digestate after biogas production. In *Biofuels-status and perspective*. IntechOpen.

[15] Leong, Y.K., Huang, C.Y. and Chang, J.S., 2021. Pollution prevention and waste phycoremediation by algal-based wastewater treatment technologies: The applications of high-rate algal ponds (HRAPs) and algal turf scrubber (ATS). *Journal of Environmental Management*, *296*, p. 113193.

[16] Devda, V., Chaudhary, K., Varjani, S., Pathak, B., Patel, A. K., Singhania, R. R., Taherzadeh, M. J., Ngo, H. H., Wong, J. W., Guo, W. and Chaturvedi, P., 2021. Recovery of resources from industrial wastewater employing electrochemical technologies: status, advancements and perspectives. *Bioengineered*, *12*(1), pp. 4697-4718.

[17] Silva, J. A., 2023. Wastewater treatment and reuse for sustainable water resources management: a systematic literature review. *Sustainability*, *15*(14), p. 10940.

[18] Rani, A., Snyder, S. W., Kim, H., Lei, Z. and Pan, S. Y., 2022. Pathways to a net-zero-carbon water sector through energy-extracting wastewater technologies. *NPJ Clean Water*, *5*(1), p. 49.

[19] Manikandan, S., Subbaiya, R., Saravanan, M., Ponraj, M., Selvam, M. and Pugazhendhi, A., 2022. A critical review of advanced nanotechnology and hybrid membrane based water recycling, reuse, and wastewater treatment processes. *Chemosphere*, *289*, p. 132867.

[20] Miller, G. W., 2006. Integrated concepts in water reuse: managing global water needs. *Desalination*, *187*(1-3), pp. 65-75.

[21] Soares, A., Czajkowska, J., Colprim, J., Gali, A., Johansson, S., Masic, A., March, A., McMeod, A., Nenov, V., Ruscalleda, M. and Siwiec, T., 2017. *Nutrients recovery from wastewater streams* (Vol. 2017, pp. 369-398). IWA Publishing, London.

[22] Krishnamoorthy, N., Dey, B., Unpaprom, Y., Ramaraj, R., Maniam, G. P., Govindan, N., Jayaraman, S., Arunachalam, T. and Paramasivan, B., 2021. Engineering principles and process designs for phosphorus recovery as struvite: A comprehensive review. *Journal of Environmental Chemical Engineering*, *9*(5), p.105579.

[23] Chowdhury, R. B., Moore, G. A., Weatherley, A. J. and Arora, M., 2017. Key sustainability challenges for the global phosphorus resource, their implications for global food security, and options for mitigation. *Journal of Cleaner Production*, *140*, pp.945-963.

[24] Kabeyi, M. J. B. and Olanrewaju, O. A., 2022. Biogas production and applications in the sustainable energy transition. *Journal of Energy*, *2022*(1), p. 8750221.

[25] Katuri, K. P., Ali, M. and Saikaly, P. E., 2019. The role of microbial electrolysis cell in urban wastewater treatment: integration options, challenges, and prospects. *Current opinion in biotechnology*, *57*, pp. 101-110.

[26] Nguyen, L. N., Aditya, L., Vu, H. P., Johir, A. H., Bennar, L., Ralph, P., Hoang, N. B., Zdarta, J. and Nghiem, L. D., 2022. Nutrient removal by algae-based wastewater treatment. *Current Pollution Reports*, *8*(4), pp. 369-383.

[27] Shamshad, J. and Rehman, R. U., 2025. Innovative approaches to sustainable wastewater treatment: a comprehensive exploration of conventional and emerging technologies. *Environmental Science: Advances*.

[28] Puyol, D., Batstone, D. J., Hülsen, T., Astals, S., Peces, M. and Krömer, J. O., 2017. Resource recovery from wastewater by biological technologies: opportunities, challenges, and prospects. *Frontiers in microbiology*, *7*, p. 2106.

[29] Achi, C. G., Snyman, J., Ndambuki, J. M. and Kupolati, W. K., 2024. Advanced Waste-to-Energy Technologies: A Review on Pathway to Sustainable Energy Recovery in a Circular Economy. *Nature Environment & Pollution Technology*, *23*(3).

[30] Ezeh, E. M., Agu, P. C. and Aworabhi, E., 2024. Biological Treatment Techniques for Sewage: Aerobic and Anaerobic Processes.

[31] Surendra, K. C., Takara, D., Hashimoto, A. G. and Khanal, S. K., 2014. Biogas as a sustainable energy source for developing countries: Opportunities and challenges. *Renewable and Sustainable Energy Reviews*, *31*, pp. 846-859.

[32] González, R., Peña, D.C. and Gómez, X., 2022. Anaerobic co-digestion of wastes: reviewing current status and approaches for enhancing biogas production. *Applied Sciences*, *12*(17), p. 8884.

[33] Liu, H. and Logan, B.E., 2004. Electricity generation using an air-cathode single chamber microbial fuel cell in the presence and absence of a proton exchange membrane. *Environmental science & technology*, *38*(14), pp.4040-4046.

[34] Choudhury, P., Prasad Uday, U.S., Bandyopadhyay, T. K., Ray, R. N. and Bhunia, B., 2017. Performance improvement of microbial fuel cell (MFC) using suitable electrode and Bioengineered organisms: A review. *Bioengineered*, *8*(5), pp. 471-487.

[35] Dhinesh, R., Aruna, S., Manikandavelu, D., Ravaneswaran, K., Kirthiga, S.S., Al-Ghanim, K.A. and Kamaraj, C., 2024. Naturally derived organic biochar as an alternative to commercially activated carbon in the oxygen removal of seafood processing wastewater. *Journal of Chemical Technology & Biotechnology*, *99*(7), pp. 1691-1702.

[36] Demirbas, A., 2005. Potential applications of renewable energy sources, biomass combustion problems in boiler power systems and combustion related environmental issues. *Progress in energy and combustion science*, *31*(2), pp. 171-192.

[37] Chanthakett, A., Arif, M. T., Khan, M. M. K. and Oo, A. M., 2021. Performance assessment of gasification reactors for sustainable management of municipal solid waste. *Journal of Environmental Management*, *291*, p. 112661.

[38] Mo, W. and Zhang, Q., 2013. Energy–nutrients–water nexus: Integrated resource recovery in municipal wastewater treatment plants. *Journal of environmental management*, *127*, pp.255-267.

[39] Talboys, P. J., Heppell, J., Roose, T., Healey, J. R., Jones, D. L. and Withers, P. J., 2016. Struvite: a slow-release fertiliser for sustainable phosphorus management?. *Plant and soil*, *401*, pp. 109-123.

[40] Kataki, S., West, H., Clarke, M. and Baruah, D. C., 2016. Phosphorus recovery as struvite from farm, municipal and industrial waste: Feedstock suitability, methods and pre-treatments. *Waste Management*, *49*, pp. 437-454.

[41] Chen, T. L., Chen, L. H., Lin, Y. J., Yu, C. P., Ma, H. W. and Chiang, P. C., 2021. Advanced ammonia nitrogen removal and recovery technology using electrokinetic and stripping process towards a sustainable nitrogen cycle: A review. *Journal of Cleaner Production*, *309*, p. 127369.

[42] Tao, W. and Ukwuani, A. T., 2015. Coupling thermal stripping and acid absorption for ammonia recovery from dairy manure: Ammonia volatilization kinetics and effects of temperature, pH and dissolved solids content. *Chemical Engineering Journal*, *280*, pp.188-196.

[43] Farghali, M., Chen, Z., Osman, A. I., Ali, I. M., Hassan, D., Ihara, I., Rooney, D. W. and Yap, P. S., 2024. Strategies for ammonia recovery from wastewater: a review. *Environmental Chemistry Letters*, pp. 1-53.

[44] Zhou, L., Liang, M., Zhang, D., Niu, X., Li, K., Lin, Z., Luo, X. and Huang, Y., 2024. Recent advances in swine wastewater treatment technologies for resource recovery: A comprehensive review. *Science of The Total Environment*, p. 171557.

[45] Ferreira, A., Reis, A., Vidovic, S., Vladic, J., Gkelis, S., Melkonyan, L., Avetisova, G., Congestri, R., Acién, G., Muñoz, R. and Collet, P., 2019. Combining microalgae-based wastewater treatment with biofuel and bio-based production in the frame of a biorefinery. *Grand challenges in algae biotechnology*, pp. 319-369.

[46] Rostam, A. B. and Taghizadeh, M., 2020. Advanced oxidation processes integrated by membrane reactors and bioreactors for various wastewater treatments: A critical review. *Journal of Environmental Chemical Engineering*, *8*(6), p.104566.

[47] Rathore, S. S., Chandravanshi, P., Chandravanshi, A. and Jaiswal, K., 2016. Eutrophication: Impacts of excess nutrient inputs on aquatic ecosystem. *IOSR Journal of Agriculture and Veterinary Science*, *9*(10), pp. 89-96.

[48] Shaddel, S., Bakhtiary-Davijany, H., Kabbe, C., Dadgar, F. and Østerhus, S. W., 2019. Sustainable sewage sludge management: From current practices to emerging nutrient recovery technologies. *Sustainability*, *11*(12), p. 3435.

[49] Obotey Ezugbe, E. and Rathilal, S., 2020. Membrane technologies in wastewater treatment: a review. *Membranes*, *10*(5), p. 89.

[50] Kakwani, N. S. and Kalbar, P. P., 2020. Review of Circular Economy in urban water sector: Challenges and opportunities in India. *Journal of Environmental Management*, *271*, p. 111010.

[51] Wu, T. Y., Mohammad, A. W., Jahim, J. M. and Anuar, N., 2007. Palm oil mill effluent (POME) treatment and bioresources recovery using ultrafiltration membrane: effect of pressure on membrane fouling. *Biochemical engineering journal*, *35*(3), pp. 309-317.

[52] Du, S., Zhao, P., Wang, L., He, G. and Jiang, X., 2023. Progresses of advanced anti-fouling membrane and membrane processes for high salinity wastewater treatment. *Results in Engineering*, *17*, p.1 00995.

[53] Ergozhin, E. E., Chalov, T. K., Begenova, B. E. and Khakimbolatova, K. K., 2019. Semi-permeable membranes for ultra-, microfiltration and reverse osmosis. *Chemical Journal of Kazakhstan*.

[54] Osman, A. I., Chen, Z., Elgarahy, A. M., Farghali, M., Mohamed, I. M., Priya, A. K., Hawash, H. B. and Yap, P. S., 2024. Membrane technology for energy saving: principles, techniques, applications, challenges, and prospects. *Advanced Energy and Sustainability Research*, *5*(5), p. 2400011.

[55] Gkotsis, P. K., Banti, D. C., Peleka, E. N., Zouboulis, A. I. and Samaras, P.E., 2014. Fouling issues in membrane bioreactors (MBRs) for wastewater treatment: major mechanisms, prevention and control strategies. *Processes*, *2*(4), pp. 795-866.

[56] Nguyen, H. T., Pham, M. T., Nguyen, T. M. T., Bui, H. M., Wang, Y. F. and You, S. J., 2022. Modifications of conventional organic membranes with photocatalysts for antifouling and self-cleaning properties applied in wastewater filtration and separation processes: A review. *Separation Science and Technology*, *57*(9), pp.1471-1500.

[57] Shalaby, S. M., Zayed, M. E., Hammad, F. A., Menesy, A. S. and Abd Elbar, A. R., 2024. Recent advances in membrane distillation hybrids for energy-efficient process configurations: Technology categorization, operational parameters identification, and energy recovery strategies. *Process Safety and Environmental Protection*.

[58] Shah, A. A., Walia, S. and Kazemian, H., 2024. Advancements in combined electrocoagulation processes for sustainable wastewater treatment: A comprehensive review of mechanisms, performance, and emerging applications. *Water research*, p. 121248.

[59] Othmani, A., Kadier, A., Singh, R., Igwegbe, C. A., Bouzid, M., Aquatar, M. O., Khanday, W. A., Bote, M. E., Damiri, F., Gökkuş, Ö. and Sher, F., 2022. A comprehensive review on green perspectives of electrocoagulation integrated with advanced processes for effective pollutants removal from water environment. *Environmental Research*, *215*, p.114294.

[60] Moreira, V. R., Lebron, Y. A. R., Couto, C. F., Maia, A., Moravia, W. G. and Amaral, M. C. S., 2022. Process development for textile wastewater treatment towards zero liquid discharge: Integrating membrane separation process and advanced oxidation techniques. *Process Safety and Environmental Protection*, *157*, pp.537-546.

[61] Kharraz, J. A., Khanzada, N. K., Farid, M. U., Kim, J., Jeong, S. and An, A. K., 2022. Membrane distillation bioreactor (MDBR) for wastewater treatment, water reuse, and resource recovery: A review. *Journal of Water Process Engineering*, *47*, p. 102687.

[62] Diaz-Elsayed, N., Rezaei, N., Guo, T., Mohebbi, S. and Zhang, Q., 2019. Wastewater-based resource recovery technologies across scale: a review. *Resources, Conservation and Recycling*, *145*, pp.94-112.

[63] Rezania, S., Darajeh, N., Rupani, P.F., Mojiri, A., Kamyab, H. and Taghavijeloudar, M., 2024. Recent advances in the adsorption of different pollutants from wastewater using carbon-based and metal-oxide nanoparticles. *Applied Sciences*, *14*(24), p.11492.

[64] Shukla, B. K., Sharma, P. K., Yadav, H., Singh, S., Tyagi, K., Yadav, Y., Rajpoot, N. K., Rawat, S. and Verma, S., 2024. Advanced membrane technologies for water treatment: utilization of nanomaterials and nanoparticles in membranes fabrication. *Journal of Nanoparticle Research*, *26*(9), p. 222.

[65] Mohanty, A., Mohanty, S. K. and Mohapatra, A. G., 2024. Real-time monitoring and fault detection in ai-enhanced wastewater treatment systems. In *The AI Cleanse: Transforming Wastewater Treatment Through Artificial Intelligence: Harnessing Data-Driven Solutions* (pp. 165-199). Cham: Springer Nature Switzerland.

[66] Kazim, H., Sabri, M., Al-Othman, A. and Tawalbeh, M., 2023. Artificial intelligence application in membrane processes and prediction of fouling for better resource recovery. *Journal of Resource Recovery*, *1*(January-December).

[67] Campos, J. L., Valenzuela-Heredia, D., Pedrouso, A., Val del Río, A., Belmonte, M. and Mosquera-Corral, A., 2016. Greenhouse gases emissions from wastewater treatment plants: minimization, treatment, and prevention. *Journal of Chemistry*, *2016*(1), p. 3796352.

[68] Mahmud, K., Panday, D., Mergoum, A. and Missaoui, A., 2021. Nitrogen losses and potential mitigation strategies for a sustainable agroecosystem. *Sustainability*, *13*(4), p. 2400.

[69] Chojnacka, K., 2023. Valorization of biorefinery residues for sustainable fertilizer production: a comprehensive review. *Biomass Conversion and Biorefinery*, *13*(16), pp. 14359-14388.

[70] Drioli, E. and Romano, M., 2001. Progress and new perspectives on integrated membrane operations for sustainable industrial growth. *Industrial & engineering chemistry research*, *40*(5), pp. 1277-1300.

[71] Zarei, M., 2020. Wastewater resources management for energy recovery from circular economy perspective. *Water-Energy Nexus*, *3*, pp. 170-185.

[72] Zhou, M., Wang, H., Hassett, D. J. and Gu, T., 2013. Recent advances in microbial fuel cells (MFCs) and microbial electrolysis cells (MECs) for wastewater treatment, bioenergy and bioproducts. *Journal of Chemical Technology & Biotechnology*, *88*(4), pp. 508-518.

[73] Wu, H. and Vaneeckhaute, C., 2022. Nutrient recovery from wastewater: A review on the integrated Physicochemical technologies of ammonia stripping, adsorption and struvite precipitation. *Chemical Engineering Journal*, *433*, p. 133664.

[74] Al Momani, D. E., Al Ansari, Z., Ouda, M., Abujayyab, M., Kareem, M., Agbaje, T. and Sizirici, B., 2023. Occurrence, treatment, and potential recovery of rare earth elements from wastewater in the context of a circular economy. *Journal of Water Process Engineering*, *55*, p. 104223.

[75] Sarpong, G., Gude, V. G., Magbanua, B. S. and Truax, D. D., 2020. Evaluation of energy recovery potential in wastewater treatment based on codigestion and combined heat and power schemes. *Energy Conversion and Management*, *222*, p. 113147.

## Chapter 15

# Bioremediation Approaches for the Removal of Heavy Metals from Wastewater

**S. Hemalatha**
**T. Manjari**
**R. Vasantha Kumar**
**M. Sathiyaraj**
**and S. Ambika***
PG and Research Department of Chemistry, Bishop Heber College, Tamil Nadu, India

### Abstract

Heavy metals (HMs) in wastewater pose a significant threat to ecosystems due to their harmful effects, persistence and tendency to bioaccumulate. Conventional removal methods, such as chemical precipitation and ion exchange, are quite expensive and generate hazardous by-products. In contrast, bioremediation offers a sustainable, cost-effective alternative that employs microorganisms or plants to remove and detoxify HMs from polluted water. This chapter explores key bioremediation mechanisms including biosorption, bioaccumulation, and biotransformation and examines the roles of diverse microbial communities in the remediation of HMs from wastewater. Factors affecting bioremediation efficiency, like pH, temperature, metal concentration, and microbial composition, are discussed. The chapter also highlights the benefits of bioremediation over conventional methods and outlines the challenges and future directions for enhancing its effectiveness and scalability.

**Keywords:** heavy metals, wastewater treatment, toxicity, phytoremediation, bioaccumulation, bioremediation, microorganisms, bio sorption

* Corresponding Author's Email: ambika.ch@bhc.edu.in

In: Advancements in Wastewater Treatment
Editors: Neha Agarwal, Maulin P. Shahl, Vijendra Singh Solankil et al.
ISBN: 979-8-90134-007-3

## 1. Introduction

Water is a fundamental natural resource vital for the survival of all life on Earth. It supports life, facilitates the growth of food, powers industries, and maintains healthy ecosystems. Clean drinking water is essential for human health and well-being. However, water sources around the world are being polluted by harmful substances especially HMs, which have become a major environmental problem.

Heavy metals (HMs) are among the most hazardous pollutants commonly found in water sources. Unlike many other pollutants, they do not degrade over time; instead, they persist in the environment and bioaccumulate in plants, animals, and humans, causing long-term adverse effects. Metals such as lead (Pb), mercury (Hg), cadmium (Cd), and arsenic (As) are typically present in contaminated water and are known for their serious health risks [1]. These metals behave differently depending on their chemical makeup, solubility in water, and interactions with living organisms. The impact of HMs in water depends on their type, concentration, and environmental conditions, including pH level, temperature, and chemical properties of the water. Some metals, like calcium and zinc, are beneficial to human health in small amounts. However, metals such as Hg and Pb are highly toxic even at low concentrations.

Prolonged exposure to these harmful metals can lead to problems with the nervous system, kidneys and reproductive system. Studies by organizations like the U.S. Environmental Protection Agency (EPA) have shown that long-term exposure to HMs in drinking water can lead to serious health problems. For example, research on barium has highlighted the importance of checking metal levels in drinking water systems. These findings show how important it is to use effective and safe methods to eliminate HMs from water. HMs enter water through both environmental processes and human interventions. Natural sources include volcanic eruptions, rock erosion, sea spray, and forest fires, which can release metals into rivers and lakes [2]. However, most of the metal pollution we see today comes from human actions like mining, smelting, manufacturing, and dumping waste into water bodies. Such processes emit large quantity of metals into the environment, often beyond what nature can handle. To clean such type of metals from water, many treatment methods have become established. These include chemical methods (like precipitation and oxidation), physical methods (such as membrane filtration and ion exchange), and biological methods [3]. While these methods can be effective, many have downsides. For example, chemical treatments may generate hazardous byproducts, and physical methods can be expensive and use a lot of energy.

As an alternative, bioremediation has become a popular, affordable, and eco-friendly way to treat HM pollution. This method uses living organisms mainly bacteria, algae, and fungi to absorb, break down, or transform HMs into less harmful forms. Bioremediation can happen naturally or be enhanced through targeted techniques to improve efficiency and effectiveness [4]. One of the main strengths of bioremediation is its adaptability to diverse environmental conditions and its effectiveness in treating water with low metal concentrations, where other methods may be less effective. It also produces less harmful waste compared to chemical methods [5]. The main mechanisms of bioremediation include biosorption (where metals stick to the outside of cells), bioaccumulation (where metals are absorbed into cells), and biotransformation (where metals are changed into safer forms [6]. This chapter explores how bioremediation helps remove HMs from wastewater. It details the processes involved, the different types of microbes used, and its practical applications. The goal is to demonstrate how bioremediation provides a practical and environmentally friendly solution to one of major prevailing pollution challenges.

## 2. HMs of Concern

Among the gravest ecological challenges is the contamination of water by these metals. This type of pollution not only affects aquatic ecosystems but also has far-reaching consequences for plants and human health [7]. For instance, mercury is released into the air and water by industries such as pharmaceuticals, paper, agriculture, and chemical production (e.g., chlorine and caustic soda manufacturing). The different sources of HMs are depicted in Figure 1. Similarly, cadmium, which is naturally found in rocks and soil, is used in batteries, pigments, textiles, and metal coatings, making it a major contributor to industrial pollution [8].

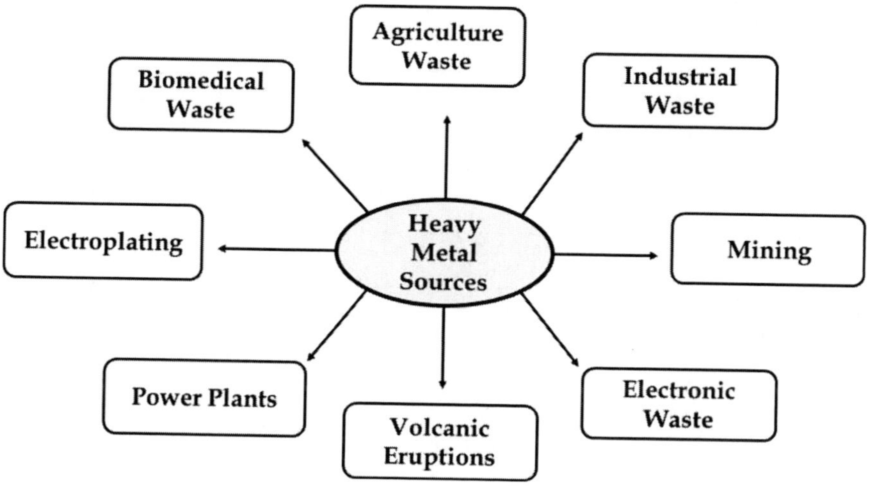


**Figure 1**. Sources of HMs.

Identifying the specific HMs present in contaminated water is only the first step; it is equally important to evaluate their concentrations within these aquatic systems. These concentration levels, often compared against established national and international regulatory limits, help determine the potential health hazards and environmental risks involved. When concentrations surpass these limits, they can lead to toxicological effects even at low doses. Thus, understanding the extent of metal contamination is crucial for prioritizing pollutants and developing effective, targeted bioremediation strategies.

### 2.1. Toxic Concentrations of HMs

Intrinsic components of the natural environment, HMs are widely recognized for their harmful effects on living organisms. Their toxicity is a major concern in contaminated environments or workplaces with a higher risk of exposure. Among the over 50 identified HMs, approximately 17 are considered highly toxic and readily released through human and natural activities [9].

These metals are generally divided into three main categories [10]:

- Highly toxic metals such as cadmium, mercury and lead which are dangerous even at very minimal concentrations.
- Moderately hazardous metals such as bismuth, indium, arsenic, thallium, and antimony, which are less toxic than the first group but still, pose significant risks.
- Essential trace elements like cobalt, iron, copper, zinc and selenium, which play key roles in biological functions but can become toxic when consumed in excess.

To limit health risks, organizations such as the World Health Organization (WHO) have established recommended safety limits for the presence of HMs in drinking water. The Table 1 lists these guidelines to help protect public health and ensure safe water quality. Beyond their concentration levels, the dangerous effects of HMs differ significantly across species and environmental contexts. Gaining a clear understanding of these varied impacts is essential for accurately assessing ecological risk and implementing effective mitigation strategies.

**Table 1.** List of some HMs present in water with their permissible limit

| HMs | Permissible Limit | Source of Contamination | Health Impact | Ref. |
|---|---|---|---|---|
| Arsenic (As) | 10 mg/L | Mining, smelting, pesticide runoff | Skin lesions, cancer, cardiovascular issues | [3] |
| Cadmium (Cd) | 0.03 mg/L | Batteries, metal plating, fertilizers | Kidney damage, bone demineralization | [11] |
| Lead (Pb) | 0.01 mg/L | Paint, pipes, gasoline | Neurotoxicity, developmental delays | [12] |
| Mercury (Hg) | 0.006 mg/L | Mining, industrial waste, fossil fuels | Nervous system disorders, cognitive impairment | [11] |
| Silver (Ag) | 0.0 mg/L | Photographic industry, electronics | Argyria (skin discoloration), liver effects | [3] |
| Magnesium (Mg) | 50 mg/L | Soil runoff, natural deposits | Laxative effect in high doses | [12] |
| Zinc (Zn) | 5 mg/L | Galvanization, mining, natural sources | Gastrointestinal issues in excess | [3] |
| Iron (Fe) | 0.3 mg/L | Corroded pipes, industrial runoff | Staining, taste issues, rarely toxic | [13] |
| Selenium (Se) | 0.01 mg/L | Coal combustion, industrial waste | Hair loss, neurological abnormalities | [14] |

## 2.2. Pathological Impacts of HM Toxicity

Toxicity level of HMs differ based on the type of metal, its biological function, and the organisms that come into contact with it. These metals significantly impact aquatic species, eventually penetrating the food chain through bio magnification and ultimately affecting humans.

### *2.2.1. Effects in Human Begins*

The harmfulness of HMs depends on the type of metal, its concentration, its role in biological systems, physiological function, and the duration of exposure. Although certain types of these metals are essential for the body in minimal amounts, they can become toxic when their levels exceed safe thresholds.

Common HMs in contaminated drinking water include lead (Pb), iron (Fe), cadmium (Cd), copper (Cu), zinc (Zn), and chromium (Cr). Long-term exposure to these metals can damage major organs such as the liver, kidneys, and nervous system, and can interfere with blood chemistry and energy levels [15]. For example, Cd and Cr are highly harmful even in small amounts and accumulate in the human body over time, with cadmium having a biological half-life of up to 33 years. A very recent study in CKD-endemic areas of Cuttack district, Odisha, revealed elevated concentrations of HMs in locally grown food crops, linking them to chronic kidney disease [16]. Chronic exposure can lead to kidney damage and other health complications. Due to its severe impact, cadmium is listed as a priority pollutant in many countries [17]. Industrial discharge of cadmium (Cd) in Japan contaminated rice crops, leading to "Itai-Itai" disease, which caused severe bone fractures and kidney dysfunction in affected communities [18]. Another incident in Santa Monica Bay, California, USA (2025), involved lead leaching into drinking water from corroded pipes after a water source switch. This exposed thousands of children to high lead levels, leading to developmental issues, lower IQ, and behavioural problems [19].

Copper, although essential for human health, becomes toxic at high concentrations, causing gastrointestinal distress and liver damage [20]. Similarly, lead is a well-known toxic metal that acts as a metabolic poison. It can interfere with enzyme activity and replace calcium in bones, leading to long-term storage in body tissues. Prolonged exposure to lead (Pb) may result in neurological disorders, such as Alzheimer's disease [17].

#### *2.2.2. Effects in Aquatic Systems*

In aquatic environments, HMs build-up gradually in living beings and can trigger long-term toxic effects. They enter the food chain through bioaccumulation and biomagnification, ultimately affecting humans and terrestrial animals reliant on aquatic resources [21].

HMs typically exist in water in either dissolved or suspended forms, eventually settling in sediments or being absorbed by aquatic life. In 2025, tests by 'Heal the Bay' found high levels of HMs in Santa Monica Bay after the Palisades Fire. In the Yamuna River, Delhi, India (ongoing), industrial effluents have caused HM accumulation, including chromium, lead, and mercury, affecting aquatic biodiversity and entering the food chain, with fish samples showing bioaccumulation [22]. These toxic metals threaten marine life and can bioaccumulate, posing risks to the food chain and human health. The uptake process in aquatic organisms happens in two stages: a rapid binding to cell surfaces, followed by a slower internal absorption [23]. Fish are particularly sensitive to metal pollution and are often used as indicators of water quality. HMs enter their bodies through three primary pathways: gills, skin, and digestive system. The gills are the primary entry point, as they are in direct contact with contaminated water. Over time, these metals build up in various fish tissues and cause changes at the biochemical, physiological, and histological levels [24]. This includes disruptions in enzyme activity and damage to essential organs. In Minamata Bay, Japan (1950s–1960s), industrial discharge of methylmercury caused Minamata disease in humans who consumed contaminated fish, leading to neurological damage and birth defects [25]. These harmful effects don't just stop at fish. They can ripple through the entire aquatic food web, impacting other freshwater species, and ultimately affecting humans who consume contaminated seafood or use affected water sources. Given the persistent and hazardous nature of HM pollution, researchers have increasingly explored environmentally friendly alternatives for their removal. Bioremediation has emerged as a compelling solution, utilizing the natural capabilities of microorganisms, plants, and other biological agents to detoxify and restore contaminated environments.

## 3. Principles of Bioremediation

### 3.1. Definition and Concept of Bioremediation

In contrast to conventional remediation methods, bioremediation offers a sustainable, cost-effective, and environmentally friendly approach to pollutant cleanup. While sometimes equated with biodegradation, which is the aging of a molecule by an organism, bioremediation encompasses diverse biotransformation processes, addressing both organic and inorganic substances through biological mechanisms. Given the diverse nature of contaminants, including agrochemicals, heavy metals (HMs), plastics, and nuclear waste, each presenting unique challenges, effective remediation must be tailored to site-specific conditions [26].

Microorganisms play a crucial role in natural biogeochemical cycles, serving as key agents in remediating polluted environments, including soil, groundwater, and industrial effluents. Sustainable remediation (SR) addresses environmental, social, and economic objectives simultaneously to minimize waste and resource consumption, while providing long-lasting benefits. Although biodegradable bio-nanoparticles are eco-friendly; their practical application in nano-bioremediation requires stringent regulatory frameworks and cost-effective production methodologies [27]. However, biologically mediated remediation processes are often slower, less predictable, and highly dependent on local physicochemical conditions, limiting their scalability and commercial adoption. While substantial research has been carried out, it remains confined to laboratory or pilot-scale studies, with significant potential for industrial application yet to be realized [27]. With the foundational principles of bioremediation established, the discussion now shifts to the specific biological processes by which HMs are removed or neutralized. These mechanisms, driven through the metabolic processes of microorganisms and plants, form the core of effective bioremediation strategies.

### 3.2. Biological Processes Involved in the Removal of Heavy Metals

#### *3.2.1. Biosorption*

Urbanization and industrialization are increasing heavy metal (HM) contamination in water bodies. The use of dead or inactive microbial biomass, including bacteria, algae, and fungi, is an effective and environmentally friendly approach to extract HMs from polluted water through biosorption [28]. HMs, a primary class of water contaminants from industrial and anthropogenic sources, have significant adverse effects on environmental and public health. Biosorption is an effective and eco-friendly alternative that is cost-effective and produces minimal secondary waste compared to conventional treatment methods. Fungi, in particular, have demonstrated great potential for metal uptake, making them promising candidates for wastewater treatment technologies [29].

The role of microbial biomass and organic or agricultural waste as biosorbents for HM removal is addressed in this section (Figure 2). This section discusses important parameters affecting biosorption including process efficiency, monitoring and control criteria, and influencing factors like pH and contact time, along with metal recovery potential, providing evidence of biosorption as a versatile and promising option for environmental clean-up [30]. New fungi strains like Penicillium chrysogenum and Aspergillus ustus have been studied to develop novel biosorbents. These fungi demonstrate a remarkable ability to remove HMs from water, dependent on optimized factors like pH, temperature, and contact. HM levels in wastewater samples collected across in Saudi Arabia ranged between 2.6–234 μg/L for $Cd^{2+}$, $Cu^{2+}$ and $Pb^{2+}$ with $Cu^{2+}$ concentrations exceeding WHO limits for safe drinking water. These findings highlight the effectiveness of biosorption [31].

The application of various sources of organic biosorbents, including agricultural waste (e.g., banana, lemon, and mango peels) has been found effective in removing metals ($Pb^{2+}$, $Cr^{3+}$, $Ni^{2+}$). A literature review from 2010–2021 confirmed their effectiveness, but some sources remain unexplored (e.g., tree tomato (lulo) peels, presenting opportunities for further exploration), representing opportunities [32]. Chitosan-based biosorbents are prominent biopolymers leading in the sorption of HMs and radionuclides from polluted aqueous environments. Forms with modifications such as chitosan composites provide improved stability and binding capacity, enhancing their suitability for industrial-scale applications [33]. In addition, synthetic materials including conducting polymers are attracting attention with their adjustable functional groups and high adsorption capacities. By tailoring these materials to target specific pollutant particles, they can offer a cost-effective solution to the degradation in quality of water by HMs, minerals [34].

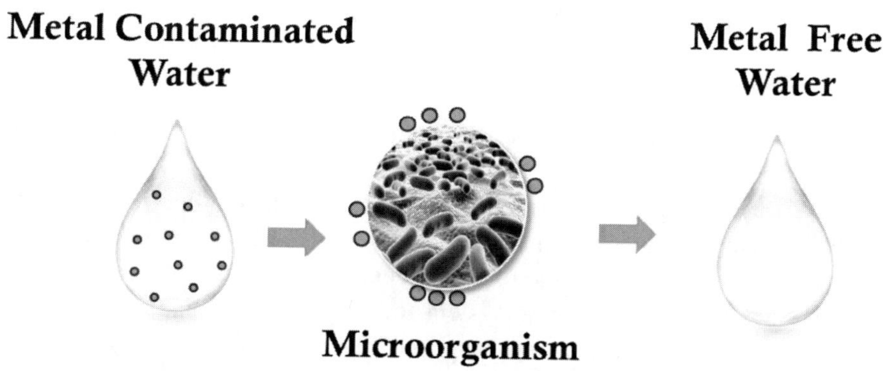


**Figure 2.** Process of Biosorption [35].

Whereas biosorption involves the passive contact of metal ions onto the surface of biological materials, bioaccumulation represents a more dynamic process wherein living cells actively transport and sequester metals internally. This distinction highlights an additional critical pathway in biological remediation strategies.

### *3.2.2. Bioaccumulation*

Bioaccumulation of pollutants in living organisms depends on their absorption, metabolism, and excretion processes. Although scientists are working to build models for understanding bioaccumulation, predicting responses remains challenging due to the complex interplay between pollutants and climate change with habitat variability, especially in sensitive ecosystems like coastal and estuarine environments [36]. Dangerous chemicals and HMs from industrial activities increasingly contaminate water bodies, threatening ecosystems and human health. Microorganisms help degrade these pollutants, offering an eco-friendly and economical alternative to conventional chemical treatment methods and filtration systems. Scaling up and optimising microbial-based solutions for broad industrial use still poses a substantial hurdle [37].

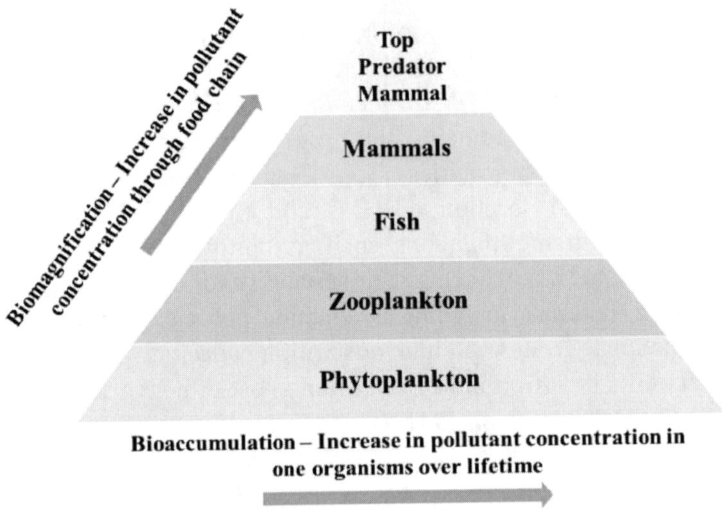


**Figure 3.** Effect of Bioaccumulation on Ecosystem [42].

In contaminant management, detoxification and bioaccumulation are important strategies used to reduce persistence of toxicity. Bioremediation can exploit GMMs, which can be outfitted with different enzymes to degrade various pollutants. In addition, emerging innovative green technologies are presented for the dual use of wastewater treatment and nanoparticle synthesis that can minimize costs while ensuring economic and environmental sustainability [38]. Hazardous HMs contaminate our ecosystem and can bioaccumulate as they enter the food chain, leading to severe risks for ecosystems and human health, represented in Figure 3. Consequently, effectively managing municipal solid waste (MSW) emerges as a critical global concern. An eco-friendly solution lies in microbial-based remediation, which employs bacteria such as Bacillus and Pseudomonas to either convert these metals into less toxic forms or sequester them. This microbial intervention not only helps in mitigating metal pollution but also enhances soil fertility and fosters plant growth. By integrating these microbial methods with recycling efforts and proper waste disposal practices, we can significantly reduce HM contamination and embark on a path toward ecosystem restoration [39]. By applying bacteria and plants through processes like phytoremediation and bioaccumulation, HMs can be sustainably removed from soil and water. The kind of metal, plant or microbial species, and soil characteristics are some of the variables that affect how effective these techniques are. Toxic metals can be converted, absorbed, or stabilised by hyper accumulator plants and metal-tolerant microbes, providing an economical and sustainable substitute for traditional mechanical and chemical remediation techniques. Environmental restoration is greatly aided by these biological tactics [40]. On the other hand, the various types of removing HMs from wastewater are frequently expensive and produce dangerous by-products. Microbial bioaccumulation, which makes use of bacteria and fungi, offers a more environmentally friendly option. However, the requirement for optimised microbial strains and reliable performance in a range of environmental conditions makes large-scale implementation difficult [41].

In addition to bioaccumulation, organisms can also immobilize metals by transforming them into stable mineral forms, a process known as biomineralization. As shown in Figure 3, bioaccumulation causes HMs to move up the food chain, starting with microorganisms and small aquatic organisms, progressing to fish, and eventually to humans and other top predators. This trophic transfer can lead to toxic metal accumulation in critical organs, affecting reproduction, growth, and overall ecosystem stability. The diagram visually emphasizes how even low concentrations of metals in water can magnify through bioaccumulation and pose severe risks to biodiversity and human health.

### *3.2.3. Biomineralization*

By employing urease-active microbial strains to transform cadmium's (Cd) toxic, easily absorbed forms into stable, less bioavailable mineral forms, biomineralisation lowers the speed and bioavailability of Cd in contaminated environments. This procedure reduces possible health hazards and ecological harm [43]. Although bacteria are the subject of the majority of microbial Biomineralisation research, fungi also contribute significantly through the generation of organic acids that form metal complexes, such as oxalic acid. Fungal involvement in calcite formation has garnered more attention recently, despite the fact that the exact mechanisms are still unknown [44]. Figure 4 illustrates the biomineralization pathway facilitated by *Sporosarcina pasteurii,* a urease-producing bacterium capable of converting soluble HM ions like $Cd^{2+}$ and $Ni^{2+}$ into stable mineral forms, such as metal carbonates. This transformation occurs via urea hydrolysis, which increases pH and facilitates the precipitation of calcite or struvite crystals. The diagram illustrates the microbial processes involved reducing metal mobility and bioavailability in contaminated water, making the method both effective and environmentally sustainable. In addition to aiding in desalination, HM removal, calcium, aluminium and phosphate recovery, and the treatment of wastewater, seawater, and groundwater, these biominerals improve the qualities of building materials like sand and cement [45]. Important environmental contaminants include toxic metals such as cadmium (Cd) and nickel (Ni). However, by creating metal carbonates, urease-producing bacteria can hydrolyse urea, making their removal easier. In one study, Cd (II) and Ni (II) were biomineralized using six of these bacterial strains; two of them demonstrated high resistance and eliminated 89% and 96% of the metals through calcite precipitation in 48 hours [45]. Additionally, protective structures that shield microbial cells from UV light, chemicals, and heat stress are formed as a result of biological metal mineralisation. In early marine environments, this process was essential for survival [46]. Biomineralisation is still fuelled by extremophiles, which are microorganisms that thrive in harsh industrial environments with high metal concentrations. In these kinds of settings, polyextremophiles are essential to maintaining this process [47]. Table 2 describes the comparison of bioremediation techniques.

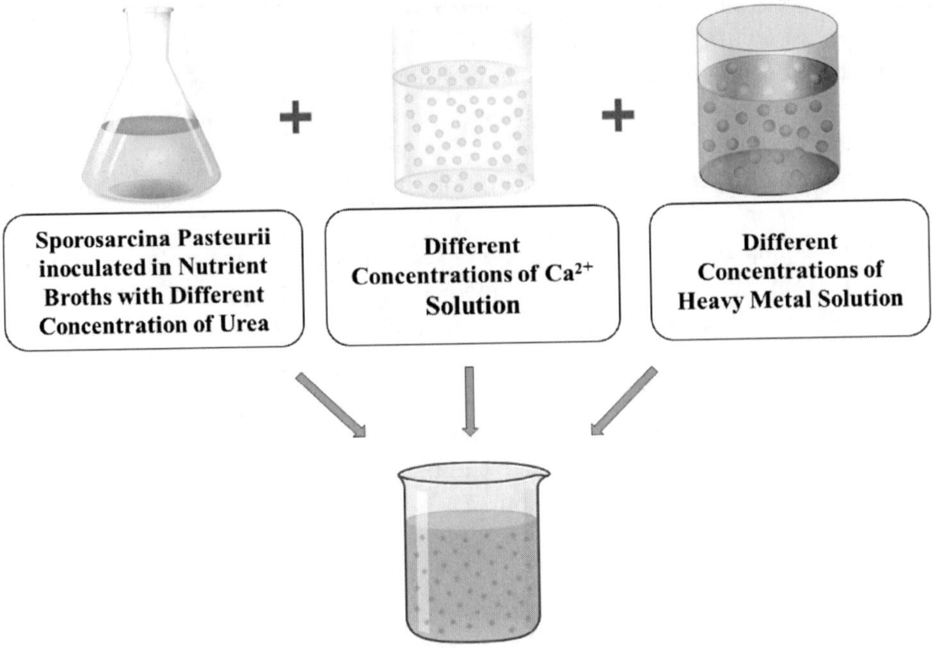


**Figure 4.** Process of Biomineralization by Sporosarcina Pasteurri [48].

**Table 2.** Comparison of bioremediation techniques

| Technique | Organism Type | Mechanism | Strengths | Ref |
|---|---|---|---|---|
| Biosorption | Bacteria, Fungi | Passive metal binding | Cost-effective, fast | [34] |
| Bioaccumulation | Bacteria, Algae | Active metal uptake | Long-term storage | [40] |
| Biomineralization | Bacteria, Fungi | Precipitation as minerals | Stabilizes metals | [49] |
| Phytoremediation | Plants | Extraction/Stabilization | Eco-friendly, large-scale | [50] |

The achievement of bioremediation, especially through biomineralization, determined largely on the microorganisms engaged. Each microbial group contributes distinct advantages to the process. In the next section, the various microorganisms utilized in bioremediation will be explored.

## 4. Microorganisms Used in Bioremediation

### 4.1. Bacteria

Bacteria play a critical role in the bioremediation of environments polluted with HMs. More than 170,000 genes linked to HM resistance have been discovered in the genomes of bacteria, with Proteobacteria and Terrabacteria accounting for a significant proportion of these genes. These genes are primarily involved in essential metabolic functions like the synthesis of amino acids and energy. Interestingly, *Rhodobacter sphaeroides* has 375 of these genes, indicating its potential for bioremediating metal-contaminated environments [51].

A combination of bacterial strains was found to be more efficient than single strains in removing HMs, such as lead, copper, and cadmium, from contaminated soil. In comparison to single-strain cultures, these mixed cultures showed higher enzymatic activity, faster growth rates, and increased resistance to HMs [52]. By providing tools to engineer bacteria with improved metal-removal capabilities, the field of synthetic biology has brought new approaches to microbial bioremediation. These modified microorganisms have improved efficiency and specificity by quickly detecting and neutralising HMs. Combining physical, chemical, and biological methods is crucial for thorough remediation. Engineered microorganisms offer a more effective alternative to conventional remediation techniques, designed to detect and remove pollutants [53]. Activated sludge systems and resistant bacterial strains collaborate effectively to remove organic matter and HMs from wastewater. These bioaugmented systems exhibit improved removal efficiency and increased tolerance to hazardous metals. This economical, effective, and ecologically friendly method can be used to treat mixed, residential, or commercial wastewater [54]. The growth and metabolic activity of maize plants irrigated with sewage water were markedly boosted by the application of silver nanoparticles and plant-growth-improving rhizobacteria (PGPR) together. The isolated PGPR strains demonstrated antifungal and antibacterial qualities, solubilised insoluble phosphate, and generated the enzymes catalase and oxidase [55].

In addition to facilitating the removal of harmful metals from irrigation water, this combined treatment helps to stimulate plant growth (Natural Attenuation) represented in Figure 2. Unlike standard physicochemical methods, which are frequently more harmful to the environment and human health, the use of genetically modified bacteria in bioremediation is significant due to their eco-friendliness and low health risks [56]. Naturally occurring bacteria and algae serve as biocatalysts, removing HMs through processes such as adsorption, aggregation, metabolism, and detoxification. Metal concentrations in wastewater and contaminated sites can be considerably decreased by these biological agents to levels that are safe for the environment [56]. Four HM-resistant bacterial strains—*Escherichia coli, Salmonella typhi, Bacillus licheniformis,* and *Pseudomonas fluorescens*—isolated from textile industry effluents showed promise in effectively removing HMs from polluted wastewater [57]. While bacteria are key players in bioremediation, fungi also contribute significantly due to their unique structural adaptations and remarkable resilience in toxic environments.

### 4.2. Fungi

Bioremediation using fungi is a simple, environmentally convivial method to eliminate HMs from polluted water. Fungi have special cell structures and natural processes that help them trap and absorb toxic metals like lead, cadmium, and mercury [58]. Because of this, they are considered a promising alternative to chemical water treatment, especially for long-term environmental clean-up. Fungi are found in different environments and can survive even in extreme conditions. They can grow in polluted areas, including those with HM contamination, which makes them ideal for bioremediation [59]. The outer layer of fungal cells contains substances like chitin, chitosan, and other functional groups that allow metals to stick to their surface or get absorbed inside the cell.

Fungi remove metals primarily through two main mechanisms:

- *Biosorption* – This happens when metal ions attach to the outside of the fungal cell. It doesn't require energy from the fungus. The metals bind through natural reactions involving the cell wall.
- *Bioaccumulation* – In this process, metals actually enter the fungal cell. Inside, they are stored or neutralized using small compounds like organic acids or sulphur-based molecules.

Some types of fungi produce antioxidants to safeguard themselves from the harmful effects of metals. Others release enzymes that help break down pollutants in the environment. Species like Aspergillus, Rhizopus, Saccharomyces, and white-rot fungi work very well in absorbing and processing these metals from water and soil [60].

It also plays other important roles in nature, such as breaking down dead materials and forming beneficial relationships with plants. Humans have used fungi for thousands of years in food and medicine and now, they're also being used to treat polluted ecosystem [61]. They grow quickly, are inexpensive to produce on a large scale, adapt easily to different conditions and play a key role in bioremediation by supplying nutrients to the surrounding environment (Figure 2). Because of these advantages, they are often chosen for bioremediation projects. Although bacteria, algae, and plants can also be used, fungi are especially good at this job due to their strong metal-binding abilities and resilience [62]._In addition to fungi, algae are also crucial in bioremediation, functioning as effective biosorbents and bioaccumulators, especially in aquatic environments.

### 4.3. Algae

Algae have shown great potential in the cleansing of HMs from water via two main processes: biosorption and bioaccumulation. Biosorption is a passive process in which metal ions attach to the surface of algal cells [63]. This occurs through several mechanisms, including ion displacement, electric charge interaction and complexation. Among these, ion exchange plays a particularly important role. Algal biomass, especially from marine sources, contains surface groups that interact with metal ions effectively, making it a cost-effective and renewable option for water treatment [64]. Traditional ion exchange systems use synthetic chemical resins, which are often expensive. However, researchers such as Volesky et al. have shown that marine algae can function as natural ion exchangers [63]. Different algal species exhibit different affinities for metal ions, which makes them suitable for targeting specific contaminants in wastewater. Bioaccumulation, on the other hand, is an effective process where metal ions are taken up into algal cells. It usually occurs in two phases: a fast initial binding to the cell surface (similar to biosorption), followed by a slower internal absorption into the cell. This process is more complex and energy-dependent compared to biosorption. Many algal species have been studied for their efficiency in HM removal. These include *Chlorella vulgaris, Chlorella fuscas*, etc. These are widely available, grow quickly, and have high binding capacities, which makes them well-suited for wastewater treatment [65].

Although algae are highly effective, other microorganisms such as bacteria also play an essential role in bioremediation. Marine bacteria are particularly useful because of their ability to survive and adapt to changing environmental conditions, including variations in temperature, salinity, and pH. Due to their resilience, marine bacteria are often isolated and studied for their potential to assist in the detaching of HMs in complex environmental settings [64]._Beyond microorganisms, certain plants have also shown remarkable abilities to absorb, stabilize, or degrade HMs in water, positioning phytoremediation as a promising approach in bioremediation.

### 4.4. Green Remediation of HM Pollutants in Water

Pollutants are removed using plants in an environmentally friendly and economical approach known as phytoremediation especially harmful metals in water. This technique has become popular because it provides an alternative to traditional methods, which are often expensive, needs a lot of energy, and use chemicals that can harm the environment. Phytoremediation helps remove toxic metals like lead, mercury, arsenic, and cadmium from polluted water, making it safer for both humans and wildlife [66]. HMs in water are a growing problem worldwide, caused mainly by human actions like industrial operations, mining, farming runoff, and improper waste disposal. These metals are dangerous to aquatic life and can affect human health if they enter the food chain. Since HMs are difficult to decompose and remain in the environment for prolonged durations and can build up, causing lasting harm. Phytoremediation offers a way to address this issue by using plants to remove or reduce the concentration of these metals from the water [66].

Plants have a remarkable ability to absorb metals through their roots. The metals they take in can be kept in the plant's tissues, changed into less harmful substances, or kept in the root area, preventing further spread. This process can work in different ways, like phytoextraction, where the plant takes up metals and stores them, or phytostabilization, where the metals are trapped in the root zone, preventing their movement into other areas [67]. One of the main benefits of phytoremediation is that it operates directly in contaminated water, eliminating the need to excavate soil or relocate water. This reduces the amount of energy and resources needed. Phytoremediation also helps protect the environment since it doesn't disturb the soil or the surrounding area. Plants slowly remove the pollutants, while also preventing soil erosion and improving water quality. Additionally, the plants used can be harvested and processed, turning them into useful products like biomass [68].

Phytoremediation is especially effective in areas with warm climates, like tropical regions, where plants grow quickly and thrive. These conditions enhance process efficiency. Plants are naturally good at adapting to different environments, making them an excellent option for cleaning polluted water, particularly in large contaminated areas where other methods may be less practical or affordable [69]. Phytoremediation is a promising and environmentally friendly method for remediating HM-polluted water (Table 3). Using plants to absorb, store, and transform these harmful metals offers an effective solution that benefits the environment while remaining cost-efficient.

**Table 3.** Some Plant Species Used for HM Removal

| Metal Removed | Plant Species | Plant Family | Mechanism | Optimal Environment | Ref |
|---|---|---|---|---|---|
| Cadmium (Cd) | Matricaria recutita, Hypericum perforatum | Asteraceae | Phytoextraction | Moist, neutral soil | [70] |
| Zinc (Zn) | Brassica juncea | Brassicaceae | Phytoextraction | Sunny, fertile loam | [71] |
| Nickel (Ni) | Psychotria douarrei | Asteraceae | Phytoextraction | Tropical, acidic soil | [72] |
| Uranium (U) | Brassica chinensis, Amaranthus spp. | Brassicaceae, Amaranthaceae | Phytoextraction | Loamy, well-drained soil | [66] |
| Thallium (Tl) | Iberis intermedia | Brassicaceae | Phytostabilization | Alkaline soils | [73] |
| Mercury (Hg) | Eichhornia crassipes | Pontederiaceae | Rhizofiltration | Aquatic environments | [74] |

Although phytoremediation requires time and careful planning, its benefits, including low cost, minimal equipment, and applicability to large areas, make it a valuable tool for addressing water pollution. As more research is conducted, phytoremediation is likely to become an even more important method for solving water pollution challenges [75]. However, there are some challenges to using phytoremediation. The process can take a long time, sometimes several years, depending on the level of contamination, the type of metal and environmental conditions. Various plants used for removing different metals, along with their family names, are listed in Table 3 [76]. Also, most plant roots are shallow, limiting the depth of water that can be treated. Introducing new plants into a local ecosystem may also have unintended effects, so it's crucial to select the right plants. Despite the promise of various bioremediation methods, their success is heavily influenced by several environmental and biological factors, which must be carefully optimized. The following section examines these critical factors that impact bioremediation effectiveness.

## 5. Factors Regulating the Effectiveness of HM Bioremediation in Water Systems

Bioremediation is a process whereby various microorganisms and plants are used to break down, remove, immobilize, or detoxify environmental contaminants, including HMs in water. This remediation method relies on the enzymatic activity of these organisms, to degrade or transform harmful substances. For bioremediation to be effective, several factors must be considered, ranging from the chemical properties of the pollutants to environmental conditions that influence microbial activity. Understanding and optimizing these factors is crucial for improving the productivity and accomplishment of HM remediation in aquatic systems [50].

### 5.1. Key Factors Affecting Bioremediation Strategies for Detoxifying HMs in Water

- *Characteristics of Metal Ions*: The chemical properties of metal ions, such as their concentration, oxidation state, and chemical form, play a significant role in determining how microorganisms interact with and neutralize them. Some metal ions are more bioavailable and easier for microorganisms to process, while others may be more toxic or less soluble, limiting microbial degradation [77].

- *Environmental Parameters*: The physical environment significantly influences bioremediation. Factors like pH, temperature, redox potential, and moisture levels directly impact microbial processes and the solubility of metals. Maintaining an optimal pH range (typically between 6.5 and 8.5) and a suitable temperature range enhances microbial activity, promoting effective bioremediation. Extremes in these parameters can inhibit microbial metabolism and slow down the degradation of pollutants [78].
- *pH*: The pH of the environment significantly influences microbial activity and the efficiency of bioremediation. Most microorganisms involved in bioremediation thrive in a pH range of 6.5 to 8.5, which is considered optimal for degradation processes. Extreme pH values (either too acidic or too alkaline) can inhibit microbial growth, reduce enzymatic activity, and slow down the degradation of contaminants. pH influences the solubility of HMs, with certain metals being more bioavailable under specific pH conditions [77].
- *Temperature*: Temperature has a critical impact on the metabolism of microorganisms. Microbial processes, including the breakdown of HMs, occur most efficiently within an optimal temperature range. While each microbial species has a preferred temperature range, extreme temperatures either excessively high or low can disrupt microbial metabolism and enzyme activity. In colder environments, microbial activity may slow down significantly, while elevated temperatures can cause enzyme denaturation, halting the remediation process [79].
- *Microbial Aspects*: The success of bioremediation largely depends on the microbe's capacity to transform HMs. Microbial competition for resources, adaptation to environmental conditions, and their ability to produce enzymes for metal transformation are critical. The existence of essential nutrients like carbon, nitrogen, and phosphorus supports microbial growth and activity, further improving the efficiency of metal removal [80].
- *Soil and Sediment Properties*: The physical properties of soil and sediment, including texture, composition, and particle size, influence the availability and mobility of HMs. Well-aerated, well-drained soils generally promote better microbial activity, aiding the bioremediation process. The mobility of metals in the soil or water affects how easily they can be accessed by microbes for degradation [81].
- *Presence of Inhibitory Substances*: Compounds that inhibit microbial activity can severely impact the bioremediation process. Toxic substances such as high metal concentrations or other pollutants may prevent microbes from effectively remediating contaminants. Understanding the nature and concentration of such inhibitors is essential for optimizing bioremediation strategies.
- *Biological Factors:* Biological factors include microbial competition for limited nutrients, antagonistic interactions, and predation by other microorganisms, such as protozoa or bacteriophages. The remediation rate depends on the pollutant concentration, the microorganism's ability to metabolize it, and the quantity of enzymes produced. Specific enzymes play a critical role in breaking down contaminants, and their effectiveness depends on the microbial population's composition and their affinity for the pollutants [82].

- *Environmental Conditions*: Environmental factors such as pH, temperature, moisture, and oxygen levels all shape microbial growth and pollutant remediation. For instance, pH levels between 6.5 and 8.5 are ideal for most bioremediation processes. Extreme temperatures or extreme moisture content can either hinder or enhance the remediation of pollutants. Similarly, oxygen availability affects whether aerobic or anaerobic degradation processes occur, with aerobic conditions typically leading to faster removal of contaminants [82].
- *Nutrient Availability*: Microorganisms require essential nutrients for growth and bioremediation. An appropriate balance of these nutrients is necessary for optimal microbial activity. In nutrient-deficient environments, the addition of these nutrients can enhance microbial efficiency and increase the rate of breakdown of pollutants [83].
- *Contaminant Properties*
  - *Metal:* Trace metals are vital for microbial metabolism but can inhibit cell functions at higher concentrations, limiting bioremediation efficiency [84].
  - *Toxic Compounds*: High concentrations of toxic substances can adversely affect microorganisms and hinder the bioremediation process.

Once the key influencing factors are understood, the next step is to select the most suitable bioremediation technique, whether applied onsite or offsite. The following section explores these techniques and approaches in detail.

## 6. Techniques and Approaches in Bioremediation

### 6.1. Ex Situ Bioremediation: Removal of Contaminated Water to Treat Offsite

In *ex-situ* bioremediation, polluted material (soil, water, or sediment) is removed for treatment at a separate site. This method enables more effective pollutant removal under controlled settings. One innovative approach involves creating sustainable building materials, such as concrete and paving stones, from contaminated sediment. This process highlights the importance of selecting appropriate remediation methods based on sediment properties, safety, environmental impact, and cost-effectiveness [85].

Fenton's reaction and activated persulphate oxidation are examples of advanced oxidation processes (AOPs) that have been investigated for the treatment of urban river sediments. These techniques, which use fewer chemicals, offer a more economical alternative to marine disposal by significantly reducing pollutant levels and enabling the use of treated sediments as infill material [86]. Bacillus sp. based biological treatments have also demonstrated promise in removing tannery waste pollutants and Cr (VI). Temperature, pH, and bacterial concentration influenced the process's effectiveness, with glucose serving as a co-substrate to enhance Cr (VI) reduction. In a study, continuous treatment was supported by sand in a packed bed reactor, and performance was evaluated by mathematical modelling [87].

Eighty to ninety percent of the harmful elements were successfully eliminated by soil washing with ethylene Diamine-N, N′-disuccinic acid (EDDS). Although metal-EDDS complexes differed in their toxicity and biodegradability, EDDS itself was extremely biodegradable. The technique worked well, especially in acidic environments, despite some loss of nutrients like iron and aluminium [88]. Ex-situ remediation approaches have significantly aided the management of hazardous pollutants affecting over 100 million people worldwide. Within the context of soil, groundwater, and air purification, multiple conventional techniques have been evaluated, highlighting their advantages and disadvantages [89]. Advances in metagenomics and next-generation sequencing have enhanced our understanding of microbial roles in HM detoxification. In ex-situ applications, microbe-induced bioremediation is becoming a viable way to deal with metal contamination [90]. Although ex-situ bioremediation is effective in controlled environments, in-situ methods offer a more practical and cost-effective approach by treating contamination directly at its source.

### 6.2. *In Situ* Bioremediation: Bioremediation Implemented On-Site at the Contamination Area

*In-situ* bioremediation involves applying treatment techniques directly at the contamination site, offering a cost-effective and minimally disruptive alternative to excavation or transportation. HM contamination in groundwater is a big global concern. In-situ approaches utilising site-specific microorganisms are becoming more popular for real-time pollution removal. The efficacy of such approaches is determined on microbial species, site circumstances, and pollutant availability [91].

Nanotechnology has been applied to site cleanup, using materials such as titanium dioxide, carbon nanotubes, and graphene. Due to their nanoscale dimensions and big surface area, these materials can treat a wide range of pollutants in a variety of environments [92]. Natural microbial communities, such as periphyton, have showed potential for cleansing surface waters. Recent advances in immobilisation technology have increased their applicability, making them suitable for habitat restoration and sustainable water treatment [93]. Biological soil remediation techniques, like vermiremediation, phytoremediation, and bacterial bioremediation, have shown to be more environmentally benign than traditional treatments. A study conducted in Gernika-Lumo found that mixing earthworms, bacteria, and plants improved the elimination of toxin such as dieldrin and HMs [93]. Additionally, investigations on Fe (III)-mediated coagulation-precipitation have provided insights into the cleaning of HMs. Cu (II) was discovered to co-precipitate more successfully with Iron than Nickel, and Copper - EDTA complexes degraded at basic pH, releasing free Cu ions. These findings help to improve our understanding of metal-ligand interactions during in-situ removal procedures [94]. Complementing these techniques are phytoremediation systems, which leverage the natural abilities of plants to clean polluted water and soil.

### 6.3. Phytoremediation Systems: Using Constructed Wetlands and Cleaning Contaminated Water with Aquatic Plants

The effectiveness of phytoremediation and the mechanisms involved are influenced by factors such as the type of pollutant, its bioavailability, and the properties of the soil or water. Plants address contamination through various processes. Contaminants are primarily absorbed by the root system, which is a vital role in preventing toxicity. The roots provide a high surface area that enables the absorption of water, essential nutrients, and non-essential contaminants [95]. Phytoremediation employs various plant mechanisms to remove, immobilize, or contain contaminants in soils, water, sediments, and sludges. The four primary mechanisms are phytoextraction, phytostabilization, phytovolatilization, and rhizofiltration, visual representation is given in Figure 5; each with distinct processes and applications. The figure illustrates key phytoremediation mechanisms, including phytoextraction, phytostabilization, phytovolatilization, phytodegradation, and rhizofiltration, demonstrating how plants absorb, stabilize, or transform HMs in water and soil, highlighting the eco-friendly and versatile nature of phytoremediation.

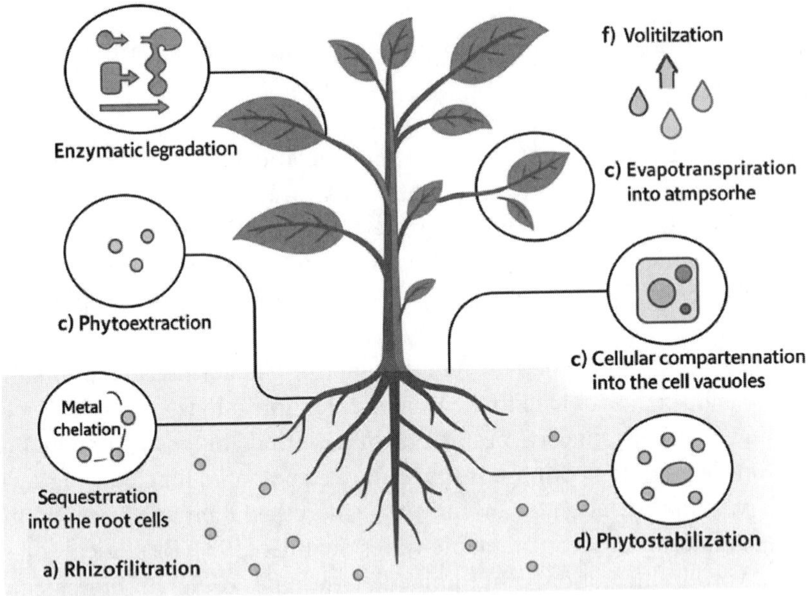


**Figure 5.** Steps involved in Mechanism of Phytoremediation [74].

### *6.3.1. Rhizofiltration*

Rhizofiltration uses plant roots to capture and remove contaminants dissolved in water. It is effective in treating surface water and wastewater by absorbing or adsorbing pollutants like HMs and organic compounds. Preferred plant characteristics for rhizofiltration include:

- Resistance to toxic metals.
- Extensive root systems with a large surface area for contaminant absorption.
- Ability to survive in low-oxygen environments.
- Terrestrial plants are often favoured due to their longer and more fibrous roots

These processes reduce the volume, mobility, and toxicity of pollutants which is the primary objective of phytoremediation. By selecting appropriate plants with specific characteristics, these phytoremediation mechanisms can effectively address environmental challenges in an affordable and environmentally sustainable manner [96].

#### *6.3.2. Phytoextraction*

Phytoextraction is a process whereby plants absorb contaminants through their roots and transport them to above-ground tissues for collection. It is effective for extracting pollutants, particularly in soils, sediments, water, and sludges [97] Essential plant traits for phytoextraction are:

- Resistance to elevated metal concentrations.
- Ability to accumulate large amounts of metals or other contaminants.
- Rapid growth and significant biomass yield.
- Efficient transfer of pollutants from roots to shoots.
- Broad root systems for accessing a large soil volume.
- Adaptability to environmental conditions and ease of cultivation.
- Resistance to pests, diseases, and herbivores to prevent contamination of the food chain.

While this mechanism primarily targets contaminants in soil, it can be applied to water if plants with sufficient bioaccumulation capacity are used [98].

#### *6.3.3. Phytostabilization*

Phytostabilization limits the movement and availability of contaminants in the environment by stabilizing them in soils or sediments through plant roots. This mechanism is primarily used for managing HMs and chlorinated solvents. Key characteristics of plants for phytostabilization are:

- Extensive and dense root systems to anchor and stabilize pollutants.
- Minimal transfer of pollutants from roots to above-ground tissues.

This mechanism is less commonly applied in water since it focuses on immobilizing contaminants in soil or sediment. It is more suitable for stabilizing HMs in sediments rather than free-floating contaminants in water [96].

#### *6.3.4. Phytodegradation*

Phytodegradation is a process where pollutants are either completely mineralized into harmless inorganic compounds or broken down into durable, harmless intermediates. These intermediates are then either absorbed by the plant's cell wall or stored in its vacuoles. Within plant tissues, enzymes catalyze the breakdown of these pollutants. In the rhizosphere, the breakdown of pollutants is enhanced by increased microbial activity, stimulated by the exudates released by the plant roots. This interaction between plants and microorganisms plays a crucial contribution in the breakdown of organic pollutants. Studies on phytodegradation have expanded our understanding of the physiological dynamics and metabolic pathways involved [72].

#### *6.3.5. Phytovolatilization*

In phytovolatilization, plants take up pollutants, change them into less harmful, and emit them into the environment through their leaves. This mechanism is particularly suited for removal of chlorinated solvents and inorganic compounds from groundwater, soil, sediment, and sludge [95].

Despite the benefits of phytoremediation systems, bioremediation as a whole faces limitation that can affect its effectiveness in certain situations, necessitating further refinement and support.

## 7. Challenges and Limitations

Although bioremediation is recognized for its eco-friendly approach, it faces several limitations and challenges. It is not suitable for all compounds, as its effectiveness is largely limited to those that are biodegradable. The efficiency of biosorption often decreases due to inhibited microbial growth, which is commonly caused by damage to microbial cell walls from toxic HMs. When occurring in large amounts, these metals are capable of kill the microbes, significantly reducing the effectiveness of wastewater treatment [99]. Furthermore, the lack of selectivity among microorganisms can lead to competition between different metal ions, lowering the overall efficiency of removing targeted HMs, as many microbes are unable to distinguish between them [100]. Bioremediation is also highly dependent on environmental conditions; even slight changes in temperature or oxygen levels can negatively impact microbial activity or lead to microbial death. Similarly, fluctuations in pH whether more acidic or alkaline, can adversely affect microbial survival and performance [101]. In addition to these issues, the process is relatively slow; for instance, one study reported that the degradation of cadmium from contaminated water took approximately 180 days, highlighting a major limitation in scenarios requiring rapid remediation [102]. To address these challenges, on-going research is exploring innovative strategies to enhance the effectiveness of bioremediation techniques.

## 8. Future Directions and Innovations

### 8.1. Genetic Engineering: Enhancing Microbial Capabilities for HM Removal

Contaminated soil and water containing inorganic contaminants can be remediated through the genetic engineering of plants. Phytoremediation is effective but relies on mechanisms of metal accumulation and resistance to facilitate toxin removal. Genetic enhancement of fast-growing plants can optimize their pollution-absorbing capabilities, leading to more effective environmental remediation. Small genetic modifications in microbes, such as fungi, bacteria, or algae, can significantly improve HM removal efficiency, enabling them to survive toxic metal exposure. This approach is safe and cost-effective. Environmental pollution is often characterized by a mix of organic and inorganic pollutants, with inorganic compounds being notably persistent due to their resistance to natural breakdown [103]. Although challenges exist, researchers are actively addressing these issues to advance wastewater treatment. This integrated approach represents a strategy for responsible wastewater management that prioritizes resource conservation, minimizes waste, and ensures a healthier environment for future generations. Another promising approach is the integration of nanotechnology, which can complement genetic engineering by working synergistically with biological systems to enhance remediation results.

### 8.2. Nanotechnology in Bioremediation: Use of Nanoparticles to Enhance Metal Removal Efficiency

Nanotechnology significantly contributes to environmental management through treatment, remediation, sensing, detection, and pollution prevention. Nano-bioremediation (NBR) employs nanomaterials for contamination treatment, applicable in both in-situ and ex-situ settings. These materials effectively break down toxic substances, aiding in the restoration of damaged ecosystems [104]. Nanoparticles can be used to remediate organic contaminants via adsorption or chemical modification, either by retaining contaminants to facilitate microbial breakdown or by stimulating microbes to produce enzymes that degrade pollutants. Researchers are drawing inspiration from nature to develop new nanoparticles, studying how organisms create complex structures and replicating these processes in laboratories to achieve stable results [105]. A promising research area involves using microbes to produce nanomaterials, offering significant potential. These organisms have evolved unique abilities that allow them to create intricate materials [106]. For example, some bacteria produce complex minerals, while certain types of algae can create delicate glass-like structures. By understanding how these organisms work, scientists hope to develop new methods for creating nanomaterials. Collectively, these advanced technologies pave the way for more effective and sustainable bioremediation solutions for water treatment.

## Conclusion

Growing concerns about public health and pollution from toxic HMs highlight bioremediation as an environmentally friendly and resilient technique for removing HMs from wastewater. Unlike conventional methods, bioremediation utilizes the natural abilities of microorganisms and plants for biosorb, bioaccumulate, biotransform, or biomineralize HMs. Despite challenges such as slow processing, metal toxicity to microbes, and optimization issues, the prospects are promising, with genetic engineering, nanotechnology, and hybrid systems enhancing efficiency and scalability. Bioremediation offers a flexible and reliable wastewater management solution, meeting the growing demand for clean water while demonstrating significant potential for further development and research.

## References

[1] Brenniman G, Kojola W, Levy P, Carnow B, Namekata T: High barium levels in public drinking water and its association with elevated blood pressure. *Archives of Environmental Health: An International Journal* 1981, 36(1):28-32.

[2] Sonone SS, Jadhav S, Sankhla MS, Kumar R: Water contamination by HMs and their toxic effect on aquaculture and human health through food Chain. *Lett Appl NanoBioScience* 2020, 10(2):2148-2166.

[3] Kannan L, Gokulprasath M, Gurusamy R, Selvam R, Palaniswamy R: Analysis of HMs contamination in water: A review. *SSRN Electron J* 2021, 8(4):201-213.

[4] Dwivedi S: Bioremediation of HM by algae: current and future perspective. *J Adv Lab Res Biol* 2012, 3(3):195-199.

[5] Bhatnagar S, Kumari R: Bioremediation: a sustainable tool for environmental management–a review. *Annual Review & Research in Biology* 2013, 3(4):974-993.

[6] Kulshreshtha A, Agrawal R, Barar M, Saxena S: A review on bioremediation of HMs in contaminated water. *IOSR Journal of Environmental Science, Toxicology and Food Technology* 2014, 8(7):44-50.

[7] Muhammad S, Shah MT, Khan S: Health risk assessment of HMs and their source apportionment in drinking water of Kohistan region, northern Pakistan. *Microchemical journal* 2011, 98(2):334-343.

[8] Zaynab M, Al-Yahyai R, Ameen A, Sharif Y, Ali L, Fatima M, Khan KA, Li S: Health and environmental effects of HMs. *Journal of King Saud University-Science* 2022, 34(1):101653.

[9] Mahurpawar M: Effects of HMs on human health. *Int J Res Granthaalayah* 2015, 530(516):1-7.

[10] Bhat UN, Khan AB: HMs: An Ambiguous Category of Inorganic Contaminants. *Research Journal of Environmental Sciences* 2011, 5(8):682-690.

[11] Edition F: Guidelines for drinking-water quality. *WHO chronicle* 2011, 38(4):104-108.

[12] Wu X, Cobbina SJ, Mao G, Xu H, Zhang Z, Yang L: A review of toxicity and mechanisms of individual and mixtures of HMs in the environment. *Environmental Science and Pollution Research* 2016, 23:8244-8259.

[13] Khan SA, Sagar SM, Sagar AK, Kadiya VN, Patel VS, Patel VS: Quality Assessment of Drinking Water in Modasa Town, Aravalli District, Gujarat, India. *International Journal of Advanced Engineering, Management and Science* 2024, 10:4.

[14] Okonji SO, Achari G, Pernitsky D: Environmental Impacts of Selenium Contamination: A Review on Current-Issues and Remediation Strategies in an Aqueous System. *Water* 2021, 13(11):1473.

[15] Sankhla MS, Kumari M, Nandan M, Kumar R, Agrawal P: HMs contamination in water and their hazardous effect on human health-a review. *Int J Curr Microbiol App Sci* 2016, 5(10):759-766.

[16] Mohanty S, Nayak RK, Jena B, Padhan K, Mohapatra KK, Sahoo SK, Dash PK, Das J, Behera SK, Sahu A: HM contamination in rice, pulses, and vegetables from CKDu-endemic areas in Cuttack district, India: a health risk assessment. *Frontiers in Environmental Science* 2023, 11:1248373.

[17] Mohod CV, Dhote J: Review of HMs in drinking water and their effect on human health. *International Journal of Innovative Research in Science, Engineering and Technology* 2013, 2(7):2992-2996.

[18] Rahimzadeh MR, Rahimzadeh MR, Kazemi S, Moghadamnia A-a: Cadmium toxicity and treatment: An update. *Caspian journal of internal medicine* 2017, 8(3):135.

[19] Lobo G, Gadgil A: Preventing leaching from lead water pipes with electrochemistry: an exploratory study. *Environmental Science: Water Research & Technology* 2021, 7(7):1267-1278.

[20] Teschke R: Copper, Iron, Cadmium, and Arsenic, All Generated in the Universe: Elucidating Their Environmental Impact Risk on Human Health Including Clinical Liver Injury. *International Journal of Molecular Sciences* 2024, 25(12):6662.

[21] Srivastava AK, Srivastava AK: Effects of chlordecone on the gonads of freshwater catfish, Heteropneustes fossilis. *Bulletin of environmental contamination and toxicology* 1994, 53:186-191.

[22] Kumar R, Goyal MK, Surampalli RY, Zhang TC: River pollution in India: exploring regulatory and remedial paths. *Clean Technologies and Environmental Policy* 2024, 26(9):2777-2799.

[23] Rahman MS, Molla AH, Saha N, Rahman A: Study on HMs levels and its risk assessment in some edible fishes from Bangshi River, Savar, Dhaka, Bangladesh. *Food chemistry* 2012, 134(4):1847-1854.

[24] Roméo M, Siau Y, Sidoumou Zn, Gnassia-Barelli M: HM distribution in different fish species from the Mauritania coast. *Science of the total environment* 1999, 232(3):169-175.

[25] Harada M: Minamata disease: methylmercury poisoning in Japan caused by environmental pollution. *Crit Rev Toxicol* 1995, 25(1):1-24.

[26] Ajona M, Vasanthi P: Bioremediation of petroleum contaminated soils–A review. *Materials Today: Proceedings* 2021, 45:7117-7122.

[27] Vázquez-Núñez E, Molina-Guerrero CE, Peña-Castro JM, Fernández-Luqueño F, de la Rosa-Álvarez MG: Use of nanotechnology for the bioremediation of contaminants: A review. *Processes* 2020, 8(7):826.

[28] Mudhoo A, Garg VK, Wang S: Removal of HMs by biosorption. *Environmental Chemistry Letters* 2012, 10:109-117.

[29] Gouda SA, Taha A: Biosorption of HMs as a New Alternative Method for Wastewater Treatment: A Review. *Egyptian Journal of Aquatic Biology and Fisheries* 2023, 27(2):135-153.

[30] Ramírez Calderón OA, Abdeldayem OM, Pugazhendhi A, Rene ER: Current updates and perspectives of biosorption technology: an alternative for the removal of HMs from wastewater. *Current Pollution Reports* 2020, 6(1):8-27.

[31] Alothman ZA, Bahkali AH, Khiyami MA, Alfadul SM, Wabaidur SM, Alam M, Alfarhan BZ: Low cost biosorbents from fungi for HMs removal from wastewater. *Separation Science and Technology* 2020, 55(10):1766-1775.

[32] Gómez-Aguilar DL, Rodríguez-Miranda JP, Salcedo-Parra OJ: Fruit peels as a sustainable waste for the biosorption of HMs in wastewater: A review. *Molecules* 2022, 27(7):2124.

[33] Wang J, Chen C: Chitosan-based biosorbents: modification and application for biosorption of HMs and radionuclides. *Bioresource technology* 2014, 160:129-141.

[34] Jadoun S, Fuentes JP, Urbano BF, Yáñez J: A review on adsorption of HMs from wastewater using conducting polymer-based materials. *Journal of Environmental Chemical Engineering* 2023, 11(1):109226.

[35] de Sa Costa HP, da Silva MGC, Vieira MGA: Biosorption of aluminum ions from aqueous solutions using non-conventional low-cost materials: A review. *Journal of Water Process Engineering* 2021, 40:101925.

[36] Wang WX: Bioaccumulation and biomonitoring. In: *Marine ecotoxicology.* edn.: Elsevier; 2016: 99-119.

[37] Filote C, Roşca M, Hlihor RM, Cozma P, Simion IM, Apostol M, Gavrilescu M: Sustainable application of biosorption and bioaccumulation of persistent pollutants in wastewater treatment: Current practice. *Processes* 2021, 9(10):1696.

[38] Chugh M, Kumar L, Bhardwaj D, Bharadvaja N: Bioaccumulation and detoxification of HMs: an insight into the mechanism. In: *Development in wastewater treatment research and processes.* edn.: Elsevier; 2022: 243-264.

[39] Sharma P, Dutta D, Udayan A, Nadda AK, Lam SS, Kumar S: Role of microbes in bioaccumulation of HMs in municipal solid waste: Impacts on plant and human being. *Environmental Pollution* 2022, 305:119248.

[40] Nnaji ND, Onyeaka H, Miri T, Ugwa C: Bioaccumulation for HM removal: a review. *SN Applied Sciences* 2023, 5(5):125.

[41] Diep P, Mahadevan R, Yakunin AF: HM removal by bioaccumulation using genetically engineered microorganisms. *Frontiers in bioengineering and biotechnology* 2018, 6:157.

[42] Streit B: Bioaccumulation processes in ecosystems. *Experientia* 1992, 48(10):955-970.

[43] Zhao X, Wang M, Wang H, Tang D, Huang J, Sun Y: Study on the remediation of Cd pollution by the biomineralization of urease-producing bacteria. *International Journal of Environmental Research and Public Health* 2019, 16(2):268.

[44] Bindschedler S, Cailleau G, Verrecchia E: Role of fungi in the biomineralization of calcite. *Minerals* 2016, 6(2):41.

[45] Arias D, Cisternas LA, Rivas M: Biomineralization mediated by ureolytic bacteria applied to water treatment: A review. *Crystals* 2017, 7(11):345.

[46] Tasleem M, Hussein WM, El-Sayed A-AAA, Alrehaily A: An *In Silico* Bioremediation Study to Identify Essential Residues of Metallothionein Enhancing the Bioaccumulation of HMs in Pseudomonas aeruginosa. *Microorganisms* 2023, 11(9):2262.

[47] Ehrlich H, Bailey E, Wysokowski M, Jesionowski T: Forced Biomineralization: A Review. *Biomimetics* 2021, 6(3):46.

[48] Li X, Wang Y, Tang J, Li K: Removal behavior of HMs from aqueous solutions via microbially induced carbonate precipitation driven by acclimatized sporosarcina pasteurii. *Applied Sciences* 2022, 12(19):9958.

[49] Khadim HJ, Ammar SH, Ebrahim SE: Biomineralization based remediation of cadmium and nickel contaminated wastewater by ureolytic bacteria isolated from barn horses soil. *Environmental Technology & Innovation* 2019, 14:100315.

[50] Abo-Alkasem MI, Hassan NmH, Abo Elsoud MM: Microbial bioremediation as a tool for the removal of HMs. *Bulletin of the National Research Centre* 2023, 47(1):31.

[51] Johnson H, Cho H, Choudhary M: Bacterial HM resistance genes and bioremediation potential. *Computational Molecular Bioscience* 2019, 9(1):1-12.

[52] Kang CH, Kwon YJ, So JS: Bioremediation of HMs by using bacterial mixtures. *Ecological engineering* 2016, 89:64-69.

[53] Thai TD, Lim W, Na D: Synthetic bacteria for the detection and bioremediation of HMs. *Frontiers in Bioengineering and Biotechnology* 2023, 11:1178680.

[54] Bestawy EE, Helmy S, Hussien H, Fahmy M, Amer R: Bioremediation of HM-contaminated effluent using optimized activated sludge bacteria. *Applied water science* 2013, 3:181-192.

[55] Khan N, Bano A: Role of plant growth promoting rhizobacteria and Ag-nano particle in the bioremediation of HMs and maize growth under municipal wastewater irrigation. *International Journal of Phytoremediation* 2016, 18(3):211-221.

[56] Verma S, Kuila A: Bioremediation of HMs by microbial process. *Environmental Technology & Innovation* 2019, 14:100369.

[57] Basha SA, Rajaganesh K: Microbial bioremediation of HMs from textile industry dye effluents using isolated bacterial strains. *Int J Curr Microbiol Appl Sci* 2014, 3:785-794.

[58] Carlile MJ, Watkinson SC, Gooday GW: The fungi: Gulf Professional Publishing; 2001.

[59] Blackwell M: The Fungi: 1, 2, 3… 5.1 million species? *American journal of botany* 2011, 98(3):426-438.

[60] Stephenson S: The kingdom fungi. *Mycoagroecology* 2022:35-49.

[61] Bano A, Hussain J, Akbar A, Mehmood K, Anwar M, Hasni MS, Ullah S, Sajid S, Ali I: Biosorption of HMs by obligate halophilic fungi. *Chemosphere* 2018, 199:218-222.

[62] Diez M: Biological aspects involved in the degradation of organic pollutants. *Journal of soil science and plant nutrition* 2010, 10(3):244-267.

[63] Zeraatkar AK, Ahmadzadeh H, Talebi AF, Moheimani NR, McHenry MP: Potential use of algae for HM bioremediation, a critical review. *Journal of environmental management* 2016, 181:817-831.

[64] Ma X-k, Ding N, Peterson EC, Daugulis AJ: HMs species affect fungal-bacterial synergism during the bioremediation of fluoranthene. *Applied microbiology and biotechnology* 2016, 100:7741-7750.

[65] Flouty R, Estephane G: Bioaccumulation and biosorption of copper and lead by a unicellular algae Chlamydomonas reinhardtii in single and binary metal systems: a comparative study. *Journal of environmental management* 2012, 111:106-114.

[66] Ali H, Khan E, Sajad MA: Phytoremediation of HMs—concepts and applications. *Chemosphere* 2013, 91(7):869-881.

[67] Sharma S, Rana S, Thakkar A, Baldi A, Murthy R, Sharma R: Physical, chemical and phytoremediation technique for removal of HMs. *Journal of HM Toxicity and Diseases* 2016, 1(2):1-15.

[68] Vasavi A, Usha R, Swamy P: Phytoremediation–an overview review. *J Ind Pollut Control* 2010, 26(1):83-88.

[69] Lone MI, He ZL, Stoffella PJ, Yang XE: Phytoremediation of HM polluted soils and water: progresses and perspectives. *J Zhejiang Univ Sci B* 2008, 9(3):210-220.

[70] Khandare RV, Watharkar AD, Pawar PK, Jagtap AA, Desai NS: Hydrophytic plants Canna indica, Epipremnum aureum, Cyperus alternifolius and Cyperus rotundus for phytoremediation of fluoride from water. *Environmental Technology & Innovation* 2021, 21:101234.

[71] Alsafran M, Usman K, Ahmed B, Rizwan M, Saleem MH, Al Jabri H: Understanding the phytoremediation mechanisms of potentially toxic elements: a proteomic overview of recent advances. *Frontiers in plant science* 2022, 13:881242.

[72] Delgado-González CR, Madariaga-Navarrete A, Fernández-Cortés JM, Islas-Pelcastre M, Oza G, Iqbal HM, Sharma A: Advances and applications of water phytoremediation: A potential biotechnological approach for the treatment of HMs from contaminated water. *International Journal of Environmental Research and Public Health* 2021, 18(10):5215.

[73] Wei Z, Van Le Q, Peng W, Yang Y, Yang H, Gu H, Lam SS, Sonne C: A review on phytoremediation of contaminants in air, water and soil. *Journal of hazardous materials* 2021, 403:123658.

[74] Laghlimi M, Baghdad B, El Hadi H, Bouabdli A: Phytoremediation mechanisms of HM contaminated soils: a review. *Open journal of Ecology* 2015, 5(8):375-388.

[75] Al-Thani R, Yasseen B: Phytoremediation of polluted soils and waters by native Qatari plants: Future perspectives. *Environmental Pollution* 2020, 259:113694.

[76] Etim E: Phytoremediation and its mechanisms: a review. *Int J Environ Bioenergy* 2012, 2(3):120-136.

[77] Medfu Tarekegn M, Zewdu Salilih F, Ishetu AI: Microbes used as a tool for bioremediation of HM from the environment. *Cogent Food & Agriculture* 2020, 6(1):1783174.

[78] Bala S, Garg D, Thirumalesh BV, Sharma M, Sridhar K, Inbaraj BS, Tripathi M: Recent strategies for bioremediation of emerging pollutants: a review for a green and sustainable environment. *Toxics* 2022, 10(8):484.

[79] Whitacre DM: Reviews of environmental contamination and toxicology, vol. 202: Springer; 2008.

[80] Alvarez A, Saez JM, Costa JSD, Colin VL, Fuentes MS, Cuozzo SA, Benimeli CS, Polti MA, Amoroso MJ: Actinobacteria: current research and perspectives for bioremediation of pesticides and HMs. *Chemosphere* 2017, 166:41-62.

[81] Ayilara MS, Babalola OO: Bioremediation of environmental wastes: the role of microorganisms. *Frontiers in Agronomy* 2023, 5:1183691.

[82] Abatenh E, Gizaw B, Tsegaye Z, Wassie M: The role of microorganisms in bioremediation-A review. *Open Journal of Environmental Biology* 2017, 2(1):038-046.

[83] Thavasi R, Jayalakshmi S, Banat IM: Application of biosurfactant produced from peanut oil cake by Lactobacillus delbrueckii in biodegradation of crude oil. *Bioresource technology* 2011, 102(3):3366-3372.

[84] Pande V, Pandey SC, Sati D, Bhatt P, Samant M: Microbial Interventions in Bioremediation of HM Contaminants in Agroecosystem. *Front Microbiol* 2022, 13:824084.

[85] Zhang Y, Labianca C, Chen L, De Gisi S, Notarnicola M, Guo B, Sun J, Ding S, Wang L: Sustainable ex-situ remediation of contaminated sediment: a review. *Environmental Pollution* 2021, 287:117333.

[86] Yan DY, Liu T, Lo IM: Treatment of urban river contaminated sediment with ex situ advanced oxidation processes: technical feasibility, environmental discharges and cost-performance analysis. *Environmental Technology* 2015, 36(16):2060-2068.

[87] Kathiravan MN, Karthick R, Muthukumar K: Ex situ bioremediation of Cr (VI) contaminated soil by Bacillus sp.: batch and continuous studies. *Chemical Engineering Journal* 2011, 169(1-3):107-115.

[88] Fabbricino M, Ferraro A, Del Giudice G, d'Antonio L: Current views on EDDS use for ex situ washing of potentially toxic metal contaminated soils. *Reviews in Environmental Science and Bio/Technology* 2013, 12:391-398.

[89] Kuppusamy S, Palanisami T, Megharaj M, Venkateswarlu K, Naidu R: Ex-situ remediation technologies for environmental pollutants: a critical perspective. *Reviews of Environmental Contamination and Toxicology Volume 236* 2016:117-192.

[90] Sharma P, Kumar S, Pandey A: Bioremediated techniques for remediation of metal pollutants using metagenomics approaches: a review. *Journal of Environmental Chemical Engineering* 2021, 9(4):105684.

[91] Sharma P, Pandey AK, Kim SH, Singh SP, Chaturvedi P, Varjani S: Critical review on microbial community during in-situ bioremediation of HMs from industrial wastewater. *Environmental Technology & Innovation* 2021, 24:101826.

[92] Hussain A, Rehman F, Rafeeq H, Waqas M, Asghar A, Afsheen N, Rahdar A, Bilal M, Iqbal HM: *In-situ, Ex-situ,* and nano-remediation strategies to treat polluted soil, water, and air–A review. *Chemosphere* 2022, 289:133252.

[93] Wu Y, Xia L, Yu Z, Shabbir S, Kerr PG: In situ bioremediation of surface waters by periphytons. *Bioresource technology* 2014, 151:367-372.

[94] Yang Z, Ma J, Liu F, Zhang H, Ma X, He D: Mechanistic insight into pH-dependent adsorption and coprecipitation of chelated HMs by in-situ formed iron (oxy) hydroxides. *Journal of Colloid and Interface Science* 2022, 608:864-872.

[95] Kafle A, Timilsina A, Gautam A, Adhikari K, Bhattarai A, Aryal N: Phytoremediation: Mechanisms, plant selection and enhancement by natural and synthetic agents. *Environmental Advances* 2022, 8:100203.

[96] Salt DE, Smith R, Raskin I: Phytoremediation. *Annual review of plant biology* 1998, 49(1):643-668.

[97] Singh N, Santal AR: Phytoremediation of HMs: the use of green approaches to clean the environment. *Phytoremediation: Management of Environmental Contaminants, Volume 2* 2015:115-129.

[98] Pulford ID, Watson C: Phytoremediation of HM-contaminated land by trees—a review. *Environment international* 2003, 29(4):529-540.

[99] Zhou B, Zhang T, Wang F: Microbial-based HM bioremediation: Toxicity and eco-friendly approaches to HM decontamination. *Applied Sciences* 2023, 13(14):8439.

[100] Karnwal A, Martolia S, Dohroo A, Al-Tawaha ARMS, Malik T: Exploring bioremediation strategies for HMs and POPs pollution: the role of microbes, plants, and nanotechnology. *Frontiers in Environmental Science* 2024, 12:1397850.

[101] Goswami RK, Agrawal K, Shah MP, Verma P: Bioremediation of HMs from wastewater: a current perspective on microalgae-based future. *Letters in Applied Microbiology* 2022, 75(4):701-717.

[102] Vardhan KH, Kumar PS, Panda RC: A review on HM pollution, toxicity and remedial measures: Current trends and future perspectives. *Journal of Molecular Liquids* 2019, 290:111197.

[103] Fasani E, Manara A, Martini F, Furini A, DalCorso G: The potential of genetic engineering of plants for the remediation of soils contaminated with HMs. *Plant, cell & environment* 2018, 41(5):1201-1232.

[104] Patel HK, Kalaria RK, Khimani MR: Nanotechnology: a promising tool for bioremediation. *Removal of toxic pollutants through microbiological and tertiary treatment* 2020:515-547.

[105] Sastry M, Ahmad A, Khan MI, Kumar R: Biosynthesis of metal nanoparticles using fungi and actinomycete. *Current science* 2003:162-170.

[106] Rajendran P, Gunasekaran P: Nanotechnology for bioremediation of HMs. In: *Environmental bioremediation technologies.* edn.: Springer; 2007: 211-221.

## Chapter 16

# The Role of Biochar as a Catalyst in Advanced Oxidation Processes for Wastewater Treatment

**Anjali Mehta**[1]
**Tanisha Kathuria**[1]
**Sudhanshu Sharma**[1]
**and Sudesh Kumar**[2,*]

[1]Department of Chemistry, Banasthali Vidyapith, Tonk, Rajasthan, India
[2]Department of Education in Science and Mathematics, National Council of Educational Research and Training (NCERT), Delhi, India

## Abstract

Biochar (BC) has been widely used in environmental remediation due to its large surface area, cost-effectiveness, and diverse range of raw material. As an adsorbent for contaminant removal, BC exhibits limited adsorption capacity over time. To expand its applications, recent efforts have focused on improving the catalytic activity of BC and BC-based catalysts. As BC contains oxygen-containing functional groups (OFGs) and persistent free radicals (PFRs), it functions as a catalyst. Moreover, the incorporation of metallic and non-metallic components to BC has a synergistic effect on the production of reactive oxygen species (ROS), electron transfer, adsorption, and oxidation of pollutants. This chapter explores BC's role in wastewater treatment, covering feedstock selection, preparation, physicochemical properties, and modifications. It also examines the mechanisms underlying the effectiveness of BC-based catalysts in catalytic processes. This chapter investigates the role of BC and BC-based catalysts in advanced oxidation processes (AOPs), including the activation of persulfate (PS), peroxymonosulfate (PMS), and hydrogen peroxide ($H_2O_2$). The objective of this chapter is to highlight the critical role that BC plays in developing environmentally friendly and sustainable technologies for environmental remediation.

**Keywords:** biochar, pyrolysis, catalyst, advance oxidation process, wastewater treatment

---

* Corresponding Author's Email: sudeshneyol@gmail.com

In: Advancements in Wastewater Treatment
Editors: Neha Agarwal, Maulin P. Shahl, Vijendra Singh Solankil et al.
ISBN: 979-8-90134-007-3

## 1. Introduction

Environmental pollution has emerged as a complex, multidisciplinary challenge in recent years as a result of the ongoing release of new chemicals into the environment [1,2]. Examples of contaminants found in water, harmful to both human health and the environment, include dyes, antibiotics, pesticides, endocrine disruptors, pharmaceuticals, volatile organic compounds, and personal care products [3,4]. To address water pollution, a number of technologies have been employed, including biological treatment [5], photocatalytic degradation [6], adsorption [7], and AOPs [8]. AOPs have become a prominent strategy among water treatment methods due to their ability to effectively degrade pollutants at the molecular level [9]. In practice, AOPs typically employ highly reactive species with strong oxidizing power to completely degrade persistent pollutants in water into simpler, harmless substances Zhou. Growing water environmental concerns require the development of effective and environmentally friendly water treatment techniques.

A porous carbonaceous material known as biochar (BC) is produced through the thermochemical decomposition of biomass in an oxygen-limited environment [10]. BC can be made from almost any biomass, such as wood chips, manures, sewage sludges, organic municipal solid wastes, food wastes, agricultural leftovers, algas, and many more [11,12]. BC has been used to eliminate pollutants from wastewater due to its extensive biomass, extensive specific surface area (SSA), numerous surface OFGs, porous composition, and affordability. Moreover, numerous modification methods were applied to improve the performance of BC and expand its range of applications [13]. Typically, BC's performance can be enhanced through physical or chemical activation methods that modify its surface area and pore structure [14]. BC materials have demonstrated exceptional activity as a catalyst or catalyst support in numerous applications addressing a range of processes, such as photocatalysis and PS/PMS oxidation, along with several other advanced applications. Incorporating BC into advanced oxidation processes offers a range of advantages.

To begin with, BC acts as a stable and inert support for different catalysts, increasing their catalytic activity and preventing agglomeration [15]. Furthermore, BC's surface is rich in various functional groups, like carboxyl groups, hydroxyl, and phenolic, which can act as adsorption and activation sites for oxidants, thereby promoting the formation of •OH and other potent ROS [16]. These radicals can efficiently break down a variety of organic contaminants because of their high oxidizing ability. In addition to its other benefits, BC's excellent adsorption properties enable it to capture and concentrate organic compounds from water within its porous network, thereby minimizing operational costs and environmental burdens [17].

This chapter aims to demonstrate that integrating BC with AOPs provides an effective approach for the mineralization and degradation of organic contaminants in water. It examines the thermochemical processes for BC production. The final section discusses the application of BC in AOPs for organic contaminant removal, highlighting its benefits, practical transformations, and methods for oxidant activation (persulfate, peroxymonosulfate, and $H_2O_2$).

## 2. Biochar (BC) Preparation

The category of biomass employed significantly influences both the amount of BC produced and its characteristics, partly due to catalytic effects during pyrolysis [18]. Numerous methods for producing BC from various feedstocks have been explored (Figure 1). The primary thermochemical conversion technologies are pyrolysis, hydrothermal carbonization, and torrefaction.

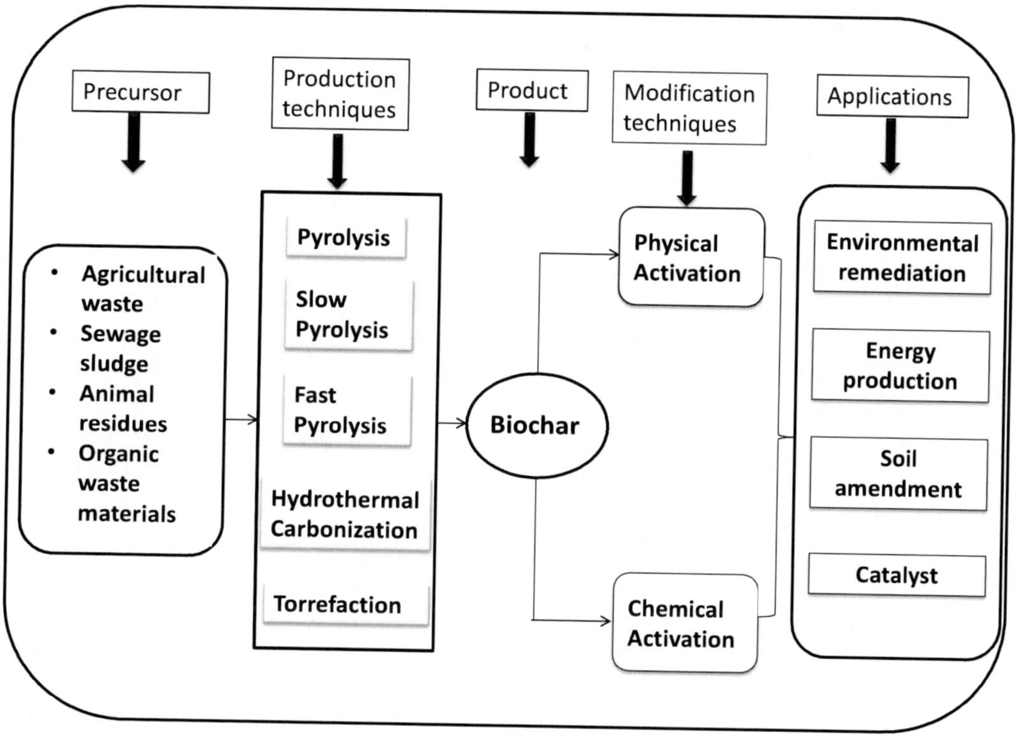


**Figure 1.** Preparation of BC/modified BC and its applications [24].

### 2.1. Pyrolysis

Pyrolysis, also known as devolatilization, involves the thermal decomposition of feedstock at temperatures between 250 and 900 °C in an oxygen-restricted or inert environment. A major advantage of this process is its ability to convert feedstock into BC, while also enabling the production of syngas and biofuel when operated at elevated temperatures or energy levels [19]. The characteristics of BCs, such as their porosity, SSA, total pore volume, and functional groups, are known to be affected by the conditions under which pyrolysis is carried out, including residence time, temperature, and heating rate. Slow pyrolysis, fast pyrolysis, hydrothermal carbonization, and torrefaction are the primary pyrolysis-related techniques, as summarized in Table 1.

Slow pyrolysis (historically used to produce charcoal) involves converting biomass into BC with a residence time exceeding two hours. The outputs of this method are BC, bio-oil, and syngas. A high BC yield is often obtained under slow pyrolysis processes.

**Table 1.** Summary of pyrolysis methods [27]

| Parameters | Slow pyrolysis | Fast pyrolysis | Hydrothermal carbonization | Torrefaction |
|---|---|---|---|---|
| Temperature | 450–500°C | 500°C–1000°C | 180–350°C | 200–300°C |
| Heating period | Very slow <10 °C/ min | Very fast 1000°C/s | Relatively slow <10°C/min | <10°C/min |
| Retention time | Hours | Seconds | Hours | Hours |
| BC yield (%) | 35 | 10–15 | 30–60 | 70–80 |

In the fast pyrolysis process, the feedstock undergoes rapid heating. The temperature typically ranges between 500 °C and 1000 °C, and the reaction occurs within a very short time frame of just 1 to 5 seconds. To facilitate fast pyrolysis, biomass is typically pre-dried to a water content of less than 10%. This process yields less BC and large amounts of bio-oil and noncondensable gasses in contrast to slow pyrolysis [20].

### 2.2. Hydrothermal Carbonization

Hydrothermal carbonization involves heating biomass in an aqueous solution within an autoclave at temperatures ranging from 180 to 350°C for several hours to produce hydrochar. Sewage sludge and other high-moisture-content feedstocks are suitable, as they do not require drying. It is also known as wet torrefaction [21] or wet pyrolysis [22]. In contrast to BC, hydrochar generally exhibits lower carbon and ash content, as well as a lower SSA (SBET) as well as reduced pore volume [23].

### 2.3. Torrefaction

Torrefaction is a mild pretreatment process conducted before pyrolysis, typically involving temperatures between 200 and 300°C, a slow heating, and a short retention period. Torrified biomass is the solid substance that results from torrefaction. It cannot be defined as 'original BC' since it possesses both biomass and BC properties. Torrefaction-char may include more oxygen-containing functional groups than BC since the thermal process was carried out at substantially lower temperatures. The kind of raw materials, temperature, retention period, and furnace atmosphere, all affect BC yield [25]. For instance, one prototype process produced solid yields for hardwood and switchgrass pellets of more than 77 wt % [26].

## 3. BC-Based Catalysts

Since the introduction of biochar to catalysis, researchers have explored various strategies to enhance its catalytic efficiency. The most commonly used methods to improve BC's catalytic performance include metal loading and heteroatom doping. These are often applied alongside conventional techniques such as pyrolysis and chemical activation.

### 3.1. Pristine BC

The products of pyrolyzing biomass at a specific temperature without the addition of any compounds are referred to as "pristine BC." Due to its distinctive physicochemical characteristics, pristine BC has become a viable catalyst in AOPs for wastewater treatment. BC, a product of biomass pyrolysis, exhibits a large specific surface area (SSA), numerous functional groups, and a porous structure, contributing to its strong catalytic potential. Because of these properties, BC can activate oxidants like hydrogen peroxide, persulfate, or peroxymonosulfate, producing ROS like hydroxyl and sulfate radicals that break down persistent organic pollutants. BC can decompose organic pollutants directly without the need for oxidants [28]. BC contains PFRs that can transfer electrons to $O_2$, resulting in the creation of superoxide radical anions and $H_2O_2$, which subsequently interact with PFRs to generate $\cdot OH$ [29]. Additionally, BC's surface functional groups and natural mineral content support electron transfer processes and radical production routes, making it an economical, sustainable, and efficient substitute for traditional catalysts [29]. Integrating raw BC into AOPs not only helps in the removal of pollutants but also supports the recovery of valuable resources and promotes a circular economy framework for wastewater management. Chen et al. [30] produced BC using hydrothermal carbonization, which, when exposed to sunshine, could break down 69% of sulfadimidine in 6 hours. Research suggests that the photocatalytic capabilities of BC can be boosted by employing a combination of pyrolysis and hydrothermal carbonization techniques [31].

### 3.2. Metal Loading

BC's surface properties can be controlled during biomass preparation, BC formation, and post-synthesis treatments. One of the best methods for increasing BC's catalytic potential is to load it with metal. Additionally, BC has the potential for dispersing metal particles and prevents metal leaching. When compared to alternative catalytic supports, BC demonstrates superior efficiency and cost-effectiveness.

Zero-valent iron supported by BC was produced at 900°C from a lignin magnetite pellet [32]. A BC-$Co_3O_4$ catalyst made from rice straw was shown to be efficient in the oxidative degradation of ofloxacin through peroxymonosulfate (PMS) (approximately 90% removal in 10minutes) [33]. Additionally, it has been found that BC made from wheat straw can accelerate electron transfer, which will increase the redox cycling of $Co_3O_4$ supported on the BC [34]. The use of zero-valent iron supported by BC was also found to be a successful approach for decomposing hydrogen peroxide and subsequently degrading sulfamethazine [35] and trichloroethylene [36]. Sono-Fenton-like reactions further demonstrated the effectiveness of BC-supported catalysts. For the sono-Fenton-like oxidation of bisphenol A, Jung et al. [37] first derived BC from rice husks, followed by the synthesis of $MnO_2$–BC composites. Within 20 minutes of the process, the composites completely removed the pollutant (100 μM), demonstrating their outstanding performance. Meng et al. [38] demonstrated that Fe/C BC, derived from corn cobs and corn stalks, could be used in situ to remove trichloroethylene from groundwater. While just 44% of trichloroethylene was eliminated in 20 minutes without PMS, under ideal conditions, 0.1 mM trichloroethylene was entirely eliminated in 20 minutes with 5.0 mM PMS. Researchers prepared a novel BC-Cu NP composite, utilizing BC from solid digestate, to activate $H_2O_2$ in order to degrade tetracycline (TC) [39]. In 6 hours, the BC-Cu/ $H_2O_2$ system eliminated almost 97.8% of the TC. The Cu(II)/Cu(I) redox system, along with the radical-mediated electron transport mechanism in BC, has been attributed to the TC degradation. •OH radicals can be produced by the Cu(II)/Cu(I) redox system reacting with $H_2O_2$ (Eqs. (1) and (2)).

$$Cu^{+}(aq) + H_2O_2 + H^{+} \rightarrow Cu^{2+}(aq) + \bullet OH + H_2O \quad (1)$$

$$Cu^{+}(aq) + \bullet OH + H^{+} \rightarrow Cu^{2+}(aq) + H_2O \quad (2)$$

### 3.3. Heteroatom Doping

Doping with heteroatoms is a significant modification technique. Over recent decades, researchers have developed functional catalysts with enhanced properties and efficiency through heteroatom doping [40, 41]. Non-metallic heteroatoms can alter the fundamental properties of BC catalysts, including SSA and pore diameter, and induce surface defects. They additionally have the potential to boost the number of Lewis acid sites, which expands the active sites [42]. Nitrogen (N) has emerged as a nonmetallic element of considerable interest. Due to the similar sizes of N and C atoms, nitrogen can readily integrate into the graphene sheet structure. Additionally, nitrogen's higher electronegativity compared to carbon can create Lewis base sites on adjacent carbon atoms [43]. Moreover, this electronegativity difference enhances electron transfer capacity and induces defect structures, improving the catalytic potential of carbonaceous materials [44].

In order to create N-doped "graphitic BC" stimulants for peroxydisulfate (PDS), nitrogen was introduced into a wetland plant (reed) using ammonium nitrate [45]. When BC was modified with nitrogen, its specific surface area (SBET) dramatically increased to 496.7 $m^2 g^{-1}$ (compared to 95.2 $m^2 g^{-1}$ for the original BC), resulting in a better adsorption capacity. It has also been reported that nitrogen doping can be utilized to boost the electrocatalytic efficiency of BC. As reported by Li et al., the incorporation of nitrogen into a loofah sponge-derived BC electrode led to an increase in both its SSA and the number of active sites, thereby enhancing its ability to remove bromate [46].

## 4. BC-Based Catalysts in AOPs for Wastewater Treatment

BC (BC)-supported catalysts have emerged as a significant area of research within AOPs. These catalysts facilitate the production of ROS, such as hydroxyl and sulfate radicals, which can decompose diverse organic contaminants, including dyes, pharmaceuticals, and pesticides.

### 4.1. Photocatalysis

BC serves as an excellent support material for photocatalytic nanoparticles like graphitic carbon nitride ($g\text{-}C_3N_4$), zinc oxide (ZnO), and titanium dioxide ($TiO_2$). The inclusion of BC (BC) in these photocatalysts has the potential to provide some benefits: (a) improved light absorption [47], (b) enhanced charge separation and reduced electron-hole recombination [48], and (c) due to their strong ability to adsorb substances, the photodegradation process is made more effective [49], resulting in the effective production of ROS when exposed to light. BC can produce ROS through its carbon matrix when exposed to sunlight. Additionally, BC serves as an effective alternative photocatalyst support. A BC combined with graphitic carbon nitride (g-$C_3N_4$) was developed by Jeon et al. [50] using a nitrogen-rich precursor. Orange G (OG) was oxidized using the prepared BC. Under light, more than 90% of OG was oxidized.

Using a solvent-free ball-milling technique, a ZnO-modified maize BC nanocomposite (MB-ZnO) was developed to facilitate the photocatalytic degradation of Mancozeb (MC) and Safranin O (Saf). Within 1 hour, the MB-ZnO composite degraded Saf under visible light (83.5%), UV light (81.0%), and darkness (78%). It was observed that the MB-ZnO nanocomposite degraded MC most effectively under visible light (56.5 %), with UV light resulting in lower degradation (27.5 %) and even less in darkness (25.2 %) [51]. The magnetic BC photocatalyst (MBC) synthesized by Fatimah et al. [52] from snake fruit peel involved the co-precipitation of iron oxide precursors and was used for the photodegradation of rhodamine B (RhB). In order to enhance its ability to degrade RhB under photocatalytic conditions, the band gap energy is 2.23 eV, the particle size ranges from 5 to 20 nm, and the BET-SSA is 126 $m^2/g$. Under optimal conditions (pH 7 and 0.5% $H_2O_2$ concentration) and UV light irradiation, the photodegradation of RhB fits well for second-order kinetics, achieving 99.9% degradation efficiency within 15 minutes. Liu synthesized an amino-modified BC-supported Ag/$Ag_3PO_4$ composite that showed significantly improved photocatalytic performance, degrading almost 80% of amoxicillin in 120 minutes, compared to only 65% degradation with pure $Ag_3PO_4$ [53].

## 4.2. Sonophotocatalysis

There's a growing interest in using ultrasound as the main energy source to drive chemical reactions. BC can be used as a sonocatalyst in ultrasound technology to decompose organic pollutants. A hydrothermal process for synthesis was used to produce $CeO_2$/BC, and the resulting material was utilized in sonocatalytic degradation studies of Reactive Red 84, an organic dye. Increases in catalyst dose and ultrasonic power improved catalyst efficiency, but increases in dye concentration and pH value reduced it. According to researchers, the presence of quenching effects from different scavengers indicates that the percentage of OH radicals plays a crucial part in the process [54].

The degradation efficiency of $ZrO_2$-BC was 96.8% when the initial concentration of Reactive Yellow 39 was 20 mg/L and the ultrasonic power was 300 W [55]. The remarkable sonocatalytic activity observed can be attributed to the phenomena of sonoluminescence and hot spot formation. The study reveals that $TiO_2$-BC nanocomposites can be effectively employed as sonocatalysts for dye degradation, utilizing mechanisms comparable to $ZrO_2$-BC, and achieving a high degradation efficiency of 97.5% for Reactive Blue 69 (initial concentration 20 mg·L−1) under 300W ultrasonic irradiation over 80 minutes [56]. There is still a lot to learn about the synthesis and use of materials based on metal oxides and BC, even after multiple successful attempts to achieve so. These materials seem to have a bright future and a lot of potential in the area of environmental preservation, particularly in the purification of water.

### 4.3. Persulfate, Peroxymonosulfate and Hydrogen Peroxide Activation System

PS and PMS ions are the main contributors of sulfate radicals. Apart from its conductive properties, BC's surface has OFG, which helps to activate PS and PMS. Specifically, the hydroxyl group (-OH) on the surface and the π-π* concentration of the BC are beneficial to PS activation [57]. Correspondingly, the ketonic functionalities on the BC surface are important in the process of PMS activation [58]. As a result of its abundance of OFGs, BC can facilitate the production of $SO_4^{\bullet-}$ radicals as well as the activation of PS, $H_2O_2$, and PMS (equations 7.3 and 7.4) [26-30].

$$(BC_{surface}) - -OOH + S_2O_8^{2-} \rightarrow SO_4^{\bullet-} + (BC_{surface}) - OO^{\bullet} + HSO_4^{-} \quad (7.3)$$

$$(BC_{surface}) - -OH + S_2O_8^{2-} \rightarrow SO_4^{\bullet-} + (BC_{surface}) - O^{\bullet} + HSO_4^{-} \quad (7.4)$$

Furthermore, PFRs generated within the BC were shown to be capable of activating persulfate (PS), and the extent of this activation varies based on the nature and amount of PFRs [59]. The research by Fu et al. [60] involved the hydrothermal synthesis of a novel graphitized hierarchical porous BC and $MnFe_2O_4$ magnetic composite, which was subsequently used to stimulate PMS for the degradation of Acid Orange 7 (AO7). Additionally, it was discovered that in the MnFe2O4/CS system, the surface of the BC made from maize stems (CS) was where radical-induced oxidation occurred most frequently. The process began with the generation of hydroxyl and sulfate radicals through electron donation by the CS to PMS, followed by the regeneration of CS via electron acceptance from PMS. Second, the transfer of electrons from Mn2+ to PMS led to the generation of $SO_4^{\bullet-}$ and the decomposition of contaminants. The presence of Fe2+ ions also led to enhanced activation of PMS. The hydroxyl and sulfate radicals hat generated decomposed the organic contaminants [60]. The $CuFe_2O_4$-loaded BC (BC) system developed by Zhang et al. effectively activates persulfate (PS) to degrade the antibiotic tetracycline (TC). The removal efficiency reached 96.0% under ideal circumstances [61]. A stalk BC-ZVI composite that has OH and COOH groups on its surface and has the ability to activate PS was developed by Liu et al. in 2022. Stalk BC combined with nZVI demonstrated exceptional efficiency, achieving almost 100% removal of Congo Red (10 min), New Coccine (30 min), and Direct Blue 71 (20 min) [62]. According to reports, the PS activation system's active sites include defects, pore structure, PFRs, etc. [63].

Huang et al. utilized BC produced from wheat straw to activate $H_2O_2$ in order to remove SMX [64]. Luo et al. employed sludge BC and $HNO_3$-treated BC to initiate $H_2O_2$ activation, enabling the degradation of ciprofloxacin (CIP) [65]. The investigation revealed that the $HNO_3$-treated BC combined with $H_2O_2$ demonstrated superior degradation performance (93%) compared to the unmodified BC/ $H_2O_2$system, which achieved 70.60% efficiency. The following explanation was given for the activation of $H_2O_2$: BC possesses an abundance of structural defects and OFGs, enhancing its catalytic and adsorption properties. A potential mechanism involves the transfer of an electron originating from the catalytic sites of PFRs within BC to $H_2O_2$, initiating the decomposition process. Table 2 provides a summary of the various applications of BC in advanced oxidation processes (AOPs).

**Table 2.** Application of BC in AOPs

| Pollutant | Process | Catalyst | Biomass | Degradation efficiency | Reference |
|---|---|---|---|---|---|
| Methyl Orange | Photocatalysis | $TiO_2$/BC | Walnut shells | 83.23% | [66] |
| Tetracycline | Photocatalysis | BiOCl ultrathin nanosheets with BC | Bamboo leaves | 71.8% | [67] |
| Gemifloxacin | Sonocatalysis Hydrothermal removal = | ZnO-BC | Wheat husks and paper sludge | 96.1% | [68] |
| Acetaminophen and naproxen | Sonocatalysis Pyrolysis | BC | Loblolly pine | 100% | [69] |
| p-hydroxybenzoic acid | PMS | $Fe_3O_4$/BC | Myriophyllum aquaticum | 100% | [70] |
| Tetracycline | PMS | Cobalt-BC | spent coffee ground | 100% | [71] |
| Metolachlor | PMS | N- BC supported $CoFe_2O_4$ | Rice straw | 24.1% | [72] |
| Bisphenol A | PS | nZVI-BC | Kenaf Bar | 98%, | [73] |

## Conclusion

In this chapter, we explored the diverse applications of BC as a sustainable and efficient catalyst in AOPs for wastewater remediation. We examined how BC synthesis via pyrolysis and torrefaction influences its physicochemical properties, such as surface area, porosity, and functional group availability, which are critical for catalytic activity. Further, various modification strategies that enhance BC's catalytic performance improve electron transfer capabilities, surface reactivity, and pollutant adsorption potential. These modifications tailor BC to function as a robust catalyst in a range of AOPs. The chapter also provided a comprehensive overview of BC's application in different oxidative systems such as photocatalysis, sonophotocatalysis, and the activation of oxidants (PS and PMS). In these systems, modified BC has shown significant potential in degrading a wide array of organic pollutants by generating ROS, such as hydroxyl and sulphate radicals, leading to mineralization of contaminants. Overall, BC stands out as a promising, low-cost, and eco-friendly catalyst for wastewater treatment, offering a circular solution that converts waste biomass into a value-added product. Future research should focus on optimizing modification techniques, deepening the understanding of reaction mechanisms, and scaling up processes for practical applications.

## References

[1] Pan, Y., Liu, X., Zhang, W., Liu, Z., Zeng, G., Shao, B., & Chen, M. (2020). "Advances in photocatalysis based on fullerene C60 and its derivatives: Properties, mechanism, synthesis, and applications." *Appl. Catal. B: Environ.*, *265*:118579. https://doi.org/10.1016/j.apcatb.2019.118579.

[2] Liu, Y., Huang, D., Cheng, M., Liu, Z., Lai, C., Zhang, C., & Liang, Q. (2020). "Metal sulfide/MOF-based composites as visible-light-driven photocatalysts for enhanced hydrogen production from water splitting." *Coord. Chem. Rev.*, *409*:213220. https://doi.org/10.1016/j.ccr.2020.213220.

[3] Cheng, N., Wang, B., Wu, P., Lee, X., Xing, Y., Chen, M., & Gao, B. (2021). "Adsorption of emerging contaminants from water and wastewater by modified BC:" A review. *Environ. Pollut.*, *273*:116448. https://doi.org/10.1016/j.envpol.2021.116448.

[4] Jiang, Z., Chen, M., Lee, X., Feng, Q., Cheng, N., Zhang, X., & Wang, B. (2023). "Enhanced removal of sulfonamide antibiotics from water by phosphogypsum modified BC composite." *J. Environ. Sci.*, *130*:174-186. https://doi.org/10.1016/j.jes.2022.10.023.

[5] Zhong, H., Wang, Z., Liu, Z., Liu, Y., Yu, M., & Zeng, G. (2016). "Degradation of hexadecane by Pseudomonas aeruginosa with the mediation of surfactants: relation between hexadecane solubilization and bioavailability." *Int. Biodeterior. Biodegrad.*, *115*:141-145. https://doi.org/10.1016/j.ibiod.2016.08.008.

[6] Huang, J., Liu, X., Zhang, W., Liu, Z., Zhong, H., Shao, B., & He, Q. (2021). "Functionalization of covalent organic frameworks by metal modification: Construction, properties and applications." *J. Chem. Eng.*, *404*:127136. https://doi.org/10.1016/j.cej.2020.127136.

[7] Lin, Y. Z., Zhong, L. B., Dou, S., Shao, Z. D., Liu, Q., & Zheng, Y. M. (2019). "Facile synthesis of electrospun carbon nanofiber/graphene oxide composite aerogels for high efficiency oils absorption." *Environ. Int*, *128*:37-45. https://doi.org/10.1016/j.envint.2019.04.019.

[8] Kathuria, T., Mehta, A., Sharma, S., & Kumar, S. (2024). "Review on ultrasound-enhanced activation of persulfate/peroxymonosulfate in hybrid advanced oxidation technologies." *Chem. Eng. Commun.*, *211*(10): 1645-1669. https://doi.org/10.1080/00986445.2024.2358372.

[9] Miklos, D. B., Remy, C., Jekel, M., Linden, K. G., Drewes, J. E., & Hübner, U. (2018). "Evaluation of advanced oxidation processes for water and wastewater treatment–A critical review." *Water Res.*, *139*:118-131. https://doi.org/10.1016/j.watres.2018.03.042.

[10] Aller, M. F. (2016). "BC properties: Transport, fate, and impact." *Crit. Rev. Environ. Sci. Technol*, *46*(14-15), 1183-1296. https://doi.org/10.1080/10643389.2016.1212368.

[11] Lopes, R. P., & Astruc, D. (2021). "BC as a support for nanocatalysts and other reagents: Recent advances and applications." *Coord. Chem. Rev.*, *426:*213585.. https://doi.org/10.1016/j.ccr.2020.213585.

[12] Shaheen, S. M., Niazi, N. K., Hassan, N. E., Bibi, I., Wang, H., Tsang, D. C.,& Rinklebe, J. (2019). "Wood-based BC for the removal of potentially toxic elements in water and wastewater: a critical review." *Int. Mater. Rev.*, *64*(4):216-247.. https://doi.org/10.1080/09506608.2018.1473096.

[13] Wang, R. Z., Huang, D. L., Liu, Y. G., Zhang, C., Lai, C., Wang, X., ... & Xu, P. (2019). "Recent advances in BC-based catalysts: properties, applications and mechanisms for pollution remediation." *J. Chem. Eng.*, *371*:380-403. https://doi.org/10.1016/j.cej.2019.04.071.

[14] Lee, J. E., & Park, Y. K. (2020). "Applications of modified BC-based materials for the removal of environment pollutants: A mini review." *Sustainability*, *12*(15):6112 https://doi.org/10.3390/su12156112.

[15] Lee, C. G., Hong, S. H., Hong, S. G., Choi, J. W., & Park, S. J. (2019). "Production of BC from food waste and its application for phenol removal from aqueous solution." *Water Air Soil Pollut.*, *230*: 1-13. https://doi.org/10.1007/s11270-019-4125-x.

[16] Nidheesh, P. V., Gopinath, A., Ranjith, N., Akre, A. P., Sreedharan, V., & Kumar, M. S. (2021). "Potential role of BC in advanced oxidation processes: a sustainable approach." *J. Chem. Eng.*, *405*:126582. https://doi.org/10.1016/j.cej.2020.126582.

[17] Gallego-Ramírez, C., Chica, E., & Rubio-Clemente, A. (2022). "Coupling of advanced oxidation technologies and BC for the removal of dyes in water." *Water*, *14*(16):2531.. https://doi.org/10.3390/w14162531.

[18] Kan, T., Strezov, V., & Evans, T. J. (2016). "Lignocellulosic biomass pyrolysis: A review of product properties and effects of pyrolysis parameters." *Renew. Sustain. Energy Rev.*,*57*:1126-1140. https://doi.org/10.1016/j.rser.2015.12.185.

[19] Kamali, M., Appels, L., Kwon, E. E., Aminabhavi, T. M., & Dewil, R. (2021). "BC in water and wastewater treatment-a sustainability assessment." *J. Chem. Eng.*, 129946. https://doi.org/10.1016/j.cej.2021.129946.

[20] Zhou, X. Y., Xie, F., Jiang, M., KE-ao, L., & Tian, S. G. (2021). "Physicochemical properties and lead ion adsorption of BC prepared from Turkish gall residue at different pyrolysis temperatures." *Microsc. Res. Tech.*, *84*(5):1003-1011. https://doi.org/10.1002/jemt.23661.

[21] Kambo, H. S., & Dutta, A. (2015). "A comparative review of BC and hydrochar in terms of production, physico-chemical properties and applications." *Renew. Sustain. Energy Rev.*, *45*:359-378. https://doi.org/10.1016/j.rser.2015.01.050.

[22] Libra, J. A., Ro, K. S., Kammann, C., Funke, A., Berge, N. D., Neubauer, Y., & Emmerich, K. H. (2011). "Hydrothermal carbonization of biomass residuals: a comparative review of the chemistry, processes and applications of wet and dry pyrolysis." *Biofuels*, *2*(1):71-106. https://doi.org/10.4155/bfs.10.81s.

[23] Sun, Y., Gao, B., Yao, Y., Fang, J., Zhang, M., Zhou, Y., & Yang, L. (2014). "Effects of feedstock type, production method, and pyrolysis temperature on BC and hydrochar properties." *J. Chem. Eng.*, *240*:574-578. https://doi.org/10.1016/j.cej.2013.10.081.

[24] Akpasi, S. O., Anekwe, I. M. S., Adedeji, J., & Kiambi, S. L. (2022). "BC development as a catalyst and its application." In *BC-Productive Technologies, Properties and Applications*. IntechOpen. https://doi.org/10.5772/intechopen.105439.

[25] Chen, W. H., Hsu, H. C., Lu, K. M., Lee, W. J., & Lin, T. C. (2011). "Thermal pretreatment of wood (Lauan) block by torrefaction and its influence on the properties of the biomass." *Energy*, *36*(5):3012-3021. https://doi.org/10.1016/j.energy.2011.02.045.

[26] Boateng, A. A., & Mullen, C. A. (2013). "Fast pyrolysis of biomass thermally pretreated by torrefaction." *J. Anal. Appl. Pyrolysis*, *100*:95-102. https://doi.org/10.1016/j.jaap.2012.12.002.

[27] Ambaye, T. G., Vaccari, M., van Hullebusch, E. D., Amrane, A., & Rtimi, S. J. I. J. O. E. S. (2021). "Mechanisms and adsorption capacities of BC for the removal of organic and inorganic pollutants from industrial wastewater." *Int. J. Environ. Sci. Technol.*, *18*(10):3273-3294. https://doi.org/10.1007/s13762-020-03060-w.

[28] Lyu, H., Zhang, Q., & Shen, B. (2020). "Application of BC and its composites in catalysis." *Chemosphere*, *240*: 124842. https://doi.org/10.1016/j.chemosphere.2019.124842.

[29] Luo, K., Pang, Y., Wang, D., Li, X., Wang, L., Lei, M., & Yang, Q. (2021). "A critical review on the application of BC in environmental pollution remediation: Role of persistent free radicals (PFRs)." *J. Environ. Sci.*, *108*:201-216. https://doi.org/10.1016/j.jes.2021.02.021.

[30] Chen, N., Huang, Y., Hou, X., Ai, Z., & Zhang, L. (2017). "Photochemistry of hydrochar: reactive oxygen species generation and sulfadimidine degradation." *Environ. Sci. Tech.*, *51*(19):11278-11287. https://doi.org/10.1021/acs.est.7b02740.

[31] He, X., Zheng, N., Hu, R., Hu, Z., & Yu, J. C. (2021). "Hydrothermal and pyrolytic conversion of biomasses into catalysts for advanced oxidation treatments." *Adv. Funct. Mater.*, *31*(7):2006505.. https://doi.org/10.1002/adfm.202006505.

[32] Lawrinenko, M., Wang, Z., Horton, R., Mendivelso-Perez, D., Smith, E. A., Webster, T. E., & van Leeuwen, J. H. (2017). "Macroporous carbon supported zerovalent iron for remediation of trichloroethylene." *ACS Sustain Chem Eng.*, *5*(2):1586-1593. https://doi.org/10.1021/acssuschemeng.6b02375.

[33] Lobos, M. L. N., Sieben, J. M., Comignani, V., Duarte, M., Volpe, M. A., & Moyano, E. L. (2016). "BC from pyrolysis of cellulose: an alternative catalyst support for the electro-oxidation of methanol. "*Int. J. Hydrogen Energy.*, *41*(25): 10695-10706. https://doi.org/10.1016/j.ijhydene.2016.04.041.

[34] Xu, H., Zhang, Y., Li, J., Hao, Q., Li, X., & Liu, F. (2020). "Heterogeneous activation of peroxymonosulfate by a BC-supported Co3O4 composite for efficient degradation of chloramphenicols." *Environ. Pollut.*, *257*:113610. https://doi.org/10.1016/j.envpol.2019.113610.

[35] Deng, J., Dong, H., Zhang, C., Jiang, Z., Cheng, Y., Hou, K., & Fan, C. (2018). "Nanoscale zero-valent iron/BC composite as an activator for Fenton-like removal of sulfamethazine." *Sep. Purif. Technol.*, *202*:130-137. https://doi.org/10.1016/j.seppur.2018.03.048.

[36] Yan, J., Qian, L., Gao, W., Chen, Y., Ouyang, D., & Chen, M. (2017). "Enhanced Fenton-like degradation of trichloroethylene by hydrogen peroxide activated with nanoscale zero valent iron loaded on BC." *Sci. Rep.*, *7*(1):43051. https://doi.org/10.1038/srep43051.

[37] Jung, K. W., Lee, S. Y., Lee, Y. J., & Choi, J. W. (2019). "Ultrasound-assisted heterogeneous Fenton-like process for bisphenol A removal at neutral pH using hierarchically structured manganese dioxide/BC nanocomposites as catalysts." *Ultrason. Sonochem.*, *57*:22-28.. https://doi.org/10.1016/j.ultsonch.2019.04.039.

[38] Meng, F., Li, Z., Lei, C., Yang, K., & Lin, D. (2021). "Removal of trichloroethene by iron-based BC from anaerobic water: key roles of Fe/C ratio and iron carbides." *J. Chem. Eng.*, *413*:127391. https://doi.org/10.1016/j.cej.2020.127391.

[39] Fu, D., Chen, Z., Xia, D., Shen, L., Wang, Y., & Li, Q. (2017). "A novel solid digestate-derived BC-Cu NP composite activating H2O2 system for simultaneous adsorption and degradation of tetracycline." *Environ. Pollut.*, *221*:301-310. https://doi.org/10.1016/j.envpol.2016.11.078.

[40] Schweitzer, C., & Schmidt, R. (2003). "Physical mechanisms of generation and deactivation of singlet oxygen." *Chem. Rev.*, *103*(5):1685-1758. https://doi.org/10.1021/cr010371d.

[41] Wang, S., & Wang, J. (2020). "Peroxymonosulfate activation by Co9S8@ S and N co-doped BC for sulfamethoxazole degradation." *J. Chem. Eng.*, *385*:123933.. https://doi.org/10.1016/j.cej.2019.123933.

[42] Zhu, F., Wu, Y., Liang, Y., Li, H., & Liang, W. (2020). "Degradation mechanism of norfloxacin in water using persulfate activated by BC@ nZVI/Ni." *J. Chem. Eng.*, *389*:124276.. https://doi.org/10.1016/j.cej.2020.124276.

[43] Guo, D., Shibuya, R., Akiba, C., Saji, S., Kondo, T., & Nakamura, J. (2016). "Active sites of nitrogen-doped carbon materials for oxygen reduction reaction clarified using model catalysts." *Science*, *351*(6271):361-365. https://doi.org/10.1126/science.aad083.

[44] Wang, H., Guo, W., Liu, B., Wu, Q., Luo, H., Zhao, Q., & Ren, N. (2019). "Edge-nitrogenated BC for efficient peroxydisulfate activation: An electron transfer mechanism." *Water Res.*, *160*:405-414. https://doi.org/10.1016/j.watres.2019.05.059.

[45] Zhou, C., Zeng, G., Huang, D., Luo, Y., Cheng, M., Liu, Y., & Li, Z. (2020). "Distorted polymeric carbon nitride via carriers transfer bridges with superior photocatalytic activity for organic pollutants oxidation and hydrogen production under visible light." *J. Hazard. Mater.*, *386*:121947. https://doi.org/10.1016/j.jhazmat.2019.121947.

[46] Yao, F., Yang, Q., Yan, M., Li, X., Chen, F., Zhong, Y., & Li, X. (2020). "Synergistic adsorption and electrocatalytic reduction of bromate by Pd/N-doped loofah sponge-derived BC electrode." *J. Hazard. Mater.*, *386*:121651. https://doi.org/10.1016/j.jhazmat.2019.121651.

[47] Chen, G., Sun, M., Wei, Q., Zhang, Y., Zhu, B., & Du, B. (2013). "Ag3PO4/graphene-oxide composite with remarkably enhanced visible-light-driven photocatalytic activity toward dyes in water." *J. Hazard. Mater.*, *244*:86-93. https://doi.org/10.1016/j.jhazmat.2012.11.032.

[48] Li, M., Huang, H., Yu, S., Tian, N., Dong, F., Du, X., & Zhang, Y. (2016). "Simultaneously promoting charge separation and photoabsorption of BiOX (X= Cl, Br) for efficient visible-light photocatalysis and photosensitization by compositing low-cost BC." *Appl. Surf. Sci.*, *386*:285-295.. https://doi.org/10.1016/j.apsusc.2016.05.171.

[49] García-Muñoz, P., Carbajo, J., Faraldos, M., & Bahamonde, A. (2014). "Photocatalytic degradation of phenol and isoproturon: Effect of adding an activated carbon to titania catalyst." *J. Photochem. Photobiol. A: Chem..*, *287*:8-18. https://doi.org/10.1016/j.jphotochem.2014.05.002.

[50] Jeon, P., Lee, M. E., & Baek, K. (2017). "Adsorption and photocatalytic activity of BC with graphitic carbon nitride (g-C3N4)." *J. Taiwan Inst. Chem. Eng.*, *77*:244-249. https://doi.org/10.1016/j.jtice.2017.05.010.

[51] Kamal, A., Saleem, M. H., Alshaya, H., Okla, M. K., Chaudhary, H. J., & Munis, M. F. H. (2022). "Ball-milled synthesis of maize BC-ZnO nanocomposite (MB-ZnO) and estimation of its photocatalytic ability against different organic and inorganic pollutants." *J. Saudi Chem. Soc.*, *26*(3), :101445. https://doi.org/10.1016/j.jscs.2022.101445.

[52] Fatimah, I., Purwiandono, G., Sahroni, I., Wijayana, A., Faraswati, M., Putri, A. D., & Doong, R. A. (2022). "Magnetically-separable photocatalyst of magnetic BC from snake fruit peel for rhodamine B photooxidation." *Environ. Nanotechnol. Monit. Manag.*, *17*:100669. https://doi.org/10.1016/j.enmm.2022.100669.

[53] Liu, X. Q., Chen, W. J., & Jiang, H. (2017). "Facile synthesis of Ag/Ag3PO4/AMB composite with improved photocatalytic performance." *J. Chem. Eng.*, *308*:889-896. https://doi.org/10.1016/j.cej.2016.09.125.

[54] Khataee, A., Gholami, P., Kalderis, D., Pachatouridou, E., & Konsolakis, M. (2018). "Preparation of novel CeO2-BC nanocomposite for sonocatalytic degradation of a textile dye." *Ultrason. Sonochem..*, *41*:503-513. https://doi.org/10.1016/j.ultsonch.2017.10.013.

[55] Khataee, A., Kayan, B., Gholami, P., Kalderis, D., Akay, S., & Dinpazhoh, L. (2017). "Sonocatalytic degradation of Reactive Yellow 39 using synthesized ZrO2 nanoparticles on BC." *Ultrason. Sonochem.*, *39*:540-549. https://doi.org/10.1016/j.ultsonch.2017.05.023.

[56] Khataee, A., Kayan, B., Gholami, P., Kalderis, D., & Akay, S. (2017). "Sonocatalytic degradation of an anthraquinone dye using TiO2-BC nanocomposite." *Ultrason. Sonochem.*, *39*:120-128. https://doi.org/10.1016/j.ultsonch.2017.04.018.

[57] Zhu, K., Wang, X., Geng, M., Chen, D., Lin, H., & Zhang, H. (2019). "Catalytic oxidation of clofibric acid by peroxydisulfate activated with wood-based BC: Effect of BC pyrolysis temperature, performance and mechanism." *J. Chem. Eng.*, *374*:1253-1263.. https://doi.org/10.1016/j.cej.2019.06.006.

[58] Huang, B. C., Jiang, J., Huang, G. X., & Yu, H. Q. (2018). "Sludge BC-based catalysts for improved pollutant degradation by activating peroxymonosulfate." *J. Mater. Chem. A.* *6*(19):8978-8985. https://doi.org/10.1039/C8TA02282H.

[59] Fang, G., Liu, C., Gao, J., Dionysiou, D. D., & Zhou, D. (2015). "Manipulation of persistent free radicals in BC to activate persulfate for contaminant degradation." *Environ. Sci. Tech.* *49*(9):5645-5653. https://doi.org/10.1021/es5061512s

[60] Fu, H., Ma, S., Zhao, P., Xu, S., & Zhan, S. (2019). "Activation of peroxymonosulfate by graphitized hierarchical porous BC and MnFe2O4 magnetic nanoarchitecture for organic pollutants degradation: Structure dependence and mechanism." *J. Chem. Eng.*, *360*:157-170. https://doi.org/10.1016/j.cej.2018.11.207.

[61] Zhang, C., Wang, Z., Li, F., Wang, J., Xu, N., Jia, Y.,& Shen, W. (2022). "Degradation of tetracycline by activated peroxodisulfate using CuFe2O4-loaded BC." *J. Mol. Liq.368*:120622. https://doi.org/10.1016/j.molliq.2022.120622.

[62] Liu, H., Hu, M., Zhang, H., & Wei, J. (2022). "Biosynthesis of stalk BC-nZVI and its catalytic reactivity in degradation of dyes by persulfate." *J. Surf. Interfac.*, *31*:102098. https://doi.org/10.1016/j.surfin.2022.102098.

[63] Zhou, X., Zhu, Y., Niu, Q., Zeng, G., Lai, C., Liu, S., & Liu, J. (2021). "New notion of BC: A review on the mechanism of BC applications in advannced oxidation processes." *J. Chem. Eng.*, *416*: 129027. https://doi.org/10.1016/j.cej.2021.129027.

[64] Huang, D., Wang, Y., Zhang, C., Zeng, G., Lai, C., Wan, J.,& Zeng, Y. (2016). "Influence of morphological and chemical features of BC on hydrogen peroxide activation: implications on sulfamethazine degradation." *RSC Adv.*, *6*(77):73186-73196. https://doi.org/10.1039/C6RA11850J.

[65] Luo, K., Yang, Q., Pang, Y., Wang, D., Li, X., Lei, M., & Huang, Q. (2019). "Unveiling the mechanism of BC-activated hydrogen peroxide on the degradation of ciprofloxacin." *J. Chem. Eng.*, *374*,:520-530. https://doi.org/10.1016/j.cej.2019.05.204.

[66] Lu, L., Shan, R., Shi, Y., Wang, S., & Yuan, H. (2019). "A novel TiO2/BC composite catalysts for photocatalytic degradation of methyl orange." *Chemosphere*, *222*:391-398. https://doi.org/10.1016/j.chemosphere.2019.01.132.

[67] Yan, Y., Tang, X., Ma, C., Huang, H., Yu, K., Liu, Y.,& Huo, P. (2020). "A 2D mesoporous photocatalyst constructed by the modification of BC on BiOCl ultrathin nanosheets for enhancing the TC-HCl degradation activity." *New J. Chem.*, *44*(1):79-86. https://doi.org/10.1039/C9NJ05219D.

[68] Gholami, P., Dinpazhoh, L., Khataee, A., & Orooji, Y. (2019). "Sonocatalytic activity of BC-supported ZnO nanorods in degradation of gemifloxacin: synergy study, effect of parameters and phytotoxicity evaluation." *Ultrasonics sonochemistry*, *55*:44-56. https://doi.org/10.1016/j.ultsonch.2019.03.001.

[69] Im, J. K., Boateng, L. K., Flora, J. R., Her, N., Zoh, K. D., Son, A., & Yoon, Y. (2014). "Enhanced ultrasonic degradation of acetaminophen and naproxen in the presence of powdered activated carbon and BC adsorbents." *Sep. Purif. Technol.*, *123*:96-105.. https://doi.org/10.1016/j.seppur.2013.12.021.

[70] Fu, H., Zhao, P., Xu, S., Cheng, G., Li, Z., Li, Y.,& Ma, S. (2019). "Fabrication of Fe3O4 and graphitized porous BC composites for activating peroxymonosulfate to degrade p-hydroxybenzoic acid: Insights on the mechanism." *J. Chem. Eng.*, *375*:121980.. https://doi.org/10.1016/j.cej.2019.121980.

[71] Nguyen, V. T., Nguyen, T. B., Chen, C. W., Hung, C. M., Huang, C. P., & Dong, C. D. (2019). "Cobalt-impregnated BC (Co-SCG) for heterogeneous activation of peroxymonosulfate for removal of

tetracycline in water." *Bioresour. Technol.*, *292*:121954. https://doi.org/10.1016/j.biortech.2019.121954.

[72] Liu, C., Chen, L., Ding, D., & Cai, T. (2019). "From rice straw to magnetically recoverable nitrogen doped BC: Efficient activation of peroxymonosulfate for the degradation of metolachlor." *Appl. Catal. B: Environ.*, *254*: 312-320. https://doi.org/10.1016/j.apcatb.2019.05.014.

[73] Liu, C., Chen, L., Ding, D., & Cai, T. (2019). "From rice straw to magnetically recoverable nitrogen doped BC: Efficient activation of peroxymonosulfate for the degradation of metolachlor". *Appl. Catal. B: Environ.*, *254*:312-320. https://doi.org/10.1016/j.apcatb.2019.05.014.

## Chapter 17

# Wastewater Treatment Methods for the Removal of Pesticides

**S. Priyanka**[1]
**A. Rubankumar**[1]
**N. Yuvarani**[1]
**Rajadurai Vijay Solomon**[2]
**and Y. Manojkumar**[1,*]
[1]PG and Research Department of Chemistry, Bishop Heber College, Trichy, India
[2]PG and Research Department of Chemistry, Madras Christian College, Chennai, India

## Abstract

The widespread use of pesticides in agriculture has led to significant water contamination, making effective wastewater treatment essential to mitigate environmental and human health risks. In this context, this chapter reviews both classical and advanced methods for pesticide removal, evaluating their efficacy, sustainability, and integration potential in treatment systems. Classical methods such as coagulation-flocculation, sedimentation, and adsorption are assessed for limitations like incomplete degradation and secondary pollution. Advanced techniques, including advanced oxidation processes (AOPs), bioremediation, membrane filtration, and emerging technologies like nanomaterials and microbial electrochemical systems, are explored for their superior efficiency in mineralizing pesticides residues. The chapter also examines the applicability of these methods to various pesticides and wastewater compositions, with an emphasis on recent advancements and the importance of interdisciplinary collaboration for developing sustainable wastewater treatment strategies.

**Keywords:** pesticides, organophosphates, agricultural pesticides, wastewater treatment, advanced oxidation processes (AOPs), nanomaterials, bioremediation

---

* Corresponding Author's Email: manojkumar.ch@bhc.edu.in

In: Advancements in Wastewater Treatment
Editors: Neha Agarwal, Maulin P. Shahl, Vijendra Singh Solankil et al.
ISBN: 979-8-90134-007-3

## 1. Introduction

Water is essential for all life on Earth, yet only 0.3% of the planet's total water resources are accessible for human use. Water scarcity is affecting many regions globally, leaving billions of people struggling to access clean water. This situation underscores the need for cautious water management, particularly as demand rises with population growth.

Water resources are being depleted, contaminated, and exploited without sufficient awareness due to various human activities [1]. These water resources are being polluted in various ways including agriculture, industries discharges and human waste. The water body gets demolished by mixing of different types of chemicals from industries and agricultural pesticides and consequently, living organisms are affected, including animals and plants. [2] The world population grew significantly from 1.5 billion to 6.9 billion between 1900 and 2000, leading to a substantial increase in food demand, which has placed significant pressure on the agricultural sector. To meet the growing demand for food, farmers have increasingly relied on pesticides. Agricultural pesticides are now considered some of the most significant emerging contaminants, with plant treatment residues being a major contributor to water pollution [3]. Many agricultural fields are situated near water bodies, which are often used for irrigation and as disposal sites for agricultural waste. This contamination commonly occurs through a process called leaching i.e., the movement of soluble compounds from the soil into groundwater and surface water [4, 5]. As water moves through the soil, these chemical that strongly bind to soil particles can be transported through leachate, further contributing to pollution. Pesticides play a key role in water contamination, especially in the actively cultivated agricultural land [6]. The residues of pesticides can be found in common food products such as fruit juice, wine, poultry food and especially in drinking water etc. It is very important to note that pesticides cannot be removed only by washing and peeling off. Surprisingly, studies on human breast milk have also revealed pesticides residues, which raise concerns about fetal exposure and the potential health effects on kids [7].

Pesticides, including insecticides, herbicides, fungicides, rodenticides, and nematicides, are classified based on the type of pest they target. Each category addresses specific agriculturally harmful organisms. Organophosphates, organochlorines, carbamates, and pyrethroids are among the chemical classes of pesticides. These classes differ in toxicity, structure, mode of action, and environmental persistence [8]. Over one thousand pesticides have been banned due to their harmful effects, yet many are still used, globally [6]. To eliminate the harmful effects of pesticides in water, many methods such as chemical, physical and biological methods are employed [9]. These methods have major challenges like toxicity to microbes, high Chemical Oxygen Demand (COD), at highly acidic condition. Other processes, such as sedimentation, coagulation, and adsorption, are often not suitable for these wastewater contaminants due to their high costs, substantial sludge production, inefficient removal, and generation of secondary pollutants [10]. To address these challenges, this article explores potential future technologies aimed at mitigating pesticide-related water contamination. It is essential to examine the specific types of pesticides that contribute to water quality degradation. The following section categorizes these compounds and highlights their environmental impacts.

## 2. Types of Pesticides in Wastewater

Pesticides are broadly classified into four types i.e. insecticides, herbicides, fungicides, and rodenticides. By enhancing plant protection and increasing crop yields, pesticides play a crucial role in advancing agricultural development. Around the world pesticide usage surpasses 2 million tons dominated by herbicides (47.5%), insecticides (29.5%), fungicides (17.5%) and rodenticides (5.5%) [11]. Table 1 presents different classes of pesticides according to their chemical nature, including examples such as organophosphates, pyrethroids, and carbamates. It also highlights their modes of action, which are critical for selecting appropriate wastewater treatment methods.

### 2.1. Insecticides

Insecticides are used in agriculture to protect plants from insects like thrips, whitefly, bugs, caterpillar and spider. Insecticides may be natural or synthetic, and organic or inorganic. Natural insecticides like nicotine pyrethrum and neem extract acts as the defense system for the plants against insects. Inorganic insecticides, often containing arsenate, copper compounds, fluorine compounds, or sulfur, are commonly used. Organic insecticides, typically synthetic chemicals, constitute the majority of currently available insecticides [12]. Based on their chemical composition, insecticides are classified into organochlorines, organophosphates, pyrethroids, neonicotinoids, ryanoids, and biological or bacterial insecticides. Insecticides can disperse through various pathways from their point of application, ultimately affecting the broader environment. Over time, they often make their way into rivers, lakes, and ponds, leading to increased toxicity and significant harm to natural ecosystems [13].

### 2.2. Herbicides

Herbicides are synthetic chemical compounds used to inhibit, destroy, or suppress the growth of unwanted or non-native plant species [14]. Herbicides are generally classified into two categories: selective and non-selective. Selective herbicides target and eliminate weeds without damaging desired crops. In contrast, non-selective herbicides are broad-spectrum and can affect all types of vegetation, including crops [15]. Over the past two decades, the herbicide industry has experienced significant growth. However, wastewater from herbicide manufacturing plants and wash-down water from formulation facilities often contain toxic and hazardous compounds. When these pollutants enter drinking water sources and soil, they pose serious environmental risks. One major consequence is the destruction of soil microorganisms, which are essential for natural biodegradation processes. The loss of these organisms severely impairs the environment's ability to degrade pollutants.

### 2.3. Fungicides

Fungicides are chemical compounds used to prevent fungal infections in plants. Based on their mode of action, they are classified into two categories: agricultural and horticultural. Agricultural fungicides are used to protect crops such as cereals, grains, and other large-scale farming products from fungal infections. Horticultural fungicides, on the other hand, are applied in the cultivation of fruits, vegetables, and ornamental plants like flowers [16]. Azole fungicides are commonly used in agriculture and are also found in biocidal products for material protection. Their persistence in soil and water is due to their resistance to hydrolytic, photolytic, and biodegradation processes, raising serious concerns about their harmful effects on aquatic environments when they enter water sources [17]. Metallic fungicides, such as cadmium chloride, copper acetate, copper silicate, mercuric chloride, and mercury sublimate, were among the earliest used fungicides in agriculture. While inorganic fungicides are considered protective and preventive, mercuric compounds have been banned in many countries due to their toxicity and environmental impact. Some fungicides are highly toxic to humans and mammals, exhibiting immune-toxic and neurotoxic effects [18].

### 2.4. Rodenticides

A recent report indicates that there are over 2,500 species of rodents. These rodents are highly adaptable to various ecosystems and play important roles in agriculture, such as soil aeration, mixing, vegetation succession, and seed and spore dispersal. They also serve as prey for numerous predatory animals. However, rodents can also cause significant damage to crops worldwide. Other mammals, including birds, rabbits, and ungulates, can similarly harm crops. To mitigate these threats, rodenticides are used to control rats and mice [19]. Figure 1 shows the usage distribution of pesticides: herbicides (47.5%), insecticides (29.5%), fungicides (17.5%), and others (5.5%) along with their health impacts. Herbicides affect hormones, insecticides harm the nervous system, and fungicides cause skin and respiratory issues.

**Table 1.** Classification of pesticides by chemical class

| Pesticides Class | Examples | Mode of Action | Ref |
|---|---|---|---|
| Organophosphates | Malathion, Parathion, Chlorpyrifos | Inhibit acetylcholinesterase, leading to overstimulation of the nervous system. | [21] |
| Organochlorines | DDT, Lindane, Heptachlor | Disrupt nerve function by interfering with ion channels. | [22] |
| Pyrethroids | Permethrin, Cypermethrin, Deltamethrin | Affect nerve function by disrupting sodium channels. | [23] |
| Carbamates | Carbaryl, Aldicarb, Methomyl | Inhibit acetylcholinesterase, similar to organophosphates. | [24] |
| Neonicotinoids | Imidacloprid, Acetamiprid, Thiacloprid | Nicotinic acetylcholine receptor agonists, affecting nerve Transmission. | [25] |
| Herbicides | Glyphosate, Atrazine, 2,4-D | Target plant metabolism, particularly photosynthesis. | [26] |
| Fungicides | Chlorothalonil, Copper sulfate, Mancozeb | Prevent fungal spore formation or growth. | [27] |
| Rodenticides | Warfarin, Bromadiolone, Zinc phosphide | Cause blood clotting, leading to death. | [28] |

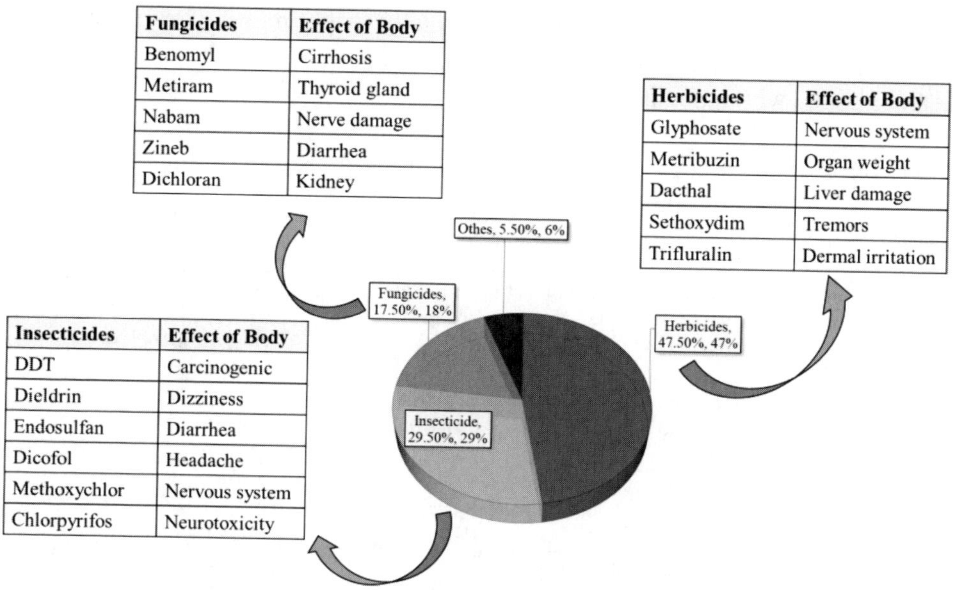


**Figure 1.** Distribution and Health Impacts of Different Types of Pesticides [20].

Common rodenticides include dicoumarol, bromadiolone, brodifacoum, chlorophacinone, coumatetralyl, flocoumafen, pindone, and difenacoum. When these anticoagulants reach water sources through heavy rainfall or agricultural runoff, they cause direct water contamination [29]. Considering the widespread presence and detrimental effects of various pesticide classes, it is essential to assess how conventional wastewater treatment methods handle these contaminants.

## 3. Conventional Wastewater Treatment Methods

Conventional wastewater treatment involves multiple stages designed to remove physical, chemical, and biological contaminants from sewage before discharge or reuse. It typically involves preliminary treatment, primary treatment, secondary treatment and tertiary treatment [30]. These stages work collectively to reduce organic matter, suspended solids, nutrients, and pathogens, ensuring compliance with environmental standards and public health protection. The key processes and objectives of each stage are outlined in Table 2.

### 3.1. Physical Treatment Methods

Physical methods play a critical role in water purification and recycling, offering some of the simplest and most accessible treatment approaches. Key physical treatment methods include sedimentation & filtration, and adsorption [37].

#### *3.1.1. Sedimentation and Filtration Method*

Sedimentation uses gravity to allow solid particles in wastewater to settle to the bottom, facilitating their removal. A coagulant is often added to help aggregate small particles into larger ones, enhancing their settling [38]. Additionally, filtration removes smaller particles from wastewater. When water passes through filter media, such as sand or gravel, the smaller particles are trapped within the media [39].

**Table 2.** Various stages of conventional wastewater treatment

| Treatment Stage | Description | Key Processes | Purpose | Ref |
|---|---|---|---|---|
| Preliminary Treatment | Involves the removal of large solids and grit to protect downstream equipment and processes. | Screening, Grit Removal | Prevents damage to equipment and reduces clogging in subsequent treatment stages. | [31] |
| Primary Treatment | Utilizes sedimentation to remove settleable organic and inorganic solids from wastewater. | Primary Sedimentation Tanks | Reduces suspended solids and organic load, facilitating | [32] |
| Secondary Treatment | Employs biological processes to degrade dissolved and colloidal organic matter. | Activated Sludge Process, Trickling Filters | Reduces biochemical oxygen demand (BOD) and suspended solids | [33] |
| Tertiary Treatment | Advanced treatment stage aimed at removing residual contaminants, including nutrients and pathogens. | Filtration, Nutrient Removal, Disinfection | Polishes effluent to meet stringent discharge or reuse standards. | [34] |
| Disinfection | Final step to eliminate pathogenic microorganisms from treated wastewater. | Chlorination, UV Radiation, Ozonation | Ensures public health safety by preventing the spread of waterborne diseases. | [35] |
| Effluent Storage | Temporary holding of treated water before its discharge or reuse, allowing for flow regulation and quality monitoring. | Storage Tanks, Lagoons | Manages effluent flow and ensures consistent water quality prior to final discharge. | [36] |

### *3.1.2. Adsorption Method*

Wastewater contains heavy metals such as Cd, Zn, Pb, Fe, Hg, Ni, Co, Mn that can be removed by various techniques including solvent extraction, chemical precipitation, ion exchange etc. Among these techniques adsorption method has more significance due to its availability and flexibility in operation. There are two types of adsorption methods i.e., physical and activated adsorption methods [40]. Activated carbon is frequently used in removal of organic and inorganic compounds in wastewater. Marine algae, such as seaweeds, are excellent sources of green nanoparticles for use in adsorption methods. Due to their strong metal-binding capacities, seaweeds provide an effective interface between the adsorbent and adsorbate, enabling the removal of a wide range of impurities [41].

## 3.2. Chemical Methods

While physical methods are straightforward and cost-effective, chemical treatments provide additional mechanisms to break down and eliminate harmful pesticides. In today's industrial landscape, many sectors consume large volumes of water daily, yet only about 50% of this water is used efficiently—the remainder is often wasted. To tackle this issue, various techniques have been developed to treat and reclaim wastewater. Among them, chemical treatment methods have proven particularly effective in removing pesticides from contaminated water [42].

#### *3.2.1. Oxidation*

Oxidation is used to oxidise organic compounds into water, $CO_2$, and some biodegradable products. The chemicals used to remove pesticides include potassium permanganate, chlorine, ozone, hydrogen peroxide, and Fenton's reagent [43]. Laboratory-scale ozonation experiments employed an ozone generator producing 10% ozone by weight in an oxygen gas stream, operating under controlled internal gas pressure. A borosilicate glass bubble column, equipped with a porous sintered glass plate diffuser at the base, was used as a semi-batch ozone reactor with a constant gas flow rate. Optimizing ozone utilization rate is crucial for designing efficient full-scale ozonation processes [44]. The efficiency of chemical oxidation depends on the oxidizing agent and factors such as pressure, temperature, and pH. This method is effective in treating pollutants like $NH_3$, $C_6H_5OH$, dyes, and other organic compounds [45].

#### *3.2.2. Precipitation*

Precipitation is another method used to remove pesticides by reducing the solubility of dissolved organic substances, transforming them into solid precipitates. This method is efficient for removing metal ions and organic substances [46]. Figure 2 illustrates a typical schematic diagram of the wastewater treatment process, showing how various stages like coagulation, sedimentation, and filtration are organized to remove pesticide contaminants. The chemicals commonly used for this method include alum, sodium bicarbonate, ferric chloride, ferrous sulphate, and lime. However, one drawback of this method is the production of sludge [47].

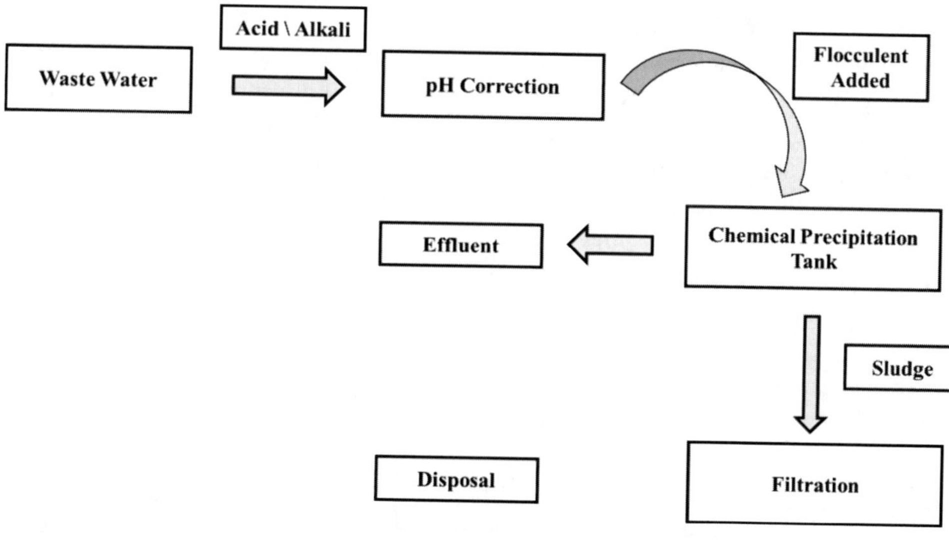


**Figure 2.** Schematic Method for the Treatment of Wastewater [47].

### 3.3. Biological Treatment Methods

#### *3.3.1. Aerobic Process*

Biological methods, using living organisms, provide an eco-friendly alternative to chemical techniques for cleaning contaminated water. Various microorganisms in activated sludge facilitate the degradation of organic contaminants in aerobic wastewater treatment. A diversified microbial community, comprising diverse species, makes the system more robust.

Not all microbes respond the same way to toxic shock, a condition that occurs when a harmful chemical unexpectedly enters the body. While some are affected, others continue functioning, helping the system recover more quickly. This microbial diversity helps maintain a stable and effective treatment process, even during unexpected disturbances [48]. Alcaligenes aquatilis, a bacterium commonly found in aerobic activated sludge, can perform heterotrophic nitrification and aerobic denitrification, enabling efficient removal of nitrogen compounds like ammonia in oxygen-rich environments. Compared to similar bacteria, it demonstrated faster and more efficient performance by removing up to 95.5% of ammonia from piggery effluent within 36 hours. This makes it a promising candidate for improving nitrogen removal in aerobic wastewater treatment systems [49].

### *3.3.2. Anaerobic Process*

Biogas is commonly produced during anaerobic treatment, which uses bacteria to break down waste in the absence of oxygen. Incorporating turbulent diffusion into digester models improves mixing simulation accuracy, enhancing reactor performance prediction and optimization [49]. Propionate, a by-product of anaerobic digestion, can inhibit methane production if it accumulates. To address this, microbial electrochemical technologies (METs) are used, employing electric currents and specific bacteria to accelerate propionate breakdown. This approach enhances methane generation, stabilizes the digestion process, and improves overall treatment system performance [50].

### *3.3.3. Bioremediation Using Bacteria and Fungi*

Microorganisms such as bacteria and fungi are used in the environmentally friendly process of bioremediation to degrade, purify, or remove contaminants from wastewater. Figure 3 illustrates how bacteria and fungi degrading harmful pesticide residues in wastewater. This bioremediation process is eco-friendly and supports sustainable water treatment. By breaking down pollutants into less harmful forms, these microorganisms help to purify water [51]. Bacteria break down dangerous chemicals in wastewater, which helps to clean it. Their metabolic adaptability allows them to effectively handle diverse pollutants, ensuring high removal efficiency.

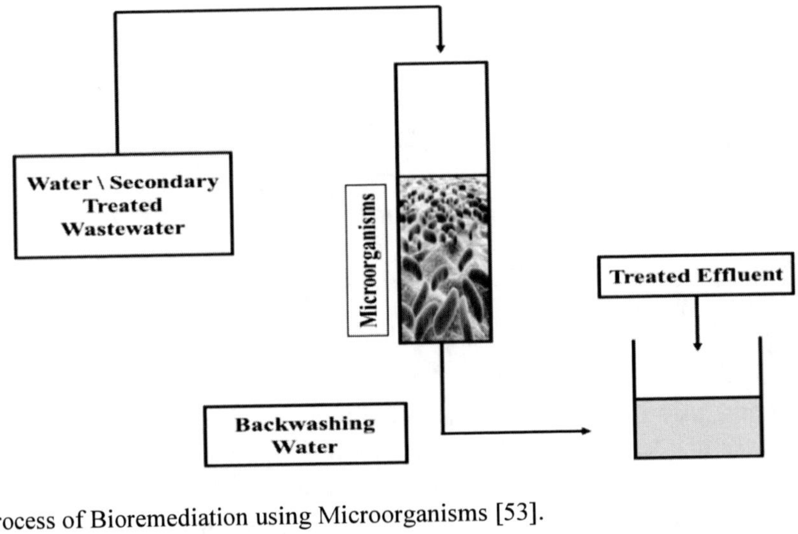


**Figure 3.** Process of Bioremediation using Microorganisms [53].

As a result, they are highly efficient at eliminating a variety of pollutants. Pesticides are among the challenging pollutants fungi especially white-rot fungi are capable of breaking down. These fungi use powerful enzymes to degrade harmful and complex compounds, making them valuable for treating wastewater containing persistent contaminants [52]. Although conventional methods provide a foundation for treatment, they often fail to completely remove complex pesticide residues. Consequently, the focus has shifted to advanced technologies, discussed in the following sections.

## 4. Advanced Wastewater Treatment Technologies

### 4.1. Advanced Oxidation Processes (AOPs)

AOPs employ highly reactive hydroxyl or sulphate radicals as primary oxidizing agents. These processes have been widely applied to treat diverse wastewaters, with powerful oxidants effectively degrading persistent organic pollutants and removing certain inorganic contaminants. The primary application of AOPs is in the degradation of organic and inorganic contaminants present in water and wastewater. In wastewater treatment, these radicals act as powerful oxidizing agents, transforming pollutants into safe or low-toxic products, ultimately offering a holistic solution for treatment of wastewater [54].

### 4.2. Photo Catalysis (UV, $TiO_2$)

One of the most studied AOPs is photocatalysis, which uses light and catalysts to degrade pollutants effectively. Photocatalysis (especially the use of titanium dioxide ($TiO_2$) with ultraviolet (UV) radiation) has become a viable advanced wastewater treatment technique. UV/$TiO_2$ system effectively eliminates numerous chemical contaminants and harmful microbes. Using model organisms like *Escherichia coli, Cryptosporidium parvum,* and *phage Qβ,* it has been effectively used to disinfect wastewater; demonstrating its capacity for pollutant breakdown and microbial inactivation [55]. Due to its benefits, such as low toxicity, high chemical stability, photoactivity, and environmental friendliness; $TiO_2$ is regarded as an ideal photocatalyst. Enhancing the photocatalytic properties of $TiO_2$ such as, through metal or non-metal doping has been shown to significantly improve its efficiency under UV light. In wastewater treatment, these modifications lead to faster degradation rates and greater adaptability to a wide range of pollutants [55].

### 4.3. Photo Fenton Process

The Photo-Fenton process, an extension of photocatalysis, enhances pollutant degradation using iron and hydrogen peroxide ($H_2O_2$). In the Photo-Fenton process, $H_2O_2$, ultraviolet (UV) light, and iron ions ($Fe^{2+}/Fe^{3+}$) combine to produce hydroxyl radicals (•OH), potent oxidants that degrade organic contaminants [56]. This advanced oxidation method is particularly effective for degrading persistent organic molecules in wastewater. This method improves overall treatment efficiency by photochemically regenerating $Fe^{2+}$ from $Fe^{3+}$, which strengthens the conventional Fenton reaction. The process also produces carboxylic acids, which may interact with iron ions, influencing the reaction mechanism. Additionally, the Photo-Fenton process effectively treats complex effluents, such as soft drink wastewater and landfill leachate, providing potent oxidative power for pollutant degradation [57].

## 4.4. Reverse Osmosis (RO)

In addition to oxidation methods, membrane-based processes like reverse osmosis (RO) play a critical role in advanced water. RO is a membrane filtration method that uses pressure to separate pollutants and dissolved salts through a semi-permeable membrane, as shown in Figure 4 [58]. Although commonly used to produce high-quality water for reuse applications, RO generates a concentrated waste stream containing effluent organic matter (EfOM) [59]. Reusing RO-treated water may be restricted by EfOM, particularly if the concentrate contains substances such as endocrine disruptors. For RO systems to be profitable and effective, a constant water and salt transfer across the membrane is essential [60].

## 4.5. Nanofiltration (NF) and Ultrafiltration (UF)

RO is often complemented by filtration techniques like nanofiltration and ultrafiltration, which offer selective separation of contaminants. Nanofiltration membranes are made from polymers, such as cellulose acetate and polyamide, or, more recently, high-temperature-tolerant ceramic materials. Figure 4 shows how different membrane filtration techniques including nanofiltration, ultrafiltration, and RO work in treating wastewater.

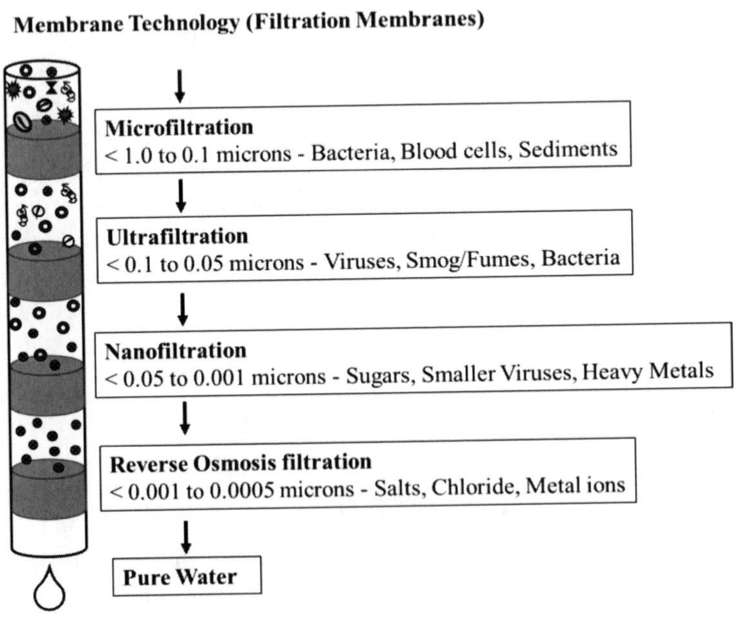


**Figure 4.** Schematic representation of Filtration Methods [61].

These advanced methods are particularly effective for removing small pesticide particles that conventional methods cannot address. NF is a promising technology in advanced water treatment due to its versatility [62]. Ultrafiltration (UF), a membrane-based, energy-efficient filtration technique, is widely used in large-scale water treatment facilities to continuously remove colloids, particulates, and microbes. UF systems frequently use periodic backwashing in dead-end filtration mode to preserve efficiency. This approach uses less energy to produce drinkable water and is used frequently [63]. These advanced treatment techniques are not only theoretical; many of them have been successfully applied across the world. The following case studies highlight how different countries implement these methods to manage pesticide-contaminated wastewater.

## 5. Case Studies

### 5.1. South Korea

A South Korean study proposed a sustainable regulatory framework for managing organophosphate pesticides (OPs) in wastewater treatment plants. The framework emphasizes the need for comprehensive monitoring and advanced treatment methods to effectively remove OPs, ensuring environmental safety and compliance with discharge standards [64].

### 5.2. Italy

A study in Italy assessed the capacity of a full-scale surface flow constructed wetland (SFCW) to treat agricultural drainage water containing high concentrations (~1 mg/L) of herbicides, including glyphosate, imidacloprid, and dimethomorph. The study demonstrated that constructed wetlands (CWs) can efficiently retain and reduce pesticide concentrations, particularly during extreme runoff events, highlighting their potential as sustainable treatment solutions in agricultural settings [65].

### 5.3. United States

Under 40 CFR Part 455, the (EPA) Environmental Protection Agency of the United States has created effluent guidelines and standards for the pesticides chemicals category. These guidelines, integrated into the National Pollutant Discharge Elimination System (NPDES), regulate pollutant discharge into U.S. waters by establishing effluent limitations and criteria for various pesticide manufacturing subcategories [66].

### 5.4. India

India's regulatory framework for pesticide management is anchored by the Insecticides Act of 1968, which governs the registration, production, distribution, transportation, and use of insecticides. The Central Insecticides Board and Registration Committee (CIBRC) oversees these regulations, ensuring pesticides meet safety and efficacy standards. In order to protect the public's health, the Food Safety and Standards Authority of India (FSSAI) also establish maximum residual limits (MRLs) for pesticides in food products [67].

### 5.5. Poland

Advanced oxidation processes (AOPs), such as UV/$H_2O_2$ and Fenton reactions, have shown high efficiency in degrading persistent pesticide chemicals. For example, a pilot-scale study in Poland that used UV/H2$O_2$ treatment to treat effluent from actual pesticides production produced a 95% of elimination [68]. These case studies underscore how various operational and environmental factors influence the effectiveness of pesticide removal techniques.

## 6. Factors Affecting Treatment Efficiency

Organophosphorus pesticide (OPP) residues in water pose risks to ecosystems and human health. Through substrate adsorption, plant absorption, and microbial activity, it has been demonstrated that CWs efficiently remove these pesticides, providing a financially feasible option. Temperature, retention duration, and pesticides dosage are significant factors influencing CW efficiency [69]. Various biological, chemical, physical, and physicochemical methods have been employed to remove pesticides from soil and water. The advantages and drawbacks of both conventional and contemporary materials such as microorganisms, clay, activated carbon, and polymers—in eliminating pesticides like carbamates, organochlorines, and organophosphorus chemicals are discussed [9]. Conventional treatment techniques face difficulties because use of seasonal pesticides frequently leads to high concentrations at water abstraction locations. For more efficient pesticides removal, new technologies such enhanced oxidation; adsorption, bioreactors, and hybrid systems are being investigated. selecting the optimum treatment approach requires a thorough investigation of the influent composition [20]. Biological wastewater treatment (BWT) aims to produce sludge containing water, pollutants, and microorganisms to transform biodegradable components into safe end products. One of the main factors affecting this process is temperature because excessive temperatures can lead to difficulties. The performance of biological treatment under such extreme conditions has been studied [70]. Advanced treatments like carbon adsorption and membrane filtration can effectively remove deformity, depending on factors such as compound type, concentration, and process conditions [71]. The primary aim of BWT is to harness microbial activity to convert biodegradable pollutants into harmless end products. However, this process generates large volumes of sludge containing water, microorganisms, nutrients such as nitrogen and phosphorus, as well as hazardous substances including pathogens and heavy metals [70]. Microbial fuel cells (MFCs) generate electricity by oxidizing organic matter in substrates like wastewater, using microorganisms that donate electrons to the anode. Substrate type, microorganisms, electrode materials, pH, temperature, and salinity are critical factors influencing MFC performance. MFCs provide a scalable, environmentally responsible alternative for groundwater bioremediation and wastewater treatment using inexpensive materials [72]. As they can be made at room temperature, geopolymers provide a sustainable substitute for cement, which uses a more energy and releases a more $CO_2$. Despite having a high organic concentration, sewage sludge can be processed with the Fenton method to remove up to 83.7% of the organics, which makes it ideal for the production of environmentally benign geopolymers [73]. The Scattering Electrical Reactor (SER) recovered up to 93.23% of dissolved organic materials from wastewater under ideal circumstances. Ru/Ir-coated electrodes used less energy than titanium ones, and a predictive model validated the deterioration process, making SER an effective and sustainable treatment technique [74]. One prominent wastewater treatment technique that is commended for its effectiveness, cost-effectiveness, and production of green energy is the upflow anaerobic sludge blanket (UASB) reactor. The potential of UASB for sustainable wastewater management was further supported by the identification of prospective future study fields and work. Given these determinants and constraints, there is an increasing drive for creative and flexible technologies that can improve treatment effectiveness, opening the door for the upcoming developments covered next.

## 7. Future Trends and Emerging Technologies

### 7.1. Nanotechnology in Pesticides Removal

Nanomaterials such as zinc oxide (ZnO), titanium dioxide ($TiO_2$), and nanoscale zero-valent iron (nZVI) have been used to degrade pesticides by photocatalysis. When exposed to light, these materials produce reactive oxygen species, which degrade pesticides molecules into less dangerous forms [75]. Carbon nanotubes (CNTs) and reduced graphene oxide (rGO), examples of advanced nanocomposites, exhibit strong adsorption and reactivity, enhancing the removal efficiency of various pesticides from aqueous solutions [76]. Research has demonstrated nano-gold particles can efficiently degrade pesticides like chlorpyrifos under ambient conditions, indicating their potential for wastewater treatment [77].

### 7.2. Green and Sustainable Methods

Organophosphate hydrolases, such as OpdA, are enzymes that may effectively break down organophosphate insecticides by immobilising themselves on substrates such as nonwoven polyester textiles. These systems have proven to be effective in breaking down pesticides like methyl parathion in water, providing a stable and reusable clean-up method [78]. Without the use of extra mediators, a variety of pesticides can also be oxidised by laccase enzymes, which are derived from fungus. According to research, commercial laccases have the ability to cleanse wastewater by eliminating up to 49.9% of specific pesticides in a 24-hour period [79].

### 7.3. Smart Monitoring and Detection of Pesticides in Water

By combining WSNs with machine learning algorithms and the Internet of Things (IoT), environmental data can be gathered and analysed, improving the precision and responsiveness of pollution monitoring systems [80]. A study addresses the combination of sensors and wireless remote sensing for surface and groundwater pesticides pollution monitoring. In order to obtain comprehensive insights into water ecosystems, it highlights the advantages of integrating IoT, machine learning, and big data analytics [80].

The use of recent advancements in smart electrochemical devices for the identification of impurities in runoff and pesticides in agricultural food has been examined. These gadgets improve agricultural sustainability by providing affordable, portable real-time monitoring options [81]. As more sustainable solutions are made possible by these cutting-edge technology and methods, it is evident that a comprehensive and progressive approach is necessary.

## Conclusion

In conclusion, there is an urgent need for effective and sustainable wastewater treatment strategies to address pesticide contamination in water. Conventional remediation methods, such as sedimentation, coagulation flocculation, and adsorption though effective but face much limitation in pesticides degradation and potential removal for secondary pollutants. Whereas AOP'S, bioremediation, membrane filtration are some of the advanced technologies, innovative methods like nanomaterials, microbial processes, electrochemical systems offer promising approaches with high efficiency and the ability to minimize pesticides residues. However, their success rate also depends on the characteristic of pesticides and wastewater contamination. Advancing these methods can help ensure a pesticides free future that safeguards both the environment and society.

## References


[1] Kılıç Z: The importance of water and conscious use of water. *International Journal of Hydrology* 2020, 4(5):239-241.

[2] Hussain M, Muhammad S, Malik RN, Khan MU, Farooq U: Status of heavy metal residues in fish species of Pakistan. *Reviews of Environmental Contamination and Toxicology volume: With Cumulative and Comprehensive Index Subjects Covered Volumes 221-230* 2014:111-132.

[3] Kılıç O, Boz I, Eryılmaz GA: Comparison of conventional and good agricultural practices farms: A socio-economic and technical perspective. *Journal of Cleaner Production* 2020, 258:120666.

[4] Morin-Crini N, Lichtfouse E, Liu G, Balaram V, Ribeiro ARL, Lu Z, Stock F, Carmona E, Teixeira MR, Picos-Corrales LA: Worldwide cases of water pollution by emerging contaminants: a review. *Environmental Chemistry Letters* 2022, 20(4):2311-2338.

[5] Pérez-Lucas G, Vela N, El Aatik A, Navarro S: Environmental risk of groundwater pollution by pesticide leaching through the soil profile. In: *Pesticides-use and misuse and their impact in the environment.* edn.: IntechOpen; 2018.

[6] Rad SM, Ray AK, Barghi S: Water Pollution and Agriculture Pesticide. *Clean Technologies* 2022, 4(4):1088-1102.

[7] Martín-Carrasco I, Carbonero-Aguilar P, Dahiri B, Moreno I, Hinojosa M: Comparison between pollutants found in breast milk and infant formula in the last decade: A review. *Science of the Total Environment* 2023, 875:162461.

[8] Kaur R, Choudhary D, Bali S, Bandral SS, Singh V, Ahmad MA, Rani N, Singh TG, Chandrasekaran B: Pesticides: An alarming detrimental to health and environment. *Science of The Total Environment* 2024, 915:170113.

[9] Marican A, Durán-Lara EF: A review on pesticide removal through different processes. *Environmental Science and Pollution Research* 2018, 25:2051-2064.

[10] Karraz FC: Presença de antibióticos em corpos hídricos: potenciais problemas e processos de remoção. *TCC (Graduação)-Curso de Engenharia Química, Universidade Federal do Rio de Janeiro, Rio de Janeiro* 2024.

[11] Sharma A, Kumar V, Shahzad B, Tanveer M, Sidhu GPS, Handa N, Kohli SK, Yadav P, Bali AS, Parihar RD: Worldwide pesticide usage and its impacts on ecosystem. *SN Applied Sciences* 2019, 1:1-16.

[12] Sabra FS, Mehana E-SE-D: Pesticides toxicity in fish with particular reference to insecticides. *Asian Journal of Agriculture and Food Sciences* 2015, 3(1).

[13] Weis GCC, de Oliveira Alves A, Assmann CE, Bonadiman BdSR, Costabeber IH: Pesticides: classifications, exposure and risks to human health. *Archives in Biosciences & Health* 2019, 1(1):29-44.

[14] Gianessi LP: The increasing importance of herbicides in worldwide crop production. *Pest management science* 2013, 69(10):1099-1105.

[15] Hazaraimi M, Goh P, Lau W, Ismail A, Wu Z, Subramaniam M, Lim J, Kanakaraju D: The state-of-the-art development of photocatalysts for the degradation of persistent herbicides in wastewater. *Science of the Total Environment* 2022, 843:156975.

[16] Zubrod JP, Bundschuh M, Arts G, Brühl CA, Imfeld G, Knäbel A, Payraudeau S, Rasmussen JJ, Rohr J, Scharmüller A: Fungicides: an overlooked pesticide class? *Environmental science & technology* 2019, 53(7):3347-3365.

[17] Stamatis N, Hela D, Konstantinou I: Occurrence and removal of fungicides in municipal sewage treatment plant. *Journal of Hazardous Materials* 2010, 175(1-3):829-835.

[18] Ballantyne B: Toxicology of fungicides. *Pesticide toxicology and international regulation* 2004:191-303.

[19] Witmer G: Rodents in Agriculture: A Broad Perspective. *Agronomy* 2022, 12(6):1458.

[20] Jatoi AS, Hashmi Z, Adriyani R, Yuniarto A, Mazari SA, Akhter F, Mubarak NM: Recent trends and future challenges of pesticide removal techniques–a comprehensive review. *Journal of Environmental Chemical Engineering* 2021, 9(4):105571.

[21] Brown JJ: Handbook of Pesticide Toxicology, Vol. 1, 2, and 3. *American Entomologist* 1991, 37(4):244-244.

[22] Jayaraj R, Megha P, Sreedev P: Organochlorine pesticides, their toxic effects on living organisms and their fate in the environment. *Interdisciplinary toxicology* 2016, 9(3-4):90.

[23] Khambay BP: Pyrethroid insecticides. *Pesticide Outlook* 2002, 13(2):49-54.

[24] Mitra S, Saran R, Srivastava S, Rensing C: Pesticides in the environment: Degradation routes, pesticide transformation products and ecotoxicological considerations. *Science of The Total Environment* 2024:173026.

[25] Thany SH: Neonicotinoide insecticides. In: *Insect Nicotinic Acetylcholine Receptors.* edn.: Landes Bioscience, Springer Science+ Business Media; 2010.

[26] Duke SO, Powles SB: Glyphosate: a once-in-a-century herbicide. *Pest Management Science: formerly Pesticide Science* 2008, 64(4):319-325.

[27] Schmahl G, Mehlhorn H, Abdel-Ghaffar F, Al-Rasheid K, Schumacher B, Jatzlau A, Pohle H: Does rain reduce the efficacy of Butox 7.5 pour on (deltamethrin) against biting midges (Culicoides specimens)? *Parasitology research* 2009, 105:1763-1765.

[28] Srilakshmi E, Ujefa S, Komali G, Priya TD, Dhanush B, Kantamaneni P: A comprehensive review on rodenticide poisoning and its management. *World J Pharm Res* 2023, 12(7):1061-1074.

[29] Regnery J, Schulz RS, Parrhysius P, Bachtin J, Brinke M, Schäfer S, Reifferscheid G, Friesen A: Heavy rainfall provokes anticoagulant rodenticides' release from baited sewer systems and outdoor surfaces into receiving streams. *Science of the Total Environment* 2020, 740:139905.

[30] Awang N, Radzi ARM, Mahat R, Fatihhi S, Johari A, Abd Wahid KA, Yajid MAM: Microbial fuel cell in industrial wastewater: treatment processes and resource recovery. In: *Resource Recovery in Industrial Wastewaters.* edn.: Elsevier; 2023: 353-363.

[31] EPA: Wastewater technology fact sheet. Screening and grit removal. In.; 2003.

[32] Camp TR: Sedimentation and the design of settling tanks. *Transactions of the American Society of Civil Engineers* 1946, 111(1):895-936.

[33] Shukla R, Ahammad SZ: Performance evaluation and microbial community structure of a modified trickling filter and conventional activated sludge process in treating urban sewage. *Science of The Total Environment* 2022, 853:158331.

[34] Hamoda M, Al-Ghusain I, Al-Mutairi N: Sand filtration of wastewater for tertiary treatment and water reuse. *Desalination* 2004, 164(3):203-211.

[35] Gomes J, Matos A, Gmurek M, Quinta-Ferreira RM, Martins RC: Ozone and photocatalytic processes for pathogens removal from water: A review. *Catalysts* 2019, 9(1):46.

[36] Jjemba P, Johnson W, Bukhari Z, LeChevallier M: Review of the leading challenges in maintaining reclaimed water quality during storage and distribution. *Journal of water reuse and desalination* 2014, 4(4):209-237.

[37] Ranjit P, Jhansi V, Reddy KV: Conventional wastewater treatment processes. In: *Advances in the Domain of Environmental Biotechnology: Microbiological Developments in Industries, Wastewater Treatment and Agriculture.* edn.: Springer; 2021: 455-479.

[38] Ghernaout D, Arabia S: Water treatment coagulation: dares and trends. *Open Access Library Journal* 2020, 7(08):1.

[39] Merkus HG, Merkus HG: Sedimentation techniques. *Particle Size Measurements: Fundamentals, Practice, Quality* 2009:319-348.

[40] Chai WS, Cheun JY, Kumar PS, Mubashir M, Majeed Z, Banat F, Ho S-H, Show PL: A review on conventional and novel materials towards heavy metal adsorption in wastewater treatment application. *Journal of Cleaner Production* 2021, 296:126589.

[41] Ghoneim MM, El-Desoky HS, El-Moselhy KM, Amer A, Abou El-Naga EH, Mohamedein LI, Al-Prol AE: Removal of cadmium from aqueous solution using marine green algae, Ulva lactuca. *The Egyptian Journal of Aquatic Research* 2014, 40(3):235-242.

[42] Gupta VK, Ali I, Saleh TA, Nayak A, Agarwal S: Chemical treatment technologies for waste-water recycling—an overview. *RSC advances* 2012, 2(16):6380-6388.

[43] Aidoo OF, Osei-Owusu J, Chia SY, Dofuor AK, Antwi-Agyakwa AK, Okyere H, Gyan M, Edusei G, Ninsin KD, Duker RQ: Remediation of pesticide residues using ozone: A comprehensive overview. *Science of the Total Environment* 2023, 894:164933.

[44] Dogruel S, Genceli EA, Babuna FG, Orhon D: An investigation on the optimal location of ozonation within biological treatment for a tannery wastewater. *Journal of Chemical Technology & Biotechnology: International Research in Process, Environmental & Clean Technology* 2006, 81(12):1877-1885.

[45] Ituma C, Etukudoh A, Abuh M, Akpomie K, Obioha C: Utilization of Nkpuma-Akpatakpa clay in ceramics: characterization and microstructural studies. *Journal of Applied Sciences and Environmental Management* 2018, 22(1):47-53.

[46] Acero JL, Von Gunten U: Characterization of oxidation processes: ozonation and the AOP O3/H2O2. *Journal-American Water Works Association* 2001, 93(10):90-100.

[47] Suja S, Almaas S, Gracy AP, Gowsika P, Jeyapradeepa K, Sri GS, Mathiya S, Asha KB: Contamination of Water by Heavy Metals and Treatment Methods–A Review. *Current World Environment* 2024, 19(1):4.

[48] Saikaly PE, Oerther DB: Diversity of dominant bacterial taxa in activated sludge promotes functional resistance following toxic shock loading. *Microbial ecology* 2011, 61:557-567.

[49] Cao X, Zhao B, Wu Y, Huang J, Wang H, Sun X, Li S: Characterization of Alcaligenes aquatilis as a novel member of heterotrophic nitrifier-aerobic denitrifier and its performance in treating piggery wastewater. *Bioresource Technology* 2022, 354:127176.

[50] Wang Z, Li D, Shi Y, Sun Y, Okeke SI, Yang L, Zhang W, Zhang Z, Shi Y, Xiao L: Recent Implementations of Hydrogel-Based Microbial Electrochemical Technologies (METs) in Sensing Applications. *Sensors* 2023, 23(2):641.

[51] Nyika J, Dinka MO: A mini-review on wastewater treatment through bioremediation towards enhanced field applications of the technology. *AIMS Environmental Science* 2022, 9(4):403-431.

[52] Dinakarkumar Y, Gnanasekaran R, Reddy GK, Vasu V, Balamurugan P, Murali G: Fungal bioremediation: An overview of the mechanisms, applications and future perspectives. *Environmental Chemistry and Ecotoxicology* 2024.

[53] Sinha P, Mukherji S: Biofiltration process for treatment of water and wastewater. *Transactions of the Indian National Academy of Engineering* 2022, 7(4):1069-1091.

[54] Deng Y, Zhao R: Advanced oxidation processes (AOPs) in wastewater treatment. *Current pollution reports* 2015, 1(3):167-176.

[55] Al-Bastaki NM: Performance of advanced methods for treatment of wastewater: UV/TiO2, RO and UF. *Chemical Engineering and Processing: Process Intensification* 2004, 43(7):935-940.

[56] Miller CJ, Wadley S, Waite TD: Fenton, photo-Fenton and Fenton-like processes. *Advanced Oxidation Processes for Water Treatment: Fundamentals and Applications* 2017, 297.

[57] Hermosilla D, Cortijo M, Huang CP: Optimizing the treatment of landfill leachate by conventional Fenton and photo-Fenton processes. *Science of the Total Environment* 2009, 407(11):3473-3481.

[58] Ridgway H, Kelly A, Justice C, Olson B: Microbial fouling of reverse-osmosis membranes used in advanced wastewater treatment technology: chemical, bacteriological, and ultrastructural analyses. *Applied and Environmental Microbiology* 1983, 45(3):1066-1084.

[59] Dialynas E, Mantzavinos D, Diamadopoulos E: Advanced treatment of the reverse osmosis concentrate produced during reclamation of municipal wastewater. *Water research* 2008, 42(18):4603-4608.

[60] Ridgway HF, Kelly A, Justice C, Olson BH: Microbial fouling of reverse-osmosis membranes used in advanced wastewater treatment technology: chemical, bacteriological, and ultrastructural analyses. *Applied and Environmental Microbiology* 1983, 45(3):1066-1084.

[61] Osman AI, Nasr M, Farghali M, Bakr SS, Eltaweil AS, Rashwan AK, Abd El-Monaem EM: Machine learning for membrane design in energy production, gas separation, and water treatment: a review. *Environmental Chemistry Letters* 2024, 22(2):505-560.

[62] Abdel-Fatah MA: Nanofiltration systems and applications in wastewater treatment. *Ain Shams Engineering Journal* 2018, 9(4):3077-3092.

[63] Chew CM, Aroua MK, Hussain MA: Advanced process control for ultrafiltration membrane water treatment system. *Journal of cleaner production* 2018, 179:63-80.

[64] Ryu H-D, Han H, Park J-H, Kim YS: Sustainable Proposal for Regulating Organophosphate Pesticides in Wastewater Treatment Plants in South Korea. *Sustainability* 2022, 14(19):11979.

[65] Braschi I, Blasioli S, Lavrnić S, Buscaroli E, Di Prodi K, Solimando D, Toscano A: Removal and fate of pesticides in a farm constructed wetland for agricultural drainage water treatment under Mediterranean conditions (Italy). *Environmental Science and Pollution Research* 2022, 29(5):7283-7299.

[66] Omoyajowo K: Sustainable Environmental Policies: An Impact Analysis of US Regulations on Pesticides and Chemical Discharges. *Available at SSRN 4943299* 2024.

[67] Bhawan F: Guidance document & standard operating procedures for fixation of maximum residue limits (MRLs) of pesticides in food commodities. *India: Food Safety & Standards Authority of India (FSSAI)* 2022.

[68] Goodwin L, Carra I, Campo P, Soares A: Treatment options for reclaiming wastewater produced by the pesticide industry. 2017.

[69] Liu T, Xu S, Lu S, Qin P, Bi B, Ding H, Liu Y, Guo X, Liu X: A review on removal of organophosphorus pesticides in constructed wetland: performance, mechanism and influencing factors. *Science of the Total Environment* 2019, 651:2247-2268.

[70] Alisawi HAO: Performance of wastewater treatment during variable temperature. *Applied Water Science* 2020, 10(4):89.

[71] Michael-Kordatou I, Michael C, Duan X, He X, Dionysiou D, Mills M, Fatta-Kassinos D: Dissolved effluent organic matter: characteristics and potential implications in wastewater treatment and reuse applications. *Water Research* 2015, 77:213-248.

[72] Aghababaie M, Farhadian M, Jeihanipour A, Biria D: Effective factors on the performance of microbial fuel cells in wastewater treatment–a review. *Environmental Technology Reviews* 2015, 4(1):71-89.

[73] Chen Y-J, Fan T-Y, Wang L-P, Cheng T-W, Chen S-S, Yuan M-H, Cheng S: Application of Fenton method for the removal of organic matter in sewage sludge at room temperature. *Sustainability* 2020, 12(4):1518.

[74] Wang H, Quan B, Bo G, Zhang Y, Liu L, Zhang J, Zhang X, Zhang C: Advanced oxidation treatment of dissolved organic matter from wastewater treatment plant secondary effluent using scattering electrical reactor. *Journal of Cleaner Production* 2020, 267:122258.

[75] Kajitvichyanukul P, Nguyen V-H, Boonupara T, Thi L-AP, Watcharenwong A, Sumitsawan S, Udomkun P: Challenges and effectiveness of nanotechnology-based photocatalysis for pesticides-contaminated water: a review. *Environmental Research* 2022, 212:113336.

[76] Yadav GK, Ahmaruzzaman M: Recent advances in the development of nanocomposites for effective removal of pesticides from aqueous stream. *Journal of Nanoparticle Research* 2021, 23(9):213.

[77] Goel P, Arora M: Remediation of wastewater from chlorpyrifos pesticide by nano-gold photocatalyst. *MRS Advances* 2020, 5:2661-2667.

[78] Gao Y, Truong YB, Cacioli P, Butler P, Kyratzis IL: Bioremediation of pesticide contaminated water using an organophosphate degrading enzyme immobilized on nonwoven polyester textiles. *Enzyme and Microbial Technology* 2014, 54:38-44.

[79] Vaithyanathan VK, Vaidyanathan VK, Cabana H: Laccase-driven transformation of high priority pesticides without redox mediators: towards bioremediation of contaminated wastewaters. *Frontiers in Bioengineering and Biotechnology* 2022, 9:770435.

[80] Mutunga T, Sinanovic S, Harrison CS: Integrating wireless remote sensing and sensors for monitoring pesticide pollution in surface and groundwater. *Sensors* 2024, 24(10):3191.

[81] Nehru R, Chen C-W, Dong C-D: A review of smart electrochemical devices for pesticide detection in agricultural food and runoff contaminants. *Science of The Total Environment* 2024:173360.

# Chapter 18

# Carbon-Based Nanoadsorbents for the Treatment of Industrial Wastewater

**Tanisha Kathuria**[1]
**Anjali Mehta**[1]
**Sudhanshu Sharma**[1]
**and Sudesh Kumar**[2,*]
[1]Department of Chemistry, Banasthali Vidyapith, Tonk, Rajasthan, India
[2]Department of Education in Science and Mathematics, National Council of Educational Research and Training (NCERT), Delhi, India

## Abstract

Carbon-based nanoadsorbents, such as biochar, carbon nanotubes, graphene, and activated carbon, are widely used for wastewater remediation. Due to their large specific surface area (SURFACE AREA), mesoporous arrangement, excellent chemical stability, high adsorption capacity, availability of various functional groups that provide easy modification, ease of regeneration, and reusability, they have attracted special attention in environmental remediation. These characteristics enable them to resist extreme conditions encountered in wastewater, including acidic, basic, and salted conditions, even at elevated concentrations or temperatures. This chapter addresses various carbon Nanomaterials (NMs), as well as their morphologies, structures, and advantages for treating wastewater. A comprehensive analysis of issues associated with scalability, affordability, and environmental impact offers perspectives on future research areas and the advancement of environmentally friendly and more efficient wastewater treatment technologies.

**Keywords:** nanoadsorbents, wastewater treatment, adsorption mechanism, industrial wastewater

---

* Corresponding Author's Email: sudeshneyol@gmail.com

In: Advancements in Wastewater Treatment
Editors: Neha Agarwal, Maulin P. Shahl, Vijendra Singh Solankil et al.
ISBN: 979-8-90134-007-3

## 1. Introduction

Over the past century, accelerated industrial development and unsustainable manufacturing practices have significantly contributed to the contamination of aquatic ecosystems [1]. Effluents from industries such as textiles, pharmaceuticals, printing, and leather tanning frequently contain toxic organic and inorganic pollutants.

When released untreated, these substances including dyes, phenols, PAHs, and heavy metals, enter aquatic environments and pose a significant risk to natural resources [2-5]. The World Health Organization has estimated that more than 1.2 billion individuals have lost their lives as a direct result of consuming contaminated water [6]. Dyes are recognized as hazardous because of their toxic, cancer-causing, and mutagenic properties [7]. Their discharge into water bodies significantly increases biological oxygen demand (BOD) and chemical oxygen demand (COD), creating significant risks to human health. Additionally, cardiovascular diseases, neurological impairment, renal injuries, and an increased risk of diabetes and cancer are all consequences of heavy metals in aqueous media [8]. Thus, it is critical to remove both organic and inorganic contaminants from effluents before their release into natural ecosystems.

Numerous techniques are employed in water remediation, such as ion exchange, photocatalysis, biological precipitation, reverse osmosis, adsorption, flocculation, membrane separation, and photoelectrocatalysis [9-12]. Among these methods, adsorption is the most advantageous for removing contaminants from wastewater due to its simple design, ease of operation, and cost-effectiveness [13]. Numerous materials have been explored for use as adsorbents in the polluted water remediation. Various NMs, including metal oxides [14], transition metal chalcogenides [15, 16], layered double hydroxides [17], gels such as xerogels, aerogels, and hydrogels, [18, 19], and carbon-based structures, have shown potential in pollutant adsorption applications. However, Carbon-based NMs are emerging as preferred nano adsorbents for water purification because of their structural versatility, eco-friendly nature, widespread availability, and practical handling [20].

### 1.1. Carbon-Based Nanoadsorbents

The utilization of carbon-derived NMs in the treatment of wastewater has increasingly drawn considerable interest. The key characteristics of carbon NMs as a special adsorbent for the elimination of both inorganic and organic pollutants during the adsorption techniques are their high surface area, enhanced adsorption capacity, affordability, adaptability of physical or chemical alteration, and significant thermal stability [21].

The broad surface area exhibited by these materials facilitates increased availability of active sites, enhancing their interaction with wide variety of chemical species in wastewater. Owing to its exceptional bonding versatility and stability, carbon is considered as most adaptable elements in the periodic table [22]. Carbon exhibits multiple allotropic forms, each with distinct structural and physicochemical properties including fullerenes, diamond, graphene, carbon nanotubes (CNTs), graphitic carbon nitrate and graphite [23]. The exceptional characteristics of carbon-based NMs have enabled their widespread implementation in generating clean and renewable energy, particularly through light-induced water splitting, as well as in the efficient removal of environmental pollutants. The adsorbent's pore size, and structure functional groups, pH, temperature, and ionic strength, as well as the adsorbate's characteristics (pKa, size, polarity, and functionality, molecular weight,), collectively govern the adsorption performance of carbon-based adsorbents toward chemical species [24]. This chapter discusses CNTs, graphene and its derivatives, activated carbon, and biochar, focusing on their physicochemical characteristics and their roles in removing inorganic and organic pollutants from wastewater.

## 2. Graphene and Graphene Oxide

Graphene possesses a large surface area, strong hydrophobicity, and electron-rich properties, enhancing its efficiency as an adsorbent [25]. The incorporation of oxygen-containing functional groups including carbonyl, hydroxyl, and carboxyl groups onto graphene enhances pollutant removal from water [26]. Graphene, a single layer of carbon atoms in a honeycomb lattice, has demonstrated significant potential in wastewater treatment due to its unique properties. It is composed of a single layer of honeycomb-shaped carbon atoms, each exhibiting $sp^2$ hybridization [27]. Graphene has shown improved adsorption for various pollutants [28–30], primarily due to its extended $\pi$-conjugated system, which allows strong $\pi$–$\pi$ and electrostatic interactions with contaminant molecules.

### 2.1. Synthesis and Modification

The preparation of graphene involves various techniques, including chemical synthesis, mechanical exfoliation, thermal decomposition of carbon sources on substrates, chemical vapor deposition (CVD), plasma-enhanced CVD (PECVD), and the longitudinal opening of CNTs (Figure 1) [31]. Moreover, the inherent hydrophobicity of graphite and graphene poses a limitation to their direct application in water filtration systems. More specifically, two-dimensional graphene variants including graphene oxide (GO) and reduced graphene oxide (rGO) have garnered significant attention in materials science owing to their excellent hydrophilicity, extended electron delocalization, ease of functionalization, and high crystallinity, which enables $\pi$–$\pi$ interactions with aromatic molecules [32, 33].

The majority of GO is produced by chemically oxidizing and exfoliating pure graphite using the Brodie, Staudenmaier, or Hummers methods, or their various modified versions [35]. Figure 1 illustrates the step-by-step process involved in the synthesis of graphene oxide. In the synthesis of graphite oxide, graphite undergoes acid-induced oxidation, followed by alkali intercalation. These treatments facilitate its exfoliation into GO. Furthermore, various reduction techniques, including hydrothermal [36], chemical [37], microwave-assisted [38], thermal [39], and photocatalytic [40] methods, can convert GO into reduced graphene oxide (rGO). Chemical reduction, the most often used method, employs different organic and inorganic reducing agents, including urea, hydrazine hydrate, pyrrole, ascorbic acid, sodium borohydride, and amino acids [41-43].

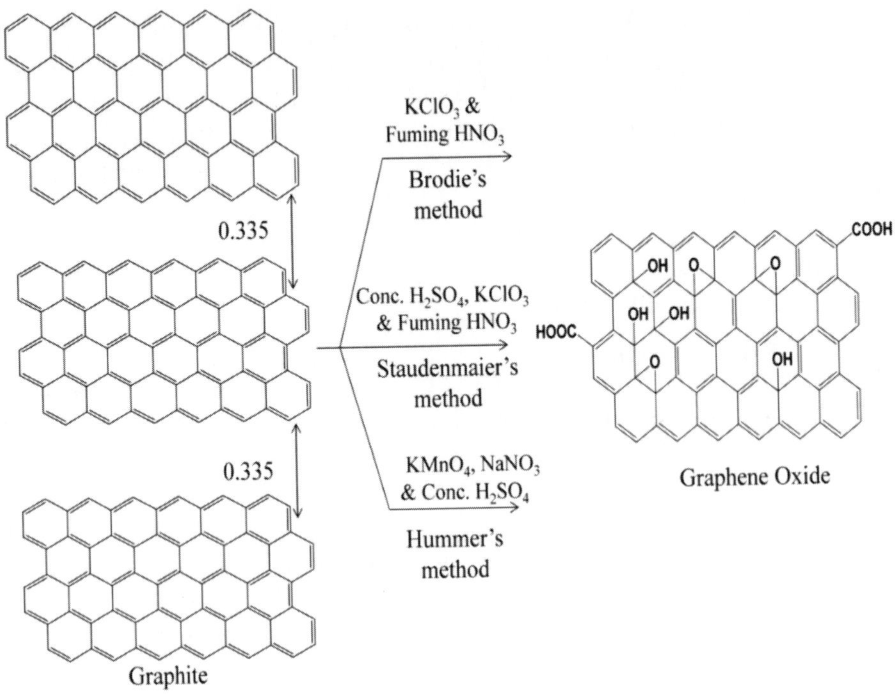


**Figure 1.** A diagrammatic representation of the synthesis process of graphene oxide [34].

(1) Brodie method

$$\text{Graphite} + KClO_3 \xrightarrow[60^{\circ}C]{+HNO_3} \xrightarrow{H_2O} \xrightarrow{\text{Exfoliation}} \text{GO}$$

(2) Staudenmaier method

$$\text{Graphite} \xrightarrow{+H_2SO_4/HNO_3} 5^{\circ}C \xrightarrow{KClO_3} \xrightarrow{H_2O} \xrightarrow{\text{Exfoliation}} \text{GO}$$

(3) Hummers-offeman method

$$\text{Graphite} \xrightarrow{+H_2SO_4} 5^{\circ}C \xrightarrow{KMnO_4} <20^{\circ}C \xrightarrow{H_2O} 90^{\circ}C \xrightarrow{H_2O_2} \xrightarrow{\text{Exfoliation}} \text{GO} \quad [44]$$

## 2.2. Applications

Several studies have documented the effectiveness of graphene and its derivatives in adsorbing various organic dyes, such as methyl orange (MO), rhodamine (RG), methylene blue (MB), Congo red (CR), and malachite green (MG), through dynamic adsorption mechanisms. Additionally, these materials have shown potential for eliminating pharmaceutical compounds, phenols, and heavy metals i.e., Fe (III), Cr (IV), Pb (II), Co (II), Cd (II), and Zn (II) [45, 46].

The surface of GO and rGO contains oxygen-containing functional groups, such as -COOH, -C-O-C-, -COO-, and -OH, which form bonds with water pollutants. GO binds with dye molecules primarily through two interactions: electrostatic interactions between its negatively charged surface and positively charged dyes, and π–π interactions between the aromatic structures of GO and the dye [47].

Direct Red 81, crystal violet, MB, Indosol SFGL direct blue, and MO, are among the dyes removed from wastewater using GO as an adsorbent. GO, modified with carboxylic and hydroxyl groups, interacts strongly with reactive functional sites in dyes during adsorption [48, 49]. Xu et al. synthesized graphene for the adsorption-based extraction of bisphenol A (BPA) from water. According to a Langmuir isotherm, graphene's maximum BPA adsorption capacity was 182 mg $g^{-1}$. The adsorption mechanism of BPA onto graphene primarily involves H-bonding and π–π bonding [50].

An adsorption capacity of 17.29 mg/g for chromium was achieved using graphene functionalized with a ferro/ferric oxide composite, as reported by Guo et al. (2014) [51]. This value exceeds than those of other magnetic materials like chitosan-coated $MnFe_2O_4$ nanoparticles (15.4 mg/g), Fe@$Fe_2O_3$ core–shell nanowires (7.78 mg/g), and $Fe_3O_4$–PEI–montmorillonite composites (8.8 mg/g) [52]. Amino-functionalized N-layer GO has been employed for the adsorption of MB and reactive black 5 from water [53]. To increase the potential of adsorption and selectivity of GO for lead (II) ions, dithiocarbamate (DTC) was used to functionalize the GO surface in different research. It is apparent that lead (II) ions can bind with functional groups like sulfur and nitrogen in DTC to generate a stable complex [21]. Various GO-based nanocomposites have been investigated for dye adsorption. For instance, methylene blue (MB) has been removed using GO-impregnated porous wood filters, $Fe_3O_4$ nanoparticle-functionalized GO/g-$C_3N_4$ nanocomposites and $Fe_3O_4$/GO composites, Additionally, GO integrated with metal ferrites and PANI–GO–$Fe_3O_4$ hybrid nanocomposites have effectively removed CR and MO dyes [54-56].

## 3. Structural and Functional Properties of CNTs

CNTs are cylindrical macromolecules composed of rolled $sp^2$-hybridized graphite sheets. Owing to their effective surface area, porosity, tubular structure, and favorable dimensional properties, CNTs are highly efficient in adsorbing various pollutants and are extensively utilized for treating water contaminants [57]. The inherent hydrophobic nature of CNTs enables strong interactions with nonpolar substances, making them effective sorbents for removing nonpolar pollutants. Despite their hydrophobic nature, the highly porous structure of CNTs facilitates the absorption of polar and ionic substances, according to the size of the molecules. The key mechanisms facilitating pollutant adsorption on CNTs include electrostatic attraction, hydrophobic interactions, π–π interactions, and electron transfer processes [58]. CNTs are generally divided into two types based on the number of graphene layers: single-walled CNTs (SWCNTs), which are made from a single sheet of carbon atoms, and multi-walled CNTs (MWCNTs), which contain multiple concentric graphene layers. Additionally, MWCNTs can have diameters up to 100 nm, compared to 0.4–3 nm for SWCNTs [22]. In contrast to SWCNTs, MWNTs typically have a lower specific surface area and no bandgap [59]. CNTs are a promising option for water treatment due to their excellent electrical and thermal properties. However, the limited use of SWCNTs in contaminant adsorption applications is largely attributed to their higher production costs and complex synthesis compared to MWCNTs.

### 3.1. Synthesis and Modification

The production of CNTs can be achieved using various methods, typically categorized into five main types: sonochemical or hydrothermal, carbon arc discharge, laser ablation, electrolysis, and chemical vapor deposition (CVD) [31]. Among these, the arc discharge technique with graphite electrodes is the most widely used and preferred for the production of CNTs with excellent purity and structure. When current is run over graphite electrodes in a temperature-controlled environment (>1700 oC), graphite vaporizes and re-condenses onto the cathode and reaction vessel walls. The carbon that condenses on the cathode forms CNTs [60].

### 3.2. Applications

Utilizing a CVD method, Alijani et al. synthesized a magnetic and environmentally friendly sorbent composed of magnetic CNTs combined with diatomaceous earth [61]. The sorbent selectivity was evaluated by conducting a lead adsorption experiment with ions including Co(II), Ni(II), Cu(II), Pb(II), Zn(II), and Cd(II), showing that the sorbent effectively removes Pb(II) ions from complex matrices. Using CVD, Duman et al. synthesized nanocomposites of MWCNT–$Fe_3O_4$ and MWCNT–κ-carrageenan–$Fe_3O_4$, which were applied as adsorbents for extracting MB from aqueous solutions [62]. Additionally, Sui et al. conducted a study on the adsorption of MB and MO using a synthesized calcium alginate/MWCNT hybrid. They reported that the maximum adsorption capacity for MB was 606.1 mg $g^{-1}$, while for MO, it was 12.5 mg $g^{-1}$ [63]. According to Liu et al., functionalized multiwalled carbon nanotubes (F-MWCNTs) can be used as demulsifier to quickly and effectively separate emulsified oil from oily effluent. It was found that F-MWCNTs could eliminate 99.8% of the oil within minutes from emulsified oily wastewater [64]. Table 1 highlights the various uses of CNTs in the removal of organic and inorganic pollutants from wastewater.

**Table 1.** The effectiveness of CNTs and their derivatives in adsorbing organic and inorganic pollutants

| Pollutants | Sorbent | Adsorption capacity (mg/g) | Ref. |
|---|---|---|---|
| Pb | L-CNTsg | 199 | [65] |
| Bromothymol blue dye | $HNO_3$/NaClO/MWCNT nanohybrid | 55 | [66] |
| MB dye | Magnetite/MWCNT | 48.06 | [67] |
| Chromium | CNTs | 24.45 | [68] |
| MO | Unmodified MWCNTs | 5.464 | [69] |
| Yellow 81 | Unmodified MWCNTs | 35.1 | [70] |
| Ciprofoxacin | Untreated SWCNTs | 933.8 | [71] |
| Tetracycline | CNTs-C@Fe-chitosan composite | 104 | [72] |
| Toluene | MWCNTs activated with KOH | 87.12 | [73] |
| Oxytetracycline | Untreated SWCNTs | 554 | [71] |
| MO | Magnetic MWCNTs | 51.74 | [74] |

## 4. Activated Carbon

Globally, activated carbon (AC) remains one of the most widely recognized and extensively used adsorbents for treating wastewater [75]. AC-based adsorbents have been extensively studied for adsorption due to their low density, structural variety, strong chemical stability, and applicability for large-scale applications [76]. This material has high porosity and is useful for sorption and catalysis because it contains both hydrophilic surface functional groups and a hydrophobic graphene layer. Demonstrating its versatility, activated carbon can effectively adsorb various contaminants, including metal ions, humic substances, dyes, phenols, pesticides, and other chemical and biological contaminants. Its chemical structure, porosity, pore volume, internal pore morphology, surface properties, and the availability of functional groups-including their activation-from their source material are typically the sources of these distinctive properties [77, 78].

### 4.1. Synthesis

AC is derived from carbon-rich materials, including various types of coal lignite, bituminous, and subbituminous, along with wood, peat, nutshells and petroleum coke [79]. AC production typically starts with pre-treatment of carbon waste materials, which includes crushing, high-temperature drying, and screening to produce uniformly sized particles. Carbonization is then performed through high temperature pyrolysis or hydrothermal carbonization (HTC), which facilitates the removal of contaminants and volatile organic compounds. HTC involves processing carbonaceous materials in a liquid medium at high pressure and temperatures above 100°C. During high-temperature pyrolysis, carbonaceous starting materials are converted in an oxygen-deficient environment at extremely high temperatures (>600°C).

### 4.2. Applications

AC finds primary industrial applications in water treatment, solvent recovery, decolorization, military protection against toxic gases like nuclear reactors, mustard gas, and air treatment [80]. The inert and porous nature of AC allows them to capture pollutants on their extensive hydrophobic internal surface, thus providing accessibility for reactants. The research by Zhang et al., focused on the utilization of powdered activated carbon (PAC) to remove 28 antibiotics from groups including chloramphenicols, sulfonamides, tetracyclines, penicillins, macrolides, and quinolones, from aqueous media [81]. The removal of basic dyes, like rhodamine, and crystal violet from synthetic textile wastewater using activated carbon produced from *Thevetia peruviana* has been examined. The Langmuir isotherm model indicated maximum monolayer adsorption capacities of 33 mg $g^{-1}$ and 37 mg/g, respectively [82]. Using acid-activated carbon prepared from *Phoenix sylvestris* leaves, Arivoli and Thenkuzali studied the adsorption of Rhodamine B and achieved a 90% removal at 60°C [83]. For Acid Violet 17, the maximum adsorption capacities for charcoal prepared from bagasse, cow dung and groundnut shell were 38.32, 50.30 and 100.57 mg $g^{-1}$ respectively. Brick kiln and wheat straw charcoal achieved adsorption capacities of 80.88 and 72.81 mg $g^{-1}$ for Acid Red 119, respectively. For Acid Red

119, adsorption capacity of 44.48 mg/g was obtained using pea shell charcoal, while for Acid Blue 15, adsorption capacity of 126.53 mg $g^{-1}$ was obtained using spent tea leaves [82]. A removal efficiency of 68.53% was attained when ibuprofen was removed using groundnut shell activated carbon (GNSAC) under optimal conditions of 0.62 g $L^{-1}$ adsorbent dose, 30°C, and 30 min. The adsorbent demonstrated a maximum capacity of 40.82 mg $g^{-1}$ for the removal of the target compound [84]. Employing KOH as an oxidizing agent, Tu et al. investigated the adsorption properties of AC made from Bermuda grass (BGAC). The experiments revealed that BGAC had an adsorption capacity of 403.23 mg $g^{-1}$ for Cr(VI), and the data fit the Langmuir isotherm model well [85]. Using the impregnation approach, α-$Fe_2O_3$ NPs doped AC was assessed for its capacity to adsorb Cr (VI) ions, according to Li et al. The maximal sorption capacity of α-$Fe_2O_3$@AC for the adsorption of Cr(VI) ions was confirmed by investigations; it fits well with pseudo-second order kinetics and was pH dependent. AC contributes to the removal of heavy metals by adsorbing Cr(VI) and subsequently reducing it to Cr(III), which then forms precipitates, such as Cr(III)-Fe(III) hydroxides, thus facilitating the overall removal process [86].

## 5. Biochar

Biochar (BC) is a carbon-rich solid byproduct derived from the pyrolysis of diverse agricultural and industrial biomass under oxygen-deficient conditions at various temperatures [87]. Its efficacy in removing emerging environmental pollutants, remediating soil, reducing greenhouse gas emissions, and generating energy is greatly influenced by characteristics like high surface area, low regeneration cost, porous structure, surface functional groups, high minerals, and microstructure [88]. Heavy metals can be adsorbed by biochar through ion exchange between the metal cations and biochar's surface protons and electrostatic interactions among the metal cations and the negative surface charge [89].

### 5.1. Synthesis

Biochar is produced as the carbon-rich solid output of thermochemical conversion processes. These include: (a) pyrolysis, which is the gradual devolatilization of organic feedstock in an inert atmosphere like $N_2$, (b) biomass thermal dehydration at 200–300 °C, (c) hydrothermal carbonization- it is the process of heating biomass that is high in moisture while applying pressure, (d) gasification, the thermal breakdown of feedstock in the presence of sub-stoichiometric agents like $CO_2$, $H_2O$, or $O_2$. Slow pyrolysis is the most common method for biochar production [90]. It is mostly derived from solid householdwaste, lignocellulosic materials, and agricultural residue [91]. One significant benefit of using waste to make biochar is that it minimizes environmental stress. Wastes are also affordable and widely accessible [92].

### 5.2. Applications

Due to its unique properties, as-fabricated biochar exhibits the capacity for adsorptive removal of various pollutants, such as organic contaminants, metals, dyes, and antibiotics. Table 2 summarizes the pollutant adsorption capabilities of BC as an adsorbent.

**Table 2.** Adsorption performance of biochar for pollutant removal

| Source | Adsorbent | Pollutants | Adsorbent capacity (mg/g) | Ref. |
|---|---|---|---|---|
| Corn straw | BC/nZVI | MG | 515.77 | [93] |
| swine manure Composte | MnO2-BC | Pb(II) | 127.75 | [94] |
| Camphor leaves | Magnetic ZnO composite | Ciprofloxacin | 500 | [95] |
| Municipal solid waste | Bentonite-biochar composite | Ciprofloxacin | 286.6 | [96] |
| Wood waste | BC | PAHs | 20.2 | [97] |
| Wheat straw | Metal doped BC | PAHs | - | [98] |
| Potato leaves and stem | Magnetic composite with humic acid | Ciprofloxacin | 12 | [99] |

## Conclusion

This chapter has highlighted the pivotal of carbon-based nanoadsorbents in the treatment of industrial wastewater, with a particular focus on four key materials: graphene, CNTs, activated carbon, and biochar. These materials are distinguished by their unique structural, surface and chemical characteristics which significantly influence their adsorption capacities. Graphene and CNTs, for instance, offer exceptionally high surface area, tunable functional groups, and strong $\pi$-$\pi$ interactions, making them highly effective in capturing a wide range of organic and inorganic pollutants. Activated carbon, known for its porous structure and affordability, remains a widely used and versatile adsorbent, particularly effective for organic compounds and heavy metals. Biochar, derived from biomass through pyrolysis, offers a sustainable and cost-effective alternative, with surface functionalities that can be engineered to target specific contaminants. Collectively, these carbon-based adsorbents exhibit high selectivity, rapid kinetics, and strong regeneration potential, which contribute to their growing application in industrial wastewater treatment. Their demonstrated efficiency in removing dyes, heavy metals, pharmaceutical residues, and other toxic pollutants underscores their essential role in advancing water purification technologies and supporting sustainable wastewater management practices.

## References

[1] Muliwa, A. M., Oyewo, O. A., & Maity, A. (2023). "Recent progress on the removal of aqueous mercury by carbon-based adsorbents: A review." *Inorg. Chem. Commun.*, *156*:111207. https://doi.org/10.1016/j.inoche.2023.111207.

[2] Song, J. Y., & Jhung, S. H. (2017). "Adsorption of pharmaceuticals and personal care products over metal-organic frameworks functionalized with hydroxyl groups: quantitative analyses of H-bonding in adsorption." *J. Chem. Eng.*, *322*: 366-374. https://doi.org/10.1016/j.cej.2017.04.036.

[3] Ntakadzeni, M., Anku, W. W., Kumar, N., Govender, P. P., & Reddy, L. (2019). "PEGylated MoS2 nanosheets: a dual functional photocatalyst for photodegradation of organic dyes and photoreduction of

chromium from aqueous solution." *Bull. Chem. React. Eng.*, *14*(1):142-152. https://doi.org/10.9767/bcrec.14.1.2258.142-152.

[4] Mukwevho, N., Kumar, N., Fosso-Kankeu, E., Waanders, F., Bunt, J., & Ray, S. S. (2019). "Visible light-excitable ZnO/2D graphitic-C3N4 heterostructure for the photodegradation of naphthalene." *Desalin. Water Treat.*, *163*:286-296 .https://doi.org/10.5004/dwt.2019.24422.

[5] Rodriguez-Mozaz, S., de Alda, M. J. L., & Barceló, D. (2004). "Monitoring of estrogens, pesticides and bisphenol A in natural waters and drinking water treatment plants by solid-phase extraction–liquid chromatography–mass spectrometry." *J. Chromatogr. A*, *1045*(1-2):85-92. https://doi.org/10.1016/j.chroma.2004.06.040.

[6] Soffian, M. S., Halim, F. Z. A., Aziz, F., Rahman, M. A., Amin, M. A. M., & Chee, D. N. A. (2022). "Carbon-based material derived from biomass waste for wastewater treatment." *Environ. Adv.*, *9*:100259. https://doi.org/10.1016/j.envadv.2022.100259.

[7] Ruan, W., Hu, J., Qi, J., Hou, Y., Zhou, C., & Wei, X. (2019). "Removal of dyes from wastewater by NMs: a review." *Adv Mater Lett*, *10*(1):9-20.https://doi.org/10.5185/amlett.2019.2148.

[8] Bhattacharjee, C., Dutta, S., & Saxena, V. K. (2020). "A review on biosorptive removal of dyes and heavy metals from wastewater using watermelon rind as biosorbent." *Environ. Adv*, *2*:100007. https://doi.org/10.1016/j.envadv.2020.100007.

[9] Das, R., Vecitis, C. D., Schulze, A., Cao, B., Ismail, A. F., Lu, X., & Ramakrishna, S. (2017). "Recent advances in NMs for water protection and monitoring." *Chem. Soc. Rev.*, *46*(22):6946-7020. https://pubs.rsc.org/en/content/articlelanding/2017/cs/c6cs00921b/unauth.

[10] Mittal, H., Kumar, V., Alhasurface arean, S. M., & Ray, S. S. (2018). "Modification of gum ghatti via grafting with acrylamide and analysis of its flocculation, adsorption, and biodegradation properties." *Int. J. Biol. Macromol*, *114*: 283-294. https://doi.org/10.1016/j.ijbiomac.2018.03.131.

[11] Ama, O. M., Kumar, N., Adams, F. V., & Ray, S. S. (2018). "Efficient and cost-effective photoelectrochemical degradation of dyes in wastewater over an exfoliated graphite-MoO 3 nanocomposite electrode." *Electrocatalysis*, *9*:623-631. https://doi.org/10.1007/s12678-018-0471-5.

[12] Nataraj, S. K., Hosamani, K. M., & Aminabhavi, T. M. (2006). "Distillery wastewater treatment by the membrane-based nanofiltration and reverse osmosis processes." *Water Res.*, *40*(12):2349-2356. https://doi.org/10.1016/j.watres.2006.04.022.

[13] Abdolali, A., Ngo, H. H., Guo, W., Zhou, J. L., Zhang, J., Liang, S., & Liu, Y. (2017). "Application of a breakthrough biosorbent for removing heavy metals from synthetic and real wastewaters in a lab-scale continuous fixed-bed column." *Bioresour. Technol.*, *229*:78-87. https://doi.org/10.1016/j.biortech.2017.01.016.

[14] Hu, J. S., Zhong, L. S., Song, W. G., & Wan, L. J. (2008). "Synthesis of hierarchically structured metal oxides and their application in heavy metal ion removal." *Adv. Mater*, *20*(15), 2977-2982. https://doi.org/10.1002/adma.200800623.

[15] Massey, A. T., Gusain, R., Kumari, S., & Khatri, O. P. (2016). "Hierarchical microspheres of MoS2 nanosheets: efficient and regenerative adsorbent for removal of water-soluble dyes." *Ind. Eng. Chem. Res.*, *55*(26):7124-7131. https://doi.org/10.1021/acs.iecr.6b01115.

[16] Wang, Z., & Mi, B. (2017). "Environmental applications of 2D molybdenum disulfide (MoS2) nanosheets Environ. Sci. Technol., *51*(15):8229-8244. https://doi.org/10.1021/acs.est.7b01466.

[17] Chubar, N., Gilmour, R., Gerda, V., Mičušík, M., Omastova, M., Heister, K., & Zaitsev, V. (2017). "Layered double hydroxides as the next generation inorganic anion exchangers: Synthetic methods versus applicability." *J. Colloid Interface Sci.*, *245*:62-80. https://doi.org/10.1016/j.cis.2017.04.013.

[18] Ye, S., Liu, Y., & Feng, J. (2017). "Low-density, mechanical compressible, water-induced self-recoverable graphene aerogels for water treatment." *ACS Appl. Mater. Interfaces.*, *9*(27): 22456-22464. https://doi.org/10.1021/acsami.7b04536.

[19] Yang, Y., Deng, Y., Tong, Z., & Wang, C. (2014). "Renewable lignin-based xerogels with self-cleaning properties and superhydrophobicity." *ACS Sustain. Chem. Eng.*, *2*(7), 1729-1733 .https://doi.org/10.1021/sc500250b.

[20] Perreault, F., De Faria, A. F., & Elimelech, M. (2015). "Environmental applications of graphene-based NMs." *Chem. Soc. Rev.*, *44*(16), 5861-5896. https://doi.org/10.1039/C5CS00021A.

[21] Ghorbani, M., Seyedin, O., & Aghamohammadhasurface arean, M. (2020). "Adsorptive removal of lead (II) ion from water and wastewater media using carbon-based NMs as unique sorbents: A review." *J. Environ. Manag.*, *254*: 109814. https://doi.org/10.1016/j.jenvman.2019.109814.

[22] Thines, R. K., Mubarak, N. M., Nizamuddin, S., Sahu, J. N., Abdullah, E. C., & Ganesan, P. (2017). "Application potential of carbon NMs in water and wastewater treatment: a review." *J. Taiwan Inst. Chem. Eng.*, *72*:116-133. .https://doi.org/10.1016/j.jtice.2017.01.018

[23] Jayaraman, T., Murthy, A. P., Elakkiya, V., Chandrasekaran, S., Nithyadharseni, P., Khan, Z., & Ashokkumar, M. (2018). "Recent development on carbon based heterostructures for their applications in energy and environment: a review." *J. Ind. Eng. Chem.*, *64*:16-59. https://doi.org/10.1016/j.jiec.2018.02.029.

[24] Aguilar-Pérez, K. M., Avilés-Castrillo, J. I., & Ruiz-Pulido, G. (2020). "Nano-sorbent materials for pharmaceutical-based wastewater effluents-An overview." ***Case*** *Stud. Chem. Environ. Eng.*, *2*:100028. https://doi.org/10.1016/j.cscee.2020.100028.

[25] Zhou, Y., Cao, S., Xi, C., Li, X., Zhang, L., Wang, G., & Chen, Z. (2019). "A novel Fe3O4/graphene oxide/citrus peel-derived bio-char based nanocomposite with enhanced adsorption affinity and sensitivity of ciprofloxacin and sparfloxacin." *Bioresour. Technol*, *292*:121951. https://doi.org/10.1016/j.biortech.2019.121951.

[26] Xiang, Y., Xu, Z., Wei, Y., Zhou, Y., Yang, X., Yang, Y., & Zhou, Z. (2019). "Carbon-based materials as adsorbent for antibiotics removal: mechanisms and influencing factors." *J. Environ. Manag.*, *237*:128-138. https://doi.org/10.1016/j.jenvman.2019.02.068.

[27] Verma, S., Mungse, H. P., Kumar, N., Choudhary, S., Jain, S. L., Sain, B., & Khatri, O. P. (2011). "Graphene oxide: an efficient and reusable carbocatalyst for aza-Michael addition of amines to activated alkenes." *Chem. Commun.*, *47*(47), :12673-12675. https://doi.org/10.1039/C1CC15230K.

[28] Wang, H., Yuan, X., Wu, Y., Huang, H., Zeng, G., Liu, Y., & Qi, Y. (2013). "Adsorption characteristics and behaviors of graphene oxide for Zn (II) removal from aqueous solution." *Appl. Surf. Sci.*, *279*:432-440. https://doi.org/10.1016/j.apsusc.2013.04.133.

[29] Cheng, J. S., Du, J., & Zhu, W. (2012). "Facile synthesis of three-dimensional chitosan–graphene mesostructures for reactive black 5 removal." *Carbohydr. Polym.*, *88*(1):61-67. https://doi.org/10.1016/j.carbpol.2011.11.065.

[30] Apul, O. G., Wang, Q., Zhou, Y., & Karanfil, T. (2013). "Adsorption of aromatic organic contaminants by graphene nanosheets: Comparison with carbon nanotubes and activated carbon." *Water Res.*, *47*(4):1648-1654. https://doi.org/10.1016/j.watres.2012.12.031.

[31] Sabzehmeidani, M. M., Mahnaee, S., Ghaedi, M., Heidari, H., & Roy, V. A. (2021). "Carbon based materials: a review of adsorbents for inorganic and organic compounds." *Mater. Adv.*, *2*(2):598-627. https://doi.org/10.1039/D0MA00087F.

[32] Pal, K., Si, A., El-Sayyad, G. S., Elkodous, M. A., Kumar, R., El-Batal, A. I., & Thomas, S. (2021). "Cutting edge development on graphene derivatives modified by liquid crystal and CdS/TiO2 hybrid matrix: optoelectronics and biotechnological aspects." *Crit. Rev. Solid State Mater. Sci.*, *46*(5):385-449. https://doi.org/10.1080/10408436.2020.1805295.

[33] Wang, X. Y., Narita, A., & Müllen, K. (2017). "Precision synthesis versus bulk-scale fabrication of graphenes." *Nat. Rev. Chem*, *2*(1):0100. https://doi.org/10.1038/s41570-017-0100.

[34] Adetayo, A., & Runsewe, D. (2019). "Synthesis and fabrication of graphene and graphene oxide: a review." *Open J. Compos. Mater.*, *9*(02):207. https://doi.org/10.4236/ojcm.2019.92012.

[35] Wang, S., Sun, H., Ang, H. M., & Tadé, M. O. (2013). "Adsorptive remediation of environmental pollutants using novel graphene-based NMs." *J. Chem. Eng.*, *226*:336-347. https://doi.org/10.1016/j.cej.2013.04.070.

[36] Zhou, Y., Bao, Q., Tang, L. A. L., Zhong, Y., & Loh, K. P. (2009). "Hydrothermal dehydration for the "green" reduction of exfoliated graphene oxide to graphene and demonstration of tunable optical limiting properties." *Chem. Mater.*, *21*(13): 2950-2956. https://doi.org/10.1021/cm9006603.

[37] Chua, C. K., & Pumera, M. (2014). "Chemical reduction of graphene oxide: a synthetic chemistry viewpoint." *Chem. Soc. Rev.*, *43*(1):291-312. https://doi.org/10.1039/C3CS60303B.

[38] Voiry, D., Yang, J., Kupferberg, J., Fullon, R., Lee, C., Jeong, H. Y., & Chhowalla, M. (2016). "High-quality graphene via microwave reduction of solution-exfoliated graphene oxide." *Science*, *353*(6306):1413-1416. https://doi.org/10.1126/science.aah3398.

[39] Wang, X., Zhi, L., & Müllen, K. (2008). "Transparent, conductive graphene electrodes for dye-sensitized solar cells." *Nano Lett*, *8*(1):323-327. https://doi.org/10.1021/nl072838r.

[40] Pei, S., & Cheng, H. M. (2012). "The reduction of graphene oxide." *Carbon*, *50*(9):3210-3228. https://doi.org/10.1016/j.carbon.2011.11.010.

[41] Shin, H. J., Kim, K. K., Benayad, A., Yoon, S. M., Park, H. K., Jung, I. S., & Lee, Y. H. (2009). "Efficient reduction of graphite oxide by sodium borohydride and its effect on electrical conductance."*Adv. Funct. Mater.*, *19*(12):1987-1992. https://doi.org/10.1002/adfm.200900167.

[42] Zhang, J., Yang, H., Shen, G., Cheng, P., Zhang, J., & Guo, S. (2010). "Reduction of graphene oxide via L-ascorbic acid." *Chem. Commun.*, *46*(7):1112-1114. https://pubs.rsc.org/en/content/articlelanding/2010/cc/b917705a/unauth.

[43] Lei, Z., Lu, L., & Zhao, X. S. (2012). "The electrocapacitive properties of graphene oxide reduced by urea." *Energy & Environmental Science*, *5*(4):6391-6399.

[44] Sheoran, K., Kaur, H., Siwal, S. S., Saini, A. K., Vo, D. V. N., & Thakur, V. K. (2022). "Recent advances of carbon-based NMs (CBNMs) for wastewater treatment: Synthesis and application." *Chemosphere*, *299*:34364. https://doi.org/10.1016/j.chemosphere.2022.134364.

[45] Yusuf, M., Elfghi, F. M., Zaidi, S. A., Abdullah, E. C., & Khan, M. A. (2015). "Applications of graphene and its derivatives as an adsorbent for heavy metal and dye removal: a systematic and comprehensive overview." *RSC Adv.*, *5*(62):50392-50420. https://doi.org/10.1039/C5RA07223A.

[46] Cortés-Arriagada, D., & Toro-Labbé, A. (2015). "Improving As (iii) adsorption on graphene based surfaces: impact of chemical doping." *Phys. Chem. Chem. Phys.*, *17*(18):12056-12064.. https://doi.org/10.1039/C5CP01313E.

[47] Lee, D. W., Kim, T., & Lee, M. (2011). "An amphiphilic pyrene sheet for selective functionalization of graphene."*Chem. Commun.*, *47*(29):8259-8261. https://doi.org/10.1039/C1CC12868J.

[48] de Assis, L. K., Damasceno, B. S., Carvalho, M. N., Oliveira, E. H., & Ghislandi, M. G. (2020). "Adsorption capacity comparison between graphene oxide and graphene nanoplatelets for the removal of coloured textile dyes from wastewater." *Environ. Technol.* https://doi.org/10.1080/09593330.2019.156760.

[49] Kalita, H., Tyagi, H., & Aslam, M. (2020). "Surface-tailored graphene oxide paper: an efficient filter for dye pollutants." *Environ. Sci.:Water Res. Technol.*, *6*(4):963-975. https://doi.org/10.1039/C9EW01129C.

[50] Xu, J., Wang, L., & Zhu, Y. (2012). "Decontamination of bisphenol A from aqueous solution by graphene adsorption." *Langmuir*, *28*(22): 8418-8425. https://doi.org/10.1021/la301476p.

[51] Guo, X., Du, B., Wei, Q., Yang, J., Hu, L., Yan, L., & Xu, W. (2014). "Synthesis of amino functionalized magnetic graphenes composite material and its application to remove Cr (VI), Pb (II), Hg (II), Cd (II) and Ni (II) from contaminated water." *J. Hazard. Mater.*, *278*:211-220. https://doi.org/10.1016/j.jhazmat.2014.05.075.

[52] Larraza, I., López-Gónzalez, M., Corrales, T., & Marcelo, G. (2012). "Hybrid materials: magnetite–polyethylenimine–montmorillonite, as magnetic adsorbents for Cr (VI) water treatment." *J. Colloid Interface Sci.*, *385*(1):24-33. https://doi.org/10.1016/j.jcis.2012.06.050.

[53] Fraga, T. J. M., De Souza, Z. S. B., Marques Fraga, D. M. D. S., Carvalho, M. N., de Luna Freire, E. M. P., Ghislandi, M. G., & da Motta Sobrinho, M. A. (2020). "Comparative approach towards the adsorption of Reactive Black 5 and methylene blue by n-layer graphene oxide and its amino-functionalized derivative." *Adsorption*, *26*(2):283-301. https://doi.org/10.1007/s10450-019-00156-9.

[54] Goodman, S. M., Bura, R., & Dichiara, A. B. (2018). "Facile impregnation of graphene into porous wood filters for the dynamic removal and recovery of dyes from aqueous solutions." *ACS Appl. Nano Mater.*, *1*(10):5682-5690. https://doi.org/10.1021/acsanm.8b01275.

[55] Khajeh, M., & Barkhordar, A. (2020). "Fe3O4/graphene oxide composite for adsorption of methylene blue and methyl orange in water treatment." *J. Appl. Spectrosc*, *87*(4):701-707. https://doi.org/10.1007/s10812-020-01057-4.

[56] Bayantong, A. R. B., Shih, Y. J., Ong, D. C., Abarca, R. R. M., Dong, C. D., & de Luna, M. D. G. (2021). "Adsorptive removal of dye in wastewater by metal ferrite-enabled graphene oxide nanocomposites." *Chemosphere*, *274*:129518. https://doi.org/10.1016/j.chemosphere.2020.129518.

[57] Abbas, A., Al-Amer, A. M., Laoui, T., Al-Marri, M. J., Nasser, M. S., Khraisheh, M., & Atieh, M. A. (2016). "Heavy metal removal from aqueous solution by advanced carbon nanotubes: critical review of adsorption applications." *Sep. Purif. Technol.*, *157*:141-161. https://doi.org/10.1016/j.seppur.2015.11.039.

[58] Yan, Y., Zhang, M., Gong, K., Su, L., Guo, Z., & Mao, L. (2005). "Adsorption of methylene blue dye onto carbon nanotubes: a route to an electrochemically functional nanostructure and its layer-by-layer assembled nanocomposite." *Chem. Mater.*, *17*(13):3457-3463. https://doi.org/10.1021/cm0504182.

[59] Mohapatra, L., Cheon, D., & Yoo, S. H. (2023). "Carbon-based NMs for catalytic wastewater treatment: a review." *Molecules*, *28*(4):1805. https://doi.org/10.3390/molecules28041805.

[60] Eatemadi, A., Daraee, H., Karimkhanloo, H., Kouhi, M., Zarghami, N., Akbarzadeh, A., & Joo, S. W. (2014). "Carbon nanotubes: properties, synthesis, purification, and medical applications." *Nanoscale Res. Lett.*, *9*:1-13. https://doi.org/10.1186/1556-276X-9-393.

[61] Alijani, H., Beyki, M. H., Shariatinia, Z., Bayat, M., & Shemirani, F. (2014). "A new approach for one step synthesis of magnetic carbon nanotubes/diatomite earth composite by chemical vapor deposition method: application for removal of lead ions." *J. Chem. Eng.*, *253*:456-463. https://doi.org/10.1016/j.cej.2014.05.021.

[62] Duman, O., Tunç, S., Polat, T. G., & Bozoğlan, B. K. (2016). "Synthesis of magnetic oxidized multiwalled carbon nanotube-κ-carrageenan-Fe3O4 nanocomposite adsorbent and its application in cationic Methylene Blue dye adsorption." *Carbohydr. Polym.147*:79-88.. https://doi.org/10.1016/j.carbpol.2016.03.099.

[63] Sui, K., Li, Y., Liu, R., Zhang, Y., Zhao, X., Liang, H., & Xia, Y. (2012). "Biocomposite fiber of calcium alginate/multi-walled carbon nanotubes with enhanced adsorption properties for ionic dyes." *Carbohydr. Polym.*, *90*(1):399-406. https://doi.org/10.1016/j.carbpol.2012.05.057.

[64] Liu, J., Li, X., Jia, W., Ding, M., Zhang, Y., & Ren, S. (2016). "Separation of emulsified oil from oily wastewater by functionalized multiwalled carbon nanotubes." *J. Dispers. Sci. Technol.*, *37*(9):1294-1302. https://doi.org/10.1080/01932691.2015.1090320

[65] Li, Z., Chen, J., & Ge, Y. (2017). "Removal of lead ion and oil droplet from aqueous solution by lignin-grafted carbon nanotubes." *J. Chem. Eng*, *308*:809-817. https://doi.org/10.1016/j.cej.2016.09.126

[66] Ghaedi, M., Khajehsharifi, H., Yadkuri, A. H., Roosta, M., & Asghari, A. (2012). "Oxidized multiwalled carbon nanotubes as efficient adsorbent for bromothymol blue." *Toxicol. Environ. Chem*, *94*(5):873-883. https://doi.org/10.1080/02772248.2012.678999.

[67] Ai, L., Zhang, C., Liao, F., Wang, Y., Li, M., Meng, L., & Jiang, J. (2011). "Removal of methylene blue from aqueous solution with magnetite loaded multi-wall carbon nanotube: kinetic, isotherm and mechanism analysis." *J. Hazard. Mater.*, *198*:282-290. https://doi.org/10.1016/j.jhazmat.2011.10.041.

[68] Mubarak, N. M., Thobashinni, M., Abdullah, E. C., & Sahu, J. N. (2016). "Comparative kinetic study of removal of Pb2+ ions and Cr3+ ions from wastewater using carbon nanotubes produced using microwave heating." *C*, *2*(1): 7. https://doi.org/10.3390/c2010007.

[69] Robati, D., Bagheriyan, S., Rajabi, M., Moradi, O., & Peyghan, A. A. (2016). "Effect of electrostatic interaction on the methylene blue and methyl orange adsorption by the pristine and functionalized carbon nanotubes." *Physica E Low Dimens. Syst. Nanostruct.*, *83*:1-6. https://doi.org/10.1016/j.physe.2016.04.005.

[70] Vuono, D., Catizzone, E., Aloise, A., Policicchio, A., Agostino, R. G., Migliori, M., & Giordano, G. (2017). "Study of adsorption behavior of multi-walled carbon nanotubes towards dyes applied in textile applications."*Adv. Sci. Lett.*, *23*(6): 5851-5854.https://doi.org/10.1166/asl.2017.9048.

[71] Ncibi, M. C., & Sillanpää, M. (2015). "Optimized removal of antibiotic drugs from aqueous solutions using single, double and multi-walled carbon nanotubes." *"J. Hazard. Mater*, *298*: 102-110. https://doi.org/10.1016/j.jhazmat.2015.05.025.

[72] Ma, J., Zhuang, Y., & Yu, F. (2015). "Facile method for the synthesis of a magnetic CNTs–C@ Fe–chitosan composite and its application in tetracycline removal from aqueous solutions."*Phys. Chem. Chem. Phys.*, *17*(24):15936-15944. https://doi.org/10.1039/C5CP02542G.

[73] Yu, F., Wu, Y., Li, X., & Ma, J. (2012). "Kinetic and thermodynamic studies of toluene, ethylbenzene, and m-xylene adsorption from aqueous solutions onto KOH-activated multiwalled carbon nanotubes." *J. Agric. Food Chem.*, *60*(50):12245-12253. https://doi.org/10.1021/jf304104z.

[74] Yao, Y., Bing, H., Feifei, X., & Xiaofeng, C. (2011). "Equilibrium and kinetic studies of methyl orange adsorption on multiwalled carbon nanotubes." *J. Chem. Eng.*, *170*(1):82-89. https://doi.org/10.1016/j.cej.2011.03.031.

[75] Bhatnagar, A., & Sillanpää, M. (2010). "Utilization of agro-industrial and municipal waste materials as potential adsorbents for water treatment—a review." *J. Chem. Eng.*, *157*(2-3):277-296. https://doi.org/10.1016/j.cej.2010.01.007.

[76] Foo, K. Y., & Hameed, B. H. (2010). "An overview of dye removal via activated carbon adsorption process." *Desalin. Water Treat*, *19*(1-3):255-274. https://doi.org/10.5004/dwt.2010.1214.

[77] Sonal, S., Prakash, P., Mishra, B. K., & Nayak, G. C. (2020). "Synthesis, characterization and sorption studies of a zirconium (iv) impregnated highly functionalized mesoporous activated carbons." *RSC Adv.*, *10*(23):13783-13798. https://doi.org/10.1039/C9RA10103A.

[78] Zheng, C., Zhao, L., Zhou, X., Fu, Z., & Li, A. (2013). "Treatment technologies for organic wastewater. In *Water treatment*." IntechOpen.

[79] Akhtar, J., Amin, N. A. S., & Shahzad, K. (2016). "A review on removal of pharmaceuticals from water by adsorption." *Desalin. Water Treat*, *57*(27): 12842-12860. https://doi.org/10.1080/19443994.2015.1051121.

[80] Allen, S. J., Whitten, L., & McKay, G. (1998). "The production and characterisation of activated carbons: a review." *Dev. Chem. Eng. Miner. Process.* ,*6*(5):231-261. https://doi.org/10.1002/apj.5500060501.

[81] Zhang, X., Guo, W., Ngo, H. H., Wen, H., Li, N., & Wu, W. (2016). "Performance evaluation of powdered activated carbon for removing 28 types of antibiotics from water." *J. Environ. Manag.*, *172*:193-200. https://doi.org/10.1016/j.jenvman.2016.02.038.

[82] Bello, O. S., Adegoke, K. A., Olaniyan, A. A., & Abdulazeez, H. (2015). "Dye adsorption using biomass wastes and natural adsorbents: overview and future prospects." *Desalin. Water Treat.*, *53*(5):1292-1315. https://doi.org/10.1080/19443994.2013.862028.

[83] Arivoli, S., & Henkuzhali, M. (2008). "Kinetic, mechanistic, thermodynamic and equilibrium studies on the adsorption of Rhodamine B by acid activated low cost carbon." *J. Chem.*, *5*(2):187-200. https://doi.org/10.1155/2008/437375.

[84] Olowonyo, I. A., Salam, K. K., Aremu, M. O., & Lateef, A. Waste Management Bulletin. Synthesis, Characterization, and Adsorptive Performance of Titanium Dioxide Nanoparticles Modified Groundnut Shell Activated Carbon on Ibuprofen Removal from Pharmaceutical Wastewater," *Waste Manag. Bull.* 1, no. 4 (2024): 217–233, https://doi.org/10.1016/J.WMB.2023.11.003.

[85] Tu, B., Wen, R., Wang, K., Cheng, Y., Deng, Y., Cao, W., & Tao, H. (2020). "Efficient removal of aqueous hexavalent chromium by activated carbon derived from Bermuda grass." *J. Colloid Interface Sci.*, *560*:649-658. https://doi.org/10.1016/j.jcis.2019.10.103.

[86] Li, B., Yin, W., Xu, M., Tan, X., Li, P., Gu, J.,& Wu, J. (2019). "Facile modification of activated carbon with highly dispersed nano-sized α-Fe2O3 for enhanced removal of hexavalent chromium from aqueous solutions." *Chemosphere*, *224*:220-227. https://doi.org/10.1016/j.chemosphere.2019.02.121.

[87] Peng, H., Gao, P., Chu, G., Pan, B., Peng, J., & Xing, B. (2017). "Enhanced adsorption of Cu (II) and Cd (II) by phosphoric acid-modified biochars." *Environ. Pollut.*, *229*:846-853. https://doi.org/10.1016/j.envpol.2017.07.004.

[88] Cha, J. S., Park, S. H., Jung, S. C., Ryu, C., Jeon, J. K., Shin, M. C., & Park, Y. K. (2016). "Production and utilization of biochar: A review." *J. Ind. Eng. Chem.*, *40*:1-15. https://doi.org/10.1016/j.jiec.2016.06.002.

[89] Agrafioti, E., Kalderis, D., & Diamadopoulos, E. (2014). "Ca and Fe modified biochars as adsorbents of arsenic and chromium in aqueous solutions." *J. Environ. Manag.*, *146*:444-450. https://doi.org/10.1016/j.jenvman.2014.07.029.

[90] Patra, B. R., Mukherjee, A., Nanda, S., & Dalai, A. K. (2021). "Biochar production, activation and adsorptive applications: a review." *Environ. Chem. Lett.*, *19*:2237-2259. https://doi.org/10.1007/s10311-020-01165-9.

[91] Zhao, Y., Qamar, S. A., Qamar, M., Bilal, M., & Iqbal, H. M. (2021). "Sustainable remediation of hazardous environmental pollutants using biochar-based nanohybrid materials."*J. Environ. Manag.*, *300*:113762. https://doi.org/10.1016/j.jenvman.2021.113762.

[92] Chausali, N., Saxena, J., & Prasad, R. (2021). "Nanobiochar and biochar based nanocomposites: Advances and applications." *J. Agric. Food Res.*, *5*:100191. https://doi.org/10.1016/j.jafr.2021.100191.

[93] Eltaweil, A. S., Mohamed, H. A., Abd El-Monaem, E. M., & El-Subruiti, G. M. (2020). "Mesoporous magnetic biochar composite for enhanced adsorption of malachite green dye: Characterization, adsorption kinetics, thermodynamics and isotherms." *Adv. Powder Technol.*, *31*(3), 1253-1263. https://doi.org/10.1016/j.apt.2020.01.005.

[94] Liang, J., Li, X., Yu, Z., Zeng, G., Luo, Y., Jiang, L,& Wu, H. (2017). "Amorphous MnO2 modified biochar derived from aerobically composted swine manure for adsorption of Pb (II) and Cd (II)." *ACS Sustain. Chem. Eng.5*(6):5049-5058. https://doi.org/10.1021/acssuschemeng.7b00434.

[95] Hu, Y., Zhu, Y., Zhang, Y., Lin, T., Zeng, G., Zhang, S., & Long, H. (2019). "An efficient adsorbent: Simultaneous activated and magnetic ZnO doped biochar derived from camphor leaves for ciprofloxacin adsorption." *Bioresour. Technol.*, *288*: 121511. https://doi.org/10.1016/j.biortech.2019.121511.

[96] Ashiq, A., Adassooriya, N. M., Sarkar, B., Rajapaksha, A. U., Ok, Y. S., & Vithanage, M. (2019). "Municipal solid waste biochar-bentonite composite for the removal of antibiotic ciprofloxacin from aqueous media." *J. Environ. Manag.*, *236*:428-435 .https://doi.org/10.1016/j.jenvman.2019.02.006.

[97] Zhou, X., Shi, L., Moghaddam, T. B., Chen, M., Wu, S., & Yuan, X. (2022). "Adsorption mechanism of polycyclic aromatic hydrocarbons using wood waste-derived biochar." *J. Hazard. Mater.*, *425*:128003. https://doi.org/10.1016/j.jhazmat.2021.128003.

[98] Zhang, X., Zhang, X., An, C., & Wang, S. (2023). "Electrochemistry-enhanced peroxymonosulfate activation by CoAl-LDH@ biochar for simultaneous treatment of heavy metals and PAHs." *Sep. Purif. Technol.*, *311*: 123341. https://doi.org/10.1016/j.seppur.2023.123341.

[99] Zhao, J., Liang, G., Zhang, X., Cai, X., Li, R., Xie, X., & Wang, Z. (2019). "Coating magnetic biochar with humic acid for high efficient removal of fluoroquinolone antibiotics in water." *Sci. Total Environ.*, *688*:1205-1215. https://doi.org/10.1016/j.scitotenv.2019.06.287.

## Chapter 19

# Current Trends, Future Perspectives, and Challenges in Wastewater Treatment

**Shisak Sharma**
**Raplang Steven Umdor**
**Imotila T. Longchar**
**and Dipak Sinha***
Department of Chemistry, Nagaland University, Lumami, Nagaland, India

## Abstract

The increasing global demand for freshwater, exacerbated by population growth and climate change, has intensified the need for effective wastewater treatment solutions. Consequently, there is a pressing need to explore current trends, future perspectives, and challenges in wastewater treatment technologies. Innovative approaches, including advanced oxidation processes, adsorption methods and biological treatments should be examined for their potential to enhance wastewater quality and sustainability. This chapter highlights recent advancements in treatment methods, including hydrothermal carbonization (HTC), which offers energy-efficient solutions for organic waste management while addressing heavy metal contamination. It also discusses the emergence of pollutants such as microplastics and pharmaceuticals, which could not be treated efficiently by traditional methods of remediation. Furthermore, the chapter emphasizes the importance of regulatory frameworks and international collaborations to tackle the complexities of wastewater management and emerging contaminants. By integrating these insights, the chapter aims to provide a comprehensive overview of the evolving landscape of wastewater treatment, underscoring the necessity for ongoing research and development to ensure environmental health and resource sustainability.

**Keywords:** wastewater treatment, hydrothermal carbonization, adsorption, emerging contaminants, advanced oxidation processes

---

* Corresponding Author's Email: dipaksinha@gmail.com

In: Advancements in Wastewater Treatment
Editors: Neha Agarwal, Maulin P. Shahl, Vijendra Singh Solankil et al.
ISBN: 979-8-90134-007-3

## 1. Introduction

The growing global demand for freshwater, driven by rapid population growth and climate change, has highlighted the importance of effective wastewater treatment systems. With increasing urbanization and industrialization, environmental contamination has worsened due to the constant release of harmful compounds into water resources, causing serious public health concerns [1,2]. Moreover, rising water demand in domestic and industrial sectors has led to an increase in wastewater volume entering sewage systems. Approximately 5 to 10 billion tonnes of industrial waste are generated annually and discharged untreated into the environment. When wastewater from numerous industries is released directly without treatment, it can have severe impacts on the environment [3,4]. As a result, removing the pollutants is critical to ensure the availability of safe drinking water and the discharged wastewater should not have detrimental environmental impacts [5]. Furthermore, recycling and reusing treated wastewater is critical for reducing the usage of freshwater.

In this context, wastewater treatment has emerged as an essential priority for safeguarding environmental sustainability and public health. Traditional wastewater management methods tend to keep up with the complexities of modern effluents, which contain a wide range of contaminants such as pharmaceuticals, personal care products, microplastics, and heavy metals [6-8]. These emerging contaminants pose major threats to aquatic ecosystems and human health, necessitating the development of novel treatment technologies capable of efficiently addressing these issues [9-11]. One promising approach to treating wastewater is hydrothermal carbonisation (HTC), a thermochemical process that turns wet biomass into hydrochar, a carbon-rich solid. This method not only makes it easier to deal with organic waste, but it also addresses issues caused by heavy metals in sewage sludge [12,13]. HTC operates under mild conditions compared to other thermochemical processes, allowing for the direct treatment of high-moisture feedstocks without the need for pre-drying, thus reducing energy consumption and operational costs [14].

Besides HTC, new methods including advanced oxidation processes (AOPs), adsorption methods, and biological solutions are also being studied for their potential to enhance wastewater quality and sustainability [15-17]. AOPs utilise highly reactive species to break down persistent organic pollution that can't be removed with regular methods [18]. Adsorption methods using different materials have also shown promise in getting rid of contaminants in wastewater [19-21].

Despite technological advances, wastewater treatment presents various challenges. The increasing complexity of contaminants, the high energy and operating costs of modern treatment procedures, and the necessity for strong regulatory frameworks are some of the most pressing concerns that must be addressed. Effective measures are critical for encouraging the use of innovative technology and ensuring compliance with emerging pollutant regulations [22-24].

Therefore, the current chapter aims to provide a comprehensive overview of current trends in wastewater treatment technologies, future perspectives on their development and implementation, and the challenges that must be addressed to achieve effective solutions. By integrating insights from recent research and case studies, this chapter underscores the necessity for ongoing innovation and investment in wastewater treatment to safeguard environmental health and ensure resource sustainability.

## 2. Current Trends in Wastewater Treatment

Recent advancements in wastewater treatment technologies have significantly improved efficiency and effectiveness. These include:

### 2.1. Hydrothermal Carbonization (HTC)

One of the most notable recent trends in wastewater treatment is the development of hydrothermal carbonization (HTC), a process that offers an energy-efficient solution for managing organic waste. HTC operates at relatively low temperatures (180-250°C) under pressure, converting organic waste into hydrochar, which can be used as a fuel or soil amendment [25]. HTC also effectively removes heavy metals from wastewater, as these contaminants become immobilized in the solid hydrochar matrix, reducing their environmental impact.

HTC holds promise not only for its capacity to handle high organic loads but also for its potential in producing value-added by-products. These by-products can be used for energy generation or in agricultural applications, promoting a circular economy within the wastewater treatment process. The energy efficiency of HTC makes it particularly attractive for industrial wastewater treatment, where large volumes of organic waste are generated [26, 27].

### 2.2. Advanced Oxidation Processes (AOPs)

Advanced Oxidation Processes are innovative treatment methods that use highly reactive species, primarily hydroxyl radicals ($^{\bullet}OH$), to degrade organic pollutants in wastewater. The generation of these radicals is crucial, as they possess a high redox potential and can oxidize virtually any organic compound present in the water matrix. This oxidation occurs rapidly and non-selectively, allowing for the effective fragmentation of pollutants into smaller, less harmful inorganic molecules such as carbon dioxide and water [28-30]. The production of hydroxyl radicals can be achieved through various means, including the use of primary oxidants like ozone and hydrogen peroxide, energy sources such as ultraviolet (UV) light, or catalysts like titanium dioxide, ZnO, $CeO_2$, $ZrO_2$, and others [31-33].

#### *2.2.1. Types of AOPs*

Several types of AOPs are employed for wastewater treatment, each with distinct mechanisms for generating hydroxyl radicals:

1. *Ozonation:* Ozonation involves the introduction of ozone ($O_3$) into water, where it decomposes to form hydroxyl radicals. Ozone is highly effective in treating organic compounds, pathogens, and odor-causing substances. Additionally, it is a powerful disinfectant that breaks down pollutants through direct oxidation or by producing $^{\bullet}OH$ radicals, especially when combined with catalysts such as hydrogen peroxide or ultraviolet (UV) light. The combination of ozone with UV radiation or $H_2O_2$ enhances

the formation of radicals, significantly increasing the oxidation potential and pollutant removal efficiency [34, 35].

2. *Fenton Processes:* The Fenton process is one of the most widely used AOPs in wastewater treatment. It utilizes hydrogen peroxide ($H_2O_2$) and ferrous iron ($Fe^{2+}$) as a catalyst to generate hydroxyl radicals via the following reaction [36]:

   $$Fe^{2+} + H_2O_2 \rightarrow Fe^{3+} + \bullet OH + OH^-$$

   The process can be further enhanced by applying UV light (photo-Fenton), which regenerates $Fe^{2+}$ ions from $Fe^{3+}$, increasing the production of hydroxyl radicals. This method is particularly effective in degrading recalcitrant organic pollutants, such as aromatic compounds, dyes, and certain industrial chemicals [36].
3. *UV/$H_2O_2$:* Ultraviolet (UV) light can be combined with hydrogen peroxide to generate hydroxyl radicals through the photolysis of $H_2O_2$. The UV light cleaves the hydrogen peroxide molecules into two hydroxyl radicals. This process is especially effective for breaking down organic pollutants that absorb UV light and are susceptible to oxidative attack. The advantage of UV/$H_2O_2$ lies in its ability to treat a wide range of contaminants without the need for additional catalysts, making it a versatile and relatively straightforward AOP for municipal and industrial wastewater treatment [37, 38].

## 2.3. Adsorption Techniques

Adsorption technologies are another important tool for removing contaminants from wastewater. Adsorption processes involve the accumulation of pollutants on the surface of a solid adsorbent, typically materials like activated carbon, zeolites, or biochar. These materials have a high surface area and are effective in capturing dissolved organic compounds, metals, and emerging contaminants like microplastics [39-43].

Activated carbon has long been used in water treatment, but recent innovations in the development of biochar (a carbon-rich material produced from the pyrolysis of organic waste) have created new opportunities for low-cost, sustainable adsorption. Biochar has proven effective in removing various pollutants, including heavy metals and microplastics, making it a valuable addition to the wastewater treatment tool kit [44, 45]

Moreover, the adsorption process is increasingly integrated into hybrid treatment systems, where it is combined with biological or chemical processes to improve overall treatment efficiency. Adsorption techniques also offer flexibility, as they can be deployed in both centralized and decentralized wastewater treatment systems [46, 47].

### *2.3.1. Adsorbent Materials*

A variety of materials have been developed and used as adsorbents, each with unique properties that influence their adsorption capacity and suitability for specific pollutants. Commonly used adsorbents include:

1. *Activated Carbon:* Activated carbon is one of the most commonly used adsorbents due to its high surface area, porosity, and ability to adsorb a wide range of contaminants. It is produced by heating carbonaceous materials (such as coal, wood, or coconut shells) in the absence of oxygen, followed by activation with steam or chemicals to enhance its adsorption capacity. Activated carbon is particularly effective for adsorbing organic compounds, such as pesticides, pharmaceuticals, and dyes, as well as certain inorganic pollutants like heavy metals [48-50].
2. *Biochar:* Biochar is a carbon-rich material produced by the thermal decomposition of biomass under oxygen-limited conditions (pyrolysis). Similar to activated carbon, biochar possesses a large surface area and porosity, making it an effective adsorbent for organic pollutants and heavy metals. Biochar's production from agricultural waste, such as crop residues and wood chips, makes it a sustainable and cost-effective alternative to activated carbon. Additionally, biochar can be further modified with chemical activation or surface functionalization to enhance its adsorption performance for specific pollutants [51,52].
3. *Nanomaterials:* Nanomaterials, including carbon nanotubes, graphene, and metal-organic frameworks (MOFs), have gained attention as advanced adsorbents due to their extraordinary surface area, tunable pore structures, and unique chemical properties. These materials exhibit enhanced adsorption capacities for a wide range of emerging pollutants, including pharmaceuticals, heavy metals, and microplastics [53,54]. For example, carbon nanotubes (CNTs) have demonstrated superior efficiency in adsorbing heavy metals like lead and cadmium due to their high surface-to-volume ratio and functional groups that promote metal ion interaction [55].

The choice of adsorbent material depends on several factors, including the target pollutant, water chemistry, cost-effectiveness, and potential for reuse or regeneration. In many cases, combining different adsorbent materials or modifying their surfaces can significantly enhance their adsorption efficiency for specific applications in wastewater treatment.

### *2.3.2. Effectiveness in Treating Microplastics and Heavy Metals*

Adsorption technologies have shown significant promise in the treatment of emerging pollutants such as microplastics and heavy metals, which are increasingly recognized as critical challenges in modern wastewater management. These pollutants are not only persistent in the environment but also pose serious risks to ecosystems and human health.

1. *Microplastics*: Microplastics, which are plastic particles smaller than 5 mm in size, are pervasive in aquatic environments and difficult to remove through traditional treatment methods. Adsorption offers a potential solution for capturing microplastics from wastewater, particularly when using adsorbents with tailored surface properties [56]. The adsorption of microplastics onto modified surfaces has been explored to enhance removal efficiencies. For instance, functionalized adsorbents can increase interaction with microplastic particles, facilitating their capture during treatment processes [57]. Studies have shown that modified activated carbon and biochar can significantly reduce microplastic concentrations in wastewater through effective adsorption mechanisms [58].

2. *Heavy Metals:* Heavy metals such as lead (Pb), cadmium (Cd), arsenic (As), and mercury (Hg) are toxic pollutants that can accumulate in water bodies and enter the food chain, posing significant risks to human health. Adsorption is particularly advantageous for removing heavy metals from wastewater due to its high efficiency even at low concentrations. Various studies have demonstrated that different adsorbents can selectively bind heavy metals such as lead, cadmium, and arsenic, thereby reducing their toxicity in treated effluents [59-61] (Ali & Gupta, 2006).

## 2.4. Biological Treatment Innovations

Biological treatment innovations, particularly Anaerobic Ammonium Oxidation (Anammox) and Microbial Fuel Cells (MFCs), are revolutionizing wastewater treatment by offering energy-positive solutions that simultaneously address nutrient removal and energy generation. These technologies represent a significant advancement in the quest for sustainable wastewater management practices.

### *2.4.1. Anaerobic Ammonium Oxidation (Anammox)*

Anaerobic ammonium oxidation is a biological process that converts ammonium ($NH_4^+$) and nitrite ($NO_2^-$) into nitrogen gas ($N_2$) under anaerobic conditions. The Anammox process is advantageous because it eliminates the need for aeration, which is a significant energy cost in conventional nitrification-denitrification processes. Additionally, Anammox does not require external organic carbon sources, making it a cost-effective solution for nitrogen removal in wastewater treatment [62-64].

The Anammox process has gained traction since its discovery in the early 1990s, with the first large-scale reactor implemented in Rotterdam in 2007. This technology has demonstrated its potential to replace traditional nitrification/denitrification steps in large-scale treatment plants, resulting in reduced operational costs and greenhouse gas emissions [65]. The overall reaction can be summarized as follows:

$$NH_4^+ + NO_2^- \rightarrow N_2 + 2H_2O$$

This reaction not only contributes to effective nitrogen removal but also enhances the sustainability of wastewater treatment systems.

### *2.4.2. Microbial Fuel Cells (MFCs)*

Microbial Fuel Cells (MFCs) are another innovative biological treatment technology that harnesses the metabolic activity of microorganisms to convert organic matter into electricity while simultaneously treating wastewater [66]. In MFCs, microorganisms oxidize organic substrates, releasing electrons that are transferred to an anode, generating electrical energy in the process. The cathode then reduces terminal electron acceptors such as oxygen or nitrate, completing the circuit and allowing for energy recovery from wastewater [67, 68].

The integration of Anammox with MFC technology referred to as Anammox-MFC systems has shown great promise in achieving simultaneous nitrogen removal and energy generation. In these systems, ammonium serves as a substrate for both Anammox bacteria and electroactive microorganisms at the anode, enhancing nitrogen removal efficiency while producing electricity. Research indicates that these coupled systems can achieve higher nitrogen removal rates compared to traditional MFCs due to the faster oxidation of ammonium through the Anammox pathway [69, 70].

Both Anammox and MFC technologies have demonstrated potential for enhancing microbial degradation of complex organic pollutants. In MFCs, the presence of electrodes not only facilitates electron transfer but also promotes the growth of pollutant-degrading bacteria by providing a conductive surface for microbial attachment. This can lead to accelerated degradation rates of organic contaminants due to improved substrate availability and enhanced metabolic activity of electroactive microorganisms [71, 72].

Additionally, engineered biofilms and genetically modified microbes are being investigated to optimize these biological processes. By enhancing microbial communities through bioengineering techniques, researchers aim to improve the efficiency of pollutant degradation and nutrient recovery from wastewater. For instance, genetically modified strains of bacteria can be designed to possess superior metabolic pathways for degrading specific pollutants or to enhance their electrogenic properties in MFCs [73].

## 3. Future Perspectives

The future of wastewater treatment is increasingly focused on sustainable practices that prioritize water reuse, resource recovery, and resilience to climate change. As global water scarcity intensifies due to population growth and climate variability, innovative approaches are essential to ensure the availability of clean water and the efficient management of wastewater. This section explores key future perspectives in wastewater treatment, including water reuse and resource recovery, decentralized treatment systems, and climate resilience.

### 3.1. Indirect and Direct Potable Reuse (IPR/DPR)

Water reuse is emerging as a critical strategy for augmenting freshwater supplies, particularly in arid regions where water scarcity is prevalent. Indirect Potable Reuse (IPR) involves treating wastewater to a high standard before discharging it into natural water bodies, where it can mix with other water sources before being extracted for drinking. Direct Potable Reuse (DPR), by contrast, treats wastewater to drinking water standards for direct supply to consumers without an environmental buffer [74, 75].

The increasing importance of IPR and DPR is underscored by the need for sustainable water management practices that can alleviate pressure on dwindling freshwater resources. Advanced treatment technologies are being implemented to ensure that reused water meets stringent safety standards. These technologies include membrane filtration, advanced oxidation processes, and biological treatments that effectively remove pollutants, including pathogens, pharmaceuticals, and microplastics [76, 77].

### 3.2. Circular Economy in Wastewater Management

The concept of a circular economy is gaining traction in wastewater management, emphasizing the recovery of valuable resources from wastewater rather than viewing it solely as waste [78]. Nutrient recovery particularly phosphorus and nitrogen is a key component of this approach. Phosphorus is essential for agricultural fertilizers, while nitrogen can be recovered through processes like anaerobic ammonium oxidation (Anammox) or struvite precipitation [79].

Furthermore, energy generation from wastewater is gaining importance. Anaerobic digestion of sludge not only reduces waste volume but also produces biogas that can be utilized for electricity generation or heating. This dual benefit aligns with the principles of circular economy by transforming waste into a resource while minimizing environmental impacts [80].

### 3.3. Producing Bioenergy from Sludge via Anaerobic Digestion

Anaerobic digestion is a well-established technology for converting organic matter in wastewater in to biogas, a renewable energy source primarily composed of methane. By optimizing anaerobic digestion processes, wastewater treatment facilities can enhance biogas production while simultaneously reducing greenhouse gas emissions associated with sludge disposal [81]. This bioenergy can be harnessed to power treatment facilities or be fed into the grid, contributing to energy self-sufficiency.

Additionally, integrating anaerobic digestion with nutrient recovery technologies creates opportunities for resource-efficient wastewater management systems that align with circular economy principles [82,83]. For example, recovered nutrients can be used as fertilizers in agriculture, closing the nutrient loop and promoting sustainable farming practices [84].

### 3.4. Climate Resilience

Climate change poses significant challenges to existing wastewater management systems through the increased frequency of extreme weather events, such as floods and droughts. These events can overwhelm traditional infrastructure designed for stable conditions, leading to untreated discharges into waterways or disruptions in service delivery [85].

To adapt to these changing conditions, there is a pressing need for resilient wastewater management strategies that can withstand climate variability while maintaining service reliability.

## 4. Challenges in Wastewater Treatment

The treatment of wastewater is essential for protecting public health and the environment. However, several challenges impede the effective management of wastewater systems. These challenges include the presence of emergingpollutants, regulatory and policy hurdles, and economic and operational constraints. Addressing these issues is crucial for developing sustainable and efficient wastewater treatment practices.

## 4.1. Emerging Contaminants

### *4.1.1. Complexity of Removing Emerging Pollutants*

Emerging contaminants such as microplastics, pharmaceuticals, and endocrine-disrupting chemicals (EDCs) have become significant concerns in wastewater treatment. These pollutants are often resistant to conventional treatment methods, making their removal complex and challenging. Microplastics, for instance, are small plastic particles that can pass through standard filtration systems, leading to their accumulation in aquatic ecosystems. Pharmaceuticals and EDCs pose similar challenges; they are designed to be biologically active at low concentrations and can disrupt endocrine functions in wildlife and humans [86, 87].

The complexity of removing these pollutants necessitates advanced treatment technologies that can effectively target a wide range of contaminants. However, many existing wastewater treatment plants are not equipped to handle these emerging pollutants due to outdated technologies or insufficient capacity.

### *4.1.2. Limitations of Current Technologies*

Although advanced oxidation processes (AOPs) and other innovative technologies show promise for treating emerging contaminants, they often come with high costs and technical complexities. Upgrading traditional plants to incorporate these advanced systems can be financially prohibitive for many municipalities, particularly in developing countries where budget constraints are prevalent [88, 89]. Additionally, the need for specialized training and expertise to operate these complex systems can further complicate implementation.

### *4.1.3. Health and Environmental Risks*

The long-term impacts of untreated emerging contaminants on ecosystems and human health remain largely unknown. Studies have indicated that microplastics can accumulate in the food chain, potentially leading to adverse health effects in humans and animals [90]. Similarly, pharmaceuticals and EDCs have been linked to reproductive and developmental issues in aquatic organisms [91]. The lack of comprehensive data on the effects of these pollutants underscores the urgent need for improved monitoring and treatment strategies.

## 4.2. Regulatory and Policy Challenges

### *4.2.1. Global Variability in Standards*

One of the significant challenges facing wastewater treatment is the variability in regulatory standards across countries. Different nations have established varying thresholds for permissible levels of contaminants in treated effluent, leading to inconsistencies in water quality management practices [92, 93]. This lack of uniformity complicates international efforts to address water pollution effectively.

Furthermore, many developing countries lack cohesive national frameworks for regulating wastewater discharge. Inadequate regulations can result in insufficient treatment practices that fail to protect public health and the environment.

#### *4.2.2. Role of International Collaboration*

Considering the transboundary nature of water resources, international cooperation is vital for effective water management. Initiatives that span national borders can promote the exchange of knowledge, strengthen capacity building, and foster the establishment of best practices for wastewater treatment [94]. Collaborative efforts can also help establish harmonized standards that promote consistency in water quality across regions.

Programs aimed at enhancing technical capacity in developing regions can empower local governments to implement effective wastewater management strategies. Such initiatives may include training programs for operators, investment in infrastructure upgrades, and support for adopting advanced treatment technologies.

### 4.3. Economic and Operational Challenges

#### *4.3.1. Cost of Advanced Treatment Technologies*

The financial burden associated with implementing advanced treatment technologies poses a significant challenge for many wastewater treatment facilities. The initial capital investment required for upgrading infrastructure can be substantial, especially for municipalities with limited budgets. Balancing the need for advanced systems with financial constraints is particularly challenging in developing countries where resources are scarce.

Furthermore, ongoing operational costs, such as energy consumption and chemical inputs, can further strain budgets. As a result, many facilities may opt to continue using outdated systems that do not adequately address emerging contaminants.

#### *4.3.2. Operational Efficiency and Maintenance*

Operational efficiency is critical for ensuring that wastewater treatment facilities function effectively while minimizing costs. Reducing energy consumption is a key focus area; energy often constitutes one of the highest operational expenses for treatment plants. Implementing energy-efficient technologies, such as anaerobic digestion or solar energy systems, can reduce costs and improve sustainability.

Moreover, regular maintenance is essential to prevent system failures that could lead to untreated discharges into the environment. Aging infrastructure poses a significant challenge; many facilities operate with outdated equipment that requires frequent repairs or replacements. Ensuring adequate training for staff members is also vital; operators must possess the skills necessary to maintain complex systems effectively.

## 5. Future Outlook

The evolving role of wastewater treatment in securing global freshwater resources cannot be overstated. As water scarcity becomes increasingly prevalent due to climate change and population growth, innovative wastewater management practices will play a critical role in mitigating environmental risks. Effective treatment systems can contribute to water conservation efforts by enabling indirect and direct potable reuse, thereby augmenting available freshwater supplies.

Moreover, integrating wastewater treatment with green infrastructure initiatives such as constructed wetlands and bioswales can enhance urban resilience while providing multiple ecological benefits. These nature-based solutions not only improve water quality but also promote biodiversity and community well-being.

In conclusion, the future of wastewater treatment lies in embracing innovation, collaboration, and sustainability. By advancing technologies that enhance resource recovery and pollutant removal while fostering partnerships across sectors, societies can move toward a more sustainable water management paradigm. This proactive approach will ensure that wastewater is viewed not merely as a waste product but as a valuable resource that can contribute to environmental health, economic viability, and social equity.

## Conclusion

The landscape of wastewater treatment is undergoing significant transformation, driven by the need for sustainable practices, innovative technologies, and a commitment to addressing emerging environmental challenges. This conclusion summarizes key findings regarding trends, technologies, and challenges in wastewater treatment, emphasizes the role of ongoing research and collaboration, and presents a future outlook on the importance of effective wastewater management.

Recent advancements in wastewater treatment have highlighted several critical trends and technologies. The integration of advanced treatment methods such as anaerobic ammonium oxidation (Anammox), MFCs, and AOPs has improved the efficiency and effectiveness of pollutant removal. These technologies not only address traditional contaminants but also target emerging pollutants like microplastics and pharmaceuticals that pose significant risks to ecosystems and human health.

Furthermore, resource recovery has gained prominence, emphasizing the reclamation of valuable nutrients, such as nitrogen and phosphorus, from wastewater. This shift towards a circular economy in wastewater management is crucial for promoting sustainability and reducing reliance on synthetic fertilizers. The adoption of decentralized treatment systems has also emerged as a viable solution, particularly in rural and peri-urban areas, allowing for localized management of wastewater while minimizing transportation costs.

## References

[1] Akhtar, N., Syakir Ishak, M. I., Bhawani, S. A. and Umar, K. (2021). Various natural and anthropogenic factors responsible for water quality degradation: A review. *Water*, 13: 2660.

[2] Gomes da Silva, F. J., Gouveia, R. M., Gomes da Silva, F. J., and Gouveia, R. M. (2020). Global population growth and industrial impact on the environment. *Cleaner Production: Toward a Better Future*, 33-75.

[3] Gaur, V. K., Sharma, P., Sirohi, R., Awasthi, M. K., Dussap, C. G. and Pandey, A. (2020). Assessing the impact of industrial waste on environment and mitigation strategies: A comprehensive review. *Journal of hazardous materials, 398*: 123019.

[4] Sharma, P., Gaur, V. K., Kim, S. H. and Pandey, A. (2020). Microbial strategies for bio-transforming food waste into resources. *Bioresource technology, 299*:122580.

[5] Saravanan, A., Kumar, P. S., Jeevanantham, S., Karishma, S., Tajsabreen, B., Yaashikaa, P. R. and Reshma, B. (2021). Effective water/wastewater treatment methodologies for toxic pollutants removal: Processes and applications towards sustainable development. *Chemosphere*, *280*: 130595.

[6] Singh, B. J., Chakraborty, A. and Sehgal, R. (2023). A systematic review of industrial wastewater management: Evaluating challenges and enablers. *Journal of Environmental Management*, *348*: 119230.

[7] Li, X., Shen, X., Jiang, W., Xi, Y. and Li, S. (2024). Comprehensive review of emerging contaminants: Detection technologies, environmental impact, and management strategies. *Ecotoxicology and Environmental Safety*, *278*, 116420.

[8] Joseph, T. M., Al-Hazmi, H. E., Śniatała, B., Esmaeili, A. and Habibzadeh, S. (2023). Nanoparticles and nanofiltration for wastewater treatment: From polluted to fresh water. *Environmental Research*, 238: 117114.

[9] Yadav, D., Rangabhashiyam, S., Verma, P., Singh, P., Devi, P., Kumar, P., Hussain, C. M., Gaurav, G. K. and Kumar, K. S. (2021). Environmental and health impacts of contaminants of emerging concerns: Recent treatment challenges and approaches. *Chemosphere*, *272*: 129492.

[10] Li, X., Shen, X., Jiang, W., Xi, Y. and Li, S. (2024). Comprehensive review of emerging contaminants: Detection technologies, environmental impact, and management strategies. Ecotoxicology and Environmental Safety, 278: 116420.

[11] Rout, P. R., Zhang, T. C., Bhunia, P. and Surampalli, R. Y. (2021). Treatment technologies for emerging contaminants in wastewater treatment plants: A review. *Science of the Total Environment*, *753*: 141990.

[12] Cavali, M., Junior, N. L., de Sena, J. D., Woiciechowski, A. L., Soccol, C. R., Belli Filho, P., Bayard, R., Benbelkacem, H. and de Castilhos Junior, A. B. (2023). A review on hydrothermal carbonization of potential biomass wastes, characterization and environmental applications of hydrochar, and biorefinery perspectives of the process. *Science of The Total Environment*, 857: 159627.

[13] Xia, Y., Liu, H., Guo, Y., Liu, Z. and Jiao, W. (2019). Immobilization of heavy metals in contaminated soils by modified hydrochar: Efficiency, risk assessment and potential mechanisms. *Science of the Total Environment*, *685*: 1201-1208.

[14] Khan, T. A., Saud, A. S., Jamari, S. S., Ab Rahim, M. H., Park, J. W. and Kim, H. J. (2019). Hydrothermal carbonization of lignocellulosic biomass for carbon rich material preparation: A review. *Biomass and bioenergy*, *130*: 105384.

[15] Paździor, K., Bilińska, L. and Ledakowicz, S. (2019). A review of the existing and emerging technologies in the combination of AOPs and biological processes in industrial textile wastewater treatment. *Chemical Engineering Journal*, *376*: 120597.

[16] Giwa, A., Yusuf, A., Balogun, H. A., Sambudi, N. S., Bilad, M. R., Adeyemi, I. and Curcio, S. (2021). Recent advances in advanced oxidation processes for removal of contaminants from water: A comprehensive review. *Process Safety and Environmental Protection*, *146*: 220-256.

[17] Ponnusami, A. B., Sinha, S., Ashokan, H., Paul, M. V., Hariharan, S. P., Arun, J., Gopinath, K.P., Hoang Le, Q. and Pugazhendhi, A. (2023). Advanced oxidation process (AOP) combined biological process for wastewater treatment: A review on advancements, feasibility and practicability of combined techniques. *Environmental research*, *237*: 116944.

[18] Ghime, D. and Ghosh, P. (2019). Removal of organic compounds found in the wastewater through electrochemical advanced oxidation processes: a review. *Russian Journal of Electrochemistry*, *55*: 591-620.

[19] Ahmed, M. B., Zhou, J. L., Ngo, H. H. and Guo, W. (2015). Adsorptive removal of antibiotics from water and wastewater: progress and challenges. *Science of the Total Environment*, *532:* 112-126.

[20] El-Sayed, M. E. (2020). Nanoadsorbents for water and wastewater remediation. *Science of the Total Environment*, *739*: 139903.

[21] Singh, B. J., Chakraborty, A. and Sehgal, R. (2023). A systematic review of industrial wastewater management: Evaluating challenges and enablers. *Journal of Environmental Management*, *348*: 119230.

[22] Ateia, M., Wei, H. and Andreescu, S. (2024). Sensors for Emerging Water Contaminants: Overcoming Roadblocks to Innovation. Environmental Science & Technology, 58: 2636-2651.

[23] Saravanan, A., Kumar, P. S., Jeevanantham, S., Karishma, S., Tajsabreen, B., Yaashikaa, P. R. and Reshma, B. (2021). Effective water/wastewater treatment methodologies for toxic pollutants removal: Processes and applications towards sustainable development. *Chemosphere*, *280*: 130595.

[24] Hassan, M., Kanwal, S., Singh, R. S., Muhammed, A. SA, M. A., Anwar, M. and Zhao, C. (2024). Current challenges and future perspectives associated with configuration of microbial fuel cell for simultaneous energy generation and wastewater treatment. *International Journal of Hydrogen Energy, 50*: 323-350.

[25] Ojewumi, M. E. and Chen, G. (2024). Hydrochar Production by Hydrothermal Carbonization: Microwave versus Supercritical Water Treatment. *Biomass, 4*: 574-598.

[26] Wang, L., Chang, Y. and Li, A. (2019). Hydrothermal carbonization for energy-efficient processing of sewage sludge: A review. *Renewable and Sustainable Energy Reviews, 108*: 423-440.

[27] Owsianiak, M., Ryberg, M. W., Renz, M., Hitzl, M. and Hauschild, M. Z. (2016). Environmental performance of hydrothermal carbonization of four wet biomass waste streams at industry-relevant scales. *ACS Sustainable Chemistry & Engineering, 4*: 6783-6791.

[28] Cheng, M., Zeng, G., Huang, D., Lai, C., Xu, P., Zhang, C. and Liu, Y. (2016). Hydroxyl radicals based advanced oxidation processes (AOPs) for remediation of soils contaminated with organic compounds: a review. *Chemical Engineering Journal, 284*: 582-598.

[29] Wang, J. and Wang, S. (2020). Reactive species in advanced oxidation processes: Formation, identification and reaction mechanism. *Chemical Engineering Journal, 401*: 126158.

[30] Kurian, M. (2021). Advanced oxidation processes and nanomaterials-a review. *Cleaner Engineering and Technology, 2*: 100090.

[31] Li, X., Zhao, R., Jiang, H., Zhai, Y. and Ma, P. (2016). Preparation and catalytic properties of ZnO-$CeO_2$-$TiO_2$ composite. *Synthesis and Reactivity in Inorganic, Metal-Organic, and Nano-Metal Chemistry, 46*: 775-782.

[32] Abdelfattah, I. and El-Shamy, A. M. (2024). A comparative study for optimizing photocatalytic activity of $TiO_2$-based composites with $ZrO_2$, ZnO, $Ta_2O_5$, SnO, $Fe_2O_3$, and CuO additives. *Scientific Reports, 14*: 27175.

[33] Parvas, M., Haghighi, M. and Allahyari, S. (2019). Catalytic wet air oxidation of phenol over ultrasound-assisted synthesized Ni/$CeO_2$–$ZrO_2$ nanocatalyst used in wastewater treatment. *Arabian Journal of Chemistry, 12*: 1298-1307.

[34] Rekhate, C. V. and Srivastava, J. K. (2020). Recent advances in ozone-based advanced oxidation processes for treatment of wastewater-A review. *Chemical Engineering Journal Advances, 3*: 100031.

[35] Litter, M. I. (2005). Introduction to photochemical advanced oxidation processes for water treatment. *Environmental photochemistry part II*, 325-366.

[36] Razzaq, U., Nguyen, T. B., Saleem, M. U., Le, V. R., Chen, C. W., Bui, X. T. and Dong, C. D. (2024). Recent progress in electro-Fenton technology for the remediation of pharmaceutical compounds in aqueous environments. *Science of The Total Environment*, 174253.

[37] Manna, M. and Sen, S. (2023). Advanced oxidation process: a sustainable technology for treating refractory organic compounds present in industrial wastewater. *Environmental Science and Pollution Research, 30*: 25477-25505.

[38] Mohammed, H. A., Khaleefa, S. A. and Basheer, M. I. (2021). Photolysis of methylene blue dye using an advanced oxidation process (ultraviolet light and hydrogen peroxide). *Journal of Engineering and Sustainable Development, 25*: 59-67.

[39] Saxena, R., Saxena, M. and Lochab, A. (2020). Recent progress in nanomaterials for adsorptive removal of organic contaminants from wastewater. *ChemistrySelect, 5*: 335-353.

[40] Cheng, N., Wang, B., Wu, P., Lee, X., Xing, Y., Chen, M. and Gao, B. (2021). Adsorption of emerging contaminants from water and wastewater by modified biochar: A review. *Environmental Pollution, 273*: 116448.

[41] Sharma, S., Umdor, R. S., Longchar, I. T., Ezung, S. L. and Sinha, D. (2024). Exploring the adsorption of catechol and resorcinol onto Croton caudatus activated carbon: An integrated experimental and theoretical approach. *Groundwater for Sustainable Development, 27*: 101325.

[42] Verma, A., Sharma, G., Kumar, A., Dhiman, P., Mola, G. T., Shan, A. and Si, C. (2024). Microplastic pollutants in water: A comprehensive review on their remediation by adsorption using various adsorbents. *Chemosphere*, 141365.

[43] Zeghioud, H., Fryda, L., Djelal, H., Assadi, A. and Kane, A. (2022). A comprehensive review of biochar in removal of organic pollutants from wastewater: Characterization, toxicity,

activation/functionalization and influencing treatment factors. *Journal of Water Process Engineering, 47*: 102801.

[44] Ma, Q., Yu, Y., Sindoro, M., Fane, A. G., Wang, R. and Zhang, H. (2017). Carbon-based functional materials derived from waste for water remediation and energy storage. *Advanced materials, 29*: 1605361.

[45] Liu, W. J., Jiang, H. and Yu, H. Q. (2015). Development of biochar-based functional materials: toward a sustainable platform carbon material. *Chemical reviews, 115*: 12251-12285.

[46] Cairone, S., Hasan, S. W., Choo, K. H., Lekkas, D. F., Fortunato, L., Zorpas, A. A., Korshin, G., Zarra, T., Belgiorno, V. and Naddeo, V. (2024). Revolutionizing wastewater treatment toward circular economy and carbon neutrality goals: Pioneering sustainable and efficient solutions for automation and advanced process control with smart and cutting-edge technologies. *Journal of Water Process Engineering, 63*: 105486.

[47] Garrido-Baserba, M., Vinardell, S., Molinos-Senante, M., Rosso, D. and Poch, M. (2018). The economics of wastewater treatment decentralization: a techno-economic evaluation. *Environmental science & technology, 52*: 8965-8976.

[48] Jeirani, Z., Niu, C. H. and Soltan, J. (2017). Adsorption of emerging pollutants on activated carbon. *Reviews in Chemical Engineering, 33*: 491-522.

[49] Srivastava, A., Gupta, B., Majumder, A., Gupta, A. K. and Nimbhorkar, S. K. (2021). A comprehensive review on the synthesis, performance, modifications, and regeneration of activated carbon for the adsorptive removal of various water pollutants. *Journal of Environmental Chemical Engineering, 9:* 106177.

[50] Heidarinejad, Z., Dehghani, M. H., Heidari, M., Javedan, G., Ali, I. and Sillanpää, M. (2020). Methods for preparation and activation of activated carbon: a review. *Environmental Chemistry Letters, 18:* 393-415.

[51] Premchand, P., Demichelis, F., Chiaramonti, D., Bensaid, S. and Fino, D. (2023). Biochar production from slow pyrolysis of biomass under $CO_2$ atmosphere: A review on the effect of $CO_2$ medium on biochar production, characterisation, and environmental applications. *Journal of Environmental Chemical Engineering, 11*: 110009.

[52] Pariyar, P., Kumari, K., Jain, M. K. and Jadhao, P. S. (2020). Evaluation of change in biochar properties derived from different feedstock and pyrolysis temperature for environmental and agricultural application. *Science of the Total Environment, 713*: 136433.

[53] Sabzehmeidani, M. M., Mahnaee, S., Ghaedi, M., Heidari, H. and Roy, V. A. (2021). Carbon based materials: a review of adsorbents for inorganic and organic compounds. *Materials Advances, 2*: 598-627.

[54] Li, X., Shen, X., Jiang, W., Xi, Y. and Li, S. (2024). Comprehensive review of emerging contaminants: Detection technologies, environmental impact, and management strategies. *Ecotoxicology and Environmental Safety, 278*: 116420.

[55] Krishna, R. H., Chandraprabha, M. N., Samrat, K., Murthy, T. K., Manjunatha, C. and Kumar, S. G. (2023). Carbon nanotubes and graphene-based materials for adsorptive removal of metal ions–a review on surface functionalization and related adsorption mechanism. *Applied Surface Science Advances, 16*: 100431.

[56] Zhang, Y., Jiang, H., Bian, K., Wang, H. and Wang, C. (2021). A critical review of control and removal strategies for microplastics from aquatic environments. *Journal of Environmental Chemical Engineering, 9*: 105463.

[57] Ali, I., Tan, X., Li, J., Peng, C., Wan, P., Naz, I., Duan, Z. and Ruan, Y. (2023). Innovations in the development of promising adsorbents for the remediation of Microplastics and Nanoplastics–a critical review. *Water Research, 230*: 119526.

[58] Shen, M., Hu, T., Huang, W., Song, B., Zeng, G. and Zhang, Y. (2021). Removal of microplastics from wastewater with aluminosilicate filter media and their surfactant-modified products: Performance, mechanism and utilization. *Chemical Engineering Journal, 421*: 129918.

[59] Burakov, A. E., Galunin, E. V., Burakova, I. V., Kucherova, A. E., Agarwal, S., Tkachev, A. G. and Gupta, V. K. (2018). Adsorption of heavy metals on conventional and nanostructured materials for wastewater treatment purposes: A review. *Ecotoxicology and environmental safety, 148:* 702-712.

[60] Chakraborty, R., Asthana, A., Singh, A. K., Jain, B. and Susan, A. B. H. (2022). Adsorption of heavy metal ions by various low-cost adsorbents: a review. *International Journal of Environmental Analytical Chemistry, 102*: 342-379.

[61] Burakov, A. E., Galunin, E. V., Burakova, I. V., Kucherova, A. E., Agarwal, S., Tkachev, A. G. and Gupta, V. K. (2018). Adsorption of heavy metals on conventional and nanostructured materials for wastewater treatment purposes: A review. *Ecotoxicology and environmental safety, 148*: 702-712.

[62] Yan, X., Liu, D., de Smit, S. M., Komin, V., Buisman, C. J. and Ter Heijne, A. (2024). Oxygen-to-ammonium-nitrogen ratio as an indicator for oxygen supply management in microoxic bioanodic ammonium oxidation. *Water Research, 261:* 121993.

[63] Xi, H., Zhou, X., Arslan, M., Luo, Z., Wei, J., Wu, Z. and El-Din, M. G. (2022). Heterotrophic nitrification and aerobic denitrification process: Promising but a long way to go in the wastewater treatment. *Science of The Total Environment, 805*: 150212.

[64] Du, R., Peng, Y., Ji, J., Shi, L., Gao, R. and Li, X. (2019). Partial denitrification providing nitrite: Opportunities of extending application for anammox. *Environment International, 131*: 105001.

[65] Thakur, I. S. and Medhi, K. (2019). Nitrification and denitrification processes for mitigation of nitrous oxide from wastewater treatment plants for biovalorization: Challenges and opportunities. *Bioresource technology, 282*: 502-513.

[66] Pandya, R. S., Kaur, T., Bhattacharya, R., Bose, D. and Saraf, D. (2024). Harnessing microorganisms for bioenergy with Microbial Fuel Cells: Powering the future. *Water-Energy Nexus, 7*: 1-12.

[67] Schaetzle, O., Barrière, F. and Baronian, K. (2008). Bacteria and yeasts as catalysts in microbial fuel cells: electron transfer from micro-organisms to electrodes for green electricity. *Energy & Environmental Science, 1*: 607-620.

[68] Vijay, A., Sonawane, J. M. and Chhabra, M. (2022). Denitrification process in microbial fuel cell: A comprehensive review. *Bioresource Technology Reports, 17*: 100991.

[69] Kong, Z., Zhou, Y., Fu, Z., Zhang, Y. and Yan, R. (2022). Mechanism of stable power generation and nitrogen removal in the ANAMMOX-MFC treating low C/N wastewater. *Chemosphere, 296*: 133937.

[70] Wang, T., Chen, M., Zhu, J., Li, N. and Wang, X. (2023). Anodic ammonium oxidation in microbial electrolysis cell: Towards nitrogen removal in low C/N environment. *Water Research*, 120276.

[71] Chen, Y., Jiang, G., Sivakumar, M. and Wu, J. (2023). Enhancing integrated denitrifying anaerobic methane oxidation and Anammox processes for nitrogen and methane removal: A review. *Critical Reviews in Environmental Science and Technology, 53*: 390-415.

[72] Lin, C. W., Zhu, T. J., Lin, L. C. and Liu, S. H. (2022). Promoting biodegradation of toluene and benzene in groundwater using microbial fuel cells with cathodic modification. *Journal of Water Process Engineering, 47*: 102839.

[73] Li, M., Zhou, M., Tian, X., Tan, C., McDaniel, C. T., Hassett, D. J. and Gu, T. (2018). Microbial fuel cell (MFC) power performance improvement through enhanced microbial electrogenicity. *Biotechnology advances, 36*: 1316-1327.

[74] Cotruvo, J. A. and Bell, K. Y. (2014). Need for direct and indirect potable water reuse specifications. *Journal-American Water Works Association, 106*: 28-30.

[75] Rodriguez, C., Van Buynder, P., Lugg, R., Blair, P., Devine, B., Cook, A. and Weinstein, P. (2009). Indirect potable reuse: a sustainable water supply alternative. *International journal of environmental research and public health, 6:* 1174-1209.

[76] Krishnan, R. Y., Manikandan, S., Subbaiya, R., Karmegam, N., Kim, W. and Govarthanan, M. (2023). Recent approaches and advanced wastewater treatment technologies for mitigating emerging microplastics contamination–A critical review. *Science of the Total Environment, 858*: 159681.

[77] Gao, W., Zhang, Y., Mo, A., Jiang, J., Liang, Y., Cao, X. and He, D. (2022). Removal of microplastics in water: Technology progress and green strategies. *Green Analytical Chemistry, 3*: 100042.

[78] Abdelfattah, I. and El-Shamy, A. M. (2024). Review on the escalating imperative of zero liquid discharge (ZLD) technology for sustainable water management and environmental resilience. *Journal of Environmental Management, 351*: 119614.

[79] Vanotti, M. B., García-González, M. C., Szögi, A. A., Harrison, J. H., Smith, W. B. and Moral, R. (2020). Removing and recovering nitrogen and phosphorus from animal manure. *Animal manure: Production, characteristics, environmental concerns, and management, 67*: 275-321.

[80] Velenturf, A. P. and Purnell, P. (2021). Principles for a sustainable circular economy. *Sustainable production and consumption, 27*: 1437-1457.

[81] Kim, D., Bowen, J. D. and Ozelkan, E. C. (2015). Optimization of wastewater treatment plant operation for greenhouse gas mitigation. *Journal of Environmental Management, 163*: 39-48.

[82] Kumareswaran, K., Ranasinghe, S., Jayasinghe, G. Y. and Dassanayake, K. B. (2024). Systematic review on liquid organic waste (LOW) characteristics, processing technologies, and their potential applications: Towards circular economy and resource efficiency. *Journal of Cleaner Production*, 141286.

[83] Najar, I. N., Sharma, P., Das, R., Tamang, S., Mondal, K., Thakur, N., Gandhi, S. G. and Kumar, V. (2024). From waste management to circular economy: Leveraging thermophiles for sustainable growth and global resource optimization. *Journal of Environmental Management, 360*: 121136.

[84] Praveen, K., Abinandan, S., Venkateswarlu, K. and Megharaj, M. (2024). Synergy of eco-innovation with on-farm practices enhances circularity beyond conventional nutrient recovery framework. *Resources, Conservation and Recycling, 208*: 107735.

[85] Kurniawan, T. A., Othman, M. H. D., Liang, X., Goh, H. H. and Chew, K. W. (2023). From liquid waste to mineral fertilizer: Recovery, recycle and reuse of high-value macro-nutrients from landfill leachate to contribute to circular economy, food security, and carbon neutrality. *Process Safety and Environmental Protection, 170:* 791-807.

[86] Shanmuganathan, R., Kadri, M. S., Mathimani, T., Le, Q. H. and Pugazhendhi, A. (2023). Recent innovations and challenges in the eradication of emerging contaminants from aquatic systems. *Chemosphere, 332*: 138812.

[87] Khan, S., Naushad, M., Govarthanan, M., Iqbal, J. and Alfadul, S. M. (2022). Emerging contaminants of high concern for the environment: Current trends and future research. *Environmental Research, 207:* 112609.

[88] Bashir, Y., Raj, R., Ghangrekar, M. M., Nema, A. K. and Das, S. (2023). Critical assessment of advanced oxidation processes and bio-electrochemical integrated systems for removing emerging contaminants from wastewater. *RSC Sustainability, 1*: 1912-1931.

[89] Gong, W., Bai, L. and Liang, H. (2024). Membrane-based technologies for removing emerging contaminants in urban water systems: Limitations, successes, and future improvements. *Desalination, 590*: 117974.

[90] Al Mamun, A., Prasetya, T. A. E., Dewi, I. R. and Ahmad, M. (2023). Microplastics in human food chains: Food becoming a threat to health safety. *Science of The Total Environment, 858:* 159834.

[91] Rahman, M. F., Yanful, E. K. and Jasim, S. Y. (2009). Endocrine disrupting compounds (EDCs) and pharmaceuticals and personal care products (PPCPs) in the aquatic environment: implications for the drinking water industry and global environmental health. *Journal of water and health, 7:* 224-243.

[92] Preisner, M., Neverova-Dziopak, E. and Kowalewski, Z. (2020). An analytical review of different approaches to wastewater discharge standards with particular emphasis on nutrients. *Environmental Management, 66:* 694-708.

[93] Giri, S. (2021). Water quality prospective in Twenty First Century: Status of water quality in major river basins, contemporary strategies and impediments: A review. *Environmental Pollution, 271*: 116332.

[94] Schellenberg, T., Subramanian, V., Ganeshan, G., Tompkins, D. and Pradeep, R. (2020). Wastewater discharge standards in the evolving context of urban sustainability–The case of India. *Frontiers in Environmental Science, 8*: 30.

## Chapter 20

# A Case Study on Waste Management in the Temples of Uttar Pradesh (India) Through Social Entrepreneurship and Green Temple Initiatives

**Pratima Ghosh[1,*]**
**and Manjula Upadhyay[2]**
[1]Department of Commerce, Navyug Kanya Mahavidyalaya, Lucknow, India
[2]Principal, Navyug Kanya Mahavidyalaya, Lucknow, India

## Abstract

The Green Temple Initiative is a program launched in various religious cities of Uttar Pradesh (India) to address the environmental impact of temple waste. Flowers and other temple offerings, when dumped into rivers, harm river ecosystems. This study analyzes the various waste management programs implemented in the cities of Ayodhya, Varanasi, Lucknow, Chitrakoot, and Prayagraj during the Mahakumbh. The study is exploratory in nature, relying on secondary data from various reports, research papers, and books. The findings of the study suggest that waste management is a shared responsibility, involving both the government and the private sector. Social entrepreneurship and start-ups have introduced innovative technologies and solutions to repurpose temple waste, creating significant economic and employment opportunities. However, the programs face challenges including funding shortages, which the government needs to address by providing additional resources. Additionally, the slow pace of implementation and the lack of adequate outreach programs to educate priests and devotees on sustainable practices remain significant obstacles.

**Keywords:** circular economy, green temple, social entrepreneurship, Varanasi, waste management, India, Europe

---

* Corresponding Author's Email: rubymitraghosh@gmail.com

In: Advancements in Wastewater Treatment
Editors: Neha Agarwal, Maulin P. Shahl, Vijendra Singh Solankil et al.
ISBN: 979-8-90134-007-3

## 1. Introduction

India is a land of rich religious traditions, where people express their devotion through offerings to gods and goddesses in the form of flowers, fruits, food, milk, water, and clothing. Additionally, many worship nature, such as rivers and trees like the Peepal, Banyan, and Mango trees, offering flowers and lighting countless clay lamps (diyas). While most of the offerings, particularly food, are consumed as prasad, items such as flowers, fruit peels, and old clothes often become waste.

Many temple cities in India are located alongside rivers, with numerous temples lining the riverbanks. The disposal of temple waste, particularly floral offerings, presents a significant environmental challenge. Millions of tons of these offerings are discarded, leading to pollution of water bodies and surrounding areas. This is where the concept of a circular economy can play a transformative role by converting these wastes into valuable resources. Techniques like vermicomposting, composting, and biogas generation have proven beneficial in addressing environmental degradation while creating economic opportunities, and promoting sustainable development.

The intersection of social entrepreneurship and waste management has gained considerable attention in both academic and practical spheres. Several studies emphasize the potential of social enterprises to tackle environmental issues while driving inclusive economic growth. Globally, there is a noticeable transition toward a circular economy, which has the potential to add $4.5 trillion to the global economy by 2030. In India, however, only 20% of goods are recycled, compared to Europe's 70% recycling rate. Additionally, India is the third-largest emitter of greenhouse gases. These pressing environmental and economic challenges underscore the urgent need for adopting a circular economy model to unlock the estimated $4.5 trillion potential by 2030 [1].

### 1.1. Social Entrepreneurship

Social entrepreneurship refers to innovative ventures and start-ups that aim to address societal challenges through sustainable business models. These start-ups play an essential role in driving sustainable practices in various sectors, particularly in the temple economy. Circular economy start-ups are at the forefront of innovation, especially in waste management, recycling, sustainable agriculture, renewable energy, and other sectors crucial for environmental conservation and sustainable development.

The global transition to a circular economy is critical, with the potential to contribute $4.5 trillion by 2030. However, India faces significant challenges, as only 20% of goods are recycled compared to 70% in Europe. With India being the third-largest emitter of greenhouse gases, the adoption of a circular economy model is essential for achieving long-term sustainability and meeting global economic goals.

### 1.2. The State of Social Entrepreneurship in India

India's start-up ecosystem, recognized by the Department for Promotion of Industry and Internal Trade (DPIIT), currently houses over 117,254 start-ups. Despite the growth of the start-up sector, only about 400 active start-ups focus on sustainability, including circular economy initiatives. Interestingly, within the ecosystem of unicorns in India, ReNew Power Pvt Ltd, a renewable energy start-up, achieved a valuation of over $1 billion in 2023, showcasing the growing interest in sustainable sectors. While India's overall start-up ecosystem is valued at $141 billion, the circular economy sector has attracted only $1.8 billion in investments between 2016 and 2023. This reflects the sector's untapped potential and the need for greater investment in sustainable initiatives.

## 2. Review of Literature

Proper management of floral waste is essential due to its adverse effects on the environment and human health. With growing awareness of environmental sustainability, the utilization of floral waste for producing valuable products has become a key focus in various research studies.

A recent study highlights that flowers commonly offered in temples, such as marigold, rose, jasmine, hyacinth, and hibiscus, can be effectively managed and utilized in various value-added forms [2]. Techniques such as vermicomposting, composting, dye extraction, essential oil extraction, making of Holi colors, and biogas generation can all be employed. Since many flowers contain secondary metabolites, these can be further used in essential oil extraction and as food additives. Additionally, handmade paper can be created using these waste products. This review underscores the significant applications of floral waste, which help address both energy crises and environmental pollution. The research also connects each of the 17 Sustainable Development Goals (SDGs) with solid waste management, particularly SDG 6: Clean Water and Sanitation.

Researchers emphasize that social entrepreneurship goes beyond profit-making to solve social problems through innovative, scalable, and sustainable models. In the context of waste management, this approach has been effectively used to tackle both environmental degradation and unemployment. Another study examined waste management practices in religious places and identified a significant gap in structured, eco-friendly disposal systems [3]. This research suggested that temple waste, especially floral waste, remains an untapped resource that can be repurposed through entrepreneurial initiatives.

Another study analyzed the impact of social enterprises like "HelpUsGreen," which processes over 8.4 tons of floral waste daily from temples in North India. It also examined the concept of "Green Temples" in India, highlighting initiatives such as converting floral waste into incense sticks and compost [4]. Its findings demonstrated that these initiatives not only reduce waste but also generate local employment, particularly benefiting women from economically disadvantaged backgrounds.

A case study was conducted on temple waste recycling units in Uttar Pradesh, noting that these units provided sustainable livelihoods for over 500 families [5]. The study emphasized the community's active participation in collection, segregation, and production, showcasing strong evidence of social cohesion and economic development. The enterprise employed marginalized women, offering them stable incomes and dignified working conditions, thus transforming temple waste into a source of economic empowerment. Collectively, the literature affirms the transformative potential of social entrepreneurship in temple waste management. These models not only address environmental issues but also generate employment, uplift communities, and foster a culture of sustainability and innovation.

A report published by "The Energy and Resources Institute" (TERI) in 2018 addressed community-based approaches to waste management in India. The Green Temple Initiatives focus on eco-friendly practices within temple premises, such as source segregation (dry, wet, and floral waste), the use of biodegradable materials for offerings, and the installation of composting and biogas units for organic waste. These initiatives also promote zero-waste temple models [3].

## 3. Hypothesis

The hypothesis for the study on Social Entrepreneurship and Green Temple Initiatives is formulated to assess the effectiveness of integrating social entrepreneurship with green initiatives in managing temple waste.

- *Null Hypothesis ($H_0$):* There is no significant relationship between the implementation of social entrepreneurship and green temple initiatives and the effectiveness of temple waste management in Uttar Pradesh.
- *Alternative Hypothesis ($H_1$):* The implementation of social entrepreneurship and green temple initiatives significantly enhances the effectiveness of temple waste management in Uttar Pradesh.

This hypothesis aims to evaluate whether the adoption of sustainable practices and community-driven initiatives can lead to improved waste management outcomes in the religious cities of Uttar Pradesh.

## 4. Objectives

The primary objectives of adopting a sustainable and holistic approach to waste management in the temples of Mathura and Vrindavan, through the integration of social entrepreneurship and the Green Temple Initiatives, are as follows:

1. *Assess Existing Waste Management Practices:* Evaluate the current waste management systems in place at the temples of Uttar Pradesh, identifying inefficiencies and areas requiring improvement to enhance system effectiveness.
2. *Identify Gaps in Current Systems:* Recognize the shortcomings and challenges within the existing waste management frameworks in these temples to develop tailored, sustainable solutions.

3. *Promote Environmental Sustainability:* Implement eco-friendly practices to mitigate the environmental impact of temple waste, with a special focus on preventing pollution of river systems and surrounding ecosystems.
4. *Encourage Social Entrepreneurship:* Propose strategies to promote social entrepreneurship as a means to address temple waste issues and create economic opportunities for local communities.
5. *Implement Green Temple Initiatives:* Suggest practical methods for implementing Green Temple Initiatives to manage waste more effectively, using sustainable technologies and practices in temples.

## 5. Methodology

The present study focuses on the Green Temple initiative in several prominent temple cities of Uttar Pradesh, along with the waste management program implemented during the recently held Maha Kumbh in Prayagraj, Uttar Pradesh. The selected cities for this study include Ayodhya, Varanasi, Lucknow, Prayagraj, and Chitrakoot. Key objectives of the study are:

- *Analysis of Initiatives:* The study will analyze the various initiatives undertaken in these cities to assess their effectiveness in achieving the Sustainable Development Goals (SDGs), specifically:
  - Clean Water (SDG 6)
  - Life Below Water (SDG 14)
  - Life on Land (SDG 15)
- *Waste Assessment:* Focus will be placed on biodegradable and chemical waste generated through temple activities, which is often discarded into rivers, contributing to ocean pollution. Special attention will be given to initiatives aimed at mitigating this waste and its environmental impacts.
- *Campaigns and Interventions:* The study will evaluate the success of various campaigns (Abhiyan) introduced to tackle the temple waste problem in these areas.
- *Data Collection:* This study is exploratory in nature and will rely primarily on secondary data sourced from:
  - Official websites of the Uttar Pradesh Government
  - Research papers, reports, and books
  - Government and non-governmental organization (NGO) reports

## 6. Government Participation and Findings of the Study

In Lucknow, the discharge of untreated wastewater from towns located along the Gomti River constitutes a major source of pollution. To address this, pollution abatement measures have been implemented under the Gomti Action Plan Phase I and Phase II. These initiatives include:

- Interception and diversion of wastewater flowing into the river,
- Construction of sewage treatment plants (STPs),

- Riverfront development,
- Construction of community toilet complexes.

Under Gomti Action Plan Phase-I, projects worth Rs. 55.75 crore were sanctioned for the towns of Lucknow, Jaunpur, and Sultanpur. Under Phase-II, projects costing Rs. 263.04 crore were sanctioned specifically for Lucknow. In addition to wastewater, floral decorations, ritual offerings, and unsold flowers from markets often end up improperly discarded [6]. These organic wastes decay and become breeding grounds for disease-causing microorganisms, exacerbating environmental pollution. To assess the environmental impact and the effectiveness of existing waste management systems, various government and independent reports will be reviewed.

### 6.1. Environmental Challenges in Sacred Cities

The sacred cities of Ayodhya, Varanasi, Prayagraj, Chitrakoot, Mathura, and Vrindavan, situated along the Ganga, Yamuna, Sarju, and Gomti Rivers, are renowned for their spiritual and cultural heritage. However, the massive influx of pilgrims and tourists has led to:

- Accumulation of temple waste, including floral offerings, incense residues, and ritual materials,
- Increased pollution through the immersion of offerings in rivers and open dumping,
- Degradation of natural habitats and water bodies.

### 6.2. Emerging Sustainable Solutions

Recognizing the urgent need for sustainable temple waste management, a holistic approach has emerged that combines social entrepreneurship with green temple initiatives [7].

1. The first Green Temple Intervention was initiated in 2017 at Kalpeshwar Temple, Chennai, where a biogas plant and biocomposter were installed. This enhanced the capacity of temple authorities and local volunteers to independently manage temple waste sustainably.
2. Social enterprises such as *Help Us Green* and *Phool* have pioneered "flowercycling," the collection and transformation of temple floral waste into eco-friendly products such as incense sticks, organic fertilizers, and biodegradable materials. These initiatives:
   - Mitigate environmental pollution,
   - Generate employment opportunities, especially for marginalized communities,
   - Promote circular economy models.
3. The Green Temple Initiative promotes environmentally conscious practices within temples, such as:
   - Waste segregation,
   - Composting,
   - Use of sustainable, biodegradable materials in rituals.

By integrating traditional religious customs with modern sustainability principles, these initiatives aim to preserve both the sanctity and ecological balance of sacred spaces.

## 6.3. Key Concerns

- *Environmental:* Traditional disposal practices have led to severe water pollution and ecosystem degradation. The buildup of temple waste threatens biodiversity and reduces the ecological health of rivers and surrounding habitats.
- *Socio-Economic:* Improper waste management affects public health, diminishes the visual and spiritual appeal of sacred sites, and risks reducing tourism revenue critical to local economies.

### *6.3.1. Need for Social Entrepreneurship*

Innovative solutions are required to transform temple waste into valuable resources. Social enterprises like *"HelpUsGreen"* have demonstrated the potential of converting floral waste into eco-friendly products, simultaneously addressing environmental issues and providing employment opportunities to marginalized communities.

Environmentally conscious practices within temple premises, such as waste segregation, composting, and the use of sustainable materials in rituals, can significantly reduce the ecological footprint. These initiatives promote environmental stewardship while preserving religious values.

### *6.3.2. Research and Policy Development*

There is a pressing need for comprehensive studies to assess current waste management practices in these cities and to develop policies that integrate traditional religious practices with modern sustainability principles [8]. Such research can guide the implementation of effective waste management strategies, ensuring the preservation of cultural heritage and environmental integrity.

# 7. Challenges of Temple Waste Management

Temple waste, which includes high volumes of flowers, fruits, and coconuts, contributes significantly to the moisture content of partially degraded waste. It was found that the initial moisture content of temple waste was around 78%. Regular experimental measurements indicated that the moisture content of the finished compost remained slightly above 60%, whereas the optimal pile moisture content for effective composting should be between 50–70% (as moist as a squeezed-out sponge) [5]. A slight decrease in moisture content was observed over time, as water serves as a critical source of nutrition for earthworms and microbes [9].

The growth of tourism has resulted in an increasing number of pilgrims and visitors, contributing substantially to temple waste generation and making waste management more challenging. The Nagar Nigams (municipal corporations) in various cities have initiated plans to address these issues, focusing on eliminating dumping sites, promoting door-to-door collection of segregated waste, and establishing waste recycling centres to enhance solid waste management. Each city's waste management plan prioritizes the elimination of dumping sites and the promotion of segregated waste collection. Traditional practices, such as the offering of lamps and waste materials into water bodies, further aggravate pollution problems, as many of these items (especially lamps) are non-biodegradable.

### 7.1. Proposed Measures

- *Adopt Sustainable Practices:* Temples should implement sustainable operations, including waste segregation, composting, and the use of biodegradable materials, setting a precedent for environmentally conscious religious practices.
- *Enhance Community Awareness and Participation:* It is crucial to educate devotees, temple authorities, and local communities about the importance of sustainable waste management and encourage their active involvement in eco-friendly initiatives.
- *Develop Policy Recommendations:* Based on research findings, formulate guidelines and policies to replicate successful waste management models in other regions.

By achieving these objectives, the initiative aims to create a replicable model that harmonizes religious traditions with modern sustainability practices, ensuring the preservation of cultural heritage while safeguarding the environment.

### 7.2. Integrated Waste Management Initiatives

The Lucknow Swachhata Abhiyan (LSA) is responsible for the entire waste management cycle from collection and transportation to processing and final disposal of garbage within the city. To ensure effective and responsible waste management, the Government of Uttar Pradesh, in collaboration with GIZ India (the German agency for international cooperation in sustainable development), is working towards establishing a circular economy for plastics.

*5R Framework:* To combat pollution and promote sustainable development, the 5R principle Rethink, Refuse, Reduce, Reuse, and Recycle is being actively implemented.

Additionally, a Command and Control Center has been established to monitor and coordinate waste management activities in real-time, improving efficiency and rapid response capabilities.

## 8. Technological Developments and Sustainable Practices in Waste Management

To enhance the efficiency, accountability, and transparency of waste management operations, the LSA employs a Command and Control Center that monitors activities in real time.

- *Sustainable Practices:* Efforts are underway to encourage sustainable production and consumption patterns in urban centers, with an emphasis on composting and waste recycling to reduce the environmental impact of temple waste.

- *Partnerships and Cooperation:* The Government of Uttar Pradesh acknowledges that waste management in Mathura and Vrindavan presents a complex challenge. A comprehensive approach involving community participation, private sector engagement, and government initiatives is critical for achieving effective waste management outcomes.

*Sustainable solutions being explored include:*

- Composting biodegradable waste,
- Recycling floral offerings and ritual materials,
- Promoting eco-friendly alternatives within temple practices.

It is crucial to address these issues to maintain the cleanliness and environmental health of these sacred cities.

## 9. Challenges in Solid Waste Management: Case Studies

### 9.1. Varanasi

- The current status of Solid Waste Management (SWM) in Varanasi remains unsatisfactory.
- Best practices for waste collection, segregation, and disposal are yet to be fully implemented.
- There is a shortage of qualified waste management professionals, and public awareness about SWM practices is limited.
- Biomedical waste is often disposed of alongside municipal solid waste, posing serious public health risks [10].
- Monitoring mechanisms at dumping sites are absent.
- Apathy among local residents, along with lack of strict punitive measures against improper waste disposal, exacerbates the problem.
- Municipal workers frequently display irresponsibility, lack of accountability, and are often unaware of specific waste management guidelines issued by the government.
- Floral waste from temples commonly ends up contaminating rivers or is discarded as regular trash.

### 9.2. Chitrakoot

A waste composition analysis of Chitrakoot Nagar Panchayat revealed the following:

1. Polythene (10.12%), Plastic (4.78%), Rubber (3.70%), Metals (2.84%), Glass (4.08%),
2. Wood (5.55%), Cotton & Clothes (4.29%), Paper & Cardboard (7.77%),
3. Vegetable Waste (11.66%), Soil & Construction Waste (29.87%),
4. Garden Waste (9.73%), Rags (2.29%), and Ash (3.32%).

5. The biodegradable fraction of the waste was found to be 44.61%.
6. Municipal bins and a structured waste collection system are almost non-existent; households and commercial establishments often dispose of waste openly on streets or vacant plots.
7. Composting technology was identified as a suitable treatment method for Chitrakoot's waste.
8. Nutrient values (N, P, K) of composted waste were improved with the use of activators, showing increases from:
   - Nitrogen (N): 1.04% to 1.51%,
   - Phosphorus (P): 0.63% to 0.94%,
   - Potassium (K): 0.92% to 1.09% (Singh et al.).

#### *9.2.1. Awareness and Capacity Building Initiatives*

The Department of Environment, Government of Uttar Pradesh, in collaboration with GIZ NSCM India, organized public awareness campaigns to educate citizens about sustainable waste management practices, the importance of segregation at source, and the environmental consequences of improper waste disposal.

### 9.3. Maha Kumbh 2025: Sustainable and Circular Waste Management

The waste management strategy for Maha Kumbh 2025 highlights a Circular Economy Model, where temporary infrastructure such as modular STPs (Sewage Treatment Plants) and Faecal Sludge Treatment Plants (FSTPs) can be dismantled and reused in rural areas after the event.

*Key Highlights:*

- Temporary STPs & FSTPs reduce land use by 70% and eliminate the need for chemical treatments.
- *Sludge Recycling:* The treated sludge from temporary FSTPs can be repurposed as organic fertilizers, supporting sustainable agricultural practices.
- *Minimized Carbon Footprint:* Reduced transportation of sludge and decentralized waste processing contribute to a lower overall carbon footprint.

## 10. Social Entrepreneurship

The government is collaborating with organizations such as GIZ India on projects related to the circular economy. These initiatives not only contribute to environmental preservation but also create social value. One such example is the production of herbal colors from temple waste, which are safe for both the environment and skin. These projects provide employment opportunities for women, particularly in the production process. The project manager for the mission emphasized that each district has a target of 5 to 10 lakh in sales. Additionally, the floral waste will be sold online through platforms like Amazon and Flipkart, promoting wider market access.

In Kannauj, temple waste is converted into raw materials for the local perfume industry. Studies have shown a positive correlation between the success of the perfume industry and the financial assistance provided to businesses, highlighting the economic potential of converting floral waste [11]. This process not only enhances economic efficiency but also supports entrepreneurship and job creation. Moreover, the residual floral waste is utilized in composting, contributing to sustainable waste management practices. This model demonstrates the power of floral waste conversion in generating both economic and environmental benefits.

## 11. Policy Recommendation

The Green Temple Solid Waste Management Program is a closed-loop waste management system that processes temple waste into usable products, such as biogas and compost. These products can be used on the temple campus itself, promoting sustainability within sacred spaces.

*Key Policy Recommendations:*

a. *Clear Operational Guidelines:* The government should establish clear guidelines for the operation, recycling, processing, and refurbishing of temple waste to ensure consistency and compliance. These guidelines should outline the necessary steps for sustainable waste management practices.
b. *Incentivizing Circular Economy Business Models:* The government should incentivize business models that promote the circular economy and sustainability, especially in sectors like waste management, resource recovery, and sustainable product manufacturing.
c. *Dedicated Funds for Circular Economy:* Each state government should consider setting up dedicated funds to promote circular economy initiatives, particularly in regions experiencing rapid population growth, industrial development, and increased consumption. This funding would support sustainable practices across industries and communities.
d. *Expansion of Waste Management Programs:* Government waste management programs and guidelines should not be restricted to specific areas but should be expanded to include a wider range of waste management practices and technologies that can be scaled and adapted to different regions.
e. *Collaborative Circular Economy Zones:* In alignment with the Public-Private Partnership (PPP) model, the government and industry stakeholders should collaborate to identify circular sustainable corridors [12-14]. These corridors should include dedicated zones or hubs for circular economy start-ups, where resources can be shared, and facilities can be accessed. These zones would foster innovation and provide a platform for start-ups to collaborate and scale their sustainable solutions.

## Conclusion

In conclusion, the Green Temple Initiative holds the potential to create a sustainable and resource-efficient environment. By involving social entrepreneurs such as NGOs and SHGs, along with government support, temple waste can be transformed into eco-friendly products. The government can further promote the use of digital and biodegradable offerings to reduce environmental impacts. This initiative not only fosters sustainability but also creates employment and income opportunities for local communities. It can further strengthen its impact by developing outreach programs aimed at training priests, devotees, and the broader community on the importance of sustainable practices and effective waste management.

## References

[1] World Economic Forum. (2019). Making the $4.5 *Trillion Circular Economy Opportunity a Reality*. WEF Article. Available at: https://www.weforum. org/impact/helping-the-circular-economy-become-a-reality/#:~:text=A%20circular%20economy%20is%20%20a,additional%20economic%20output%20by%202030.

[2] Kumar, H., & Sen, R. Management of temple floral waste and utilization of value-added floral waste products: A review, *International Journal of Advances in Engineering and Management*, 2 (1), (2022), 759-764, https://doi.org/10.35629/5252-45122323.

[3] Sharma, V. Green temples of India: A path to sustainable devotion, Journal of Religious Studies and Ecology, 12(1), (2018), 32–47.

[4] Reddy, S., & Kumar, R. Social Entrepreneurship in Waste Management: Creating Value for Community and Environment. *Indian Journal of Social Work*, 80(2), (2019) ,159-176.

[5] Singh, A., & Gupta, G. Solid waste management. *Indian Journal of Science and Technology*, 4(7), (2011), 6846. https://doi.org/10.17485/ijst/2011/v4i7/68008.

[6] Ministry of Environment, Forest and Climate Change (MoEFCC). (2016). *Solid Waste Management Rules*, 2016. Government of India.

[7] Srivastava, A., & Tripathi, R. (2021). Circular economy and waste valorization in temple towns. *Environmental Economics and Policy Studies*, 23(1), 55–73.

[8] UNDP India. (2020). Waste management and circular economy in India: A sectoral overview. Retrieved from https://www.undp.org/india-G

[9] Hujuri, S. S., et al. (2023). Microbial composting and vermicomposting of temple waste using Eisenia fetida. *Materials Today*, 72(5), 2780–2784.

[10] Shouvk, B. (2023). Waste management system and problems of *Varanasi* Municipal Corporation area, Varanasi, UP. In *Futuristic trends in social sciences* (Vol. 3, Book 18, Part 2, Chapter 5). IIP Series. e-ISBN: 978-93-5747-624-9.

[11] Shukla, P., & Ali, M. (2025). Utilizing temple floral waste to address the resource crunch in the perfume industry: A case study of *Kannauj* district, Uttar Pradesh, India. *Agro Environmental Sustainability*, 3, 58–63. https://doi.org/10.59983/s2025030107.

[12] Council on Energy, Environment and Water (CEEW). (2023). India's circular economy potential: *Opportunities and challenges*. Retrieved from https://www.ceew.in/sites/default/files/india-circular-economy-potential-webceew16dec.pdf.

[13] National Centre for Sustainable Coastal Management (NCSCM). (2023). Circular economy solutions preventing marine litter in three ecosystems.

[14] Chase India. (2024). Startups accelerating circular economy for sustainable Bharat: Survey report. Retrieved from https://www.chase-india.com/media/qqfc0xyh/startup-accelerating-circular-economy-report-revised.pdf.

# About the Editors

**Dr. Neha Agarwal** is presently serving as an Assistant Professor in the Department of Chemistry at *Navyug Kanya Mahavidyalaya*, affiliated with the University of Lucknow, India. She earned her Doctorate (PhD.) from the University of Lucknow and holds prestigious qualifications, having cleared both the UGC-CSIR-NET and GATE examinations in Chemical Sciences. She has made significant scholarly contributions, with numerous research papers and review articles published in esteemed Scopus-indexed journals. She is also an accomplished innovator, having secured four patents, and has played a pivotal role in academic publishing—editing multiple volumes for reputed publishers including Nova Science Publishers, Bentham Science, Springer Nature, CRC Press, and Anu Books International. Additionally, she has authored three undergraduate textbooks. An active member of several national and international scientific bodies, Dr. Agarwal also serves on the editorial boards of various international, peer-reviewed journals. Her research primarily centres on exploring the environmental impacts of organic pollutants—such as pharmaceuticals, dyes, and pesticides—and developing cost-effective and eco-friendly remediation techniques to degrade these contaminants into harmless, non-toxic substances. Her ultimate aim is to contribute toward a green and sustainable future.

**Dr. Vijendra Singh Solanki** is currently serving as an assistant Professor in the Department of Chemistry at the Institute of Science and Research, IPS Academy, Indore, India. He holds both MPhil and PhD. degrees, with his doctoral research focused on the degradation of persistent dyes in wastewater, contributing to sustainable environmental solutions. Dr. Solanki has published numerous research articles in Scopus-indexed journals, primarily centered on dye degradation, nanoparticle synthesis, characterization, and their environmental applications. He has been granted several Indian patents and has also edited three books published by Bentham Science, CRC Press and Springer Nature. His research interests include the photocatalytic degradation of organic pollutants, wastewater remediation, green chemistry, and the synthesis and characterization of nanomaterials. An active member of several scientific organizations, Dr. Solanki is frequently invited to deliver expert talks in his field of expertise.

**Dr. Sankara Rao Miditana** serves as an Assistant Professor of Chemistry at Government Degree College, Puttur, Andhra Pradesh. With over 15 years of dedicated teaching experience, he is a highly regarded member of the academic community. He earned his PhD. in photocatalysis from Andhra University, Visakhapatnam. He also serves as a member of Board of Studies (BoS) at Government College (Autonomous), Nagari, Andhra Pradesh, where he plays an active role in curriculum development and maintaining academic standards. Dr. Miditana is an accomplished researcher with 11 publications in reputed peer-reviewed journals and also published 3 book chapters published by international publishers. His research focuses on heterogeneous photocatalysis and organic-inorganic nano hybrid materials, where he continues to contribute to advancing the frontiers of chemistry.

**Dr. Maulin P. Shah** is a Scientist in the Industrial Wastewater Research Lab at Enviro Technology Ltd., Ankleshwar, Gujarat, India. He earned his PhD. in Environmental Microbiology from Sardar Patel University, Gujarat, in 2005, and previously served as an Assistant Professor at Gujarat University. With a strong background in Microbial Biotechnology, his research spans Applied Microbiology, Environmental Biotechnology, Bioremediation, and Industrial Liquid Waste Management. Dr. Shah's primary focus is on understanding microbial adaptation to toxic substances and their applications in environmental restoration, including the development of bacterial bioreporters. His extensive work includes isolating and genetically engineering microbes for hazardous material degradation, reflected in over 350 publications in esteemed national and international journals. At Enviro Technology Ltd., he directs research programs and has supervised over 100 postgraduate students. Dr. Shah also plays active roles as an Editorial Board Member for over 100 Journals of high repute including Elsevier, Springer, Taylor & Francis, Wiley, De Gruyter, KeAi, Frontiers. He has edited more than 230 books in the area of Industrial Wastewater Microbiology and Waste Management.

# List of Contributors

**Agarwal, Neha**
Department of Chemistry,
Navyug Kanya Mahavidyalaya (University of Lucknow),
Lucknow, Uttar Pradesh, India.

**Ambika, S.**
PG & Research Department of Chemistry,
Bishop Heber College, Trichy -17.

**Ansari, Almas**
PG Department of Biotechnology,
St. Xavier's College, Mumbai.

**Ansari, Muazzama**
PG Department of Biotechnology,
St. Xavier's College, Mumbai.

**Arora, Unik**
Department of Chemistry,
Meerut College, Meerut (U.P.) 250003 India.

**Bisht, Yash**
PG Department of Biotechnology,
St. Xavier's College, Mumbai.

**Chaudhary, Archana**
Centre for Translational Research,
Office of Research & Development,
Indian Institute of Technology, Indore 453552, India.

**Cheema, Harveer Singh**
Department of Botany,
Meerut College,
Meerut (U.P.) 250003 India.

**Dhali, Ashruta M.**
PG Department of Biotechnology,
St. Xavier's College, Mumbai.

**Dhinesh R.**
Department of Aquatic Environment Management,
Faculty of Fisheries Science,
Kerala University of Fisheries and Ocean Studies, Kochi, Kerala, India.

**Dokka, Winnie Teja**
Department of Chemistry,
Government Degree College,
Chintalapudi, Andhra Pradesh-534460, India.

**Ghosh, Pratima**
Department of Commerce,
Navyug Kanya Mahavidyalaya,
Lucknow, India.

**Hemalatha, S.**
PG & Research Department of Chemistry,
Bishop Heber College, Trichy -17.

**Inwati, Gajendra Kumar**
School of Chemical Sciences,
Devi Ahilya Vishwavidyalaya,
Indore, M P 452001, India.

**Jain, Punita**
PG Department of Biotechnology,
St. Xavier's College, Mumbai.

**Kathuria, Tanisha**
Department of Chemistry,
Banasthali Vidyapith, Tonk,
Rajasthan-304022, India.

**Kavuluko, Jacinta Muthini**
Department of Science,
School of Science, Health and Engineering,
Daystar University, P.O Box 44400 -00100
GPO, Ngong Rd Nairobi, Kenya.

**Kiarii, Ephraim Muriithi**
Department of Physical Sciences,
University of Embu, P.O Box 6-60100, Embu.

**Korlam, Saivenkatesh**
Department of Botany,
S.V.A. Government College,
Srikalahasti, Tirupati Dt., Andhra Pradesh.

**Kori, Deepak K. K.**
Department of Chemistry,
Indian Institute of Technology,
Indore, 453552, India.

**Kumar, Raj**
Department of Biotechnology and Bioinformatics,
Jaypee University of Information Technology,
Waknaghat, Himachal Pradesh, India.

**Kumar, R. Vasantha**
PG & Research Department of Chemistry,
Bishop Heber College, Trichy -17.

**Kumari, Sheetal**
Amity Institute of Environmental Sciences,
Amity University, Noida-201313, India.

**Kumar, Sudesh**
Department of Chemistry,
National Council of Educational Research and Training (NCERT), Delhi.

**Longchar, Imotila T.**
Department of Chemistry,
Nagaland University, Lumami-798627,
Nagaland, India.

**Manjari, T.**
PG & Research Department of Chemistry,
Bishop Heber College, Trichy -17.

**Manojkumar, Y.**
PG & Research Department of Chemistry,
Bishop Heber College, Trichy.

**Marwaha, Lovleen**
Department of Zoology,
School of Bioengineering and Biosciences,
Lovely Professional University,
Phagwara, Punjab, India.

**Mehta, Anjali**
Department of Chemistry,
Banasthali Vidyapith, Tonk,
Rajasthan-304022, India.

**Miditana, Sankara Rao**
Department of Chemistry,
Government Degree College, Puttur,
Tirupati, Andhra Pradesh, India-517583.

**Patel, Prabhat Kumar**
School of Sciences,
Harbin Institute of Technology Shenzhen,
Shenzhen-518055, China.

**Patel, Yamini**
School of Chemical Sciences,
Devi Ahilya Vishwavidyalaya,
Indore, M P 452001, India.

**Priyanka, S.**
PG & Research Department of Chemistry,
Bishop Heber College, Trichy.

**Pushpa Latha, A H D**
Department of Botany,
Dr V S Krishna Government Degree College (A),
Visakhapatnam-530013, Andhra Pradesh, India.

**Rana, Rishi**
Department of Civil Engineering,
Jaypee University of Information Technology,
Waknaghat, Himachal Pradesh, India.

**Raghuvanshi Urmila**
Department of Chemistry,
Shri G S Institute of Technology and Science,
Indore, M P 452003.

**Ramesh Babu, A**
Department of Chemistry,
SVA Government Degree College(M),
Srikalahasti, Andhra Pradesh, India.

**Rubankumar, A.**
PG & Research Department of Chemistry,
Bishop Heber College, Trichy.

**Sarkar, Joy**
Department of Chemistry
University of Lucknow
Lucknow, India

**Kirthiga S. S.**
Department of Aquatic Environment Management,
Faculty of Fisheries Science,
Kerala University of Fisheries and Ocean Studies,
Kochi, Kerala, India.

**Sathiyaraj, M.**
PG & Research Department of Chemistry,
Bishop Heber College, Trichy -17.

**Sharma, Apoorav**
Department of Zoology,
School of Bioengineering and Biosciences,
Lovely Professional University,
Phagwara, Punjab, India.

**Sharma, Pratibha**
School of Chemical Sciences,
Devi Ahilya Vishwavidyalaya,
Indore, M P 452001, India.

**Sharma, Richa**
Department of Civil Engineering,
Jaypee University of Information Technology,
Waknaghat, Himachal Pradesh, India.

**Sharma, Shisak**
Department of Chemistry,
Nagaland University, Lumami-798627,
Nagaland, India.

**Sharma, Sudhanshu**
Department of Chemistry,
Banasthali Vidyapith, Tonk,
Rajasthan-304022, India.

**Simma, Nalini**
Department of Chemistry,
Government Degree College(M),
Srikakulam, Andhra Pradesh-532001, India.

**Sinha, Dipak**
Department of Chemistry,
Nagaland University, Lumami-798627,
Nagaland, India.

**Pandey, Jaya**
Department of Chemistry,
Amity School of Applied Sciences
Amity University, Uttar Pradesh, India

**Solomon, Rajadurai Vijay**
PG & Research Department of Chemistry,
Madras Christian College, Chennai.

**Srivastava, Amrita**
Department of Chemistry,
University of Lucknow,
Lucknow, India.

**Steve, Dembede**
Department of Science, School of Science,
Health and Engineering, Daystar University,
P.O Box 44400 -00100
GPO, Ngong Rd Nairobi, Kenya.

**Steven Umdor, Raplang**
Department of Chemistry,
Nagaland University, Lumami-798627,
Nagaland, India.

**Todiwala, Nilufer**
PG Department of Biotechnology,
St. Xavier's College, Mumbai.

**Tomar, Ashish**
Department of Chemistry,
Meerut College,
Meerut (U.P.) 250003 India.

**Trivedi, Sudakshina**
Department of Chemistry,
University of Lucknow,
Lucknow, India.

**Upadhyay, Manjula**
Department of Commerce,
Navyug Kanya Mahavidyalaya,
Lucknow, India.

**Wanale, Shivraj Gangadhar**
Netaji Subhashchandra Bose, Art's, Commerce, and Science College,
Nanded, 431601, Maharashtra, India.

**Yuvarani, N.**
PG & Research Department of Chemistry,
Bishop Heber College, Trichy.

# Index

## #



## A



## B

## C



## D



## E

## F



## G



## H



## I

## L



## M



## N



## O



## P

## R



## S

## T



## U



## V



## W